Saas-Fee Advanced Course 34
2004

J. D. Haigh   M. Lockwood   M. S. Giampapa

The Sun, Solar Analogs and the Climate

Saas-Fee Advanced Course 34
2004

Swiss Society for Astrophysics and Astronomy

Edited by I. Rüedi, M. Güdel and W. Schmutz

With 288 Illustrations, 32 in Color

Springer

Joanna Dorothy Haigh
Space and Atmospheric Physics
Blackett Laboratory
Imperial College
Prince Consort Road
London SW7 2AZ, UK

Mark S. Giampapa
National Solar Observatory/NOAO
950 N. Cherry Ave.
P.O. Box 26732
Tucson, AZ 85726-6732, USA

Michael Lockwood
Space Science and Technology Department
Rutherford Appleton Laboratory
Chilton
Didcot, Oxfordshire, OX11 0QX, UK
and
School of Physics and Astronomy
University of Southampton
Highfield
Southampton, SO17 1BJ, UK

Volume Editors:

Isabelle Rüedi
Werner Schmutz
Physikalisch-Meteorologisches Observatorium Davos
World Radiation Center
Dorfstr. 33
7260 Davos Dorf, Switzerland

Manuel Güdel
Paul Scherrer Institut
Würenlingen and Villigen
5232 Villigen PSI, Switzerland

This series is edited on behalf of the Swiss Society for Astrophysics and Astronomy:
Société Suisse d'Astrophysique et d'Astronomie
Observatoire de Genève, ch. des Maillettes 51, 1290 Sauverny, Switzerland

Cover picture: Composite showing the Earth from Apollo 17 and the Sun observed in soft X-rays by the Yohkoh satellite. Credit: US National Aeronautics and Space Administration and the Japanese Institute of Space and Astronautical Science (ISAS/JAXA).

ISBN 978-3-642-06279-7 e-ISBN 978-3-540-27510-7

Springer is a part of Springer Science+Business Media
springeronline.com

Printed in The Netherlands

Data conversion by TechBooks
Cover design: *design & production* GmbH, Heidelberg

Printed on acid-free paper 55/3141/mh - 5 4 3 2 1 0

Preface

The 34th Saas-Fee advanced course of the Swiss Society of Astronomy and Astrophysics (SSAA) took place from March 15 to 20, 2004, in Davos, on the subject of *The Sun, Solar Analogs and the Climate.*

Presently the Swiss mountain resort of Davos is probably most well known for hosting an event on globalization. However, it is because Davos also happens to be the seat of the Physikalisch-Meteorologisches Observatorium Davos and World Radiation Center, that this course on a "global" subject was hosted here.

Exceptionally, the topic of this course was not purely astrophysical, but the members of the SSAA decided to support it all the same due to the timely topic of global warming and its possible link to solar variations.

In these times of concern about global warming, it is important to understand solar variability and its interaction with the atmosphere. Only in this way can we distinguish between the solar and anthropogenic contributions to the rising temperatures. Therefore, this course addressed the observed variability of the Sun and the present understanding of the variability's origin and its impact on the Earth's climate. Comparing the solar variability with that of solar analog stars leads to a better understanding of the solar activity cycle and magnetic activity in general, and helps us to estimate how large the solar variations could be on longer time scales.

In spite of the fantastic weather and snow conditions which reigned during this week, the participants assiduously took part in the lectures. This is proof of the high quality of the lectures that the three speakers, Joanna Haigh, Mike Lockwood and David Soderblom, delivered. We deeply thank them for their contributions and efforts and hope that the readers will enjoy the book as much as we enjoyed their lectures.

We particularly appreciated the cooperation of the lecturers in coping with an unexpected situation: Mark Giampapa took ill the day before the beginning of the course, and we found an excellent replacement lecturer in David Soderblom, who agreed at very short notice to fly to Switzerland and to give the lectures prepared by Mark Giampapa. We owe him very special thanks for his outstanding effort. The chapter on stars in this book was, however, prepared entirely by Mark Giampapa.

We also have the pleasure of thanking Sonja Degli Esposti and Annika Weber for their help in the practical organization of the course. Finally, the financial support of the Swiss Academy of Natural Sciences and of the Swiss Society of Astrophysics and Astronomy is greatly acknowledged, without which the course could not have taken place.

Davos and Zürich,
January 2005

Isabelle Rüedi
Manuel Güdel
Werner Schmutz

Contents

List of Previous Saas-Fee Advanced Courses

!! 2004 The Sun, Solar Analogs and the Climate
J.D. Haigh, M. Lockwood, M.S. Giampapa

!! 2002 The Cold Universe
A.W. Blain, F. Combes, B.T. Draine

!! 2000 High-Energy Spectroscopic Astrophysics
S.M. Kahn, P. von Ballmoos, M. Güdel

!! 1999 Physics of Star Formation in Galaxies
F. Palla, H. Zinnecker

* 1998 Star Clusters
B.W. Carney, W.E. Harris

* 1997 Computational Methods for Astrophysical Fluid Flow
R.J. LeVeque, D. Mihalas, E.A. Dorfi, E. Müller

* 1996 Galaxies Interactions and Induced Star Formation
R.C. Kennicutt, F. Schweizer, J.E. Barnes

* 1995 Stellar Remnants
S.D. Kawaler, I. Novikov, G. Srinivasan

* 1994 Plasma Astrophysics
J.G. Kirk, D.B. Melrose, E.R. Priest

* 1993 The Deep Universe
A.R. Sandage, R.G. Kron, M.S. Longair

* 1992 Interacting Binaries
S.N. Shore, M. Livio, E.J.P. van den Heuvel

* 1991 The Galactic Interstellar Medium
W.B. Burton, B.G. Elmegreen, R. Genzel

* 1990 Active Galactic Nuclei
R. Blandford, H. Netzer, L. Woltjer

! 1989 The Milky Way as a Galaxy
G. Gilmore, I. King, P. van der Kruit

! 1988 Radiation in Moving Gaseous Media
H. Frisch, R.P. Kudritzki, H.W. Yorke

! 1987 Large Scale Structures in the Universe
A.C. Fabian, M. Geller, A. Szalay

!	1986	Nucleosynthesis and Chemical Evolution *J. Audouze, C. Chiosi, S.E. Woosley*
!	1985	High Resolution in Astronomy *R.S. Booth, J.W. Brault, A. Labeyrie*
!	1984	Planets, Their Origin, Interior and Atmosphere *D. Gautier, W.B. Hubbard, H. Reeves*
!	1983	Astrophysical Processes in Upper Main Sequence Stars *A.N. Cox, S. Vauclair, J.P. Zahn*
*	1982	Morphology and Dynamics of Galaxies *J. Binney, J. Kormendy, S.D.M. White*
!	1981	Activity and Outer Atmospheres of the Sun and Stars *F. Praderie, D.S. Spicer, G.L. Withbroe*
*	1980	Star Formation *J. Appenzeller, J. Lequeux, J. Silk*
*	1979	Extragalactic High Energy Physics *F. Pacini, C. Ryter, P.A. Strittmatter*
*	1978	Observational Cosmology *J.E. Gunn, M.S. Longair, M.J. Rees*
*	1977	Advanced Stages in Stellar Evolution *I. Iben Jr., A. Renzini, D.N. Schramm*
*	1976	Galaxies *K. Freeman, R.C. Larson, B. Tinsley*
*	1975	Atomic and Molecular Processes in Astrophysics *A. Dalgarno, F. Masnou-Seeuws, R.V.P. McWhirter*
*	1974	Magnetohydrodynamics *L. Mestel, N.O. Weiss*
*	1973	Dynamical Structure and Evolution of Stellar Systems *G. Contopoulos, M. Hénon, D. Lynden-Bell*
*	1972	Interstellar Matter *N.C. Wickramasinghe, F.D. Kahn, P.G. Metzger*
*	1971	Theory of the Stellar Atmospheres *D. Mihalas, B. Pagel, P. Souffrin*

---

* Out of print

! May be ordered from Geneva Observatory
Saas-Fee Courses
Geneva Observatory
CH-1290 Sauverny
Switzerland

!! May be ordered from Springer

The Earth's Climate and Its Response to Solar Variability

J.D. Haigh

Imperial College London
j.haigh@imperial.ac.uk

1 The Climate System

At periods of higher solar activity the Earth is subject to enhanced solar irradiance, a greater incidence of solar energetic particles and fewer galactic cosmic rays. The aim of this series of lectures is to assess to what extent solar variability affects the Earth's climate and to consider how any apparent response is brought about.

The absorption of solar radiation determines the Earth's mean temperature and radiation budget, while the latitudinal distribution of the absorbed radiation is the primary driver for atmospheric circulations. At any point in the atmosphere the radiative heating rate is the net effect of solar heating and infrared cooling, the latter being intrinsically related to the atmospheric composition and temperature structure. But solar radiation (and solar energetic particles) are also key to determining the composition of the atmosphere so that variations in solar output have the potential to affect the atmospheric temperature structure in a complex and non-linear fashion.

Clouds play a key role in the radiation budget and it has been proposed that variations in cosmic ray intensity may influence cloud cover and properties. By these means solar variability may also result in significant changes in climate.

The first lecture provides a descriptive background of the Earth's climate system while the factors determining atmospheric circulations are discussed in lecture 2. Lecture 3 presents evidence for the influence of solar variability on climate and subsequent lectures present material on the physical and chemical processes determining the structure of the Earth's lower and middle atmospheres. Finally lectures 8 and 9 discuss the mechanisms whereby solar activity may influence the atmospheric radiation balance, composition, cloud cover and circulation.

1.1 The Earth's Climate System

The temperature, motion and composition of the atmosphere, and how these properties vary in space and time, are determined by many complex interactions between the atmosphere itself and the oceans, cryosphere and biosphere;

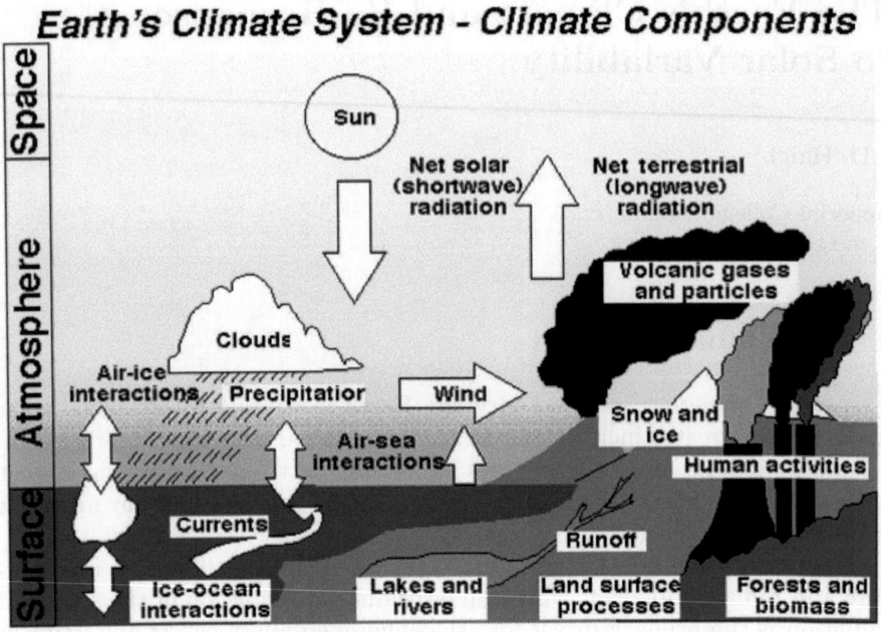

Fig. 1.1. Components of the Earth's climate system

Fig. 1.1 illustrates the major contributing factors. The main energy source is solar irradiance and energy loss is largely through the emission of heat radiation to space. The transport and conversion of the radiant energy depends on the composition of the atmosphere, a key component of which is water, present as vapour or (in clouds) in liquid form or as ice. The water in the atmosphere is a small component of the Earth's hydrological cycle which involves evaporation from the oceans, condensation into clouds, transport of the clouds by winds and deposition of rain or snow over land and sea. Large amounts of fresh water are stored in the polar ice caps and the deep ocean circulations are determined by the temperature and salinity of polar water. Forests and land cover are important in the radiation budget (because of its dependence on surface albedo) and the hydrological cycle. Volcanic eruptions import large quantities of gases and aerosols into the atmosphere, as do human industrial and agricultural activities. The winds, which move these gases etc. around, themselves are mainly driven by the need to transport heat from low to high latitudes.

1.2 Temperature

The temperature of the atmosphere varies with height according to the physical, chemical and dynamical processes taking place and traditionally names are attached to different layers according to their vertical temperature gradients. Figure 1.2 identifies these layers in a global mean plot of the temperature of the atmosphere below 100 km. The lowest portion is called the troposphere; here temperature decreases with height up to around 12 km where a change in gradient, called the tropopause, is present. Above this lies the stratosphere which is a stable layer in which the temperature increases to an altitude of about 50 km, the altitude of the stratopause where the gradient reverses again into the mesosphere. Above the mesopause lies the thermosphere. It is common in meteorology to measure altitude not by height but by atmospheric pressure. The pressure at any level indicates the weight of the atmosphere above so, from Fig. 1.2, it can be seen that the troposphere, which contains the weather and climate as experienced by humankind, contains about 80% of the entire atmospheric mass.

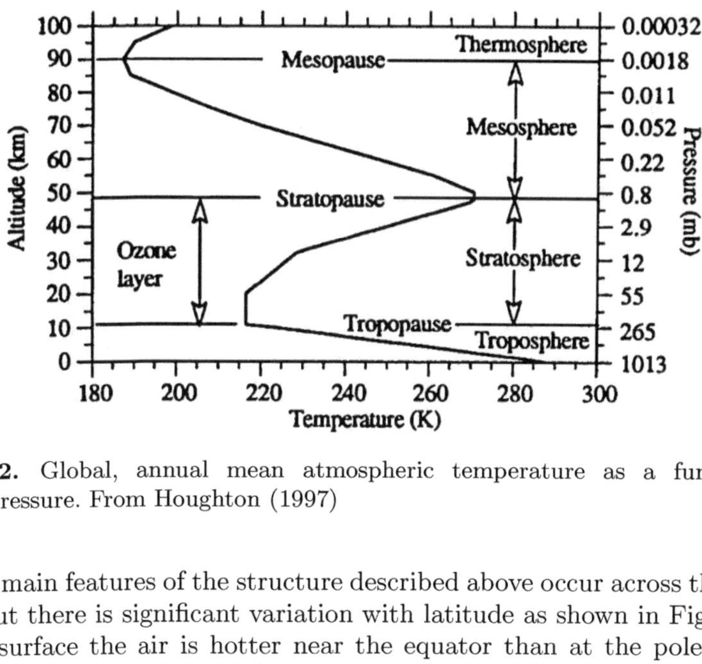

Fig. 1.2. Global, annual mean atmospheric temperature as a function of height/pressure. From Houghton (1997)

The main features of the structure described above occur across the whole globe but there is significant variation with latitude as shown in Fig. 1.3(a). At the surface the air is hotter near the equator than at the poles but at the tropopause (which is higher at low latitudes) the reverse is the case. The stratopause decreases in temperature monotonically from the summer to the winter pole while the opposite is true at the mesopause. The physical and chemical factors responsible for some of these features are discussed in lectures 2, 4 and 7.

1.3 Winds

The zonal mean thermal structure is closely related (see Sect. 2.1) to the zonal winds shown in Fig. 1.3(b). In the latter figure positive values indicate winds from the west (called westerlies by meteorologists) and negative winds from

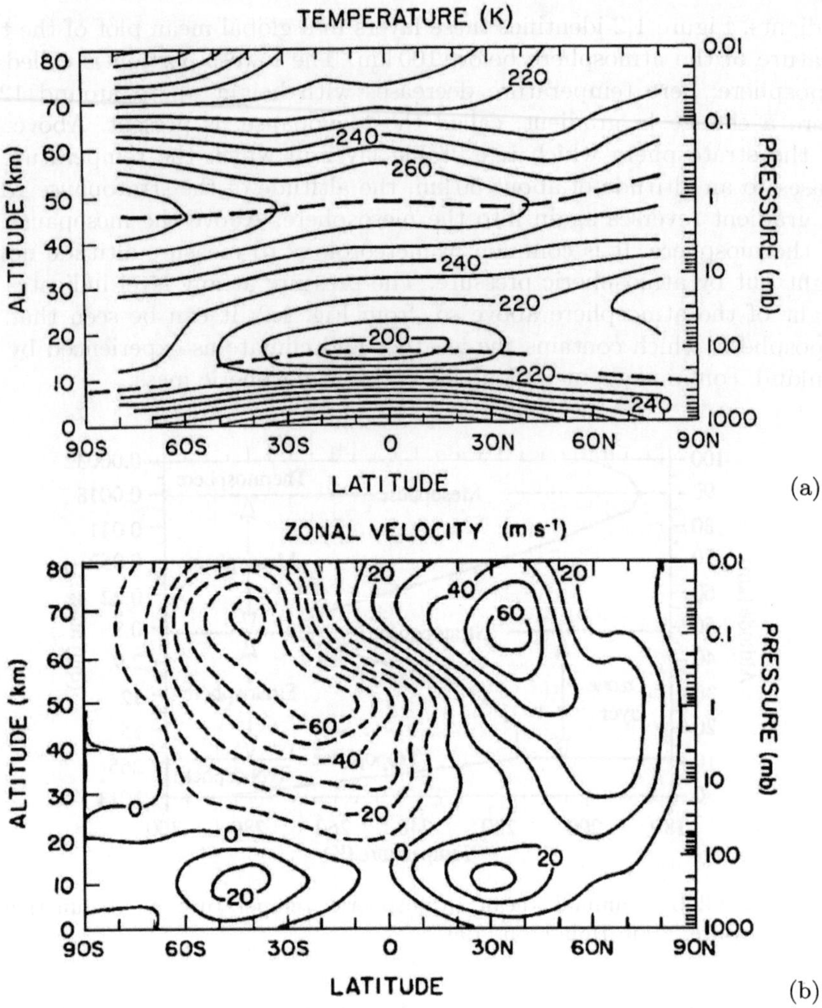

Fig. 1.3. (**a**) Zonal mean temperature (K) during northern hemisphere winter as a function of latitude and altitude/pressure. The *dashed* lines indicate the positions of the tropopause and stratopause. (**b**) As (**a**) but for zonal mean zonal wind (ms^{-1}); westerly winds *solid* lines, easterly winds *dashed* lines. From Salby (1996)

the east. The troposphere is dominated by the two sub-tropical westerly jets which have peak values at the tropopause. Between them is a region of weak easterlies which strengthens with height at low latitudes and into the summer stratosphere and mesosphere . The winter hemisphere middle atmosphere has a strong westerly jet.

The mean meridional circulation of the troposphere (Fig. 1.4(a)) is dominated by the winter Hadley cell in which air rises to the summer side of the

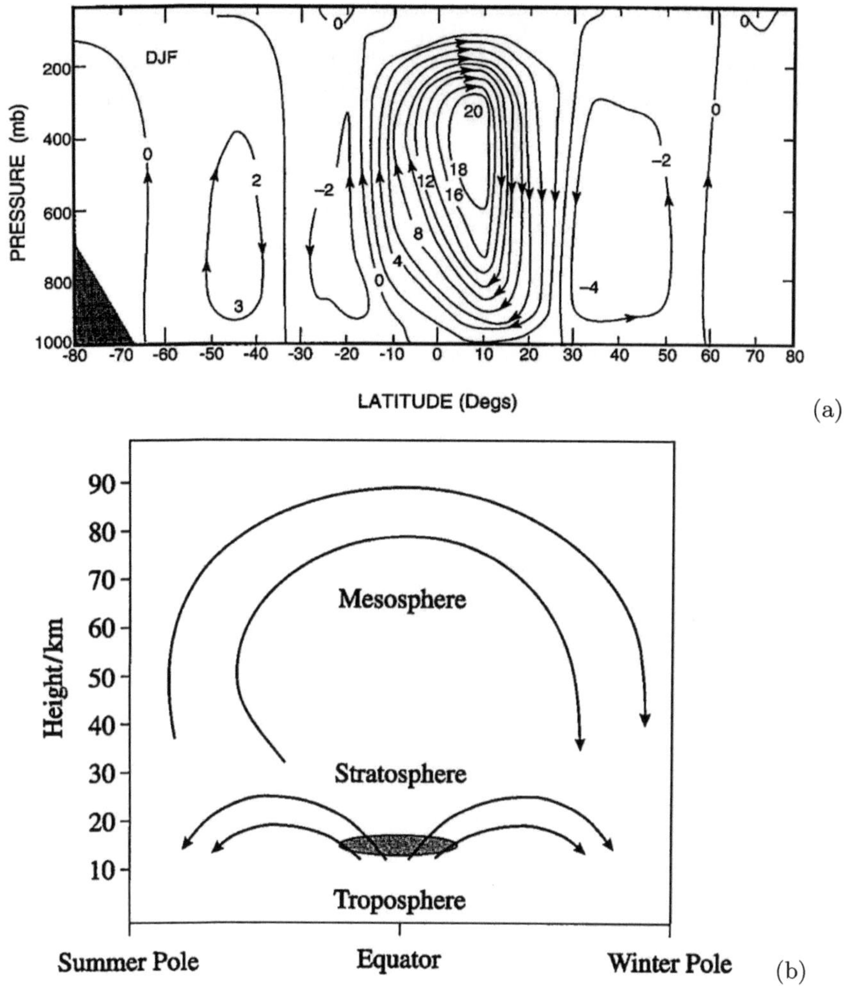

Fig. 1.4. (**a**) Mean meridional circulation of the troposphere during northern hemisphere winter (note linear pressure scale). From Salby (1996). (**b**) Schematic plot of mean meridional circulation of the stratosphere and mesosphere. From Andrews (2000)

equator, flows towards the winter pole near the tropopause, sinks in the winter sub-tropics and returns equatorwards near the surface. A much weaker summer Hadley cell exists in the other hemisphere. In mid-latitudes in both hemispheres there are cells circulating in the opposite direction; these Ferrel cells are thermally indirect (i.e. transport warm air up the temperature gradient) and are driven by zonally asymmetric eddies (see Sects. 1.5 and 2.2).

In the lower stratosphere air is transported from the tropics towards the poles by the Brewer-Dobson circulation while in the upper stratosphere and

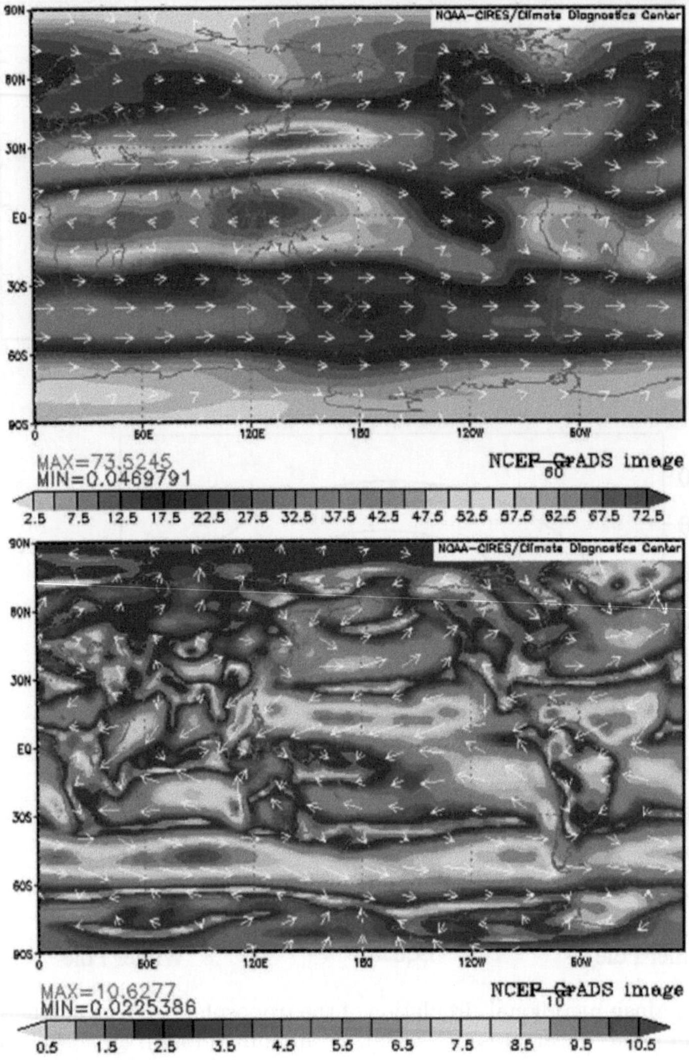

(a)

(b)

Fig. 1.5. Wind strength (contours, ms^{-1}) and direction (arrows) for DJF as a function of longitude and latitude at (**a**) 200 hPa and (**b**) 1000 hPa. Data from http://www.cdc.noaa.gov/ncep_reanalysis/

mesosphere there is a solstitial circulation with upward motion in the summer hemisphere, a summer-to-winter transport in the mesosphere and descent near the winter pole. Both these circulations are wave-driven: the former by planetary waves and the latter by gravity waves (see Sect. 1.5).

These zonal mean figures show the winds averaged over longitude; the variation with longitude produces a more complex picture. Figure 1.5 shows the mean wind fields for December/January/February (DJF) at the surface and near the tropopause. The surface trade winds can be seen as easterly/equatorward flow in the tropics. In mid-latitudes the surface flow is predominantly westerly/poleward but this flow is disrupted by the presence of high topography in the northern hemisphere. At 200 hPa the flow is more zonal, the strong jets east of North America and Asia mark the North Atlantic and North Pacific storm tracks respectively.

1.4 Surface Pressure

The surface winds are closely related to surface pressure. Figure 1.6 shows a map of mean sea level pressure in DJF. The low pressure near the equator and high pressure bands near 30° latitude are signatures of the rising and sinking portions, respectively, of the Hadley cells. The Siberian High and Aleutian Low are persistent features of the northern hemisphere winter, associated with the topography and responsible for steering depressions to the north and south respectively.

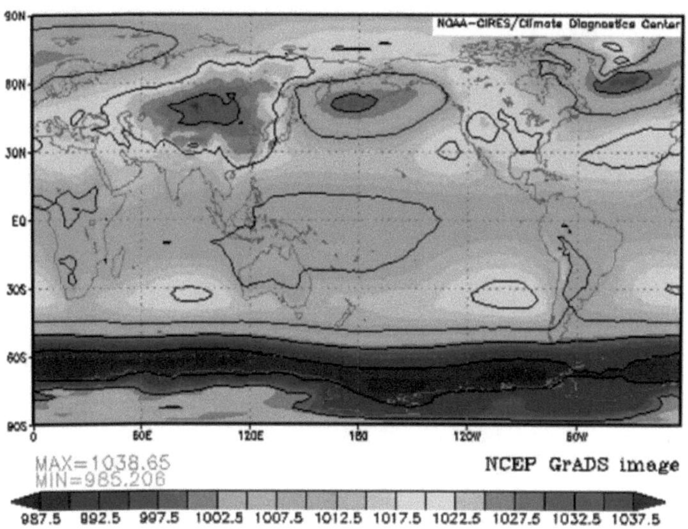

Fig. 1.6. Mean sea level pressure (hPa) for DJF as a function of longitude and latitude. Data from: http://www.cdc.noaa.gov/ncep_reanalysis/

On any individual day the surface pressure map is far much more disturbed and shows, for example, the individual weather patterns in mid-latitudes seen Fig. 2.1.

1.5 Waves in the Atmosphere

The mid-latitude cyclones discussed in the previous section are a manifestation of baroclinic instability in which air moving across a horizontal temperature gradient results in a release of potential energy which is converted into the kinetic energy of the waves. These baroclinic waves have horizontal wavelengths typically of order 1–2000 km. Other instabilities in the atmosphere result in waves of much greater and much smaller wavelength.

Rossby (or planetary) waves depend on the rotation and spherical geometry of the Earth and are several thousand kilometres long. Waves generated by large scaled weather disturbances in the troposphere propagate upwards but only through regions of westerly winds and only those of the longest wavelengths reach into the stratosphere. Figure 1.7 gives an example of a Rossby wave in atmospheric temperature in the lower stratosphere. This is essentially a wavenumber 2 case, with two positive and two negative anomalies around a line of latitude.

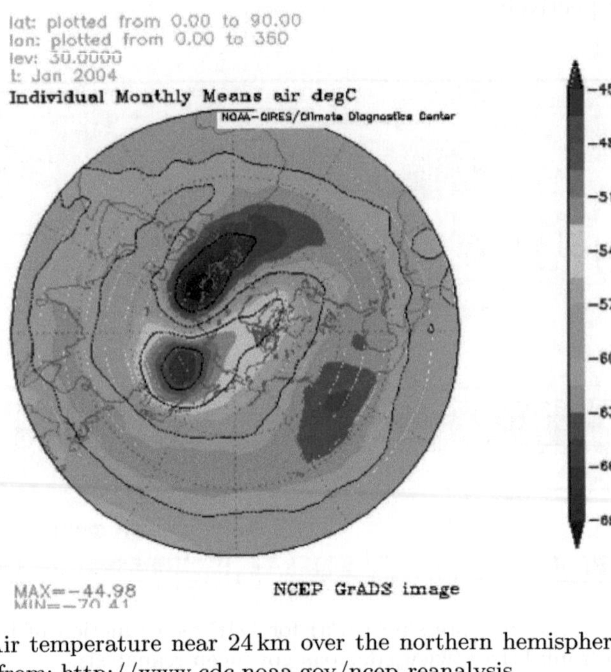

Fig. 1.7. Air temperature near 24 km over the northern hemisphere for January 2004. Data from: http://www.cdc.noaa.gov/ncep_reanalysis

Gravity waves, for which the restoring force is buoyancy, can be generated by flow over mountains or by convective activity and have horizontal wavelengths of perhaps 100 km and vertical wavelengths of about 15 km. A common signature of gravity waves is the pattern of lee wave clouds – bands of cloud lying perpendicular to the flow downstream of a mountain range.

Gravity waves can penetrate high into the atmosphere; lee wave clouds are sometimes seen in the stratosphere and the momentum and energy deposited by gravity waves near the mesopause are responsible for the solstitial circulation and the reverse temperature gradient at these levels (see Fig. 1.4).

1.6 Composition

Below about 100 km altitude the atmosphere is well-mixed because turbulent mixing occurs on shorter timescales than molecular diffusion (which tends to sort lighter from heavier molecules). Higher up the atmosphere is so rarefied (at 100 km the pressure is about one millionth its value near the surface) that vertical mixing is controlled by diffusion. Thus at these high altitudes the free electrons produced by the sun's ionising radiation have long lifetimes leading to the presence of a charged layer called the ionosphere which plays a key role in determining the Earth's electric field. This will be discussed further in lecture 9.

The composition of the lower/middle atmosphere is given in Table 1.1. It is dominated by nitrogen which, however, plays a negligible role in atmospheric radiative and chemical processes. Oxygen is chemically inert but in the middle atmosphere is photodissociated into highly reactive oxygen atoms and thus plays a key role in determining the chemical composition of the stratosphere. In Table 1.1 a range of concentrations is given for water vapour and ozone because of their variability with altitude. For these, and other minor constituents not included in the table, the deviation from a uniform profile is due to local sources and/or sinks. Some examples are given in Fig. 1.8; the chemical processes governing these concentrations are discussed in lecture 7.

Table 1.1. Composition of the atmosphere below 100 km

Constituent	Concentration (fraction by number of molecules)
Nitrogen	0.78
Oxygen	0.21
Argon	0.0093
Water vapour	0–0.04
Carbon dioxide	370 ppm
Neon	18 ppm
Helium	5 ppm
Krypton	1 ppm
Hydrogen	0.5 ppm
Ozone	0–12 ppm

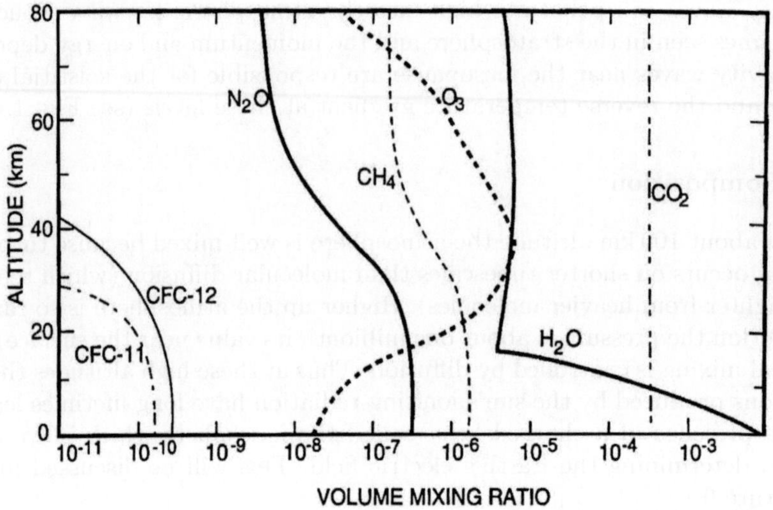

Fig. 1.8. Concentration profiles of radiatively active species (Salby, 1996)

Human activity is perturbing the chemical composition of the atmosphere, both locally and globally. Figure 1.9 shows measurements of the CO_2 concentration made in Hawaii, far from any sources of industrial pollution. An annual cycle, reflecting the hemispheric asymmetry in biomass and the seasonal cycle in photosynthesis, is superposed on a clear upward trend. The present-day value is far higher than any present over at least the past 420,000 years (see Fig. 3.3).

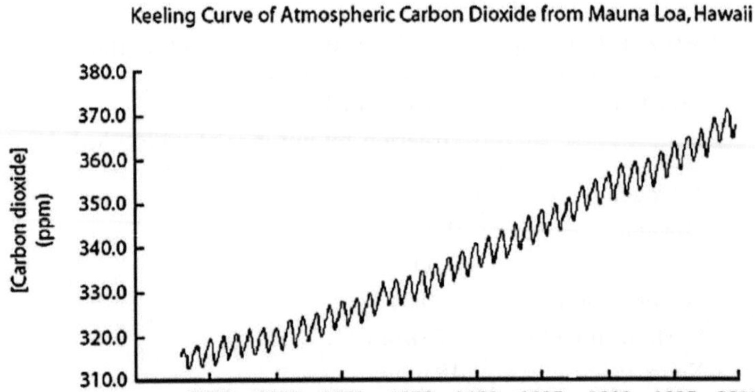

Fig. 1.9. Carbon dioxide concentrations at Mauna Loa. Keeling and Whorf (2004)

1.7 Clouds and Precipitation

Clouds play a crucial role in the radiation, heat and water vapour budgets of the planet. Their distribution is variable but generally occur in regions of the ascending branches of the Hadley and Ferrel cells in the tropics and mid-latitudes respectively. Figure 1.10 shows that precipitation rate is largely associated with the high (deep) cloud in the tropics and mid-latitude storm tracks. Large areas of low cloud are present over the sub-tropical oceans; these do not produce much rainfall but are very important in the Earth radiation budget. The processes of cloud formation and the radiative properties of cloud are the subjects of lecture 6.

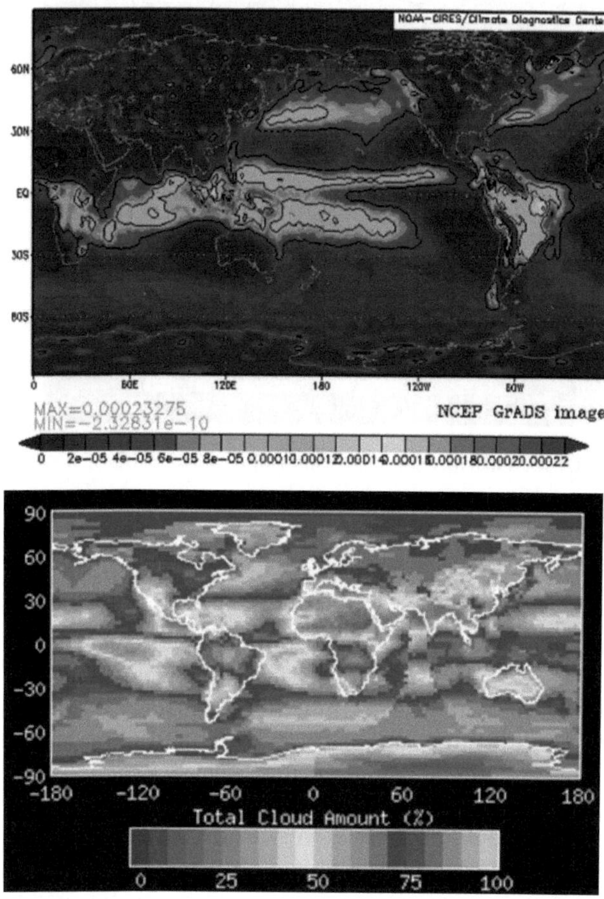

(a)

(b)

Fig. 1.10. (**a**) Mean DJF precipitation rate (kg m^{-2} s^{-1}) from http://www.cdc.noaa.gov/ncep_reanalysis/. (**b**) Mean DJF percentage total cloud cover (1983–2001) and the components of (**c**) high and (**d**) low cloud. From: http://isccp.giss.nasa.gov/

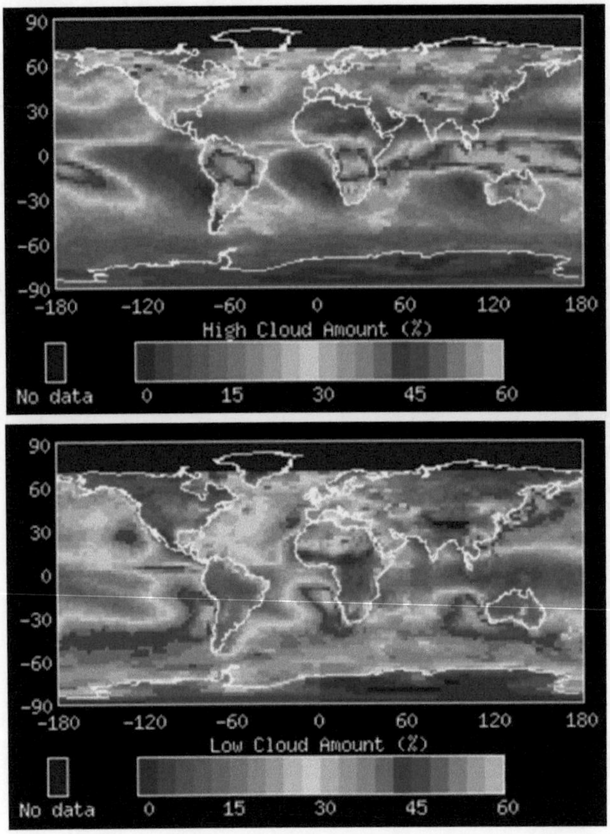

(c)

(d)

Fig. 1.10. for caption see previous page (note shift of longitude)

1.8 Oceans

The role of the oceans in the climate system is to transport heat from low to high latitudes as well as to act as the source and sink of atmospheric water vapour. Ocean surface currents are wind-driven so that Fig. 1.11 shows a clear signature of the trade winds either side of the equator. In each of the major ocean basins the flow is predominantly in an anticyclonic gyre with poleward flow on the west side of the basin and equatorward flow to the east. The western boundary currents bring relatively warm water to mid-latitudes thus ameliorating the climate in these regions. Figure 1.11 also shows regions of surface divergence and convergence indicating where water may be rising and sinking respectively. The deep ocean circulation operates through a global scale "conveyor belt", Fig. 1.12, which is driven by gradients of heat and salinity. Surface water in the Pacific is heated by the sun and flows westward across the Pacific and Indian Oceans then round the Cape of Good Hope

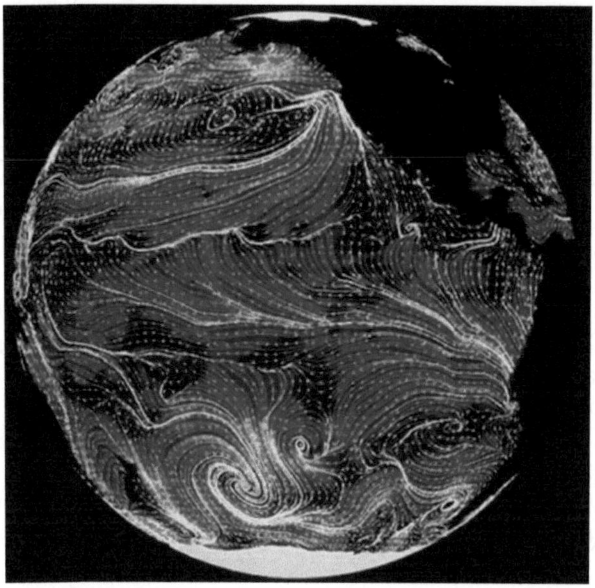

Fig. 1.11. Surface currents in the Pacific Ocean. *Red* colours indicate regions of surface divergence, *blue* convergence. From: http://www.phys.ufl.edu/

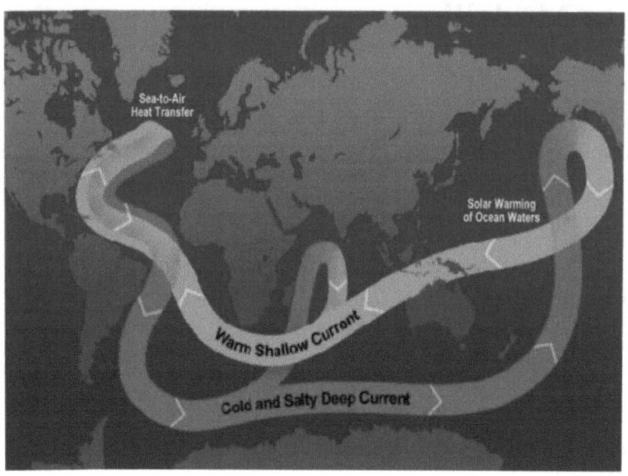

Fig. 1.12. Ocean "conveyor belt" deep sea circulation. From the US Global Change Research Program

into the Atlantic. It crosses the equator and travels northward into the North Atlantic where much of its heat is lost to the atmosphere. The cold water then sinks and returns to the Pacific at depth via the Antarctic. Some studies have indicated that climate change might alter the conveyor belt with significant potential impact especially on lands bordering the North Atlantic.

1.9 Climate and Weather

The climate of any geographical region can be described by long-term (perhaps several decade) averages of the meteorological conditions which it experiences. The weather at a particular place, however, essentially represents the instantaneous conditions which can be enormously variable. To understand climate we do not need to know the detail of every individual weather event but in assessing the effects of climate change great care has to be taken that natural variability on all timescales is taken into account. For example, the average daily maximum temperature in London in August is 26.4°C but on 6 August 2003 it reached 35.3°C. The popular press was happy to ascribe this to "global warming" although estimates suggest that the latter can be responsible for an average increase of only about 0.5°C over the past century. Nevertheless, some modelling studies have suggested that global warming will be accompanied by more extreme events so that extremely hot (or cold or wet or dry) periods may become more frequent. In any event, it is clear that great care must be taken in extracting solar signals from noisy climate records.

2 Atmospheric Dynamics, Modes of Variability and Climate Modelling

The atmosphere is a continuous, compressible fluid resting on the surface of a rotating planet. By applying some of the basic laws of physics – conservation of energy, conservation of mass, Newton's 2nd law of motion and the ideal gas law – to this fluid we can acquire an understanding of the main features of the global atmospheric circulation.

2.1 Equations of Motion

Continuity Equation

Consider the flow of air through an elemental volume. The local density, ρ, will increase if mass converges within the volume. Thus the rate of change of density with time is given by:

$$\frac{\partial \rho}{\partial t} = -\nabla . (\rho \boldsymbol{u}) \tag{2.1}$$

where $\boldsymbol{u}$ is the local velocity (bold type indicates vector variables) and the right-hand-side of the equation represents the divergence of the flux of mass crossing unit area in the direction of $\boldsymbol{u}$.

An alternative way to write this equation is:

$$\frac{1}{\rho}\frac{D\rho}{Dt} + \nabla . \boldsymbol{u} = 0 \tag{2.2}$$

where the material derivative

$$\frac{D}{Dt} \equiv \frac{\partial}{\partial t} + \boldsymbol{u}.\nabla \tag{2.3}$$

represents the rate of change with time following the motion, contrasting with $\frac{\partial}{\partial t}$ which represents the rate of change at a fixed point in space. The former approach, following air parcels, is often referred to as the Lagrangian method.

Equation of State

Assuming that air under normal atmospheric conditions can be treated as an ideal gas (an assumption which is sufficiently accurate where water vapour is not condensing into cloud) we can write:

$$pM = \rho RT \tag{2.4}$$

where p is atmospheric pressure, M the molecular weight, R the universal gas constant and T temperature.

Thermodynamic Equation

Applying the First Law of Thermodynamics to our parcel of air, again using the Lagrangian approach, we find:

$$\frac{DT}{Dt} = Q + \frac{1}{\rho C_p}\frac{Dp}{Dt} \tag{2.5}$$

where the second term on the right-hand-side represents adiabatic warming/cooling due to compression/expansion and Q is the heating rate due to diabatic factors: mainly radiative processes in the stratosphere and latent heating/cooling in the troposphere.

Navier-Stokes Equation

Newton's second law of motion states that the acceleration of a body is equal to the ratio of the net force acting on it to its mass. Applying this to a parcel of air we can write:

$$\frac{D\boldsymbol{u}}{Dt} = -g\boldsymbol{k} - \frac{1}{\rho}\nabla p + \frac{\eta}{\rho}\nabla^2\boldsymbol{u} \tag{2.6}$$

where the terms on the right-hand-side represent the forces per unit mass due to gravity, pressure gradients and friction (viscous drag). $\boldsymbol{k}$ is the unit vector in the vertical direction and η the dynamic viscosity. This equation, however, only applies in an inertial frame of reference (i.e. one which itself is not subject to external forces) and thus needs modification for use in a rotating frame of reference, such as a coordinate system fixed relative to the Earth. In a frame rotating with angular velocity $\boldsymbol{\Omega}$ the equation becomes:

$$\frac{D\boldsymbol{u}}{Dt} + 2\boldsymbol{\Omega} \times \boldsymbol{u} + \boldsymbol{\Omega} \times (\boldsymbol{\Omega} \times \boldsymbol{r}) = -g\boldsymbol{k} - \frac{1}{\rho}\nabla p + \frac{\eta}{\rho}\nabla^2 \boldsymbol{u} \qquad (2.7)$$

where the second and third terms are the Coriolis and centripetal accelerations respectively. The centripetal term is small in the atmosphere but acts to distort the shape of the solid Earth from spherical to bulging at the equator. Thus it is usual to incorporate this term in with gravity by defining:

$$g' = -g\boldsymbol{k} - \boldsymbol{\Omega} \times (\boldsymbol{\Omega} \times \boldsymbol{r}) \qquad (2.8)$$

where the effective gravity, vector $\boldsymbol{g}'$, is perpendicular to the local surface of the Earth. Thus the momentum equation becomes:

$$\frac{D\boldsymbol{u}}{Dt} = -2\boldsymbol{\Omega} \times \boldsymbol{u} + \boldsymbol{g}' - \frac{1}{\rho}\nabla p + \frac{\eta}{\rho}\nabla^2 \boldsymbol{u} \qquad (2.9)$$

The coordinate system is conventionally chosen such that local coordinates $(x,\ y,\ z)$ refer to displacements in the eastward, northward and upward directions respectively.

Scale Analysis of the Momentum Equation

By investigating the typical magnitudes of the terms in (2.9) for large-scale motion we can produce some simplifications.

Using the typical magnitudes of Table 2.1 we find that in the horizontal the Coriolis and pressure gradient terms dominate so that approximately:

$$2\boldsymbol{\Omega} \times \boldsymbol{u_h} = -\frac{1}{\rho}\nabla_h p \qquad (2.10)$$

Table 2.1. Values used for scale analysis

Scale	Typical Magnitude
Horizontal distance	10^6 m
Vertical distance	10^4 m
Horizontal velocity	$10\,\mathrm{ms^{-1}}$
Vertical velocity	$10^{-2}\,\mathrm{ms^{-1}}$
Time	10^5 s
Surface density	$1\,\mathrm{kg\,m^{-3}}$
Earth radius	6×10^6 m
Earth rotation rate	$7 \times 10^{-5}\,\mathrm{s^{-1}}$
Acceleration due to gravity	$10\,\mathrm{ms^{-2}}$

where the subscript h indicates only the horizontal components of the vector. This is referred to as geostrophic balance and applies on synoptic scales (i.e. the scale of weather patterns – a thousand kilometres in the horizontal, say), away from the Earth's surface and under conditions for which the horizontal wind is not too large. It can not be applied very close to the equator where the vertical component of the Earth's rotation becomes zero. Equation (2.10) may alternatively be written:

$$f\boldsymbol{k} \times \boldsymbol{u_h} = -\frac{1}{\rho}\nabla_h p \tag{2.11}$$

where the Coriolis parameter $f = 2\Omega \sin\phi$ and ϕ is latitude.

We can see from (2.10) that geostrophic balance will result in horizontal winds flowing along isobars (lines of constant pressure) and so can be deduced from daily weather charts, such as shown in Fig. 2.1. On that day there was a high pressure system over the British Isles and a low pressure system over the Baltic states bringing cold northerly winds from Scandinavia to central Europe.

In the vertical the dominant balance is between pressure gradient and gravity giving:

$$\frac{\partial p}{\partial z} = -g\rho \tag{2.12}$$

which is just hydrostatic balance. It applies in the atmosphere away from regions of significant vertical motion such as present in storm clouds.

Geopotential Height

Substituting an expression for density from the ideal gas law (2.4), into the hydrostatic equation we obtain an expression for pressure as a function of height:

$$p(z) = p_0 \exp\left(-\int_0^z \frac{gM}{RT}dz'\right) \tag{2.13}$$

where p_0 is the pressure at the surface ($z = 0$). Meteorologists often use pressure (or log(pressure)) as a vertical coordinate. Turning (2.13) round another way we obtain:

$$Z_i = \frac{R\bar{T}}{Mg}\ln\left(\frac{p_0}{p_i}\right) \tag{2.14}$$

where Z_i is referred to as the geopotential height of the layer at pressure p_i and $\bar{T}$ is a measure of the temperature of the layer $0 < z < Z_i$ (or $p_i < p < p_0$). Note that the mean temperature of a layer of atmosphere between two pressure levels is given by its thickness in terms of geopotential height.

Thermal Wind

Using the geostrophic (2.11), the equation of state (2.4) and hydrostatic (2.12) equations we can find a relationship between the windshear (the variation of horizontal wind with height) and the horizontal temperature gradient:

$$\frac{\partial u_h}{\partial z} \approx \frac{g}{fT}\boldsymbol{k} \times \nabla_h T \tag{2.15}$$

Where the vertical gradient of temperature has been neglected. This shows that, for example, in a situation where temperature decreases towards the pole then westerly wind will increase with height. Such a relationship can be clearly observed to exist throughout the atmosphere in the plots of zonal mean wind and temperature in Figs. 1.3. Note that this applies both in the northern and southern hemispheres as f has negative values in the latter.

2.2 Transport of Heat by the Global Circulation

The Troposphere

The energy driving the global atmospheric circulation comes from the Sun and on a global and annual average the energy of incoming solar irradiance must be balanced by thermal infrared radiation emitted by the planet (as discussed in lecture 5). The absorbed solar radiation, however, has a strong latitudinal variation, with much higher values near the equator and lower values near the poles, while the outgoing infrared radiation has only a weak latitudinal dependence so that there is a net surplus of radiation at low latitudes and deficit at high latitudes, as shown in Fig. 2.2.

This differential heating tends to set up a thermal convection cell with air rising near the equator, flowing polewards then cooling, sinking and flowing equatorwards again near the surface. At some point, however, the strength of the west winds, induced in the poleward-flowing air necessary to satisfy the thermal wind relation in the presence of the latitudinal temperature gradient, becomes dynamically unstable. The resulting waves (see below) transport heat polewards in mid-latitudes with sufficient strength to compensate for the equator-to-pole temperature gradient. Thus the mean meridional circulation consist of a tropical Hadley cell extending to about 30° latitude, a mid-latitude cell operating in the reverse direction and a third, thermally direct, cell at high latitudes as seen in Fig. 1.4. Coriolis acceleration of the surface flows associated with these cells produce the trade winds (north-easterly in the Northern hemisphere and south-easterly in the Southern hemisphere) and the mid-latitude westerlies shown in Fig. 1.5.

Mid-latitude weather systems, such as presented in Fig. 2.1, develop from the waves that are induced in the zonal flow in response to the instability described above.

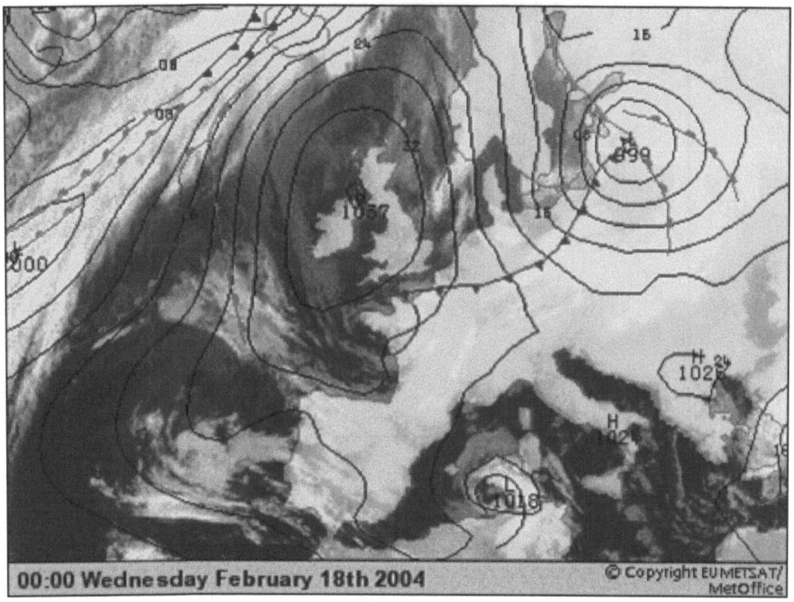

Fig. 2.1. Composite of satellite cloud imagery and surface pressure chart. From http://www.meto.gov.uk/weather/charts/composite.html

Because of the steeper temperature gradients in winter the intensity of these systems, and the amount of energy they transport polewards, is greater during that season as shown in Fig. 2.2.

The Middle Atmosphere

In the stratosphere the evaporation and condensation of water vapour, that is so important in the heat budget of the troposphere, is not important. Infrared cooling is largely balanced by heating due to the absorption of solar ultraviolet radiation (see lecture 4) so that the meridional temperature structure is quite different in the two regions. In the upper stratosphere, above about 30 hPa, the temperature increases uniformly from the winter to the summer pole (see Fig. 1.3). In thermal wind balance with this a strong westerly jet exists in the winter hemisphere and an easterly jet in the summer hemisphere (Fig. 1.3). In the lower stratosphere upper tropospheric convective processes affect the tropics so that the temperature has a minimum near the equator and maxima at the summer pole and in winter mid-latitudes. This temperature structure is associated with a mean meridional flow in which air entering the tropical lower stratosphere is transported towards both poles in the lower stratosphere, but higher up forms a simple single cell with ascent near the summer pole and descent near the winter pole (Fig. 1.4).

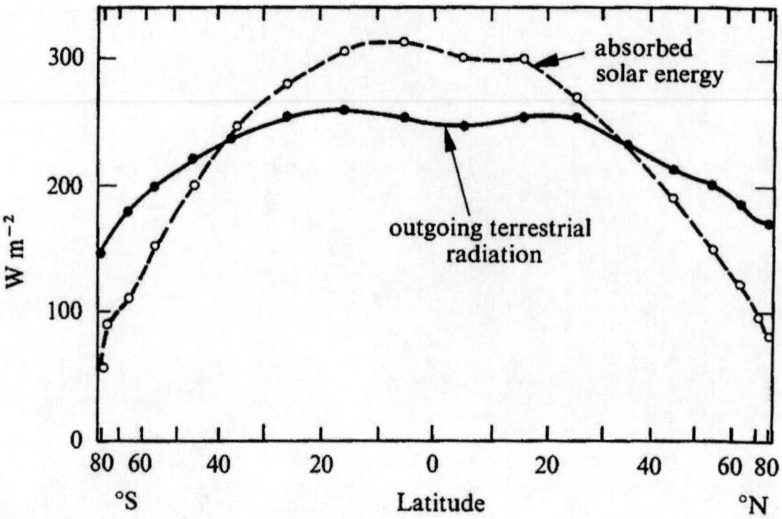

Fig. 2.2. Zonal annual mean of absorbed solar energy and emitted thermal energy. Houghton (1977)

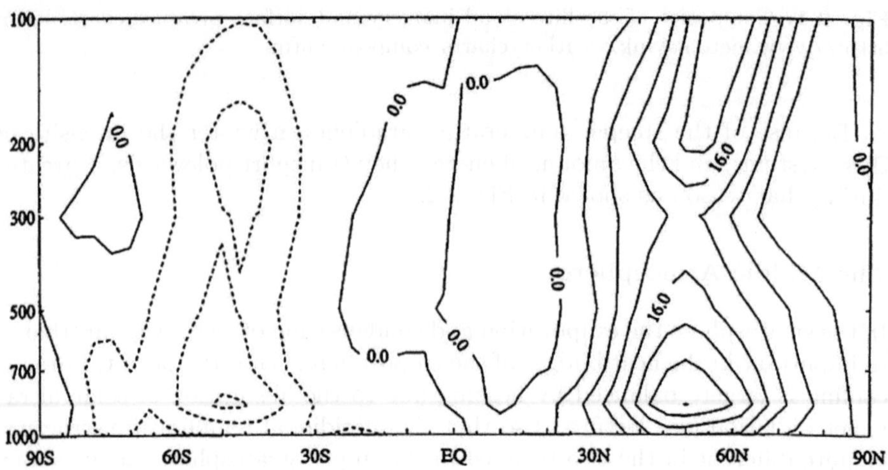

Fig. 2.3. Zonal mean eddy heat flux (°C m s^{-1}) in December as a function of latitude and pressure (hPa), positive values indicate a northward flux. Holton, 1992

2.3 Modes of Variability

The atmosphere exhibits a number of characteristic modes of variability that are important in determining the local climate in various regions. A discussion of the physical bases of these modes is beyond the scope of this course but descriptions of some of them are included as they may influence how the impact of solar variability on climate is experienced.

Identified modes of variability are several, each with different geographical influences, but all showing specific patterns of response with large regional variations. The El Niño-Southern Oscillation (ENSO) phenomenon is the leading mode in the tropics, although its influence is felt globally. At higher latitudes the leading northern winter modes are the North Atlantic Oscillation (NAO) and the Pacific-North America Oscillation (PNA), which are sometimes viewed as part of the same phenomenon, referred to as the Arctic Oscillation (AO). There is an Antarctic Oscillation (AAO) in the southern hemisphere. In the equatorial lower stratosphere a Quasi-Biennial Oscillation (QBO) modulates winds and temperatures.

It has been suggested that the impact of solar variability, as well as other climate forcing factors, may be to affect the frequency of occupation of certain phases of these modes. Below follows some short descriptions of the modes as well as how they have been proposed to respond to solar activity.

The North Atlantic Oscillation

Over the Atlantic in winter the average sea level pressure near 25–45°N is higher than that around 50–70°N (Fig. 1.6). This pressure gradient is associated with the storm-tracks which cross the ocean and determine, to a large extent, the weather and climate of western Europe. Since the 1930s it has been known that variations in the pressure difference are indicative of a large-scale pattern of surface pressure and temperature anomalies from eastern North America to Europe. If the pressure difference is enhanced then stronger than average westerly winds occur across the Atlantic, cold winters are experienced over the north-west Atlantic and warm winters over Europe, Siberia and East Asia, with wetter than average conditions in Scandinavia and drier in the Mediterranean. The fluctuation of this pattern is referred to as the NAO and the pressure difference between, say, Portugal and Iceland can be used as an index of its strength, see Fig. 2.4. Some authors regard the NAO as part of a zonally symmetric mode of variability, the AO, characterised by a barometric seesaw between the north polar region and mid-latitudes in both the Atlantic and Pacific.

Using individual station observational records of pressure, temperature and precipitation, values for the NAO index have been reconstructed back to the seventeenth century. The index shows large inter-annual variability but until recently has fluctuated between positive and negative phases with a period of approximately four or five years, see Fig. 2.4. Over the past two decades the NAO has been strongly biased towards its positive, westerly, phase and it has been suggested that this might be a response to global warming. Computer models of the general circulation of the atmosphere (GCMs) are quite successful in simulating NAO-type variability and some GCM studies do show increasing values of the NAO index with increased greenhouse gases. However, this is not true of all models' results and some studies suggest that the NAO pattern itself may be modified in a changing climate so

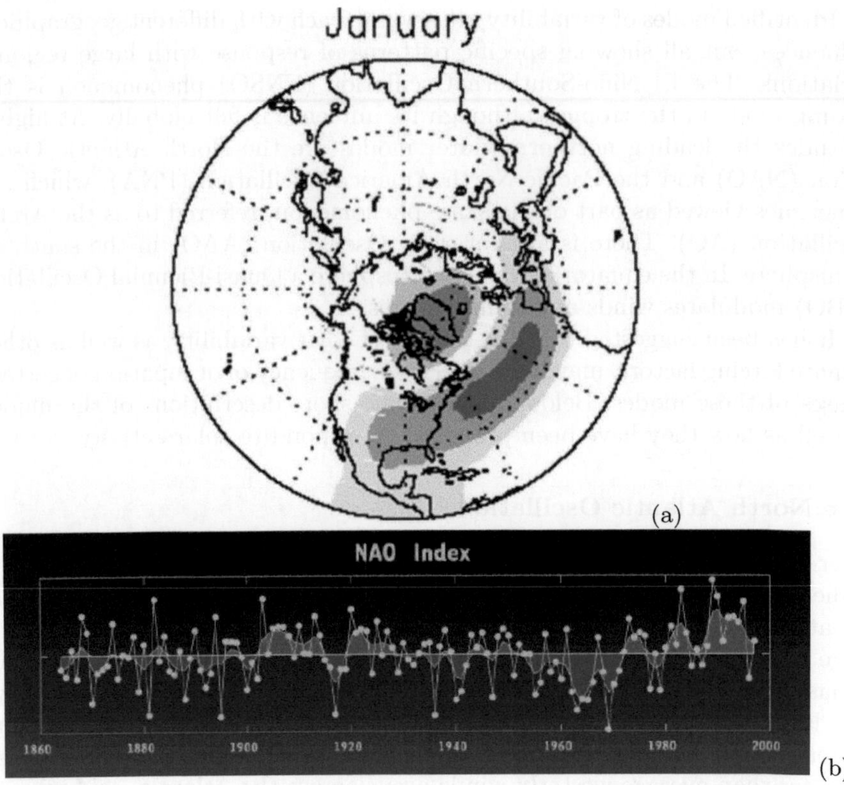

Fig. 2.4. *Top*: Pattern of anomalies in surface pressure for the positive phase of the NAO (red=+ve, blue=-ve). *Bottom*: time series of NAO index. From: http://www.ldeo.columbia.edu/res/pi/NAO/

that use of simple indices may not be appropriate. One recent GCM study (Shindell et al., 2001) of the Maunder Minimum period has suggested that the negative phase of the NAO may have been dominant during that period of low solar activity.

Another uncertainty is how significant might be coupling with either sea surface temperatures and/or the state of the middle atmosphere in producing a realistic NAO/AO pattern. Planetary-scale waves, produced in the lower atmosphere by longitudinal variations in topography, propagate upwards in winter high latitudes through the stratosphere and deposit momentum and heat which feeds into the general atmospheric circulation. Where this wave absorption takes place depends on the ambient temperature and wind structure. Thus any changes induced in the mean temperature structure of the stratosphere may result in a feedback effect on lower atmosphere climate. An analysis of zonal wind observations does suggest a downward propagation of AO patterns in many winters (Baldwin and Dunkerton, 2001). This offers a plausible mechanism for the production of NAO/AO-type signals in tropospheric climate by factors which affect the heat balance of the stratosphere (Kodera, 1995), specifically solar variability. Data and modelling studies have already shown such a response to heating in the lower stratosphere by volcanic eruptions (Robock, 2001).

The El Niño-Southern Oscillation

The El Niño was the name given to the periodic warming of the ocean waters near the coast of Peru and Ecuador which adversely impacts the fishing industry in this region due to the suppression of upwelling nutrient-rich deep waters. The Southern Oscillation was the term used to describe a periodic variation observed in the east-west gradient of surface atmospheric pressure across the equatorial Pacific. It is now realised that these two phenomena are both parts of the same complex interaction of the ocean and atmosphere, subsequently referred to as ENSO. In an ENSO event positive anomalies of sea surface temperature initially appear on the east side of the equatorial Pacific and over a period of a few months spread westward until they cover most of the ocean at low latitudes, as shown in Fig. 2.5.

Associated with this the region of maximum rainfall, normally over the maritime continent, shifts eastward into the Pacific. A wide range of interannual climate anomalies, in the extra-tropics as well as near the equator, appear to be associated with ENSO but there is not, as yet, a complete theory which can explain the complex coupling of the atmospheric and oceanic dynamics and thermodynamics which is taking place. ENSO events occur every 2–5 years but seem to have been increasing in frequency over the past two decades, see Fig. 2.5.

The Quasi-Biennial Oscillation

In the equatorial lower stratosphere an oscillation in zonal winds occurs with a period of approximately 28 months. A given phase (east or west) starts in the upper stratosphere and moves downward at a rate of about 1km per month to be replaced by winds of the opposite phase (see Fig. 2.6). The largest amplitude in the zonal wind variation occurs at about 27 km altitude. The QBO comes about because of interactions between vertically propagating waves and the mean flow. When the wind blows from the west (QBO west phase) westward moving waves can propagate freely but eastward moving waves are absorbed and deposit their momentum, thus strengthening the existing westerlies and moving the westerly peak downwards. Somewhat above this absorption layer the westward moving waves are dissipated and weaken the west wind, eventually changing the direction to easterly. The absorption of the westward waves then starts to propagate downwards, reversing the phase of the QBO. Baldwin and Dunkerton (2001) give a good review of current understanding of the QBO.

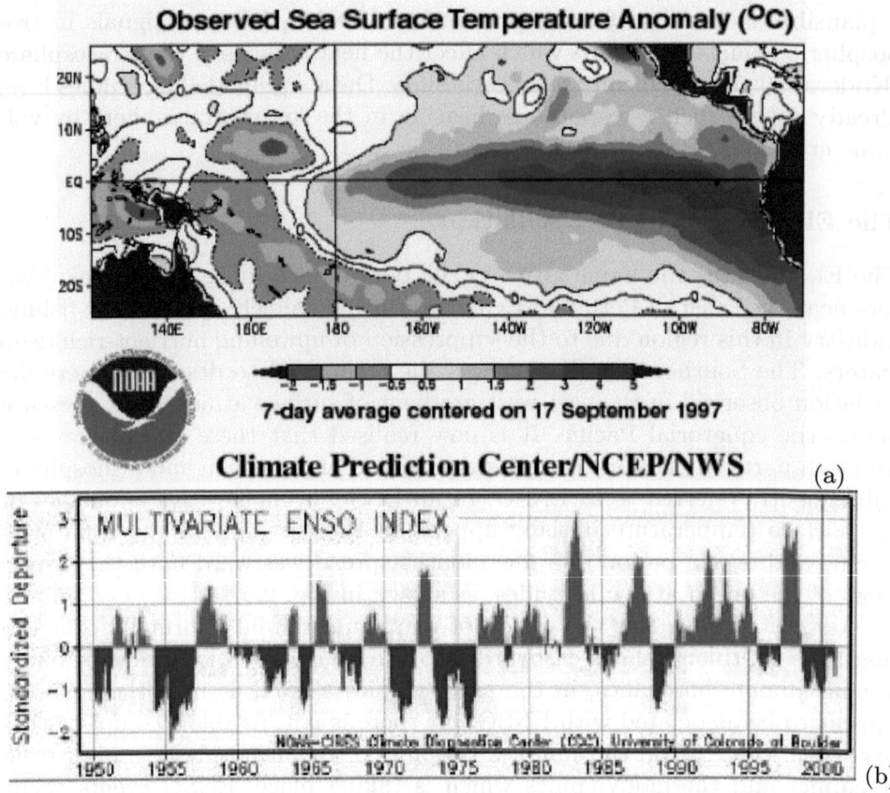

Fig. 2.5. *Top*: Anomalies in Pacific sea surface temperatures associated with ENSO SST. From the U.S. National Oceanic and Atmospheric Climate Prediction and Diagnostics Centers

The effects of the QBO are not restricted to equatorial regions. The transport of heat, momentum and ozone to high latitudes are all modulated. On average the westerly phase of the QBO is associated with colder winter temperatures at the north pole in the lower stratosphere. This can be explained by an enhanced ability of mid-latitude planetary waves to propagate into the equatorial westerlies leaving the cold winter pole undisturbed. However, this relationship only appears to hold when the Sun is at lower levels of activity. Near 11-year cycle maxima the relationship breaks down and possibly reverses. The absorption of solar radiation is greater in the stratosphere than lower down and its modulation by solar activity quite significantly larger (see Fig. 7.7). The potential for solar modulation of stratospheric temperatures and winds to modify planetary wave propagation, and hence tropospheric climate, is an active area of research.

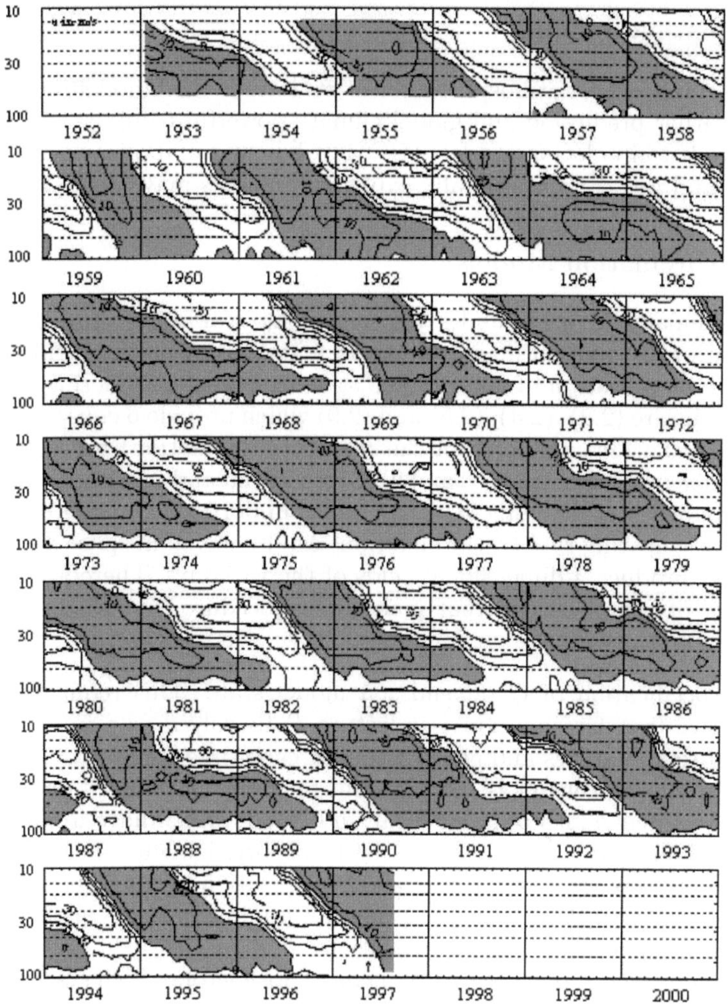

Fig. 2.6. A time series of the zonal wind profile above Singapore. The contour interval is 10 ms^{-1}) and negative (easterly) winds are shaded. The downward propagation of alternating easterly and westerly wind is the QBO

Signals with a periodicity of 2–3 years have also been found in records of surface temperature and precipitation and, recently, in the NAO index. This interaction between modes presents a much more complex picture but again suggests a mechanism whereby the response of climate parameters to solar variability could have a marked geographical distribution.

2.4 Climate Modelling

An important tool for atmospheric scientists is numerical modelling. For studies of weather prediction and global climate powerful computers are needed to run global circulation models (GCMs). For many applications, however, simpler, and less computationally expensive, models may be appropriate.

Global Circulation Models

GCMs simulate the state and evolution of the atmosphere through solving the equations of motion at all points on a three-dimensional grid (or other discrete representation of space). The starting point, considering at present only dry air, are (2.2), (2.4), (2.5) and (2.9) which provide 6 equations (taking account of the 3 components of the momentum equation) for 6 unknowns, *viz* $\boldsymbol{u}$ (three components), p, ρ and T. Given initial and boundary conditions, some properties of dry air (M, C_p, η) and, importantly, a grid of values of the diabatic heating rate Q, the equations can be solved in space and iterated in time to produce 4-dimensional fields of the variables. There are, however, a number of complications that make this procedure somewhat less simple than may first appear:

1. The specification of Q depends on model state. This requires that the radiative transfer equation (see lecture 4) be solved as a function of the model temperatures and composition at each time-step.
2. The presence of water is fundamental to the atmospheric condition providing sources/sinks of latent heat and clouds which have a huge impact on Q. Thus water vapour has to be transported within the model and a scheme to represent convection and condensation needs to be included.
3. The discretisation in space and time results in rounding errors and also, possibly, numerically unstable solutions.
4. Some representations need to be included for factors which occur on scales smaller than the grid size but nevertheless impact the large scale (e.g. turbulence).
5. Constraints on computer resources (processing speed, memory, storage) place limitations on what it is possible.

Models of Reduced Complexity

Due to the computational constraints, mentioned above, it is not feasible to include the best possible representations of all the known relevant physical and chemical processes in a GCM. It may not even be desirable – sometimes interpreting GCM output is almost as difficult as data from the real atmosphere! Under such circumstances it is often more useful to use simplified models which focus on one (or more) aspects of a problem at the expense of the treatment of other features which are deemed to be less important in the particular context.

A common approach is to reduce the dimensionality of the model, at the expense of a complete representation of dynamical processes. For example, 2D (latitude-height) models, in which zonal mean quantities are considered, have been successfully used in studies of stratospheric chemistry (see Sect. 7.1). 1D (height only) Energy Balance Models, which have no transport apart from vertical diffusion, can provide a useful first order estimate of global mean response to radiative forcing perturbations (an example is given in Sect. 8.2).

Alternatively, if the focus of attention is on a process occurring in a particular region of the atmosphere, it may be useful to restrict the spatial extent of the model, although care must then be taken with boundary conditions. Some of the work on wave-mean flow interactions in the middle atmosphere, discussed in Sect. 8.3, was carried out with a model in which the lower boundary was set at the tropopause, thus avoiding the necessity to simulate the complex cloud-radiation and boundary layer effects within the troposphere and allowing long integrations to take place.

Another approach is to simplify the treatment of certain processes in order to focussing on others. For example, the studies of coupling between the tropical lower stratosphere and tropospheric circulations discussed in Sect. 8.3 were carried out with a model in which all diabatic processes were reduced to a Newtonian relaxation to a reference equilibrium temperature. This maintained the full representation of dynamical processes but avoided the necessity for detailed calculations of Q thus allowing numerous experiments to be carried out.

3 Climate Records

In this lecture we will consider how information on past climates is derived and look at some of the (mainly circumstantial) evidence that variations in solar activity have affected climate on a wide range of timescales.

3.1 Measurements and Reconstructions

Observational Records

Assessment of climate variability and climate change depends crucially on the existence and accuracy of records of meteorological parameters. Ideally records would consist of long time series of measurements made by well-calibrated instruments located with high density across the globe. In practice, of course, this ideal can not be met. Measurements with global coverage have only been made since the start of the satellite era about 25 years ago. Instrumental records have been kept over the past few centuries at a few locations in Europe. For longer periods, and in remote regions, records have to be reconstructed from indirect indicators of climate known as proxy data.

The longest homogeneous instrument-based temperature series in the world is the Central England Temperature record dating back to 1659. It was first constructed in the 1970s from an accumulation of measurements made by amateur meteorologists in central lowland England. The construction of a homogeneous record requires knowledge of how, and at what time of day, the measurements were made and how local conditions may account for regional variations. Other similar temperature records dating back to the mid 18th century are available for Munich, Vienna, Berlin and Paris among other sites in Europe and at least one in the Eastern United States. These datasets have been extensively analysed for periodic variations in temperature. Generally they show clear indications of variations on timescales of about 2.2 to 2.4 years (QBO?) and 2.9 to 3.9 years (ENSO?) but on longer timescales individual records show peaks at different frequencies with little statistical significance.

The most complete record of rainfall comes from eastern China where careful observations of floods and droughts date back to the fifteenth century. Again spectral analysis results in periodicities which vary from place to place.

Figure 3.1 shows instrumental measurements of surface temperature compiled to produce a global average dating back to 1860. Much of the current concern with regard to global warming stems from the obvious rise over the twentieth century. A key concern of contemporary climate science is to attribute cause(s) to this warming. This is discussed further in lecture 8.

Other climate records suggesting that the climate has been changing over the past century include the retreat of mountain glaciers, sea level rise, thinner Arctic ice sheets and an increased frequency of extreme precipitation events.

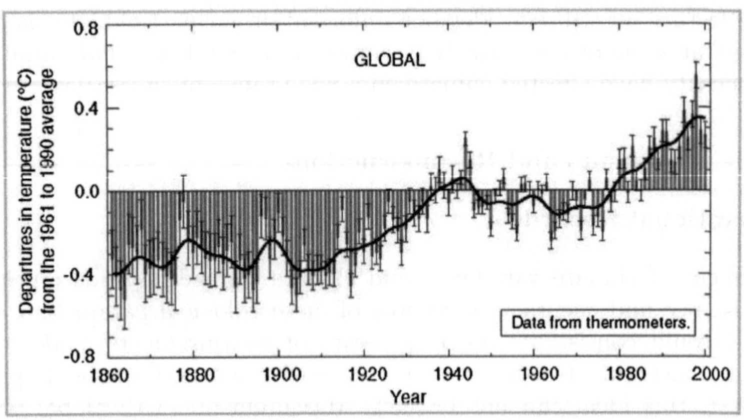

Fig. 3.1. Variations in the global annual average surface temperature over 140 years from instrumental records. From IPCC (2001)

Proxy Records

Proxy data provide information about weather conditions at a particular location through records of a physical, biological or chemical response to these conditions. Some proxy datasets provide information dating back hundreds of thousands of years which make them particularly suitable for analysing long term climate variations and their correlation with solar activity.

One well established technique for providing proxy climate data is dendrochronology, or the study of climate changes by comparing the successive annual growth rings of trees (living or dead). It has been found that trees from any particular area show the same pattern of broad and narrow rings corresponding to the weather conditions under which they grew each year. Thus samples from old trees can be used to give a time series of these conditions. Felled logs can similarly be used to provide information back to ancient times, providing it is possible to date them. This is usually accomplished by matching overlapping patterns of rings from other trees. Another problem that arises with the interpretation of tree rings is that the annual growth of rings depends on a number of meteorological variables integrated over more than a year so that the dominant factor determining growth varies with location and type of tree. At high latitudes the major controlling factor is likely to be summer temperature but at lower latitudes humidity may play a greater role. Figure 3.2 shows a 1000-year surface temperature record reconstructed from proxy data, including tree rings. It shows that current temperatures are higher than they have been for at least the past millennium.

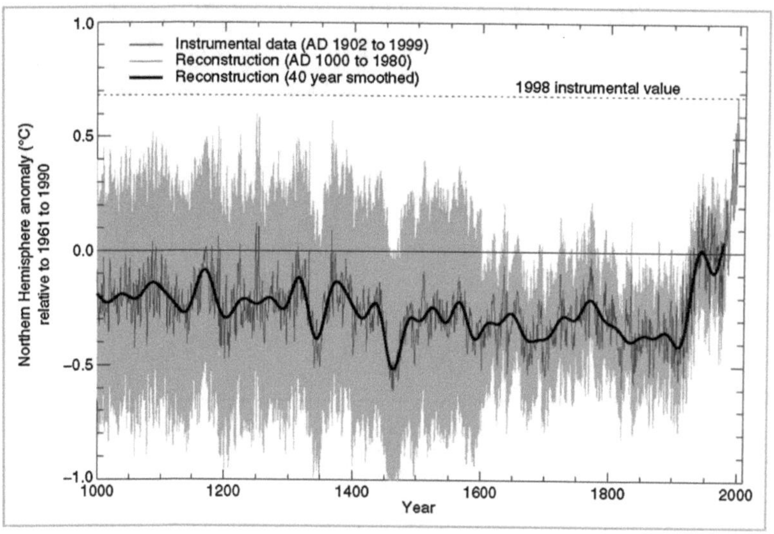

Fig. 3.2. Variations in Northern Hemisphere surface temperature over the past millennium from proxy records (tree rings, corals, ice cores). From IPCC (2001)

Much longer records of temperature have been derived from analysis of oxygen isotopes in ice cores obtained from Greenland and Antarctica. The ratio of the concentration of ^{18}O to that of ^{16}O, or ^{2}H to ^{1}H, in the water molecules is determined by the rate of evaporation of water from tropical oceans and also the rate of precipitation of snow over the polar ice caps. Both these factors are dependent on temperature such that greater proportions of the heavy isotopes are deposited during periods of higher global temperatures. As each year's accumulation of snow settles the layers below become compacted so that at depths corresponding to an age of more than 800 years it becomes difficult to precisely date the layers. Nevertheless, variations on timescales of more than a decade have been extracted dating back over hundreds of thousands of years.

Figure 3.3 shows the temperature record deduced from the deuterium ratio in an ice core retrieved from Vostok in East Antarctica. The roughly 100,000 year periodicity of the transitions from glacial to warm epochs is clear and suggests a relationship with the variations in eccentricity of the Earth's orbit around the Sun (see the discussion of Milankovitch cycles in lecture 8) although this does not explain the apparently sharp transitions from cold to warm seen in Fig. 3.3. The figure also presents the concentrations of methane

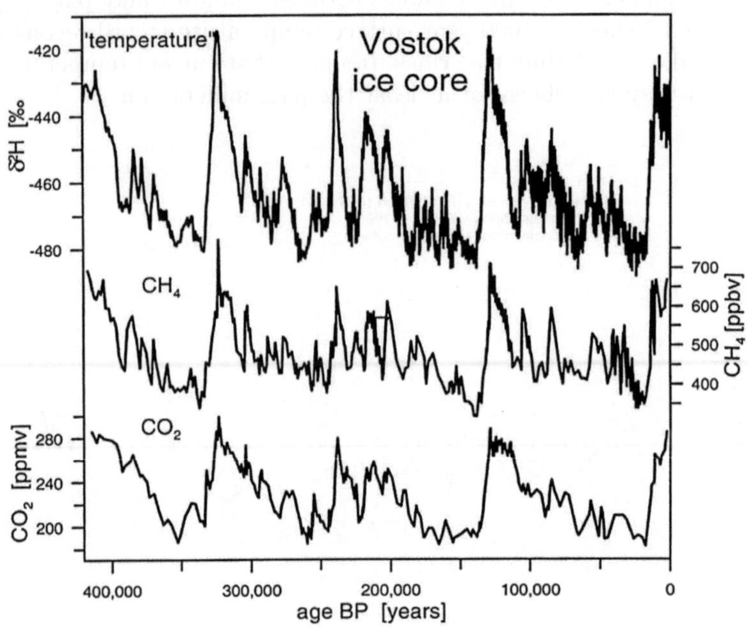

Fig. 3.3. Records derived from an ice core taken from Vostok, East Antarctica, showing variations in temperature (derived from deuterium measurements) and the concentrations of methane and carbon dioxide over at least 400,000 years. From Stauffer (2000)

and carbon dioxide preserved in the ice core showing a strong correlation between these and temperature. Note that neither concentration is as high as the present day values. One theory (Petit et al, 1999) suggests that the warming of southern high latitudes caused by the orbital variations is amplified by the release of CO_2 from the southern oceans and this warming is then further amplified through a reduction in albedo resulting from the melting of Northern Hemisphere ice sheets. Such positive feedback mechanisms might explain the sharp increases in temperature seen in the record.

Evidence of very long term temperature variations can also be obtained from ocean sediments. The skeletons of calciferous plankton make up a large proportion of the sediments at the bottom of the deep oceans and the ^{18}O component is determined by the temperature of the upper ocean at the date when the living plankton absorbed carbon dioxide. The sediment accumulates slowly, at a rate of perhaps 1m every 40,000 years, so that changes over periods of less than about 1,000 years are not detectable but ice age cycles every 100,000 years are clearly portrayed.

Ocean sediments have also been used to reveal a history of temperature in the North Atlantic by analysis of the minerals believed to have been deposited by drift ice (Bond et al, 2001). In colder climates the rafted ice propagates further south where it melts, depositing the minerals. An example of such an analysis is presented in Sect. 3.2 below.

3.2 Solar Signals in Climate Records

Many different approaches have been adopted in the attempt to identify solar signals in climate records. Probably the simplest has been the type of spectral analysis mentioned above, in which cycles of 11 (or 22 or 90 etc) years are assumed to be associated with the sun. In another approach time series of observational data are correlated with time series of solar activity. In an extension of the latter method simple linear regression is used to extract the response in the measured parameter to a chosen solar activity forcing. A further sophistication allows a multiple regression, in which the responses to other factors are simultaneously extracted along with the solar influence. Each of these approaches gives more faith than the previous that the signal extracted is actually due to the sun and not to some other factor, or to random fluctuations in the climate system, and many interesting results emerge. It should be remembered, however, that such detection is based only on statistics and not on any understanding of how the presumed solar influence takes place. Some of the mechanisms which have been proposed to explain how changes in the Sun affect the climate will be discussed in lecture 8.

Millennial, and Longer, Timescales

Section 3.1 mentioned how temperature records may be extracted from ice cores and ocean sediments. These media may also preserve information on cosmic ray flux, and thus solar activity, in isotopes such as ^{10}Be and ^{14}C. Thus simultaneous records of climate and solar activity may be retrieved. An example is given in Fig. 3.4 which shows fluctuations on the 1,000 year timescale well correlated between the two records, suggesting a long-term solar influence on climate.

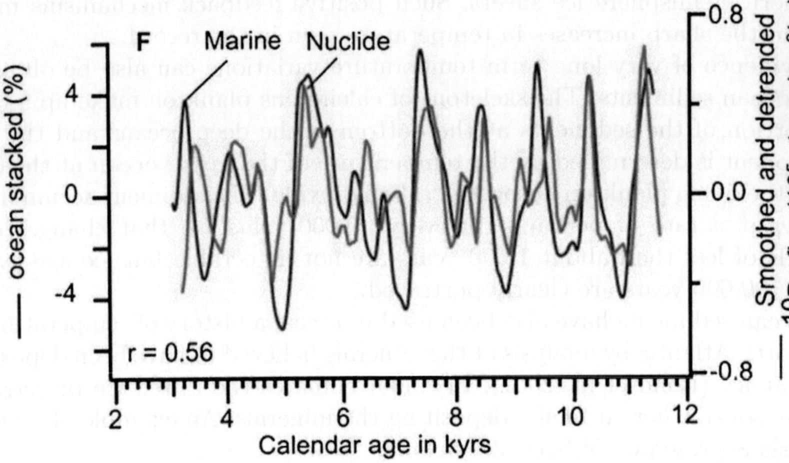

Fig. 3.4. Records of ^{10}Be and ice-rafted minerals extracted from ocean sediments in the North Atlantic. Bond et al. (2001)

On somewhat shorter timescales it has frequently been remarked that the Maunder Minimum in sunspot numbers in the second half of the seventeenth century coincided with what has become known as the "Little Ice Age " during which western Europe experienced significantly cooler temperatures. Figure 3.5 shows this in terms of winter temperatures measured in London and Paris compared with the ^{14}C ratio found in tree rings for the same dates. Similar cooling has not, however, been found in temperature records for the same period across the globe so attribution of the European anomalies to solar variability may be unwarranted.

Century Scale

A paper published by Friis-Christensen and Lassen in 1991 caused considerable interest when it appeared to show that temperature variations over the observational period could be ascribed entirely to solar variability. The measure of solar activity used was the length of the solar cycle (SCL) and, as can be seen in Fig. 3.6, this value coincides almost exactly with the Northern Hemisphere land surface temperature record. The SCL values were, however, smoothed with a (1, 2, 2, 2, 1) running filter so that at each point the value

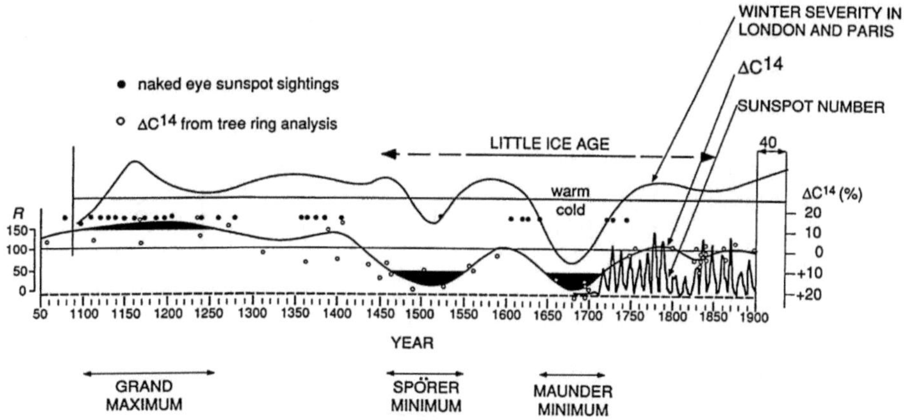

Fig. 3.5. From a paper by Eddy (1976) suggesting that winter temperatures in NW Europe are correlated with solar activity. Note the coincidence of the "Little Ice Age" with the Maunder Minimum in sunspots

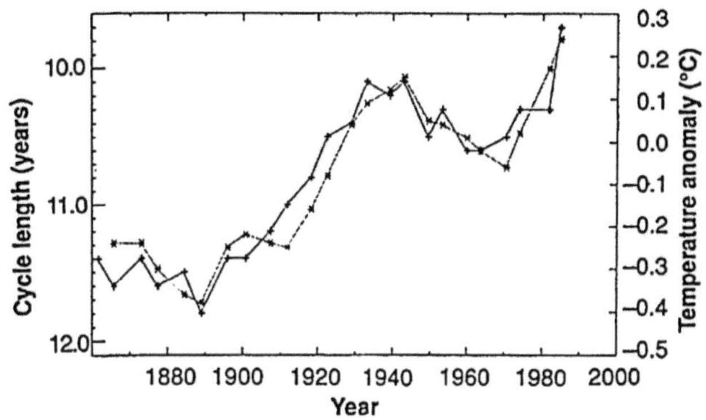

Fig. 3.6. Records of northern hemishere land temperature (*stars*) and length of the solar cycle (inverted, pluses)

given has contributions from four solar cycles both forwards and back in time, i.e. a total of ~88 years. This means that the values given for dates more recent that about 1950 required some extrapolation, and the more recent the more the result depended on assumptions about future behaviour. Thus the upturn in values in the latter part of the twentieth century was essentially construed. The authors later (Lassen and Friis-Christensen, 2000) corrected this, and extended their analysis back over 500 years using a temperature reconstruction, as shown in Fig. 3.7(a) and the fit still shows a strong correlation between SCL and temperature at least up to about 1950. This work has, however, been reanalysed by Laut and Gundermann (2000) who argue that the higher weightings in the fit given to more recent data were not justified and presented a new version in which equal weightings were given to

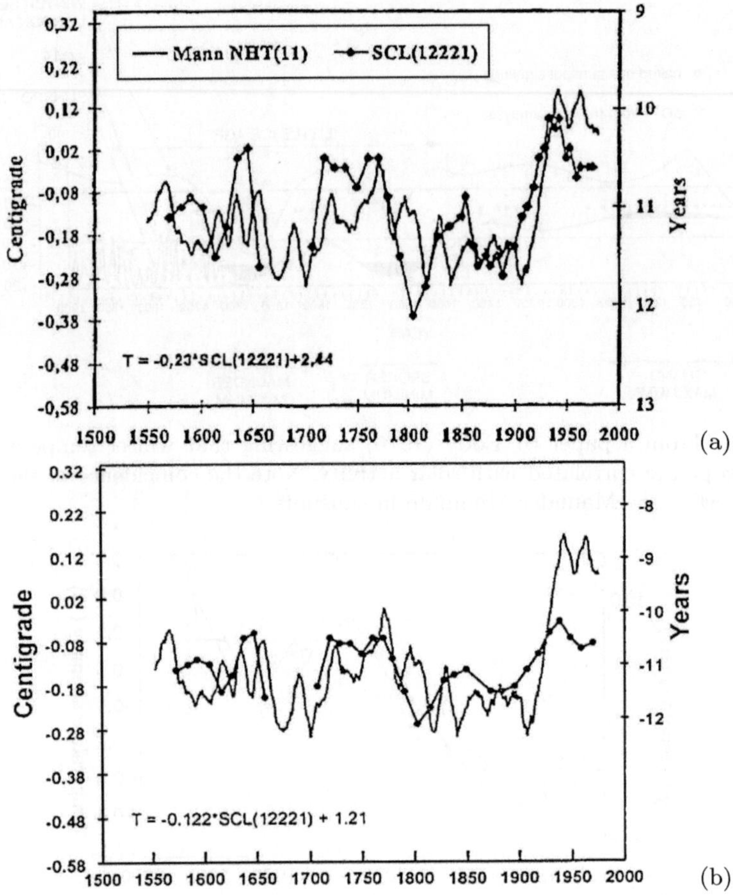

Fig. 3.7. Time series of smoothed solar cycle lengths as fitted by: (**a**) Lassen and Friis-Christensen (2000) and (**b**) Laut and Gundermann (2000) to the Mann et al. (1999) Northern Hemisphere temperature record

all data points, as shown in Fig. 3.7(b). The correspondence between the two records in the twentieth century now appears much less marked. This saga presents a salutary lesson in the care which must be taken in data correlation studies and the conclusions which can be drawn from them.

The detection and attribution of the causes of twentieth century climate change is still not fully resolved, see lecture 8.

Solar Cycle

Many studies have purported to show variations in meteorological parameters in phase with the "11-year" solar cycle. Some of these are statistically not robust and some show signals that appear over a certain period only to disappear, or even reverse, over another period. There is, however, considerable evidence that solar variability on decadal timescales does influence climate.

Figure 3.8 shows the annual mean, over the sub-tropical Pacific Ocean, of the geopotential height of the 30 hPa pressure surface – a measure of the mean temperature of the atmosphere below about 24 km altitude. It varies in phase with the solar 10.7 cm radio wave flux over three and a half solar cycles with an amplitude suggesting that the lower atmosphere is 0.5–1.0 K warmer at solar maximum than at solar minimum. This is a large response but from this figure alone it is not clear whether it applies locally or globally or how the temperature anomaly is distributed in the vertical.

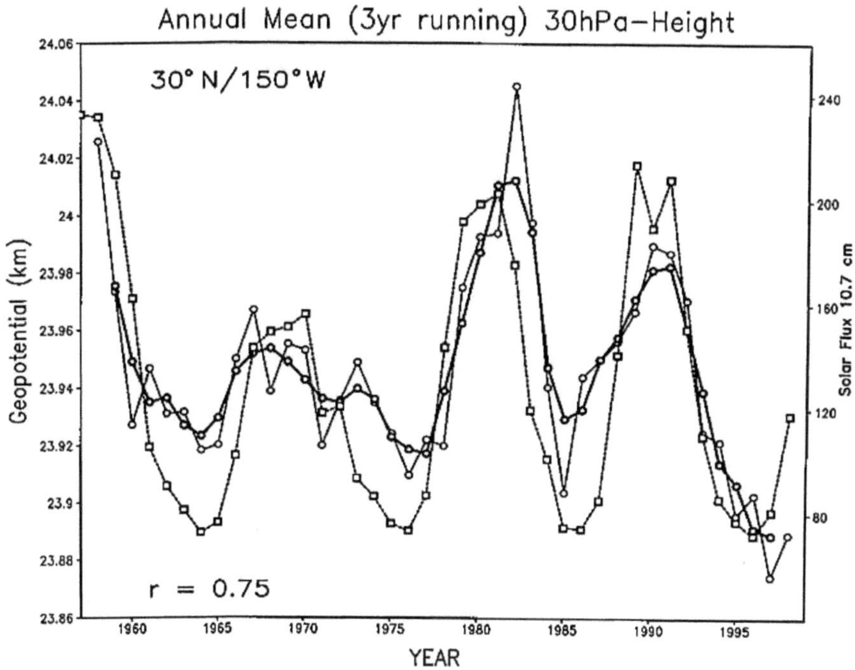

Fig. 3.8. Time series of annual mean 30 hpa geopotential height (km) at 30°N, 150°W (*thin line with circles*), its three-year running mean (*thick line with circles*) and the solar 10.7 cm flux (*dashes line with squares*). Labitzke and van Loon (1995)

Figure 3.9 shows the mean summer time temperature of the upper troposphere (between about 2.5 and 10 km) averaged over the whole northern hemisphere. Again this parameter varies in phase with the solar 10.7 cm index although with a smaller amplitude, 0.2–0.4 K, than suggested by the single sub-tropical station discussed above.

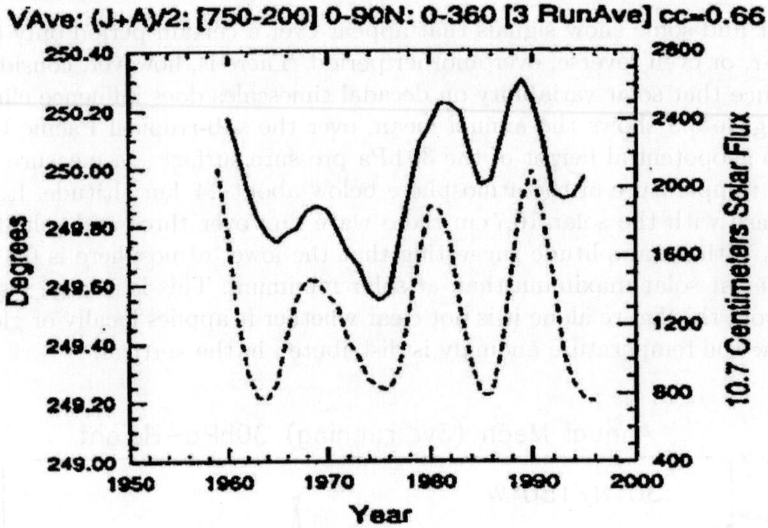

Fig. 3.9. Time series of mean temperature of 750–200 hpa layer in the northern hemisphere summer (*solid line*) and the solar 10.7 cm flux (*dashes line*). Van Loon and Shea (2000)

Solar signals have also been detected in sea surface temperatures (SSTs). Figure 3.10 presents some results from an EOF analysis of SSTs with the upper panel showing the time varying amplitude of the pattern of response shown in the lower panel. Two interesting features emerge from this work, one is that SSTs do not warm uniformly in response to enhanced solar activity: indeed, the pattern shows latitudinal bands of warming and cooling, and secondly that the amplitude of the change is larger than would be predicted by radiative considerations alone, given the known variations in TSI over the same period.

A similar pattern is shown for the solar signal in the results of a multiple regression analysis of NCEP/NCAR Reanalaysis zonal mean temperatures (Fig. 3.11). In this work data for 1978-2002 were analysed simultaneously for ten signals: a linear trend, ENSO, NAO, solar activity, volcanic eruptions, QBO and the amplitude and phase of the annual and semi-annual cycles. The patterns of response for each signal are statistically significant and separable from the other patterns. The solar response shows largest warming in the stratosphere and bands of warming, of >0.4 K, throughout the troposphere in mid-latitudes. The mechanisms whereby this takes place are discussed in lecture 8 but it is worth noting here that the analysis was carried out independently at each grid-point so that the hemispheric symmetry of the solar signal and the confinement of the NAO pattern to northern mid-high latitudes provides support for the validity of these two results.

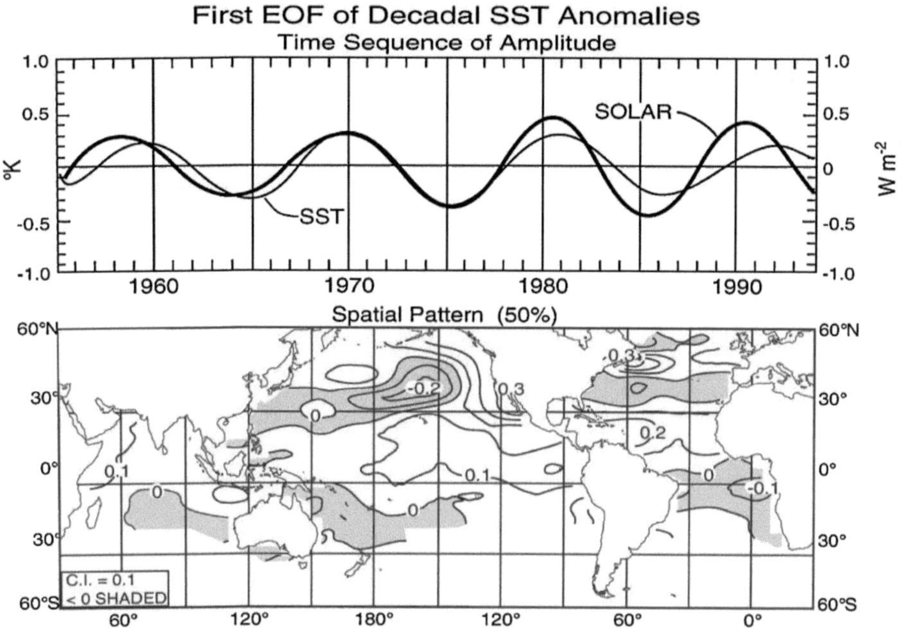

Fig. 3.10. *Bottom*: pattern of response in sea surface temperatures derived for solar variability. *Top*: time series of the magnitude of this pattern. White et al. (1997)

Also shown in Fig. 3.11 are some results from a similar multiple regression analysis of zonal mean zonal winds. These show that the effect of solar activity is to weaken the sub-tropical jets and to move them polewards. The effect of a volcanic eruption is also to weaken the jets but to move them equatorwards. In both cases the predominant direct effect is heating of the lower stratosphere with the tropospheric signal showing a dynamical response. How this takes place is discussed in lecture 8.

The solar effects seen in the NCEP temperature and wind data have been reproduced by GCM simulations of the effects of increased solar variability (Haigh, 1996, 1999; Larkin et al. 2000). These studies also predicted a weakening and expansion of the tropical Hadley cells in response to solar activity. Recent analysis of NCEP vertical velocity data has confirmed that this effect is also seen in the real atmosphere (Fig. 3.12).

Another interesting study of correlations between solar activity and climate was carried out by Marsh and Svensmark (2000) who showed that over about one and a half solar cycles low latitude low cloud cover varied in phase with galactic cosmic ray (GCR) intensity (see Fig. 3.13). Physical mechanisms whereby this might take place are discussed in lecture 9 but it is worth noting here that, if there is a link between cloud and solar activity, then this result alone does not show that it is due to GCRs any more than any other

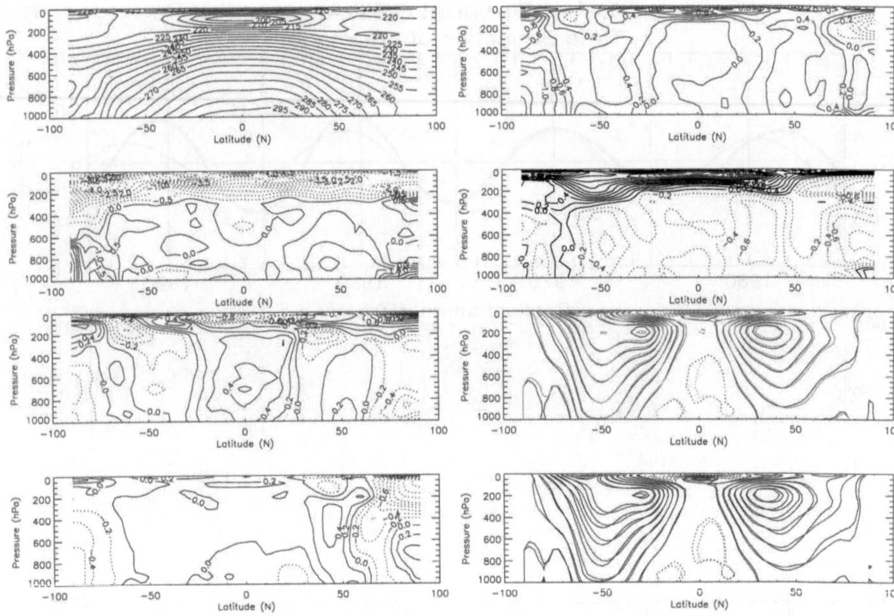

Fig. 3.11. Results from multiple regression analysis of NCEP zonal mean temperatures and zonal winds (Haigh, 2003). *Left-hand-side* from top: mean temperature; temperature trend (K/decade); ENSO max-min temperature (K); NAO max-min temperature (K). *Right-hand-side*: solar cycle max-min temperature (K); effect of Piña Tubo eruption on temperature (K); zonal mean zonal wind solar max (*grey*), solar min (*black*), positive contours *solid* lines, negative dashed, contour interval 5 ms^{-1}; after Piña Tubo (*grey*), before (*black*)

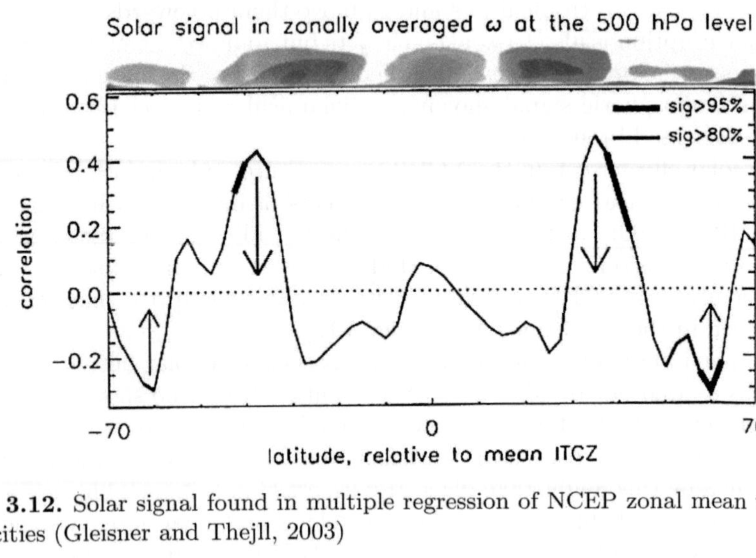

Fig. 3.12. Solar signal found in multiple regression of NCEP zonal mean vertical velocities (Gleisner and Thejll, 2003)

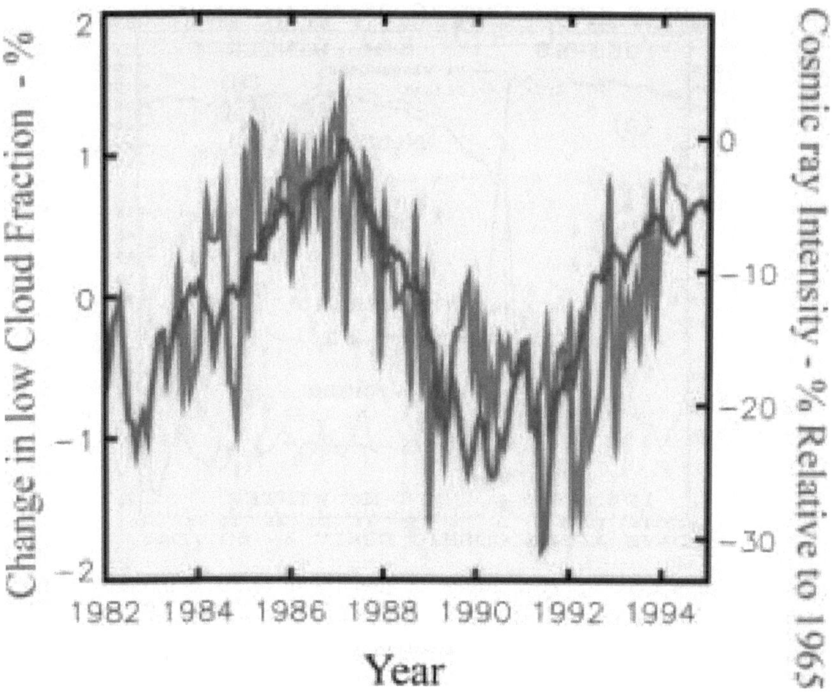

Fig. 3.13. Anomaly in low latitude, low level cloud over the ocean (*darker curve*) and galactic cosmic rays (paler), from Marsh and Svensmark (2000)

solar-modulated input to the Earth. Furthermore, an update to Fig. 3.13 (Haigh, 2003) suggests that the correlation weakens after 1994 so it remains to be seen if this result is robust.

Shorter Timescales

Various studies have suggested that the atmosphere responds to solar activity effects on timescales much shorter than the solar cycle. A response to the solar 27-day rotation has been clearly observed in middle atmosphere composition and temperature, which will be discussed in lecture 7. On even shorter timescales correlations have been observed between decreases in GCRs associated with solar coronal mass ejections and the areas of cyclonic storms, see Fig. 3.14.

4 Radiative Processes in the Atmosphere

Solar radiation is the fundamental energy source for the atmosphere. In this lecture we will consider the input of solar radiation to the Earth and how this

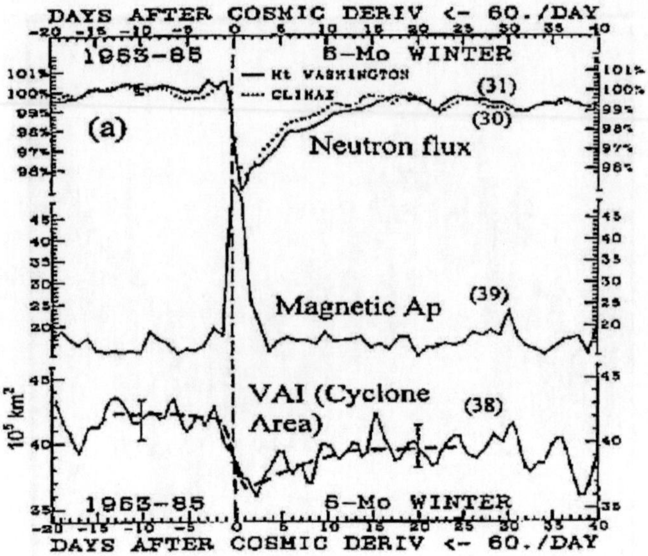

Fig. 3.14. Superposed epoch analysis of changes in neutron monitor count rate (associate with GCRs), and vorticity area index (a measure of the area covered by cyclonic storm in the northern hemisphere in winter). Tinsley (2000)

is balanced by thermal emission. We will investigate the radiative processes which determine the atmospheric temperature structure and some of the theoretical basis of the scattering of radiation, including the influence of cloud droplet size on albedo.

4.1 Earth Radiation Budget

Global Average

The global average equilibrium temperature of the Earth is determined by a balance between the energy acquired by the absorption of incoming solar radiation and the energy lost to space by the emission of thermal infrared radiation. The amount of solar energy absorbed depends both on the incoming irradiance and on the Earth's reflective properties. If either of these changes then the temperature structure of the atmosphere-surface system tends to adjust to restore the equilibrium. In order to understand how these processes affect climate it is important to investigate in more detail the solar and infrared radiation streams. Figure 4.1 shows the components of the global annual average radiation budget and how much radiation is absorbed, scattered and emitted within the atmosphere and at the surface. The value for the incoming radiation, 342 $\mathrm{Wm^{-2}}$, is equivalent to a total solar irradiance at the Earth of 1368 $\mathrm{Wm^{-2}}$ averaged over the globe. Of this 31% (107 $\mathrm{Wm^{-2}}$) is reflected back to space by clouds, aerosols, atmospheric molecules and

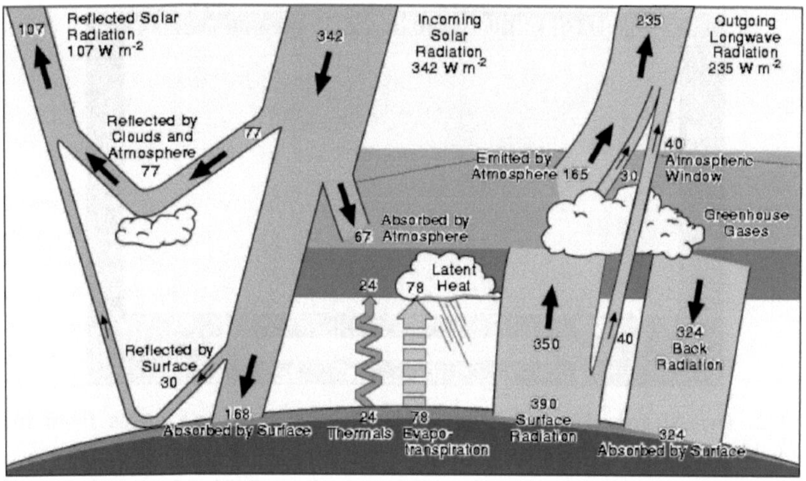

Fig. 4.1. Globally averaged energy budget of the atmosphere. Figure from http://asd-www.larc.nasa.gov/ceres/brochure/clouds-and-energy based on data from Kiehl and Trenberth, 1997

the surface, with the clouds playing the most important role, so that only 235 Wm^{-2} is absorbed by the Earth system. 20% (67 Wm^{-2}) of the incident radiation is absorbed within the atmosphere leaving 49% (168 Wm^{-2}) to reach and heat the surface.

The temperature and emissivity of the surface are such that 390 Wm^{-2} of infrared energy are emitted into the atmosphere. Only 40 Wm^{-2} of this, however, escapes to space with the remainder being absorbed by atmospheric gases and cloud. The atmosphere returns 324 Wm^{-2} to the surface. The energy balance at the surface is achieved by non-radiative processes such as evaporation and convection. The radiation balance at the top of the atmosphere is achieved by the 195 Wm^{-2} of heat energy emitted to space by the atmosphere and clouds. The figure also gives an indication of how atmospheric and surface properties may affect the vertical temperature structure of the atmosphere. This is discussed further in Sect. 4.3.

Geographical Distribution

Figure 4.1 considers only the global annual average situation. Hidden within this energy balance, however, there are wide geographic and seasonal variations in the radiation budget components. Figure 4.2 presents data acquired by the ERBE instrument (on the ERBS and NOAA-9 satellites) which give an indication of the spatial distributions. Figure 4.2 (a) shows the absorbed solar radiation in July. The incident radiation is greatest at the sub-solar point (near 20°N), and no radiation is incident in the polar night (south of 70°S), but the pattern of absorbed radiation is complicated by the presence

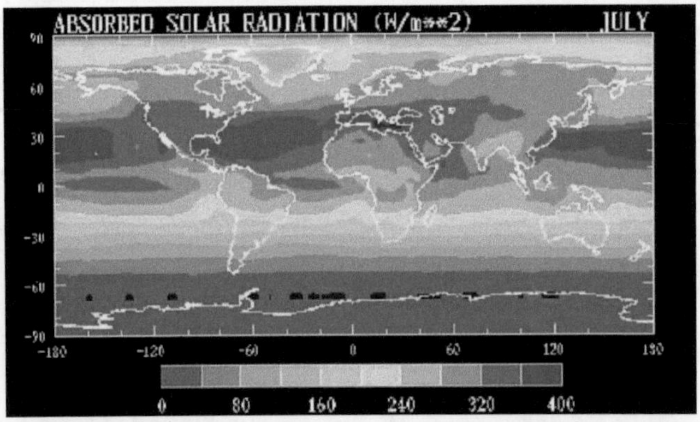

Fig. 4.2. (**a**) Absorbed solar radiation in July (Wm^{-2}) (Figures from http://rainbow.ldeo.columbia.edu/ees/data/)

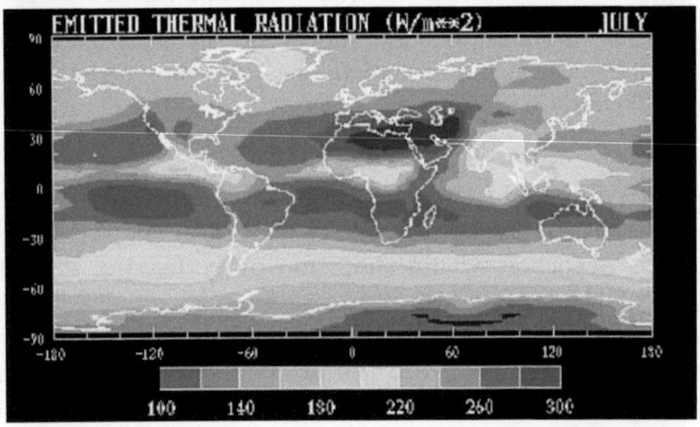

(b) As Fig. 4.2(**a**) but for emitted thermal radiation

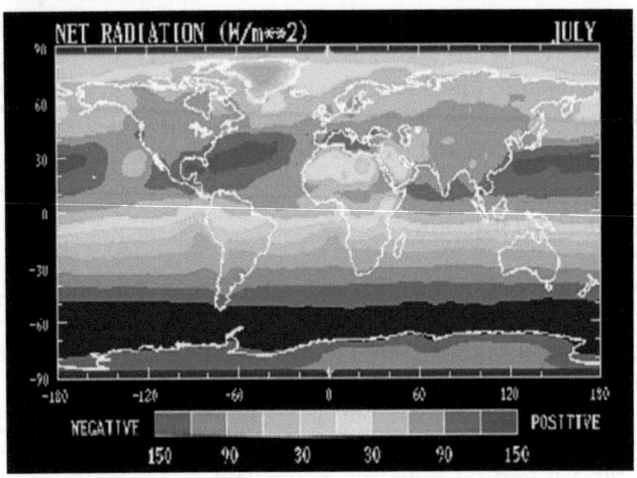

(c) As Fig. 4.2(**a**) but for net incoming radiation

of tropical cloud and reflective surfaces. These enhance the albedo and thus result in patches of reduced absorption.

The map of thermal radiation (Fig. 4.2 (b)) indicates the temperature of the emitting surface with, in general, more being emitted from the warmer low latitudes and summer hemisphere. However, the presence of cloud near 10°N and in the Indian summer monsoon provide colder radiating surfaces and thus patches of lower emission. The distribution of net radiation (Fig. 4.2 (c)) shows generally positive values (more absorbed than emitted) in the summer hemisphere and negative values south of 10°S and near the north pole. There are significant exceptions, however, e.g. negative values over the Sahara and the Arabian peninsular, due to a combination of hot surface temperatures and high albedo. The low latitude excess of energy must be transported, by either the atmosphere or oceans, to make up the deficit at high latitudes. Thus the radiation balance is intrinsically linked with large scale atmospheric and oceanographic circulations.

4.2 Absorption of Solar Radiation by the Atmosphere

Solar Irradiance

Absorption by the atmosphere of solar radiation depends on the concentrations and spectral properties of the atmospheric constituents. Figure 4.3 shows a blackbody spectrum at 5750 K, representing solar irradiance at the top of the atmosphere, and a spectrum of atmospheric absorption. Absorption features due to specific gases are clear with molecular oxygen and ozone being the major absorbers in the ultraviolet and visible regions and water vapour and carbon dioxide more important in the near-infrared.

The direct solar flux (i.e. ignoring diffuse radiation scattered into the beam), in the direction of the beam, at wavelength λ and altitude z is given by:

$$F_\lambda(z) = F_{0\lambda} \exp(-\tau_\lambda(z)) \tag{4.1}$$

where F_0 is the flux incident at the top of the atmosphere and the optical depth, τ, depends on the air density, ρ, the mass mixing ratio, c, the extinction coefficient, k, of the radiatively-active gas and the solar zenith angle, ζ:

$$\tau_\lambda(z) = \int_z^\infty k_\lambda c(z')\rho(z') \sec\zeta dz'. \tag{4.2}$$

Clearly the flux at any point depends on the properties and quantity of absorbing gases in the path above. The altitude at which most absorption takes place at each wavelength can be seen in Fig. 4.4 which shows the altitude of unit optical depth for an overhead Sun. At wavelengths shorter than 100 nm most radiation is absorbed at altitudes between 100 and 200 km by atomic and molecular oxygen and nitrogen, mainly resulting in ionized products. Between about 80 and 120 km oxygen is photodissociated as it absorbs in the

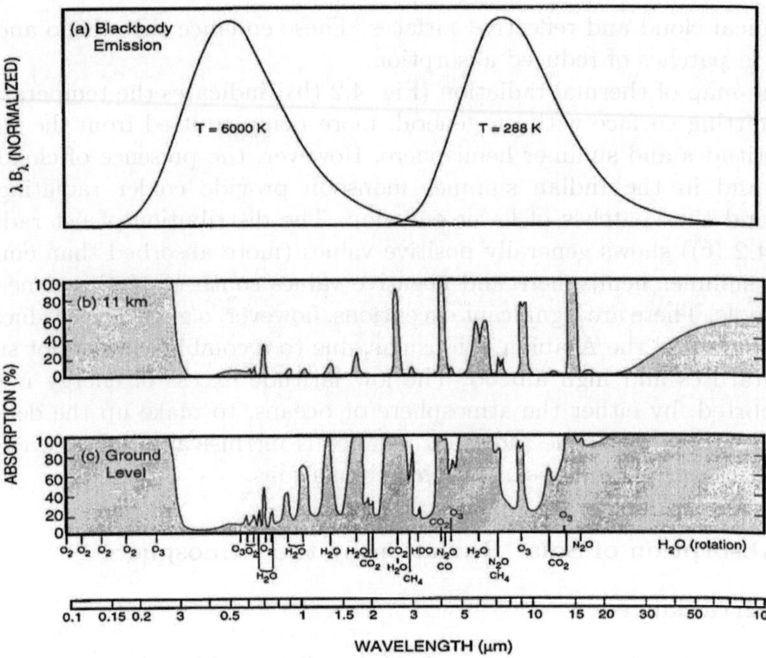

Fig. 4.3. (**a**) Black body functions at the emitting temperatures of the Sun and the Earth. The functions are scaled to have equal area to represent the Earth's radiation balance. (**b**) Absorption by a vertical column of atmosphere. The most important gases responsible for absorption are identified below, near the appropriate wavelengths (Houghton, 1977)

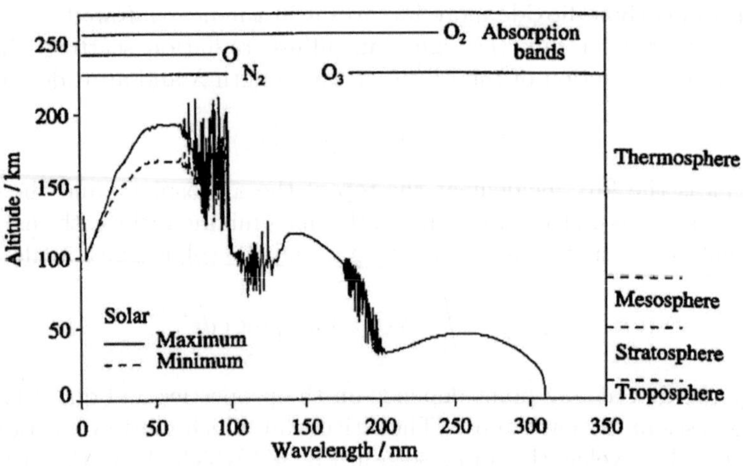

Fig. 4.4. Wavelength dependence of the altitude of one optical depth for absorption of solar radiation with an overhead Sun. After Andrews (2000)

Schumann-Runge continuum between 130 and 175 nm. The Schumann-Runge bands, 175–200 nm, are associated with electronic plus vibrational transitions of the oxygen molecule and are most significant between 40 and 95 km altitude. The oxygen Herzberg continuum is found in the range 200–242 nm and is overlapped by the ozone Hartley-Huggins bands between 200 and 350 nm which are responsible for the photodissociation of ozone below 50 km. The ozone Chappuis bands, in the visible and near-infrared, are much weaker than the aforementioned bands but, because they absorb near the peak of the solar spectrum, the energy deposition into the atmosphere is significant. Furthermore, this deposition takes place in the lower atmosphere and so is particularly relevant for climate. The absorption of solar near-infrared by carbon dioxide and water vapour is smaller but makes an important contribution to the heat budget of the lower atmosphere (see Sect. 4.2).

Solar irradiance varies with solar activity, as discussed by Prof Lockwood. The spectral composition of the variation determines which parts of the atmosphere respond most in terms of heating rates (see below). Note, however, that if the composition of the atmosphere were to remain unchanged, then variations in irradiance do not affect the height of unit optical depth shown in Fig. 4.4.

"Anomalous" Absorption

Observed values of the absorption of solar radiation in the atmosphere almost always exceed theoretical values. This effect has become known as "anomalous absorption" and is the subject of considerable debate (a good review is given by Ramanathan and Vogelmann, 1997). Discussion concerns not only possible physical explanations for its existence but also its magnitude (including whether it actually exists), spectral composition and whether it is a property only of cloudy skies. Some aircraft studies of the visible radiation field around cloud suggest that the apparent anomaly is an artefact of the imperfect sampling of the 3D structure. However, studies using a combination of satellite and ground-based data have confirmed the existence of a global average anomaly of 25–30 $\mathrm{W\,m^{-2}}$ (i.e., 10–12% of the total solar irradiance absorbed by the Earth), with the near-infrared region appearing to be significant. Some of the studies concluded that cloudy skies were responsible for the excess absorption but others have suggested a significant discrepancy between the results of GCM radiation schemes and observations in clear-sky absorption.

Mechanisms proposed to account for the underestimate of absorption in radiation models include problems with the formulation of the radiative transfer (band models, treatment of scattering, etc.), uncertainties in the radiative properties of water vapour (the spectral database and continuum absorption), the loading, composition and radiative properties of aerosol particles, cloud impurities, cloud drop-size distributions, the inability of models to simulate 3D radiation field in inhomogeneous cloud and the enhancement of the photon mean-free-path in 3D cloud. Work continues in investigating these avenues but, in the context of the effects of solar variability on climate it is worth noting that large uncertainties remain in quantitative estimates of the absorption of solar radiation by the atmosphere.

Solar Heating Rates

Most of the absorbed solar radiation eventually becomes heat energy so the local solar heating rate, Q (degrees per unit time), can be estimated from the divergence of the direct solar flux:

$$Q_\lambda(z) = \frac{1}{C_p\rho(z)\sec\zeta}\frac{dF_\lambda(z)}{dz} = \frac{1}{C_p}k_\lambda c(z)F_\lambda(z) \tag{4.3}$$

where Cp is the specific heat at constant pressure of air.

Figure 4.5 shows the instantaneous heating rate spectrum as a function of altitude calculated for a tropical atmosphere with an overhead sun. Clearly visible are the strong absorption in the ultraviolet (wavenumbers > 25,000 cm^{-1}) in the upper stratosphere, in the visible (~14,000–25,000 cm^{-1}) throughout the stratosphere and in several narrow bands in the near infrared.

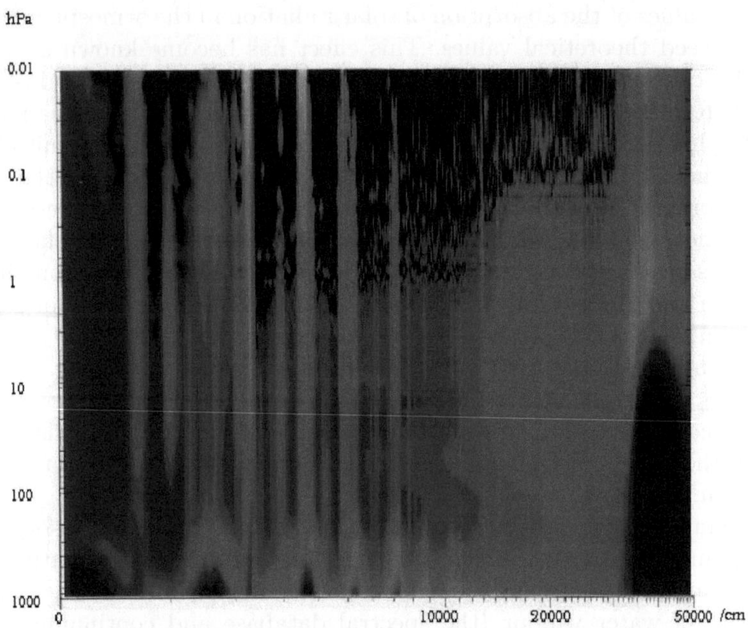

Fig. 4.5. Solar heating rates as a function of wavenumber (=1/λ(cm)) and altitude (pressure in hPa), paler colours indicate greater heating. Figure courtesy of Gail P. Anderson, U.S. Air Force Research Laboratory

Figure 4.6 presents a vertical profile of diurnally averaged solar heating rates for equatorial equinox conditions, showing the contribution of each of the UV/vis absorption bands mentioned in Sect. 4.2. This vertical structure in the absorption of solar radiation is crucial in determining the profile of atmospheric temperatures and plays an important role in atmospheric chemistry and thus composition.

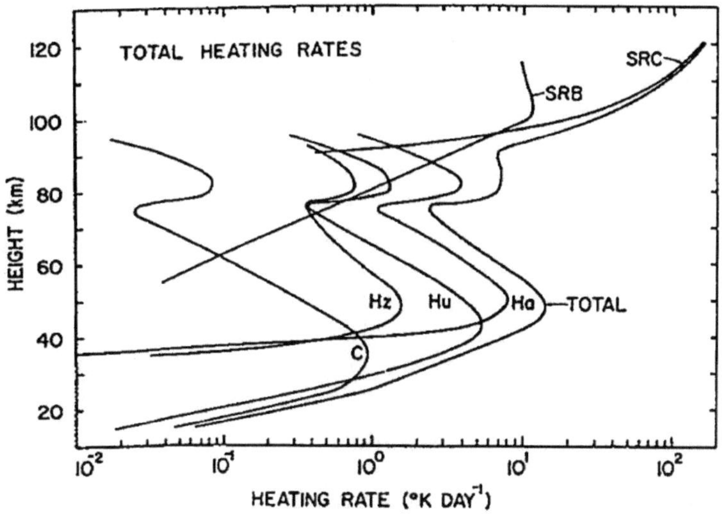

Fig. 4.6. Diurnal average solar heating rate (Kd^{-1}, log scale) as a function of altitude for equinoctial conditions at the equator showing contributions by the Schumann-Runge continuum and band (SRC and SRB), the Herzberg continuum (Hz) and the Hartley (Ha), Huggins (Hu) and Chappuis (C) bands. After Strobel (1978)

If the solar spectral irradiance varies then, without a change in composition, from the equation for the heating rate above we can see that the spectral heating rate just varies in proportion to the irradiance. If, however, as is actually the case, the atmospheric composition also responds to solar variability then this will affect both F and Q in a non-linear fashion. For example, an increase in F_0 will tend to increase F and Q. However, an increase in $c(z)$ (of ozone for example) enhances τ tending to reduce F. The sign of the change in F at any altitude depends on the competition between these two factors which is determined by the spectral composition of the change in F_0 and its effects on the photochemistry of the atmosphere. The effect on Q is then a product of the changes in F and $c(z)$. This is discussed further in Sect. 8.3.

4.3 Infrared Radiative Transfer and the Atmospheric Temperature Structure

Infrared Absorbers

The atmosphere absorbs solar radiation, as discussed above, and, while this energy may at first be used in photodissociation, molecule excitation or ionization processes, it essentially ends up as molecular kinetic energy, i.e., in raising atmospheric temperatures. To balance this the atmosphere must lose heat by radiating energy in the thermal infrared. The amount of energy radiated depends on the local temperature and on the infrared spectral properties (emissivities) of the atmospheric constituents. Figure 4.3 shows a black body spectrum at 245 K, the radiative equilibrium temperature for a planet with albedo 31% at approximately 1AU from the Sun, and the atmospheric absorption spectrum. Far infrared radiation is strongly absorbed by water vapour in its rotation bands and across the thermal infrared there are further water vapour absorption bands as well as features due to other "greenhouse" gases. There are strong carbon dioxide bands at 15 μm, 4.3 μm and 2.7 μm, water vapour bands at 6.3 μm and 2.7 μm, an ozone band at 9.6 μm as well as features due to methane, nitrous oxide and chlorofluorocarbons.

Atmospheric Temperature Profile

Where the atmosphere is optically thin in the infrared, such as in the stratosphere, radiant heat energy may be transmitted to space, causing local cooling. At lower altitudes where the atmosphere is optically much thicker, however, emitted infrared radiation is absorbed and reemitted by neighbouring layers. Thus the atmospheric temperature profile is determined by interactions between levels as well as by solar heating and direct thermal emission.

In the middle atmosphere absorption of solar ultraviolet radiation by oxygen and ozone, as described in the previous section, produces a peak in temperature near 50 km called the stratopause as shown in Fig. 1.2. This heating is counteracted by thermal emission, mainly by carbon dioxide in its 15 μm band, but also by the ozone 9.6 μm band and the water vapour 6.3 μm band. Typical profiles of infrared cooling and solar heating rates can be seen in Fig. 4.7. The lower stratosphere (between approximately 15 and 25 km) is in approximate radiative equilibrium. Here heating is due to ozone absorption both of visible radiation in its Chappuis bands and also of infrared radiation emanating from lower levels in its 9.6 μm band. Cooling is mainly by carbon dioxide.

In the troposphere radiative transfer is largely accomplished by water vapour, and solar heating is relatively small. However, radiative processes do not determine the temperature profile in this region. This is because a radiative equilibrium profile would be convectively unstable, i.e., a small upward

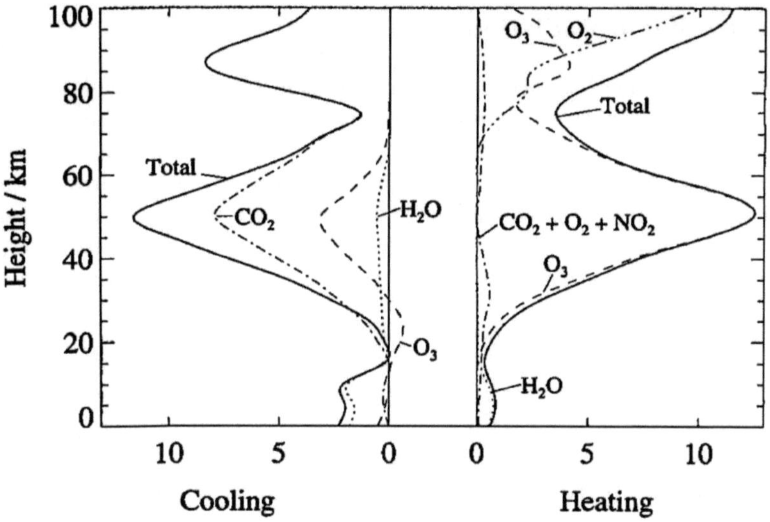

Fig. 4.7. Global mean infrared cooling rates and solar heating rates (Kd^{-1}) as a function of altitude showing contributions of the major gases. After Andrews (2000)

displacement of an air parcel would result in it remaining hotter than its environment, despite expansion and adiabatic cooling, and thus continuing to rise. The temperature profile of the troposphere is therefore limited by convective processes which result in the adiabatic lapse rate of about $-7\,\mathrm{K\,km}^{-1}$ seen in Fig. 1.2. Temperatures are locally warmer than would be the case based on radiative processes alone so that infrared emissions increase. This results in infrared cooling due to tropospheric water vapour, as shown in Fig. 4.7, rather than the warming which would come about from infrared trapping alone. Thus the tropopause marks the region where radiative processes (ozone heating and carbon dioxide cooling) take over from mainly convective processes. Note that an important factor determining the value for the temperature lapse rate is the release of latent heat from the condensation of water vapour into cloud droplets. Thus clouds play an integral part in determining the temperature structure of the lower atmosphere. The accurate representation of clouds and precipitation remains a major challenge in global climate modelling (see also lecture 6).

Above the ozone layer the effects of ultraviolet absorption by ozone are reduced and there is a minimum in temperature at the mesopause. Higher still, heating due to the absorption by molecular oxygen of far ultraviolet radiation takes over and there is a steep increase in temperature in the lower thermosphere.

4.4 Scattering of Radiation

Section 4.2, on the absorption of solar radiation by the atmosphere, omitted any discussion of scattering processes. Scattering of solar radiation by cloud droplets and aerosols is very important and scattering by ice cloud of thermal radiation can also be significant. To gain a deeper understanding of the role of scattering in atmospheric radiative transfer we will look in more detail at the radiative transfer equation. This gives the change in intensity $I_\lambda\,(s,\Omega)$ of radiation of wavelength λ in direction Ω between positions s and $s+ds$ is as:

$$dI_\lambda\,(s,\Omega) = -\{I_\lambda(s,\Omega) - (1-\omega_\lambda)B_\lambda(s,\Omega) - \frac{\omega_\lambda}{4\pi}\int I_\lambda(s,\Omega')P_\lambda(s,\Omega,\Omega')d\Omega'\}\;k_\lambda\rho(s)ds \tag{4.4}$$

where: ω_λ is the single scattering albedo, i.e. the ratio of the scattering coefficient to the extinction coefficient; $B_\lambda(s,\Omega)$ is the source function, i.e. the radiation emitted by the atmosphere (negligible at solar wavelengths but dominant in the thermal infrared) and $P_\lambda(s,\Omega,\Omega')$ is the scattering phase function, i.e. the probability that radiation incident in solid angle Ω' will be scattered into Ω.

The first term on the right-hand-side of the equation represents removal (by absorption and scattering) of radiation from the incident beam, the second term represents emission into the beam and the third scattering into the beam of radiation incident from all other directions.

The solution of this equation for radiation emerging upwards (direction Ω_0) at the top of the atmosphere (toa, at altitude z_t) is:

$$I(z_t,\Omega_0) = I(0,\Omega_0)\mathrm{T}(0,z_t) + \int_0^{z_t}\left\{(1-\omega)B(z,\Omega_0) - \frac{\omega}{4\pi}\int_{4\pi}^{0} I(z,\Omega')P(z,\Omega',\Omega_0)d\Omega'\right\}\frac{d\mathrm{T}(z,z_t)}{dz}dz \tag{4.5}$$

where the suffix λ has been dropped for simplicity but is still implied and $\mathrm{T}(z,z') = \exp(-\tau(z,z'))$ is the transmittance of the atmosphere (gases and clouds) between z and z'. The first term on the right-hand-side represents radiation starting at the Earth's surface (either emitted or reflected) being transmitted to the toa. The second term integrates the transmission of radiation emitted and scattered by all atmospheric layers.

The radiation propelled to space depends on the scattering properties of any clouds present in the atmosphere and thus on the microphysical, as well as large scale, properties of the cloud. The scattering can be described under three regimes depending on the relative magnitudes of the droplet size, r, and the wavelength of the radiation:

If $r \ll \lambda$ (e.g. sunlight & air molecules; microwaves & raindrops) Rayleigh scattering occurs which is strongly wavelength-dependent ($\propto \lambda^{-4}$) with peaks in the forward and back-scattering regions.

If $r \sim \lambda$ (e.g. sunlight by dust particles; infrared radiation by cloud droplets) it becomes an electromagnetic EM boundary-value problem which is weakly wavelength dependent and usually solved by Mie theory for spheres.

If $r \gg \lambda$ (e.g. sunlight by raindrops; infrared by ice crystals) geometric optics apply.

Some examples of phase functions for real clouds are given in Fig. 6.4. The functions are often complex but can be expanded by a sum of Legendre polynomials as a function of scattering angle θ:

$$P(\cos\Theta) \approx \sum_{l=0}^{2N-1} (2l+1)\chi_l L_l(\cos\Theta) \tag{4.6}$$

where the weighting, χ_1 of $L_1(=\cos\theta)$ is often referred to as the asymmetry parameter, g. A commonly used empirically-derived phase function is the Henyey-Greenstein:

$$P_{HG}(\cos\Theta) = \frac{(1-g^2)}{(1+g^2-2g\cos\Theta)^{3/2}} \tag{4.7}$$

Drop Size and Optical Depth

The scattering optical depth depends on the cross-sectional area of the droplets in the path of the radiation and can be expressed in terms of the droplet size distribution $n(r)$ by:

$$\tau \propto \iint r^2 n(r) dr dz \tag{4.8}$$

The liquid water path is the mass of water per unit area along the path of the radiation and so can be expressed as:

$$LWP \propto \iint r^3 n(r) dr dz \tag{4.9}$$

Thus we can relate τ to LWP by

$$\tau \propto \frac{LWP}{r_e} \tag{4.10}$$

where the droplet effective radius is defined by

$$r_e = \frac{\int r^3 n(r) dr}{\int r^2 n(r) dr} . \tag{4.11}$$

From this it becomes clear that if the total amount of liquid water available is fixed then the optical depth will be inversely related to droplet size. Thus if the number density of cloud condensation nuclei increases, the drop size will get smaller and the albedo of the cloud increase. This is the basis of the "indirect effect of aerosol" on radiative forcing discussed in Sect. 5.3 and is also related to the proposed link between cosmic rays and cloud cover which is the subject of Sect. 9.3.

5 The Greenhouse Effect and Radiative Forcing of Climate Change

In this lecture we will consider how the global mean surface temperature of the Earth responds to changes in the radiation budget. The discussion will centre around the concept of "radiative forcing" of climate change which provides a relatively simple method for estimation of the surface temperature response.

5.1 Radiative Equilibrium Temperature

The radiative equilibrium temperature of a planet may be estimated by considering how hot it needs to be in order to emit a quantity of radiative energy equal to that which it absorbs from the sun.

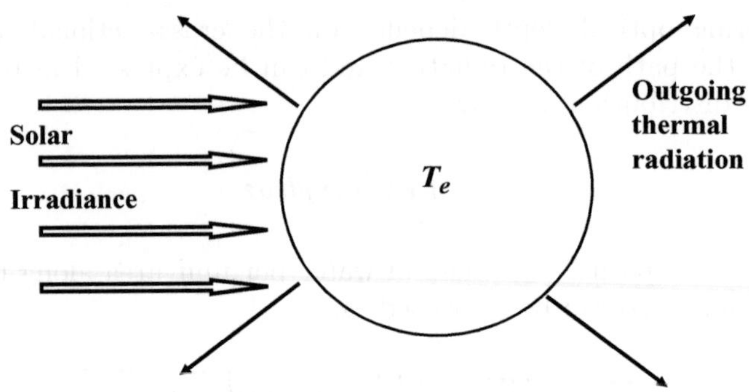

Fig. 5.1. Determination of the radiative equilibrium temperature

Assuming the planet to be spherical and of radius R it absorbs a quantity of radiation equal to $(1\text{-}\alpha)\pi R^2 S$, where S is the irradiance and α the planetary albedo. Assuming that the planet has effective emission temperature T_e then the absorbed solar radiation is balanced by emission of thermal radiation to space of σT_e^4 per unit surface area:

$$\pi R^2 (1-\alpha) S = 4\pi R^2 \sigma T_e^4 \tag{5.1}$$

thus

$$\sigma T_e^4 = (1-\alpha)\,S/4 = F_s \qquad (5.2)$$

where F_s is the global average absorbed irradiance. Taking S=1368 $\mathrm{Wm^{-2}}$ and $\alpha = 0.31$ as values appropriate to the Earth (see Fig. 4.1), we find F_s=235 $\mathrm{Wm^{-2}}$ and $T_e = 254\,\mathrm{K}$. Clearly this temperature is much colder than the global average surface temperature of the Earth. It corresponds to the temperature of the layer of atmosphere, at greater altitude, from which the majority of radiation escapes to space. The presence of the atmosphere warms the surface to a temperature greater than T_e by the so-called greenhouse effect.

5.2 A Simple Model of the Greenhouse Effect

First it should be noted that the "greenhouse" in this context is a misnomer as garden greenhouses keep warm mainly by inhibiting convection (i.e., trapping the hot air) rather than by limiting infrared emission. Nevertheless, the terminology has achieved such widespread acceptance that it is retained here.

The basic premise of the greenhouse effect is that the atmosphere is relatively transparent to solar radiation, thus allowing it to reach and warm the surface, while being absorptive in the infrared, thus trapping the heat energy at low levels. In a very simple model the atmosphere is assumed to have a single uniform temperature and grey absorption properties (i.e., not varying with wavelength except in as much as they are different for solar and thermal radiation) and to lie above a surface at a different temperature. The whole system is in radiative equilibrium. As discussed in the preceding sections, these assumptions are not accurate but this model does allow us firstly to observe the principle of greenhouse warming and secondly to establish a basis for the concept of the radiative forcing of climate change, discussed below.

Figure 5.2 shows the fundamentals of the model. Solar irradiance F_s is incident at the top of the atmosphere and a fraction T_S of this is transmitted to the surface. The surface, at temperature T_g, emits irradiance F_g and a fraction T_L of this reaches the top of the atmosphere. The suffixes $_S$ and $_L$ to the transmittances represent shortwave (solar) and longwave (infrared) properties. The atmosphere, at temperature T_a, emits irradiance F_a in both upward and downward directions. [N.B. note different notation for temperature T and transmittance T]

For radiation balance at the top of the atmosphere and at the surface respectively:

$$F_s = F_g \mathrm{T}_L + F_a F_s \mathrm{T}_S = F_g - F_a \qquad (5.3)$$

from which can be deduced that:

$$F_g = F_s(1+\mathrm{T}_S)/(1+\mathrm{T}_L) \qquad (5.4)$$

If the surface is a black body (i.e., has unit emissivity) then:

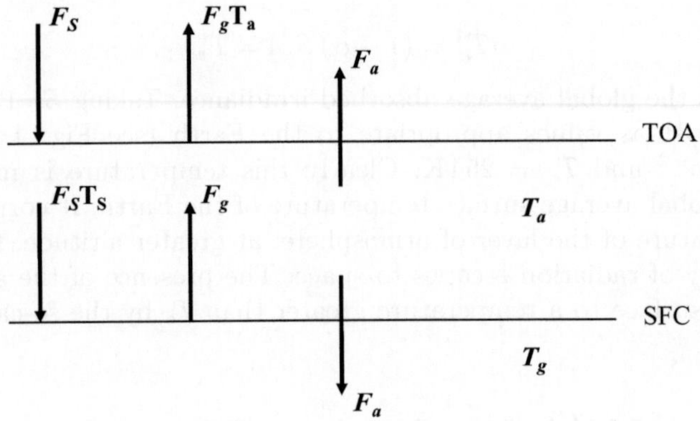

Fig. 5.2. A simple model of radiative fluxes across a slab atmosphere

$$F_g = \sigma T_g^4 \tag{5.5}$$

Furthermore, from the discussion in Sect. 5.1 the solar flux may be related to the equilibrium temperature of the Earth, T_e, by:

$$F_s = \sigma T_e^4 = (1 - \alpha)S/4 \tag{5.6}$$

Thus:

$$T_g^4 = T_e^4(1 + \mathrm{T}_S)/(1 + \mathrm{T}_L) \, . \tag{5.7}$$

If the atmosphere is more transparent to solar than thermal radiation, i.e. $\mathrm{T}_S > \mathrm{T}_L$, then the model predicts that the surface temperature will be greater than the equilibrium temperature. For example, with $\mathrm{T}_S = 0.49$ and $\mathrm{T}_L = 0.10$ (as suggested by Fig. 4.1) $T_g = 1.08T_e = 274\,\mathrm{K}$, an increase of 20K over the value calculated for the atmosphere-free planet. This is a demonstration of greenhouse warming.

In this simple model the atmospheric temperature, T_a, is given by:

$$T_a^4 = T_e^4(1 - \mathrm{T}_S\mathrm{T}_L)/(1 - \mathrm{T}_L^2) \, , \tag{5.8}$$

where it has been assumed that F_a=(1 - $\mathrm{T}_L)\sigma T_a^4$, i.e., that the emissivity plus the transmissivity of the atmosphere is unity (Kirchoff's Law). For $\mathrm{T}_S >$ T_L, T_a is less than the equilibrium temperature such that the total emitted irradiance remains σT_e^4.

Changes to any of the parameters within the expression for T_g will affect the equilibrium surface temperature. Thus variations in total solar irradiance, planetary albedo, which is determined by land cover and cloud and aerosol properties, will have an impact on climate as will the concentrations of any of the gases involved in determining atmospheric transmittances. These include ozone, water vapour and nitrogen dioxide for shortwave radiation and H_2O, CO_2, CH_4, N_2O, O_3, CFCs etc – i.e. the "greenhouse gases".

5.3 Radiative Forcing of Climate Change

The Concept of Radiative Forcing

Continuing for a little longer with the simple model of the previous section we can express the net downward flux at the top of the atmosphere as:

$$F_N^{\downarrow} = F_s - F_g \mathrm{T}_L - F_a = \sigma[T_e^4 - (T_g^4 - T_a^4)\mathrm{T}_L - T_a^4] \tag{5.9}$$

In equilibrium, as discussed above, $F_N^{\downarrow} = 0$. However, consider a situation in which T_g and T_a have their equilibrium values and some external factor acts to perturb the value of T_e or T_L. Then, before T_g and T_a have adjusted and equilibrium is re-established, the instantaneous value of $F_N^{\downarrow}$ is not zero. The simplest definition of radiative forcing (RF) is just the change in the value of $F_N^{\downarrow}$. If the RF is positive then there is an increase in energy entering the system (or equivalently a decrease in energy leaving the system) and it will tend to warm until the outgoing energy matches the incoming and the net flux is again zero. The perturbing factors might again be changes in solar irradiance, planetary albedo or the concentrations of radiatively active gases, aerosols or cloud.

The concept of radiative forcing has been found to be a useful tool in analysing and predicting the response of surface temperature to imposed radiative perturbations. This is because experiments with general circulation models (GCMs) of the coupled atmosphere-ocean system have found that the change in globally averaged equilibrium surface temperature is linearly related to the radiative forcing:

$$\Delta T_g = \lambda \Delta F_N^{\downarrow} = \lambda RF \tag{5.10}$$

where λ is the "climate sensitivity parameter" which has been found to be fairly insensitive to the nature of the perturbation and to lie in the range $0.3 < \lambda < 1.0\,\mathrm{K}\ (\mathrm{W\ m^{-2}})^{-1}$. Thus a calculation of the radiative forcing due to a particular perturbant gives a first-order indication of the potential magnitude of its effect on surface temperature without the need for costly GCM runs.

Note that this relationship between ΔT_g and RF is not consistent with the simple model above (which would suggest that the surface temperature varied with the cube root of the change in flux). This is because of the gross assumptions made in the model: an isothermal, grey atmosphere, neglect of convective adjustment and, most importantly, the role of water vapour in a positive feedback process. This comes about because as the atmosphere warms it can hold more water vapour which acts as a greenhouse gas to increase the warming. Thus the simple model is useful to introduce the fundamentals of the greenhouse effect and the concept of radiative forcing but should not be used in any quantitative assessment of potential temperature change.

The large, factor 3, range in the value of λ given above represents the spread of values given by different GCMs. This gives an indication of the uncertainties in climate prediction. It should be noted, however, that for each *particular* GCM the range of λ found using different sources of radiative forcing is much narrower. This suggests that, while absolute predictions are subject to large uncertainty, the forecast of the relative effects of different factors is much more robust.

Instantaneous and Adjusted Radiative Forcing

It has been found that the value of λ, and thus the usefulness of the radiative forcing concept, is more robust if, instead of using the instantaneous change in net flux at the top of the atmosphere, RF is defined at the tropopause with the stratosphere first allowed to adjust to the imposed changes. Thus a formal definition of radiative forcing, as used by the Intergovernmental Panel on Climate Change (IPCC, see Sect. 5.5) is the change in net flux at the tropopause after allowing stratospheric temperatures to adjust to radiative equilibrium but with surface and tropospheric temperatures held fixed. The effects of the stratospheric adjustment are complex as can be illustrated by the case of changes in stratospheric ozone. An increase in ozone masks the lower atmosphere from solar ultraviolet i.e., reducing net flux and thus RF; however, the presence of ozone in the lower stratosphere increases the downward infrared emission (and RF) both directly through the 9.6 µm band and also indirectly through the increase in stratospheric temperatures which it produces. Whether the net effect is positive or negative depends on whether the shortwave or longwave effect dominates and this is determined by the vertical distribution of the ozone change.

The direct effect of an increase in total solar irradiance is to increase the radiative forcing; the heating of the stratosphere by the additional irradiance will enhance this by increasing the downward emission of thermal radiation. However, the sign of the radiative forcing due to any solar-induced increases in ozone is not clear – published estimates show both positive and negative values – because of the uncertainties in the distribution of the ozone change (see lecture 7).

Radiative Forcing Since 1750

Figure 5.3, reproduced from the IPCC 2001 scientific assessment of climate change, shows the time-varying components or radiative forcing, relative to the values in 1750, from natural and anthropogenic sources. These give an indication of the relative magnitudes of the different components but can not be *directly* interpreted in terms of variation in surface temperature because the inertia of the climate system means that time lags in the system need to be taken into account.

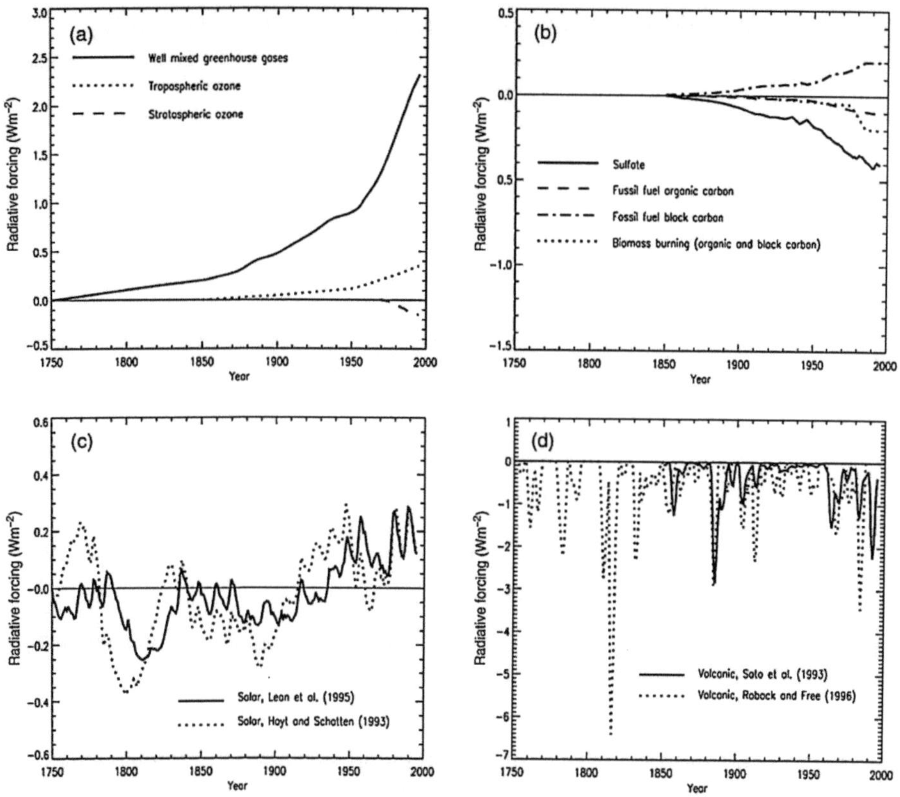

Fig. 5.3. Time evolution of global annual mean radiative forcing due to (**a**) well-mixed greenhouse gases, stratospheric and tropospheric ozone, (**b**) direct effect of sulphate aerosols, organic and black carbon arrosols from fossil fuels and biomass burning, (**c**) solar irradiance variations (two different reconstructions, see lecture 8) and (**d**) volcanic aerosols in the stratosphere. N.B. different scales are used in each panel (IPCC, 2001)

Figure 5.4 shows the RF values deduced for the period 1750 to 2000 for a range of different factors. The largest component, of 2.43 Wm^{-2}, is due to the increase in well-mixed greenhouse gas concentrations. The other components are all of magnitude a few tenths of a Wm^{-2}. For example, sulphate aerosol has produced a RF of –0.4 Wm^{-2}, negative because it enhances the albedo. Also shown in Fig. 5.4 is a bar from 0 to –2 Wm^{-2} for the "first indirect effect" of all aerosol types. This represents the process whereby an increase in aerosol concentration produces more cloud condensation nuclei so that the composition of the cloud tends to a higher number density of smaller droplets which increases albedo. The magnitude of the indirect effect is very uncertain but probably negative for the reasons outlined above (see Sects. 4.4 and 6. for further discussion). No estimates were given in the IPCC 2001 report

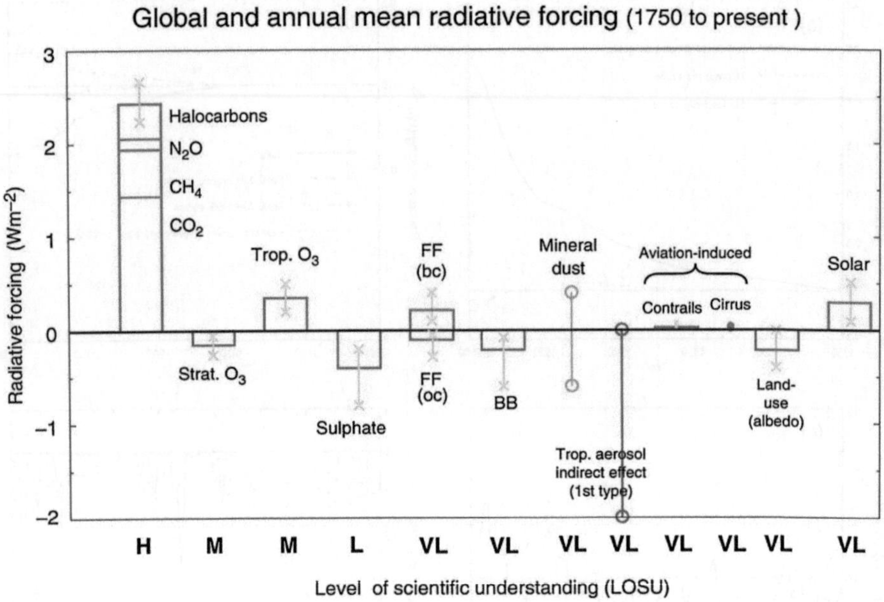

Fig. 5.4. Global, annual average radiative forcing contributions 1750–2000 from the IPCC third assessment report (IPCC, 2001)

for the "second indirect effect", in which cloud lifetime is extended due to a suppression of precipitation, as no reliable estimates were available.

The vertical bars delimited by crosses in Fig. 5.4 do not represent statistical ranges of uncertainty but just the ranges of values published in peer-reviewed scientific literature. This figure also gives an indication of the "level of scientific understanding" for each effect which is a subjective analysis intended to indicate whether the scientific processes involved were perceived to be complete and well-understood.

The solar contribution is assessed to be $0.3{\pm}0.2\,\mathrm{Wm^{-2}}$. Note that when calculating solar radiative forcing it is necessary to use the global average value (F_s from 5.2) rather than the total solar irradiance at the Earth (S in the same equation). The year 1750 was chosen by the IPCC (see Sect. 5.5) to represent the pre-industrial atmosphere but for a naturally-varying factor like the Sun this date is, of course, arbitrary. A choice of later in the 18th century would have given a slightly reduced solar RF but early in the 19th century a significantly larger one. The value of the climate sensitivity parameter λ is estimated from GCM calculations to be approximately $0.6\,\mathrm{K\,(W\,m^{-2})^{-1}}$ (but see discussion on range of values in Sect. 5.3) so that a solar radiative forcing of $0.3\ \mathrm{W\ m^{-2}}$ would indicate that a global average surface warming of only about $0.18\,\mathrm{K}$ since 1750 could be ascribed to the Sun. However, the IPCC gives the assignation "very low" to the LOSU associated with solar radiative forcing, thereby acknowledging that there may be factors as yet unknown, or not fully understood, which may act to amplify (or even diminish) its effects. Some of these are discussed in lectures 8 and 9.

Complications with the Application of Radiative Forcing

As discussed above the radiative forcing value provides a useful zeroth order estimate of the equilibrium response of global annual average surface temperature to a change in the radiation balance. However, caution needs to be exerted in its application to more complex (real) situations. Some examples follow:

1. As noted above, the equilibrium ΔT_g is predicted so that a time series of radiative forcing time series cannot be interpreted directly as temperature change record. It can, however, be used as input to energy balance models (see Sect. 8.2).
2. Confusion can occur between what is a forcing and what a feedback. A chemical or microphysical response to a radiative perturbation which results in a change in atmospheric composition or in cloud properties may produce an additional radiative forcing (as in the indirect aerosol effect). A dynamical response, however, in which atmospheric circulations are affected, is a feedback because such effects are implicitly included in the calculations by GCMs of the value of λ. In this context an interesting example is that of stratospheric water vapour (SWV). There is observational evidence that SWV has been increasing by about 1.5% per annum for at least the past 20 years but, to date, no explanation that can account for this magnitude of change has been found. If it is part of changing circulations in response to GHG increases then it can be viewed as a feedback effect but if it is due to some in situ chemical process (e.g. methane oxidation) then it should be included as an indirect radiative forcing.
3. Some forcings have very inhomogeneous spatial distributions. The question arises as to whether the radiative forcing concept, which has been derived based on global annual mean considerations, may be applied to these. The pragmatic approach to vertical distributions has already been discussed in Sect. 5.3. Examples of horizontal distributions are given in Fig. 5.5. It is not possible to simply map these distributions onto local surface temperature changes, because gradients in heating produce dynamical responses, but it is important that these distributions are accurately estimated because their relationships to such factors as surface albedo and solar zenith angle are important for proper simulation of regional climate change.

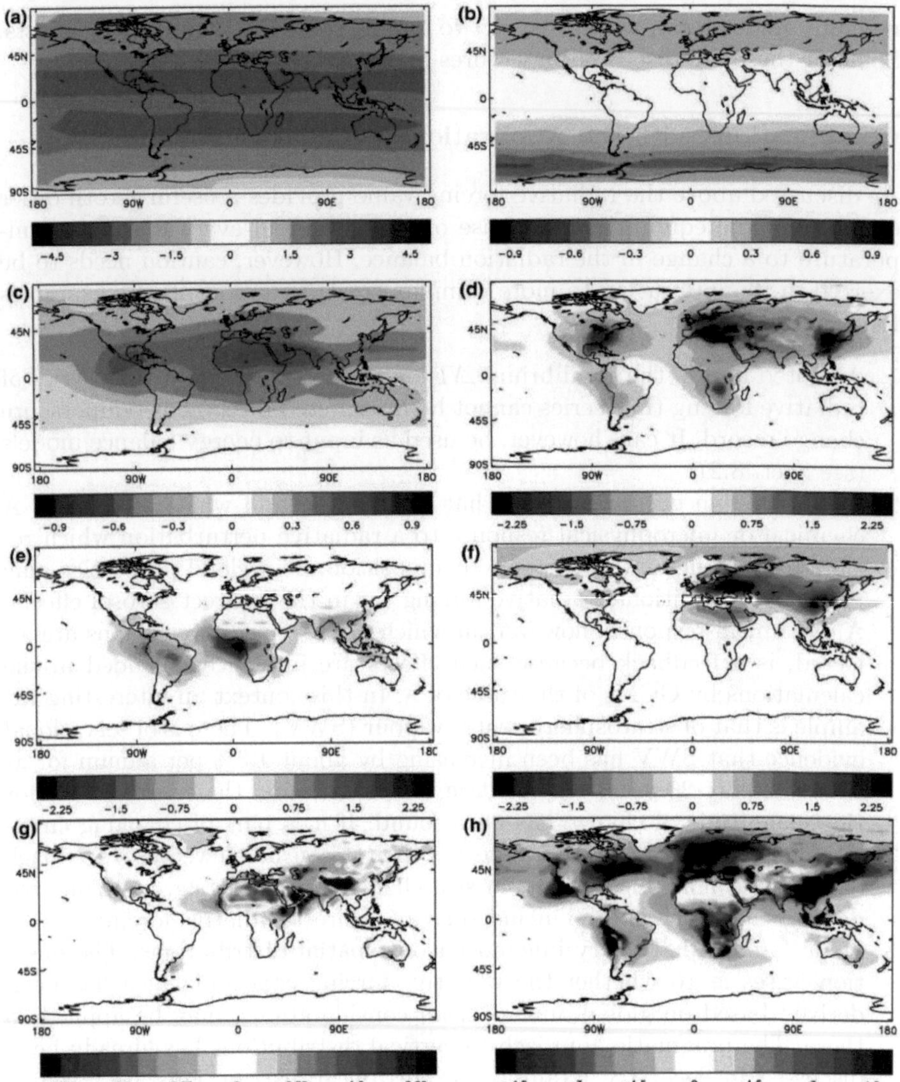

Fig. 5.5, for caption see next page

5.4 Global Warming Potentials

The radiative forcing components discussed above represent best estimates for the effect of actual changes in atmospheric concentrations. Sometimes, however, particularly where new chemical compounds are being developed for industrial applications, it is useful to compare the effectiveness of compounds on a per unit mass basis. The global warming potential (GWP) of a compound is the ratio of the time-integrated radiative forcing from the instantaneous

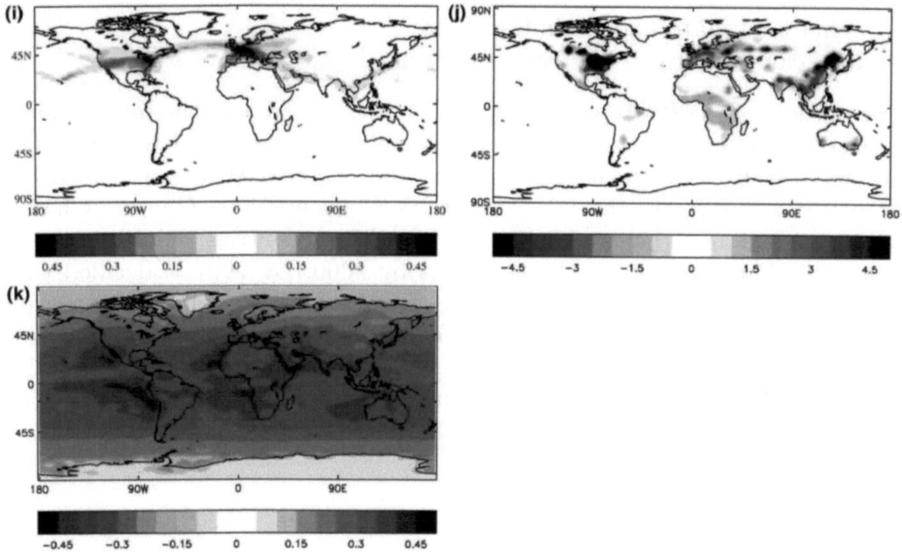

Fig. 5.5. Geographical distribution of radiative forcing (1750–2000) due to (**a**) well-mixed greenhouse gases, (**b**) stratospheric ozone depletion, (**c**) increses in tropospheric ozone, (**d**) direct effect of sulphate aerosol, (**e**) direct effect of carbon aerosol from fossil fuel burning, (**g**) direct effect of (anthropogenic) mineral dust, (**h**) first indirect effect of sulphate aerosol, (**i**) aircraft condensation trails, (**j**) albedo effects of land-use changes. Units are Wm^{-2}, note different scales in each panel. Source: IPCC (2001)

release of 1kg of a trace substance relative to 1kg of a reference gas (usually CO_2). This gives a measure of the potential radiative impact of a particular species over a given time horizon. Thus the GWP value depends not only on the radiative properties of a compound but also on its atmospheric lifetime and the time horizon over which the GWP is estimated. For a 100 year horizon some examples are:

CO_2 1; CH_4 23; CCl_2F_2 10,600; SF_6 22,200.

5.5 The Intergovernmental Panel on Climate Change (IPCC)

The IPCC was established in 1988 by the World Meteorological Organisation and the United Nations Environment Programme to assess scientific, technical and socio-economic information relevant for understanding the risk of human-induced climate change. It does not carry out research, nor does it monitor climate-related data but bases its assessments on published and peer-reviewed scientific technical literature. It seeks answers to the questions:

- Is climate changing?
- If so why?

- What is likely to happen in the future?
- How can we mitigate the effects and/or adapt to them?

The authors of the IPCC assessments are representatives of the international scientific community and are subject to changeover between the reports which are extensively reviewed by government, and other, scientists. Its work has been key to informing the UN Framework Convention on Climate Change which in 1992 agreed that OECD and Transition countries should return to 1990 emission levels of greenhouse gases by 2000. By 1997 it decided that stronger measures should be taken and the Kyoto Protocol prescribes that average emissions should be cut by 5% by 2010, although this varies between countries and there are complex arrangements for "emissions trading". By 2003 188 countries had ratified the Kyoto Protocol but it is not clear that all will actually follow its recommendations.

6 Clouds

Clouds have a major impact on the heat and radiation budgets of the atmosphere. They transport latent heat from the oceans to the atmosphere. They reflect solar radiation back to space, reducing the net incoming radiative flux, and they trap infrared radiation, acting in a similar way to greenhouse gases. The magnitudes of these effects depends on the location, altitude, time of year and also the physical properties of the cloud. In this lecture we will consider the role of cloud in the radiation budget and how this impacts radiative forcing. We will also consider how clouds are formed and how their radiative properties relate to their microphysical structure. A short discussion of how clouds are represented in GCMs completes this section.

6.1 Clouds and the Earth Radiative Budget

In lecture 4 we saw that on a global average clouds increase the planetary albedo by reflecting about 23% of the incoming solar irradiance back to space. The magnitude of the reflectance depends on the optical thickness of the cloud, the water phase (liquid or ice), the particle size distribution and also the particle shape. The degree of longwave trapping depends on the transmissivity of the cloud and also its temperature: high (cold) cloud is more effective because it emits less radiation to space while trapping the (warm) radiation from below. Thus clouds tend to reduce both the incoming solar radiation and the outgoing longwave radiation.

The net effect of cloud on the radiation budget depends on whether the shortwave or longwave effect is larger and thus on the location, height and microphysical properties of the cloud. The effect of cloud on the radiation balance is measured using a cloud radiative forcing parameter CRF. This is defined as follows:

Total radiative forcing = absorbed solar radiation – outgoing longwave radiation (as discussed in lecture 5)

i.e.

$$RF = S - L \tag{6.1}$$

Now

$$CRF = (S - L) - (S - L)_{clear} \tag{6.2}$$

is the difference between the actual RF and its value should all clouds be removed. This can be rewritten:

$$CRF = (S - S_{clear}) - (L - L_{clear}) \tag{6.3}$$

i.e. the difference between the shortwave cloud forcing and the longwave. Note that both terms in brackets are usually negative.

Studies using satellite data suggest that for the global average CRF < 0, i.e. the net effect of the presence of cloud is to reduce the net absorbed radiation and thus cool the climate (i.e., the shortwave dominates the longwave). Figure 6.1 shows the results of an analysis of two independent satellite datasets: one of cloud cover and one of radiation budget components. The curves labelled ASR and OLR are essentially scaled zonal mean values of ($S - S_{clear}$) and ($L - L_{clear}$), respectively, shown as a function of latitude. At most latitudes the ASR has a greater magnitude than the OLR and this effect is most marked in the southern hemisphere where (bright) cloud over the large areas of (dark) ocean create a particularly marked response in solar radiation. In the tropics, however, the 2 terms almost balance and between about 7°N and 20°N the longwave effect dominates.

The contributions of the geographical regions to the cloud forcing components can be seen in Fig. 6.2. The longwave plot shows – ($L - L_{clear}$) and thus the values are always greater than zero where cloud tends to exist, particularly in the tropics. The shortwave plot shows generally negative values, as discussed above, but when cloud arrives over a high albedo surface its effect may be much smaller, as can be seen over the Saharan and Australian deserts, or even of the opposite sign as seen over the Antarctic. The net cloud forcing, shown in Fig. 6.2 (c), is thus positive over sub-tropical deserts and high latitude ice caps.

The cloud radiative forcing components are further diagnosed in Table 6.1 and Table 6.2 which show that there are strong seasonal as well as geographical variations in the components and, furthermore, that the altitude and thickness of the cloud has a strong influence.

Overall cloud radiative forcing is negative as the effect of cloud in reflecting solar radiation dominates its ability to trap thermal radiation. This effect is much smaller in the winter hemisphere, however, where solar irradiance is lower.

High and mid-level thin cloud have a net warming effect on climate. Hemispheric differences are due to the much larger proportion of the surface being

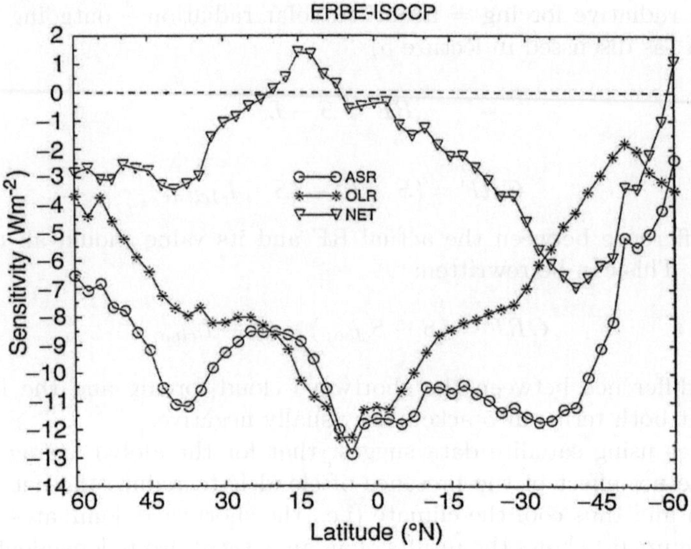

Fig. 6.1. Sensitivity of radiation fields (from ERBE data) to cloud cover (from IPCC data). See text for further details. Source: Ringer and Shine (1997)

ocean in the southern hemisphere. NB the values shown in Table 6.1 and Table 6.2 are not entirely consistent because of the use of different datasets; this gives an indication of the large uncertainties involved in such calculations.

6.2 Clouds and Radiative Forcing of Climate Change

In the global equilibrium state clouds form part of the overall radiation balance (as shown in Fig. 4.1). A factor which induces a change in cloud cover, drop size or altitude will therefore introduce a radiative forcing. If, however, the changes are brought about by another forcing factor then their effects may be viewed as a feedback on the initial forcing. For example, an increase in greenhouse gases might cause a surface warming and this may induce enhanced convection and an increase in cloud cover. The thick convective cloud produced will have a negative radiative forcing and thus reduce the potential greenhouse gas warming. Such feedback effects, however, are included in the GCMs used to assess the viability of the radiative forcing concept and are therefore implicitly included in the value of the climate forcing parameter λ.

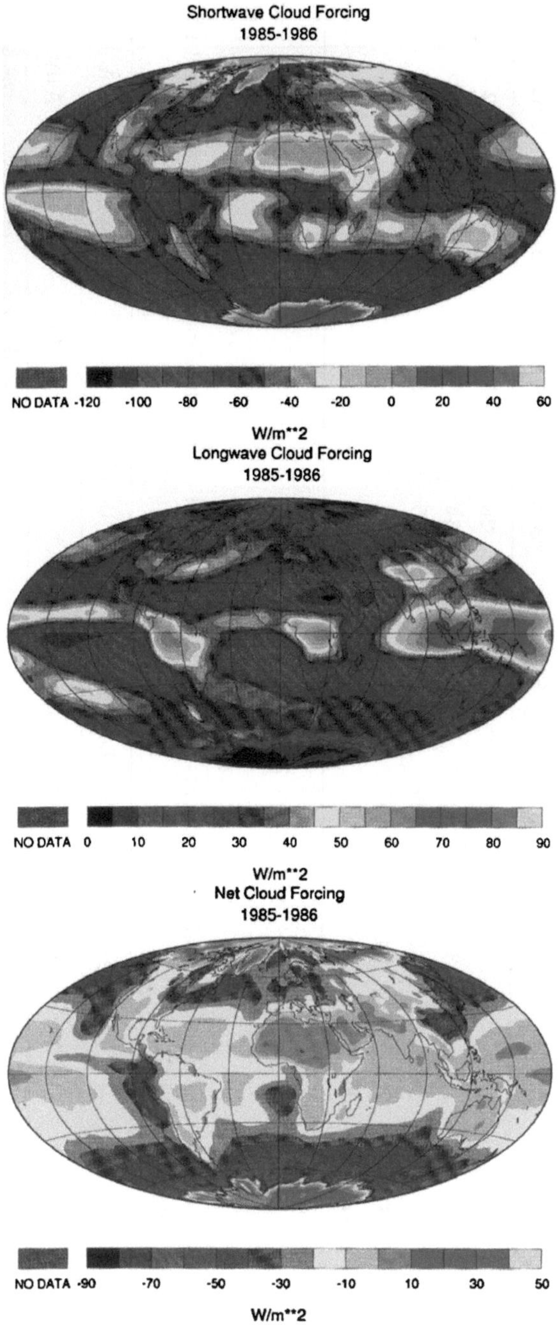

Fig. 6.2. Annual average cloud radiative forcing from ERBE data 1985-6. *Top*: shortwave; *middle*: longwave; *bottom*: net (Hobbs, 1993)

Table 6.1. Cloud radiative forcing components by season and hemisphere derived from ERBE data. See Sect. 6.1 for definitions of symbols. Hobbs (1993)

	Longwave ($W m^{-2}$)			Solar ($W m^{-2}$)			Net ($W m^{-2}$)		
	L	**L_{cl}**	**-(L-L_{cl})**	**S**	**S_{cl}**	**(S-S_{cl})**	**(S-L)**	**(S_{cl}-L_{cl})**	**CRF**
NH DJF	228	254	**26**	168	196	**−28**	−59	−57	**−2**
NH JJA	242	277	**35**	307	371	**−64**	65	95	**−30**
SH DJF	234	268	**34**	320	395	**−75**	86	127	**−41**
SH JJA	234	258	**24**	162	190	**−28**	−72	−67	**−5**
Global annual	234	264	**30**	240	288	**−48**	6	24	**−18**

Table 6.2. Global average cloud radiative forcing components derived from ERBE and ISCCP data sorted by season, cloud altitude and thickness. Source: Hobbs (1993)

	High Thin		High Thick		Mid Level Thin		Mid Level Thick		Low		All
	JJA	**DJF**	**JJA**	**DJF**	**JJA**	**DJF**	**JJA**	**DJF**	**JJA**	**DJF**	**Annual**
-(L-L_{cl})	6.5	6.3	8.8	8.8	4.8	4.9	2.4	2.4	3.5	3.5	25.8
(S-S_{cl})	−4.1	−4.0	−14.8	−16.3	−3.4	−4.1	−9.0	−10.9	−18.6	−21.7	−53.4
Net	**2.4**	**2.3**	**−6.**	**−7.5**	**1.4**	**0.8**	**−6.6**	**−8.5x**	**−15.1**	**−18.2**	**−27.6**

Thus the cloud (or indeed atmospheric humidity) produced by a dynamical response to other forcings can not be viewed as an additional forcing component. Only if changes to cloud properties are induced in situ by chemical or microphysical processes can they produce a radiative forcing, in the climate change sense, rather than a feedback. The suggestion that galactic cosmic rays may produce such an effect in discussed in lecture 9.

6.3 Cloud Radiative Properties and Microphysics

The ability of a cloud to scatter or absorb radiation depends critically on its microphysical properties. These include the phase, size, shape and density of the cloud particles. It also depends on the wavelength of the radiation. See lecture 4.4 for the theoretical background.

Figure 6.3 shows that at visible wavelengths, except in the presence of very small drops, the reflection is almost independent of droplet size and depends only on optical depth. The near infrared reflectance, however, for thick clouds (optical depth $\tau > 16$) depends on optical depth but only on particle size, with greater scattering for smaller drops. For thin clouds with small droplets the situation is more complex with no unique microphysical values for some pairs of spectral measurements.

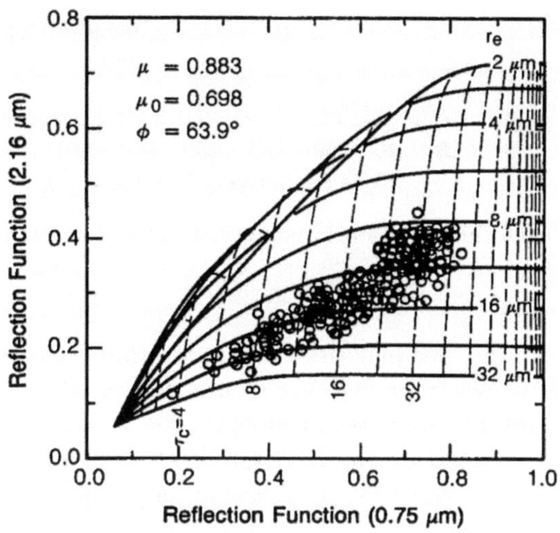

Fig. 6.3. *Curves*: Theoretical relationship between the reflection at 0.75 and 2.16 μm of a cloud composed of spherical water drops as a function of optical depth (at 0.75 μm) and droplet radius. Solar zenith angle 45.7°, observation zenith angle 28°, relative azimuth 69°. *Spots*: measurements in stratocumulus cloud. Nakajima and King (1990)

For ice cloud the situation is even more complex because the ice crystals can not be assumed to be spherical. Figure 6.4 shows the phase scattering function (see Sect. 4.4) calculated for a spherical water droplet and for ice crystals with various shapes. All have a strong forward scattering peak but vary widely at different angles. The water droplet curve shows a rainbow feature at 140° and the hexagonal column curve the 23° halo often seen round the sun in the presence of thin cirrus cloud.

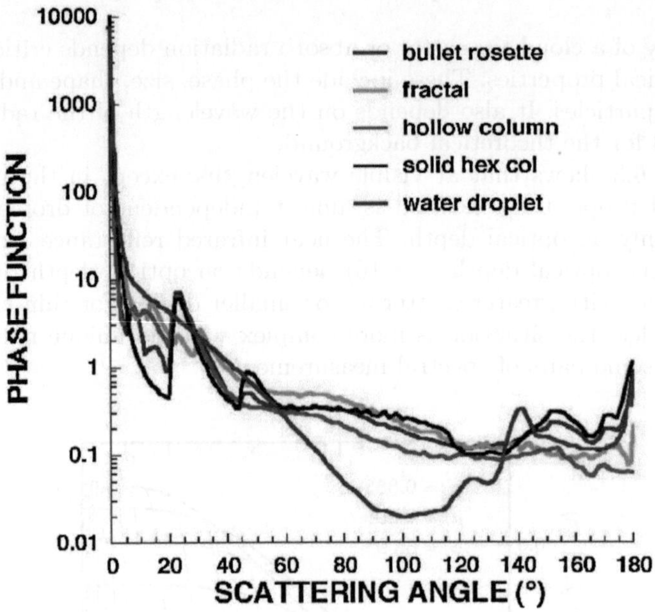

Fig. 6.4. Phase scattering function for spherical water droplets and ice crystals of various shapes

An example of the peculiarities of cloud radiation effects can be seen in Fig. 6.5. This shows radiation at 3.7 μm measured by a polar-orbiting meteorological satellite. At this wavelength the signal will comprise both thermal radiation emitted by the Earth and reflected solar radiation but at this location and time the solar component dominates. The land appears brighter than the sea, except for a strip to the west of Eire where specular reflection from the sea surface results in "sunglint". Clouds generally appear bright, as would be expected by a high reflectivity – see, for example, the low cloud lying off the north coast of Scotland. In several places, however, the cloud appears black as can be seen in the highest cloud on the weather front to the west side of the image, some of the storm clouds over Spain and some aircraft condensation trails over N. Scotland Snow lying on the Alps also appears black. This happens because the ice associated with these

Fig. 6.5. Image acquired at 3.7 μm from Advanced Very High Resolution Radiometer on a NOAA meteorological satellite at 15:12 h in April 1984. Courtesy University of Dundee Satellite Receiving Station

features is composed of crystals which are so large that instead of scattering the radiation they absorb it. Exploitation of such features in multi-spectral data allows us to retrieve information about cloud microphysical properties from satellite data.

6.4 Cloud Formation

Large Scale Processes

Clouds form when the water vapour in the air condenses. Generally this occurs in airmasses which are rising as these expand due to the reduction in pressure and therefore cool adiabatically. The cooler air has a lower saturated vapour pressure so that the airmass, with a given humidity, becomes saturated. Air masses rise either due to convection or through being forced to pass over some sort of barrier.

Convection takes place when air at the surface is heated. This can be due either to the ground being warmed by the sun or when cool air flows over a warmer surface. The warmed air becomes less dense than that of the surrounding air and starts to rise and cool; at some height (called the condensation level) the air becomes saturated and cloud forms. Within the cloud the decrease of temperature with height is generally less than that of clear air so the ascending volumes remain warmer and more buoyant than the surrounding air. Convective clouds appear as fluffy white heaps but given enough energy can develop into deep storm clouds, reaching the tropopause where the wind blows ice "anvils" off the tops.

Forced ascent may take place over surface topography (hills and mountains) or over other bodies of air. An example of the latter occurs in mid-latitude cyclonic systems which are formed when polar and sub-tropical air-masses collide, creating weather fronts and forcing the warmer air to ride over the colder.

Microphysical Processes

Cooling the air to saturation point is not, however, a sufficient condition for cloud to form. In fact it is possible for relative humidities to reach values up to 500% without any condensation having taken place. In practice the water vapour requires a suitable surface, called a condensation nucleus, on which to condense. If the condensation nucleus is not a water surface then heterogeneous nucleation is said to take place; if it is then homogeneous nucleation occurs. In the free atmosphere, however, heterogeneous nucleation is the only important process because homogeneous nucleation requires prohibitively high relative humidities. (For a comprehensive discussion of cloud formation see the classic text by Ludlam, 1980).

Particles which act as condensation nuclei include sea salt, sulphates, mineral dust and aerosols produced from biomass burning. The concentration and composition of atmospheric aerosol vary geographically with, for example, sulphate aerosol being more abundant in the northern hemisphere as it is generated in industrial regions. A region with a higher concentration of condensation nuclei will produce a larger number of smaller cloud droplets than a remote area with clean air which will produce fewer larger droplets for the same total water content. Smaller drops are more effective at scattering radiation (see Sects. 6.3 and 4.4) so that the cloud produced in the air with more aerosol has a higher albedo. An example of this can be seen in Fig. 6.6 which shows a satellite image in which emissions from ships' funnels have modified the reflectivity of a pre-existing deck of stratocumulus cloud. This effect has become known as the "first indirect effect of aerosol" in the context of the radiative forcing of climate change by anthropogenic aerosol (see Sect. 5.3). It has been suggested that ionisation of atmospheric aerosol by

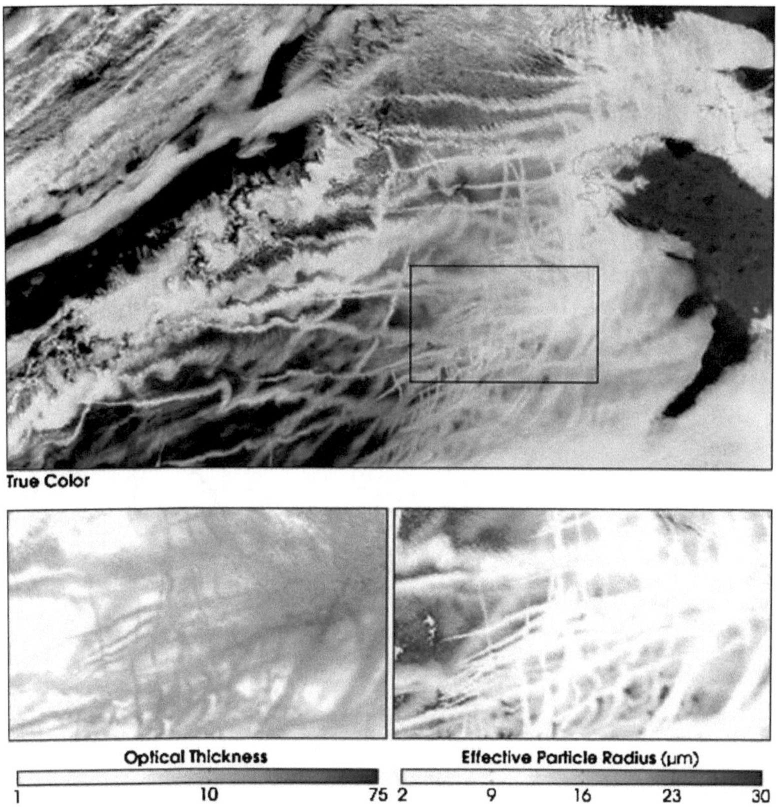

Fig. 6.6. Ship tracks observed off the west coast of France by the MODIS instrument on the Aqua satellite on 27 January 2003. The two lower panels show the optical properties of the cloud retrieved for the area enclosed by the box in the upper panel. From: http://visibleearth.nasa.gov/

galactic cosmic rays may make them more efficient cloud condensation nuclei and thus potentially have the ability to enhance global albedo (see lecture 9).

6.5 Clouds in GCMs

Uncertainties and approximations in the representation of cloud formation and cloud radiative properties remain a major cause of uncertainty in current climate prediction models, and are the main reason for the range of uncertainty of approximately a factor two in estimates of the climate sensitivity parameter, λ, as outlined in Sect. 5.3. As discussed above, cloud formation depends on factors ranging from local topography, large scale flow and the temperature and humidity of air masses to the microphysical composition of particulates in the atmosphere. It is not feasible for GCMs to take account fully of all these factors and so they include a variety of parameterisations for cloud prediction. On the largest spatial scales and annual timescales these parameterisations have limited success (see below). For local cloud on timescales of hours, however, the predictions can be very poor and, with grid sizes of the order of 200 km in the horizontal and 1 km in the vertical, there is no possibility of being able to reproduce cloud structures such as those shown in Fig. 6.5.

Figure 6.7 shows the zonal annual mean cloud cover as observed and as predicted by an international selection of climate models. Most of the models reproduce the maxima in cloud observed in the tropics and in mid-latitudes but all underestimate the amount of cloud in the tropics and sub-tropics and overestimate it in mid- to high latitudes.

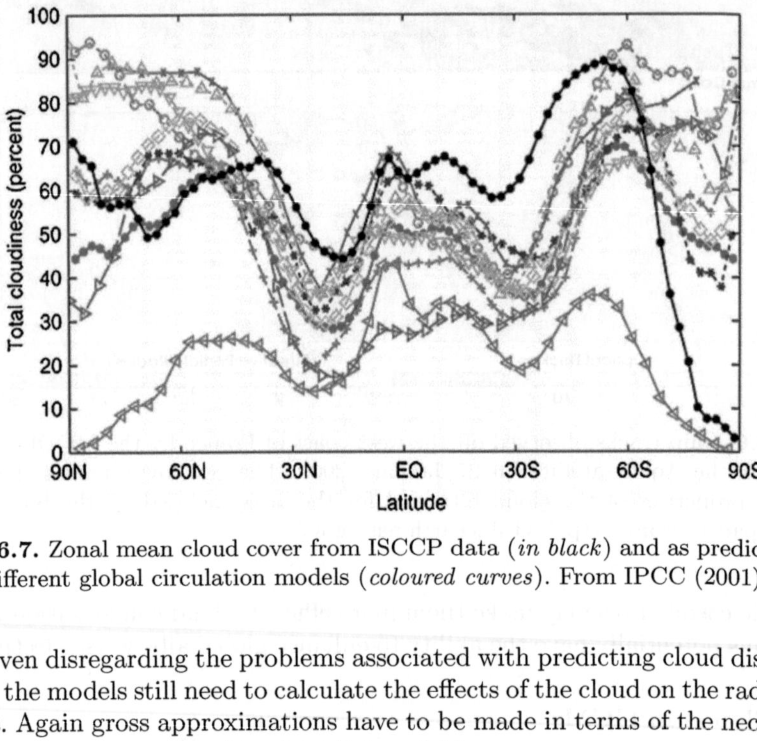

Fig. 6.7. Zonal mean cloud cover from ISCCP data (*in black*) and as predicted by ten different global circulation models (*coloured curves*). From IPCC (2001)

Even disregarding the problems associated with predicting cloud distributions the models still need to calculate the effects of the cloud on the radiation fields. Again gross approximations have to be made in terms of the necessary integrations over spectral and spatial variables.

Cloud-radiation interactions present arguably the largest uncertainty in our current understanding of climate and our ability to predict climate change.

7 Atmospheric Photochemistry

The interaction of solar radiation with the atmosphere is key to the initiation or constraint of many of the chemical processes which take place in the atmosphere. In this lecture we will consider the role of solar UV radiation in determining the budget of stratospheric ozone and how the composition and thermodynamic structure of the stratosphere are modulated by solar variability. We will also look at how radiative perturbations to the upper stratosphere may impact winter polar dynamics. The section finishes with short pieces on the impact of solar energetic particles and the photochemistry of the troposphere.

7.1 Stratospheric Photochemistry

Major Processes

In the stratosphere the main chemical reactions determining the concentration of ozone are:

$$\begin{aligned} &O_2 + h\nu \rightarrow O + O \\ &O + O_2 + M \rightarrow O_3 + M \\ &O_3 + h\nu \rightarrow O_2 + O \\ &O + O_3 \rightarrow 2O_2 \\ &O_3 + X \rightarrow XO + O_2 \\ &XO + O \rightarrow X + O_2 \end{aligned} \tag{7.1}$$

The first of these reactions represents the photodissociation of oxygen at wavelengths less than 242 nm. This process is the key step in ozone formation because the oxygen atoms produced react with oxygen molecules to produce ozone molecules, as depicted in the second reaction (the M represents any other air molecule whose presence is necessary to simultaneously conserve momentum and kinetic energy in the combination reaction). Because the short wavelength ultraviolet radiation gets used up as it passes through the atmosphere, concentrations of atomic oxygen increase with height. This would tend to produce a similar profile for ozone but the effect is counterbalanced by the need for a 3-body collision (reaction 2) which is more likely at higher pressures (lower altitudes). Thus a peak in ozone production occurs at around 50 km. The third reaction is the photodissociation of ozone, mainly by radiation in the Hartley band ($\lambda < 310$ nm), into one atom and one molecule of oxygen. This does not represent the fundamental destruction of the ozone because the oxygen atom produced can quickly recombine with an oxygen molecule. The fourth reaction represents the destruction of ozone by combination with an oxygen atom. The fifth and sixth reactions represent the destruction by any catalyst X, which may include OH, NO and Cl. The various destruction paths are important at different altitudes but the combined effect is an ozone concentration profile which peaks near 25 km in equatorial regions.

Because photodissociation is an essential component of ozone formation most ozone is produced at low latitudes in the upper stratosphere. Observations show, however, that it is also present in considerable quantities in the mid- and high latitude stratosphere due to transport by the mean meridional circulation (see Sect. 1.3). The resulting distribution of mixing ratio is shown in Fig. 7.1. The atmospheric circulations tend to move the ozone away from its source region towards the winter pole and downwards. Indeed, the quantity of ozone above unit area of the Earth's surface (referred to as the ozone column amount) is usually greater at mid-latitudes than the equator, see Fig. 7.1 In the lower stratosphere its photochemical lifetime is much longer, because of the reduced penetration of the radiation which destroys it, and its distribution is determined by transport, rather than photochemical, processes. In winter high latitudes photochemical destruction essentially ceases and the ozone accumulates until the spring.

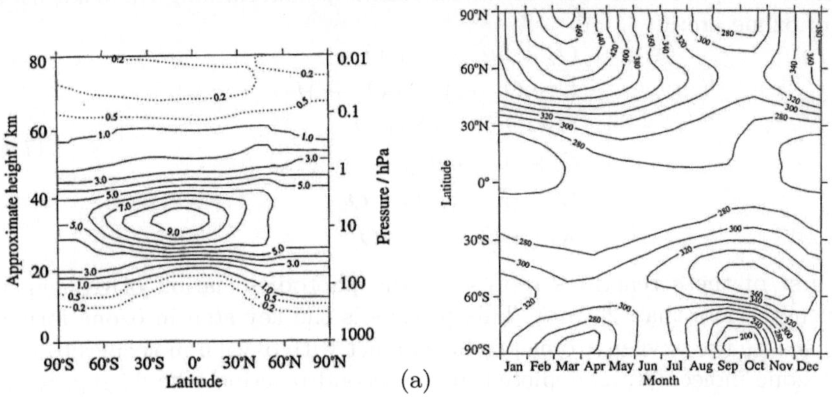

Fig. 7.1. (**a**) Zonal mean concentration of ozone (ppmv) in January. (**b**) The annual cycle in the latitudinal distribution of zonal mean ozone column amounts. The units are 10^{-3} atm cm (1 atm cm = 2.69 × 1019 molecules cm^{-2}). From: Andrews (2000)

The minimum in ozone column which occurs near the south pole in spring (October) has deepened considerably during the past thirty years into what has become known as the "ozone hole". Observations and theoretical studies have shown that the depletion occurs mainly in the lower stratosphere and is due to catalytic destruction of ozone on the surface of polar stratospheric cloud particles by active chlorine which is released as the sun rises in spring from chlorine compounds which have been stored through the winter. The source of the chlorine is CFCs which are now banned under international agreement. Because of the long atmospheric lifetime of CFCs and the natural interannual variability of the winter polar atmosphere it is difficult to see a trend over short (few year) periods but it appears that the ozone hole is now filling.

Response of Stratospheric Ozone to Solar UV Variability

We see from the above discussion that ozone is produced by short wavelength solar ultraviolet radiation and destroyed by radiation at somewhat longer wavelengths. Images of the sun acquired in the visible and ultraviolet (Fig. 7.2) suggest that it is more active at shorter wavelengths and, indeed, the amplitude of solar cycle variability is greater in the far ultraviolet (see Fig. 7.3). This means that ozone production is more strongly modulated by solar activity than its destruction and this leads to a higher net production of stratospheric ozone during periods of higher solar activity.

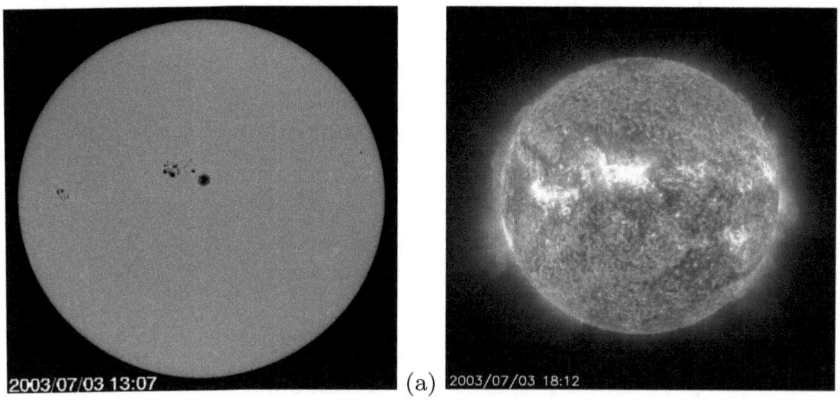

(a) (b)

Fig. 7.2. Images of the sun acquired at (**a**) visible (6768 Å) and (**b**) far UV (304 Å) wavelengths by instruments on the SOHO satellite. From: http://sohowww.nascom.nasa.gov/

Both observational records and model calculations show approximately 2% higher values in ozone columns at 11-year cycle maximum relative to minimum. However, there are some discrepancies between satellite observations and model predictions in the vertical and latitudinal distributions of the response. Figure 7.4 (lower panel) shows a typical model calculation with largest changes in the middle stratosphere and less above and below while the upper panel shows the solar cycle modulation of ozone derived from a satellite dataset.

Equatorial profiles of ozone solar cycle modulation from a selection of other model calculations and two other data analyses, are shown in Fig. 7.5. The models all show the same profile shape while the observational datasets suggest something rather different: a larger response near the stratopause and possibly a second maximum in the lower stratosphere. This remains a key area of uncertainty in solar effects on the atmosphere. There may be some factors missing in the models and their poor simulation of the lower stratospheric response suggests this might be related to ozone transport. However, full

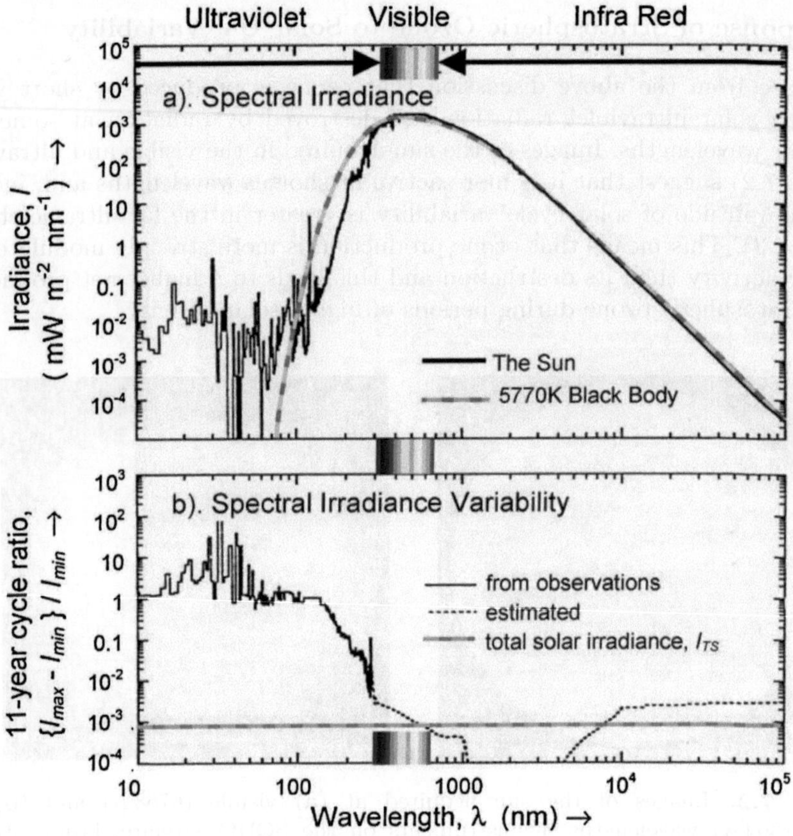

Fig. 7.3. *Top*: solar spectrum. *Bottom*: Fractional difference in solar spectral irradiance between maximum and minimum of the 11-year cycle. Adapted from Lean (1991) by M. Lockwood

3D GCMs produce very similar profiles to those of the 2D models with less complete dynamics so the mechanisms involved are not clear. It should also be borne in mind that the observational data are only available over less than two solar cycles so there remains some doubt about the statistical robustness of the signals derived from them.

Influence of Solar 27-Day Rotation Period on Stratospheric Ozone

Anisotropies on the surface of the Sun mean that the radiation it emits is not uniform in the longitudinal dimension. Thus, as it rotates on its axis there is a variation in the radiative energy received at the Earth. The magnitude of this variation depends on the phase of the solar cycle: it is much smaller at solar min when the Sun's surface is less active. Figure 7.6 shows the percentage

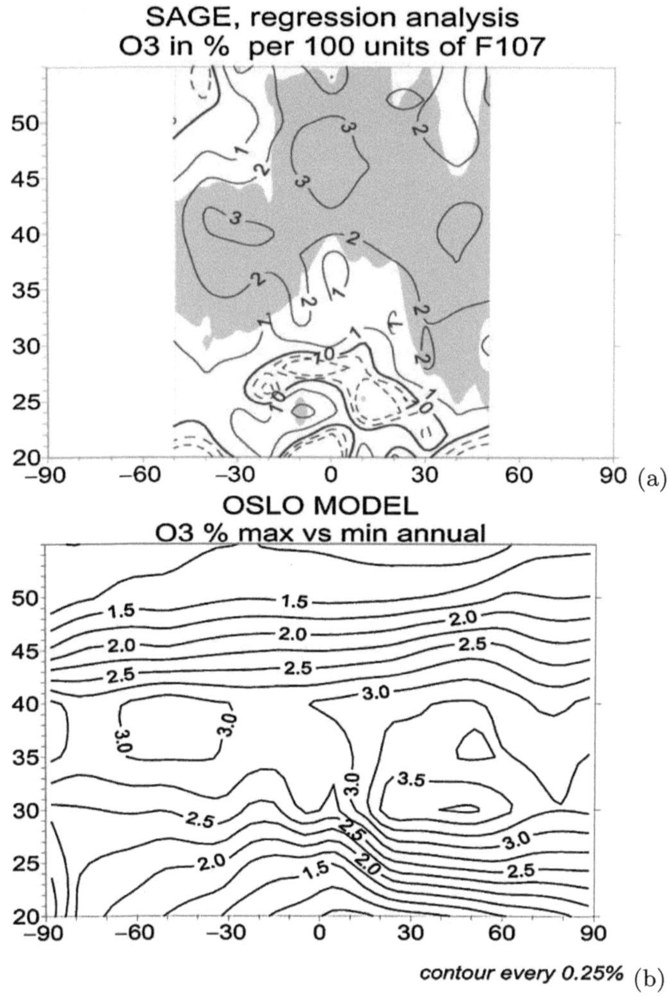

Fig. 7.4. Percentage increase in ozone from solar min to max as a function of latitude (°N) and altitude (km). *Top*: estimated from SAGE data. *Bottom*: estimated by 2D model (SOLICE, 2004)

variation as a function of wavelength derived from SOLSTICE data during early 1992 when the Sun was near maximum. Also in Fig. 7.6 is presented a calculation of the correlation between ozone concentrations and incoming solar UV. At zero time lag a peak response is seen in the lower mesosphere and this propagates downwards reaching 10 hPa about 5 days later but then appears to peter out. The magnitude of the response, of order 1–2%, compares well with available observations in the upper stratosphere.

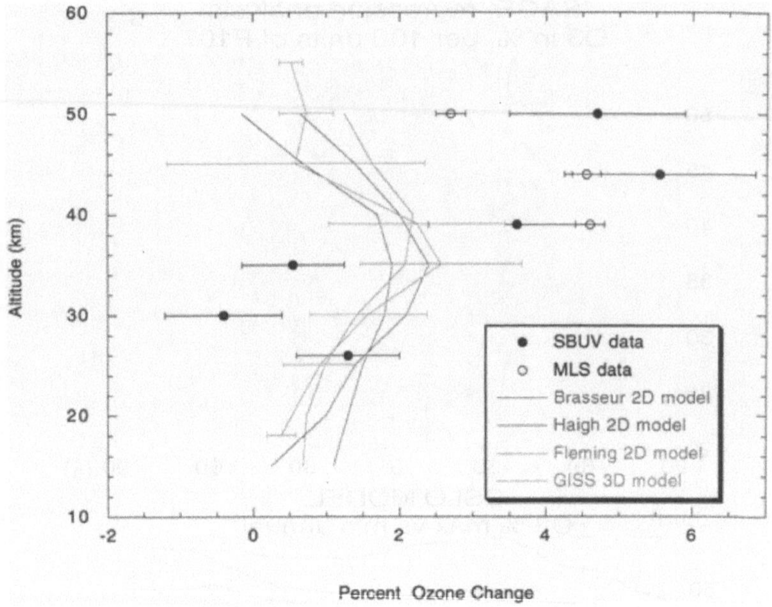

Fig. 7.5. Estimates of percentage increase in tropical ozone from solar min to solar max from satellite data (data points and horizontal bars) and models (coloured curves)

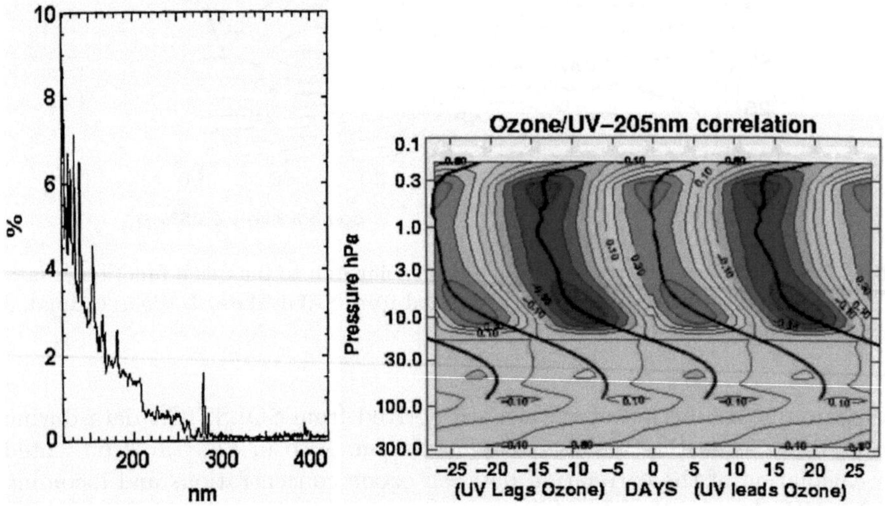

Fig. 7.6. *Left*: 27-day signal in solar spectral irradiance derived from SOLSTICE data, peak amplitude as percentage of mean irradiance. (Williams et al, 2001). *Right*. Correlation between ozone concentration and incoming solar UV as a function of time-lag and altitude. Calculated for solar max conditions using a GCM with coupled chemistry (SOLICE, 2004)

Response of Stratospheric Temperature to Solar UV Variability

Figure 7.7 shows vertical profiles of the solar fluxes and heating rates for solar minimum conditions in the ultraviolet, visible and near infrared spectral regions. The magnitude of the UV flux is much smaller than in the other two regions but its absorption by ozone causes the largest heating rates in the middle atmosphere. The weaker absorption of visible radiation in the lower stratosphere and of near infrared radiation in the troposphere give much smaller heating rates.

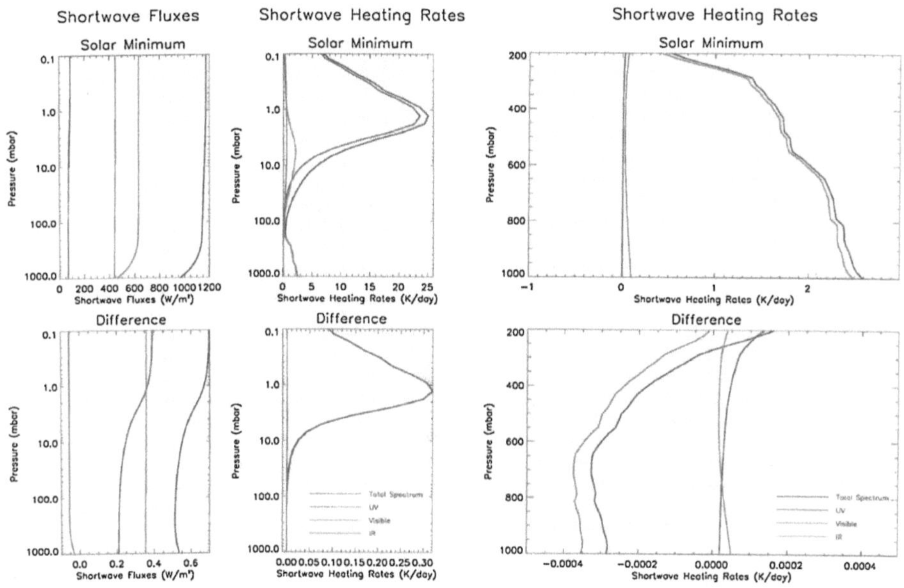

Fig. 7.7. Solar fluxes (*left*) and heating rates (*middle* and *right*). *Top*: solar minimum values; *bottom*: difference between solar min and max. The spectrum is divided into UV (220–320 nm), visible (320–690 nm) and near infrared (690–1000 nm) bands. The right hand column has a linear pressure scale for the ordinate so emphasising the troposphere (Larkin, 2000)

Also shown in Fig. 7.7 are the differences in fluxes and heating rates between 11-year solar cycle minimum and maximum conditions. At the top of the atmosphere the increases in incoming radiation in the visible and UV regions are of similar magnitude but the stronger absorption of UV produces much greater heating. The same data are shown with a linear pressure scale for the ordinate (to emphasise the troposphere) in the last two panels of Fig. 7.7. The spectral changes prescribed for these calculations were such that near-infrared radiation was weaker at solar maximum so decreases in heating rate are shown. This is contentious but the changes are anyway very small being about one part in ten thousand.

The response of atmospheric temperatures to solar variability is large in the upper atmosphere, with thermospheric variations of order 100 K being typical over the 11-year cycle, reflecting the large modulation of far ultraviolet radiation in that region. At lower altitudes the response is smaller, and less certain. Measurements made from satellites suggest an increase of up to about 1 K in the upper stratosphere at solar maximum, a minimum, or even negative, change in the mid-low stratosphere with another maximum, of a few tenths of a degree below. However, precise values, as well as the position (or existence) of the negative layer, vary between datasets: some examples are given in Fig. 7.8.

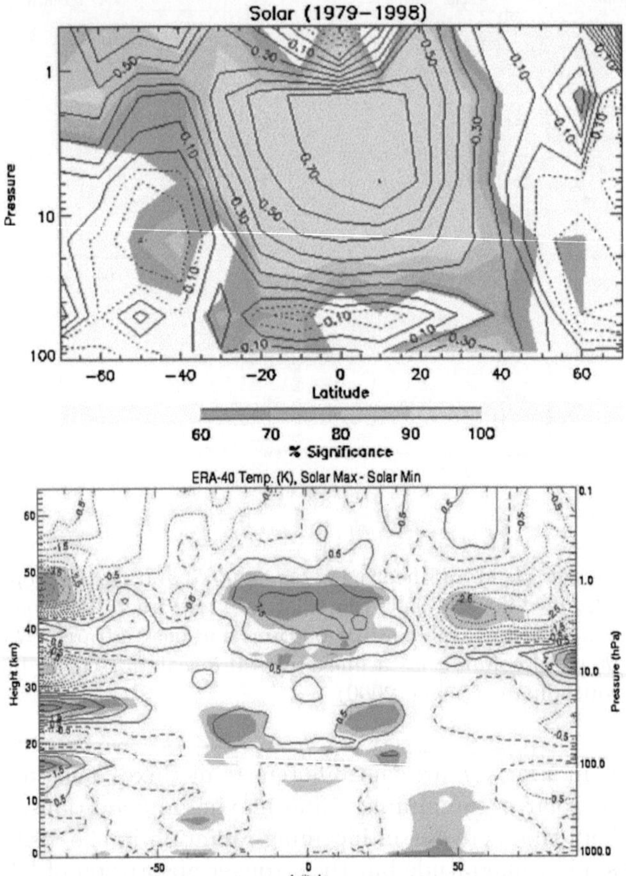

(a)

(b)

Fig. 7.8. Difference in temperature between solar maximum and solar minimum derived from observational data. (**a**) SSU/MSU satellite data (*grey shading* denotes statistical significance as shown in the legend). (**b**) ERA reanalysis data for the period 1979–2001; *light/dark shading* denotes 95% and 99% significance. Note the different height ranges in the two panels (SOLICE, 2004)

Studies of the solar influence using middle atmosphere GCMs also show a range of responses. Figure 7.9 presents the temperature change predicted by five different GCMs with three different approaches to implementing the necessary change in ozone. The best agreement is shown between the ERA data and the GCM with coupled chemistry but there remain large uncertainties in the nature of the response.

Response of NH Winter Polar Stratosphere to Solar UV Variability

During the winter the high latitude stratosphere becomes very cold and a polar vortex of strong westerly winds is established. The date in spring when this vortex finally breaks down is very variable, particularly in the northern hemisphere, but plays a key role in the global circulation of the middle atmosphere (see lecture 1). In a series of studies Kuni Kodera has suggested that, because variations in solar UV input change the latitudinal temperature gradient in the upper stratosphere, the evolution of the winter polar vortex will be affected. The first column in Fig. 7.10, derived from satellite data, shows that the vortex strengthens in November and December in response to solar activity and this positive anomaly in zonal mean zonal winds then propagated polewards and downwards until by February it is replaced by an easterly anomaly.

The other columns in Fig. 7.10 show the results of various GCM runs. None is successful in reproducing the apparent observational signal. It remains to be seen whether this disparity is due to fundamental flaws in the models, or to the design of the experiments, to insufficiently long integrations or even lack of statistical significance in the observational data analysis.

A Link to the QBO

Accepted theory has it that a west phase in the tropical QBO is linked with cold temperatures in the winter lower stratosphere, with vertically propagating planetary waves being channelled equatorwards. Karin Labitzke has pointed out, however, that while this relationship holds well during periods of lower solar activity it tends to break down near solar maximum. Figure 7.11, which contains data for the years 1956–1991, shows that warm polar temperatures tend to occur during the east phase of the QBO at solar min and west phase at solar max. The reason for this is subject of current research and is discussed in Sect. 8.3.

Solar High Energy Particle Effects

As described by the other lecturers, solar activity is manifest in not only variations in output of solar electromagnetic radiation but also in a range

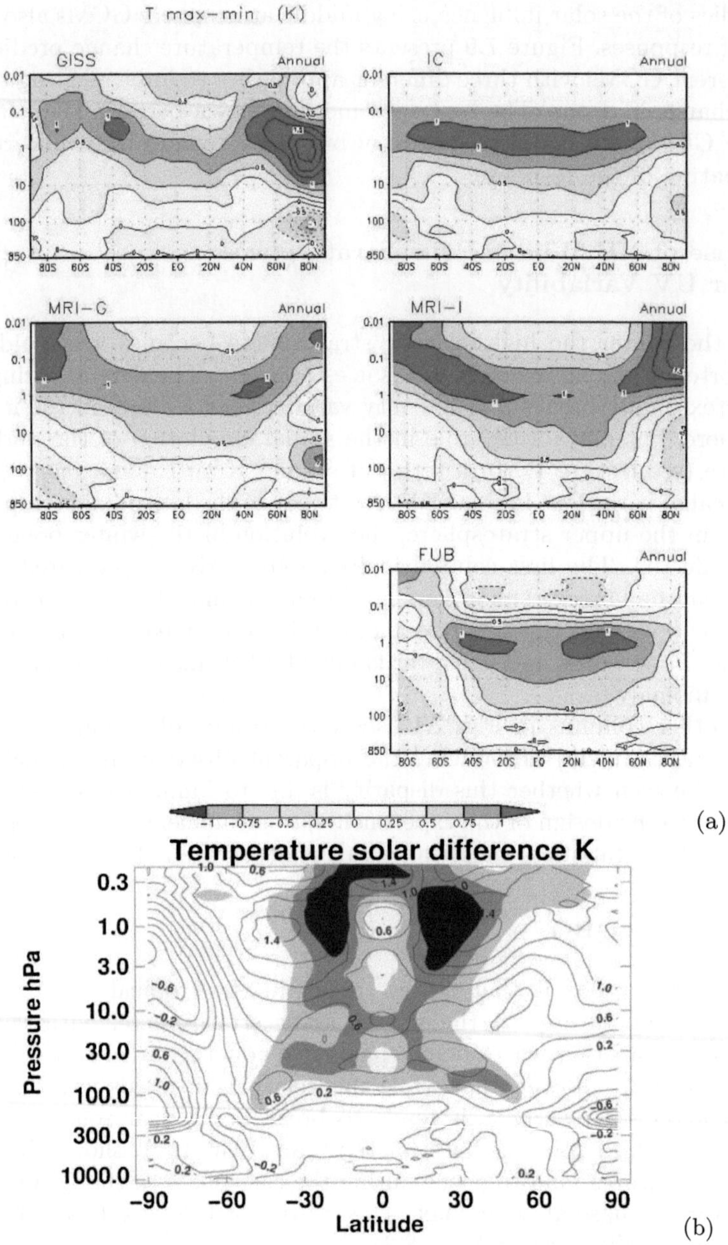

Fig. 7.9. Results from GCM simulations of the response of middle atmospheric temperatures to the 11-year solar cycle. The first column shows runs of two different GCMs which have both specified a solar-induced ozone response. The second column contains similar experiments but with a different specified ozone field (SOLICE, 2004). The panel in the third column comes from a GCM with a coupled chemistry scheme and thus an interactively-produced ozone response

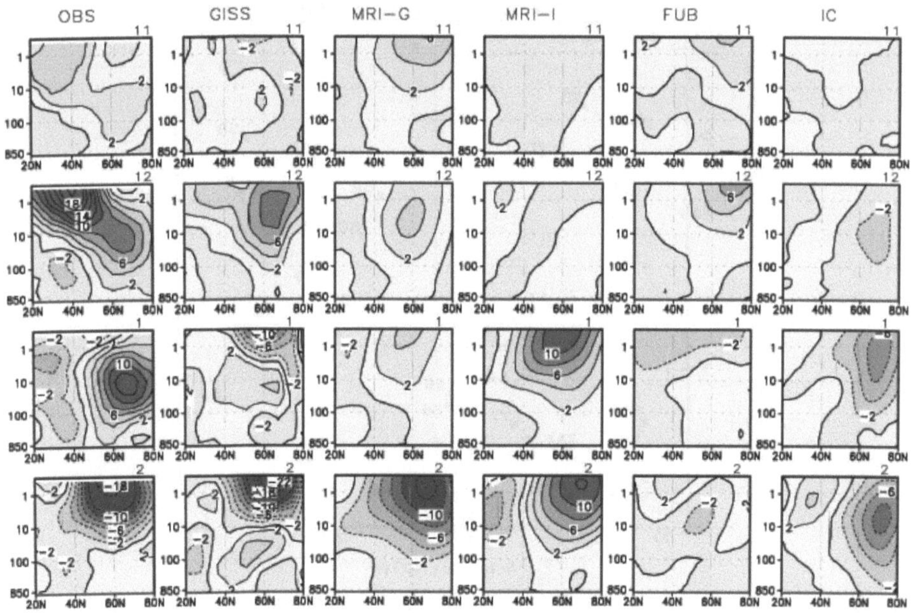

Fig. 7.10. Difference between solar maximum and solar minimum in Northern Hemisphere zonal mean zonal wind from November (*top*) to February (*bottom*). First column from satellite measurements; other columns different GCMs (Matthes et al. 2003)

of other parameters. One of these is the occurrence and severity of coronal mass ejections which result in the emission of energetic particles, some of which reach the Earth. The highest energy particles penetrate well into the stratosphere and affect the chemical composition of the atmosphere. Precipitating electrons and solar protons affect the nitrogen oxide budget of the middle atmosphere through ionization and dissociation of nitrogen and oxygen molecules. NO catalytically destroys ozone, as discussed above, and reductions in ozone concentration may occur down to the middle stratosphere for a particularly energetic event (Jackman et al, 2001). As the solar particles follow the Earth's magnetic field lines these effects have greatest initial impact at high latitudes but ozone depletion regions may propagate downwards and equatorwards over the period of a few weeks. Figure 7.12 shows calculations of the effects on NOy and O3 of a large solar proton event, which took place during October 1989. In the polar lower mesosphere nitrogen oxides are more than doubled in concentration leading to an ozone reduction in the polar stratosphere of over 4%. It is interesting to note that the effect of energetic particle events on ozone is in the opposite sense to that of enhanced ultraviolet irradiance. As particle events are more likely to occur when the Sun is in an active state the combined effect on ozone may be complex in its geographical, altitudinal and temporal distribution.

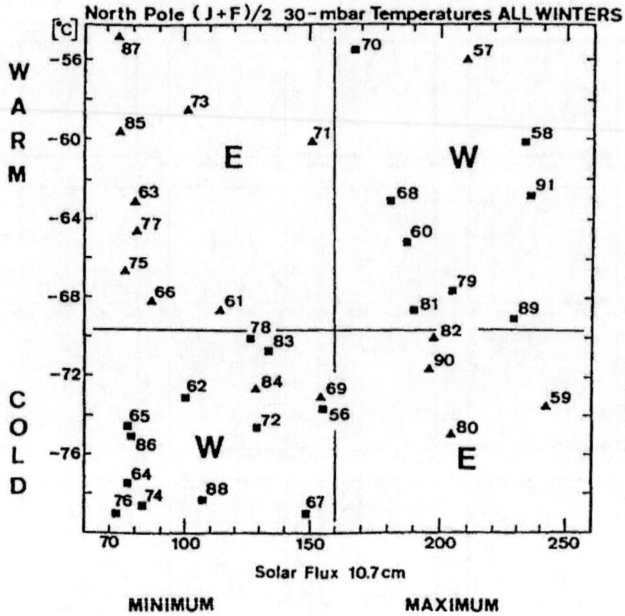

Fig. 7.11. A scatter plot showing for each year values of 30 hPa temperature at the north pole in Jan/Feb (ordinate), solar 10.7 cm flux (abscissa) and phase of the QBO (symbols, triangle for east and square for west). The *horizontal* and *vertical* lines, and the E & W labels, have been drawn to indicate regions of the diagram in which certain phases of the QBO predominate (Labitzke and van Loon, 1992)

7.2 Photochemistry of the Troposphere

Solar radiation is also fundamental in determining the composition of the troposphere. The daytime chemistry of the troposphere is dominated by the hydroxyl radical, OH, because its high reactivity leads to the oxidation and chemical conversion of most other trace constituents. OH is formed when an excited oxygen atom, $O(^1D)$, reacts with water vapour. The source of the $O(^1D)$ is the photodissociation (at wavelengths less than about 310 nm) of ozone; thus the presence of ozone is fundamental to the system. A major source of tropospheric ozone is transport from the stratosphere, but it is also formed through the photolysis (at wavelengths less than 400 nm) of nitrogen dioxide, which can be catalytically regenerated. Because OH is photolytically produced its concentration drops at night and the dominant oxidant becomes the nitrate radical, NO_3, itself photochemically destroyed during the day.

Thus any variation in the intensity, or spectral composition, of solar radiation may affect the lower atmosphere not only through direct heating but also potentially through modifying its chemical composition.

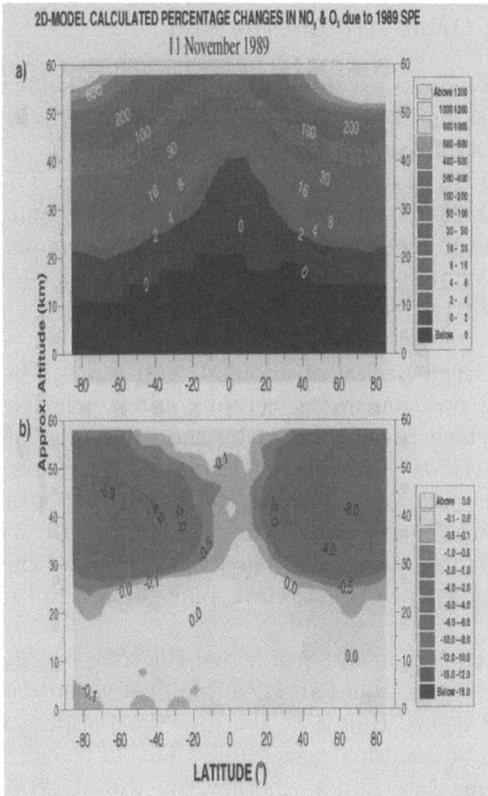

Fig. 7.12. Latitude-height section of percentage change in NO_y (*top*) and O_3 (*bottom*) calculated using a 2D model for November 1989, following the major solar proton event of October 1989 (Vlachogiannis and Haigh, 1998)

8 Response of Climate to Variations in Solar Irradiance

Sections 3 and 7 (part) have presented evidence of the influence of solar variability on various meteorological parameters. The other lectures have described some of the physical and chemical processes involved in the interaction of solar radiation with the atmosphere. In this lecture we try to bring these together to investigate mechanisms whereby variations in the solar radiative energy entering the Earth's atmosphere may influence climate. Three different processes are considered: first changes in irradiance due to orbital variations; second changes in total solar irradiance and third changes in solar spectral irradiance, concentrating on the role of UV.

8.1 The Earth's Orbit Around the Sun

If the Sun's output were constant then the amount of solar radiation reaching the Earth would depend only on the distance between the two bodies. This distance, R_e, varies during the year due to the ellipticity of the Earth's orbit which is measured by the value of its eccentricity (e). The value of e itself, however, varies in time with periods of around 100,000 and 413,000 years due to the gravitational influence of the Moon and other planets.

At any particular point on the Earth the amount of radiation striking the top of the atmosphere also depends on two other orbital parameters. One of these is the tilt (θ_t) of the Earth's axis to the plane of its orbit which varies cyclically with a period of about 41,000 years. The other parameter is the longitudinal position of the vernal equinox relative to the perihelion of the orbit (p), which is determined by the precession of the Earth's axis. This varies with periods of about 19,000 and 23,000 years. Figure 8.1 shows calculated values of e, θ_t and p over several hundred thousand years. Cyclical variations in climate records with periods of around 19, 23, 41, 100 and 413 kyr are generally referred to as Milankovitch cycles after the geophysicist who made the first detailed investigation of solar-climate links related to orbital variations.

Averaged over the globe and over a year the solar energy flux at the Earth depends only on e but seasonal and geographical variations of the irradiance depend on θ_t and $e\sin p$. But it is not just the temperatures of individual seasons that are at stake: the intensity of radiation received at high northern latitudes in summer determines whether the winter growth of the ice cap will recede or whether the climate will be precipitated into an ice age. Thus changes in seasonal irradiance can lead to much longer term shifts in climatic regime.

One approach to investigating a link between orbital parameters and climate is illustrated in Fig. 8.2. In that work the orbital effects of Fig. 8.1 were combined as an input to an empirical model of surface ice volume. The parameters of the model, specifically the growth and decay times of ice sheets and the lag between orbital forcing and climate response, were tuned to produce the best match of the output to oxygen isotope records taken from deep sea cores. The figure shows a good fit over the past 150 kyr suggesting that a good part of the observed variability can be explained in this way. The lack, however, of any physical mechanisms linking the forcing and response in the model, and also the poorer simulation of the earlier period, left many unanswered questions which are now beginning to be addressed by coupled atmosphere-ocean-ice GCMs.

8.2 Variation of Total Solar Irradiance

The electromagnetic radiative energy emitted by the Sun varies with solar activity, as discussed by Prof. Lockwood. In lecture 5 we saw estimates of

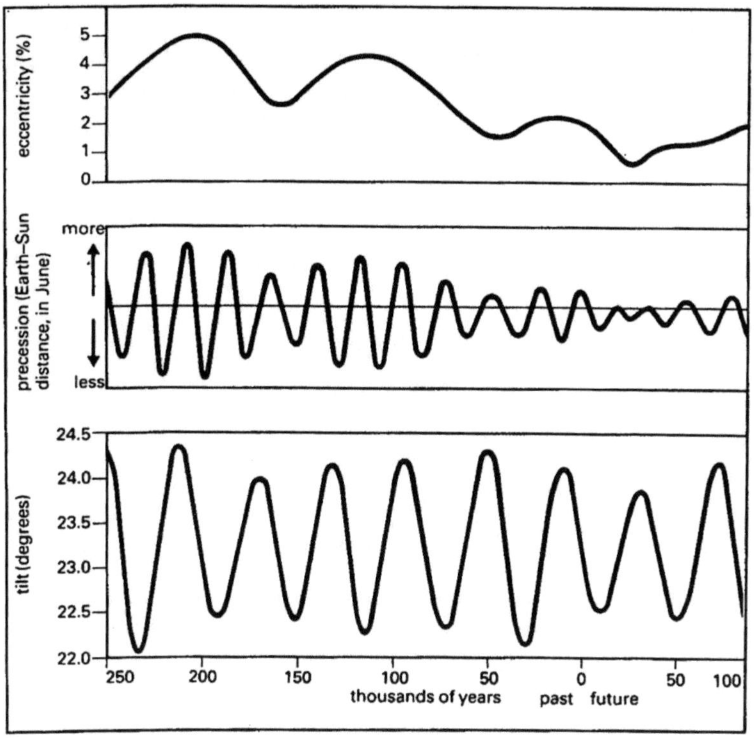

Fig. 8.1. The eccentricity, precession and tilt of the Earth's orbit calculated to take place over 350,000 years. Burroughs (1992)

how these values of change in total solar irradiance converted into radiative forcing. The next stage is to consider how the solar radiative forcing impacts climate and a relatively easy approach to this is to use a globally-averaged energy balance model (see Sect. 2.4). An example of some EBM estimates of the evolution of global average surface temperature over the past 1000 years in response to various forcing factors is given in Fig. 8.3. The upper panel shows the separate responses to changes in greenhouse gases, tropospheric aerosol, stratospheric (volcanic) aerosol and three different TSI series. The lower panel shows the response estimated to a combination of forcing factors, along with observational and reconstructed measurements of surface temperature.

The results suggest that the gross features of the temperature record are determined by volcanic and solar drivers until the twentieth century when human-induced factors, especially greenhouse gases, dominate.

Figure 8.4 shows estimates of global mean surface temperature calculated for the past 150 years using a full GCM. These results suggest that a good match between modelled and observed values can be obtained by including both natural (solar and volcanic) as well as anthropogenic forcings with the

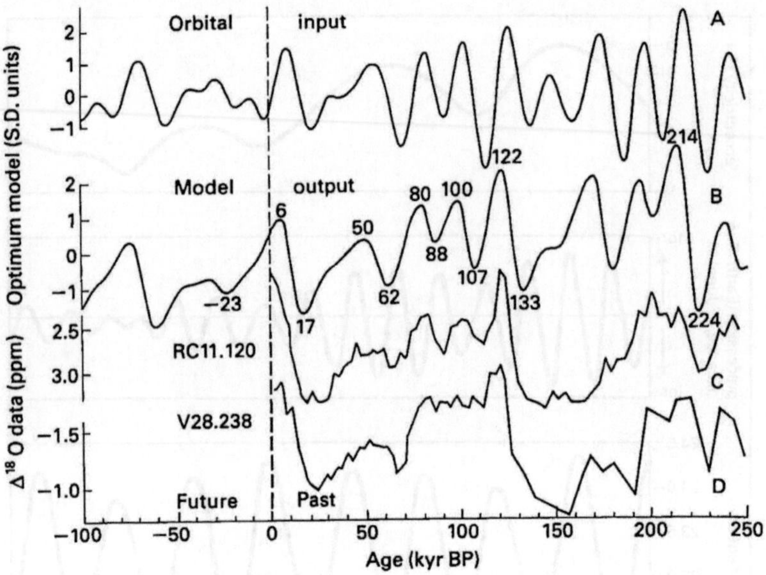

Fig. 8.2. Land ice volume best fit of a model (**B**) with input orbital forcing (**A**) to deep sea core oxygen isotope variations obtained from the southern Indian Ocean (**C**) and the Pacific Ocean (**D**). From Burroughs (1992)

increase in solar irradiance being particularly important in producing the warming over the first half of the twentieth century.

However, there is a large amount of noise in both model and observational datasets. An alternative approach to detecting component causes of climate change, is "optimal fingerprinting". For this it is assumed that the geographical patterns of response to particular factors are known, that the time-dependences of the forcing factors are known but that the amplitudes of the responses are unknown. The task is then essentially to perform a multiple regression analysis on a dataset to find which weighted combination of the response patterns best matches the data, taking into account known errors/uncertainties in both the data and patterns. An example of the results of one such analysis, using a dataset of surface temperature observations on a latitude-longitude grid over the twentieth century, is shown as global averages in Fig. 8.5(a). The black curve is the observations with the grey band representing measurement uncertainty; the red curve shows the result of only using anthropogenic forcing factors, the green only natural factors and the blue both together. A good fit is obtained when both types of forcing are included but Fig. 8.5(b) shows the derived magnitudes of the forcings. Here the value 1 indicated that the derived magnitude equals that the model gives using standard radiative forcing estimates. The model appears to be underestimating the solar influence by a factor of 2 or 3 implying that some amplification factors of the solar influence are not incorporated into the model's representation.

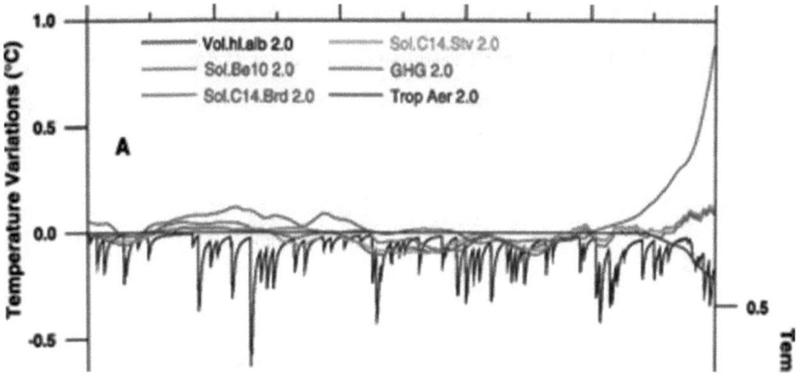

Fig. 8.3. (**a**) Global average surface temperature calculated using an EBM: results for individual forcings

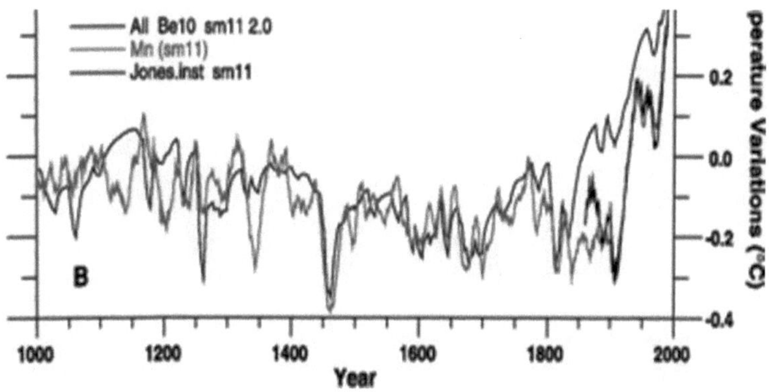

(**b**) Result for combined forcings (*in blue*) compared with the observational record (*in black*) and a series reconstructed from proxy data (*red*) (Crowley, 2000)

Furthermore, GCM runs which only include variations in TSI are unable to reproduce the distribution of temperature response as shown, for example, in Fig. 3.11 which confirms that something is lacking in our ability to simulate the response of climate to solar activity.

Figure 8.6 shows the results of a similar analysis on dataset consisting of zonal mean temperatures on a latitude-height grid from 1958 to 1996. Thus in this case there is information on the vertical structure of the response of temperature to climate change factors. A clear signal of the two solar cycles is detected and interestingly the amplitude factor β is about a factor 3 larger than is found directly using solar radiative forcing in a GCM, suggesting a significant underestimate of the modelled response to solar variability.

To explain these underestimates it is necessary to find some factor(s) which amplify the effect from that derived simply by consideration of total solar irradiance as the primary driving mechanism behind the impact of so-

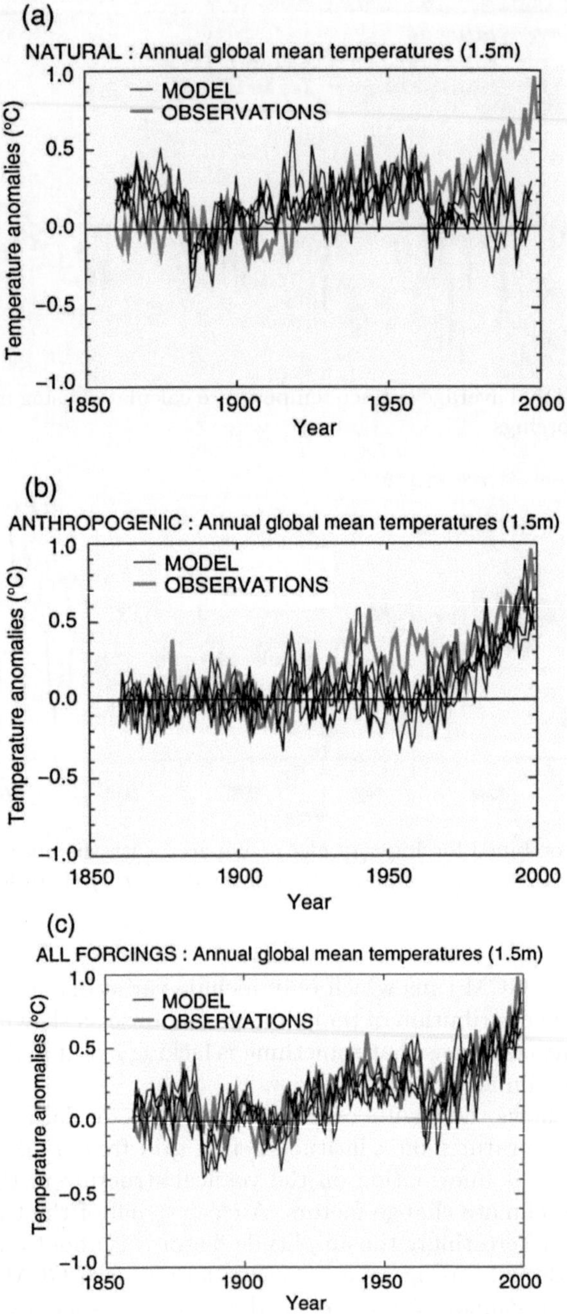

Fig. 8.4. Global annual mean surface temperature calculated using a GCM. Each panel shows 4 model integrations, each with the same forcing but with slightly different initial conditions, to give an indication of intrinsic natural variability (IPCC, 2001)

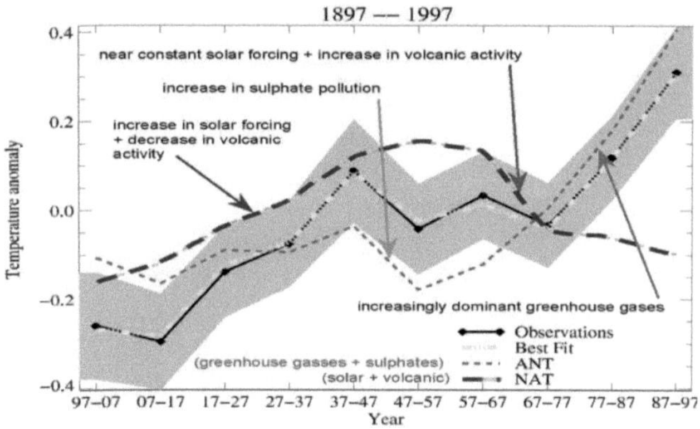

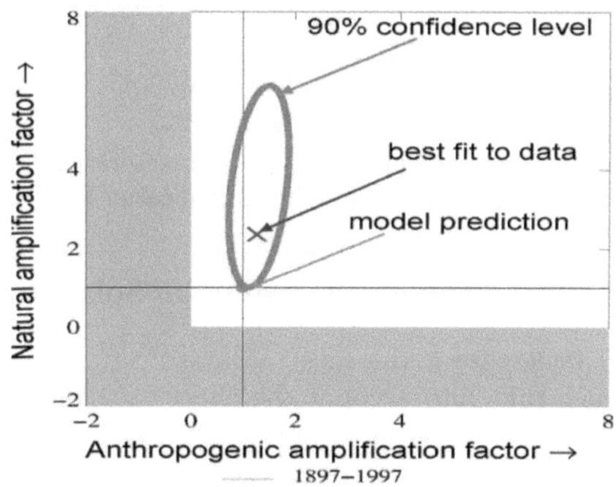

Fig. 8.5. Optimal fingerprinting technique in which geographical patterns of surface temperature change for different forcings are fitted to the observed time series. (**a**) results for different forcing factors, (**b**) derived magnitude of natural and anthropogenic forcings relative to that found from standard model runs. Figure courtesy Peter Stott, U.K. Meteorological Office

lar variability on climate. Potentially one such amplification mechanism is through the effects of variations in solar UV radiation on the stratosphere.

8.3 Variation of Solar Ultraviolet Radiation

The GCM studies discussed in the previous section represented variations in solar activity only by changes in TSI but, as noted in lecture 7, the effects of the larger variations in solar UV variation particularly impact the temperature and ozone structure of the middle atmosphere. Here we consider how these changes may impact the climate of the troposphere.

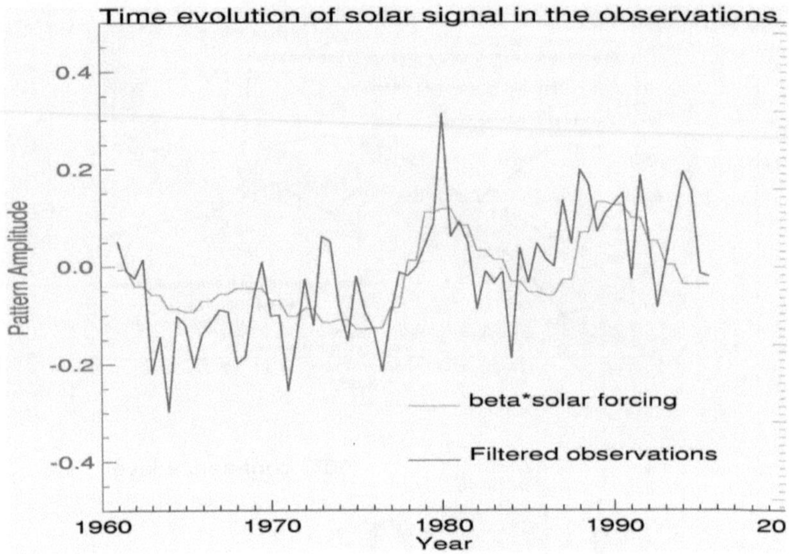

Fig. 8.6. Optimal fingerprinting applied to a time series of latitude-height patterns of temperature (Myles Allen, personal communication)

Effect of the Stratospheric Response on Solar Radiative Forcing

As discussed in lecture 5 the most accurate estimates of radiative forcing are those which take into account the effect of changes in the stratosphere on the radiative flux at the tropopause. The larger variations that take place in the UV part of the solar spectrum have a significant impact on stratospheric temperature and composition and so have a knock-on effect on the radiation reaching the tropopause (as first noted by Haigh, 1994). Figure 8.7(a) shows solar irradiance at winter mid-latitudes as a function of wavelength and altitude calculated using a 2D atmospheric model with accurate representations of photochemical and radiative processes. At the top of the atmosphere most energy is at visible wavelengths and this is transmitted almost unaffected through to the surface; at wavelengths shorter than ~300 nm, however, most radiation is absorbed by the time it reaches an altitude of ~40 km. There is also some absorption at longer visible wavelengths.

Figure 8.7(b) shows the difference in spectral irradiance between maximum and minimum periods of the 11-year solar cycle. At the top of the atmosphere there is more energy at all wavelengths but this is not perpetuated throughout the depth of the atmosphere. At wavelengths <330 nm and >500 nm there is actually less radiation reaching the troposphere at solar maximum than solar minimum because the enhanced concentrations of stratospheric ozone are resulting in greater absorption at these wavelengths. This is a strongly non-linear effect which varies with latitude and season and thus its impact on the value of solar radiative forcing is not easy to predict.

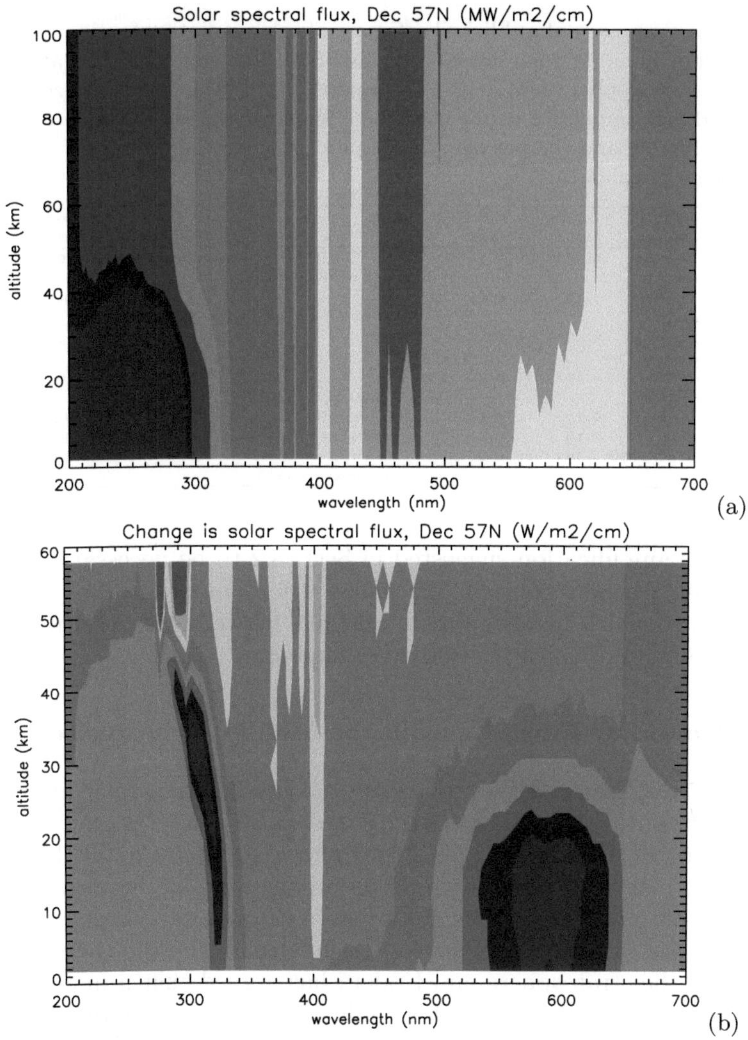

Fig. 8.7. (**a**) Solar spectral irradiance as a function of wavelength and altitude for 57°N at the winter solstice. Values range from $<10^{-4}$ (*violet*) to >6 (*red*) MW m^{-2} cm^{-1}. (**b**) As (**a**) but showing the difference between values at maximum and minimum periods of the 11-year solar cycle, contour interval is 500 W m^{-2} cm^{-1} ranging from <-1500 (*violet*) to >2500 (*red*). From Haigh (1994)

As discussed in lecture 5, changes to the temperature of the stratosphere also affect radiative forcing, adding a further complication. Estimates of the net effect of solar-induced ozone increases on solar radiative forcing vary widely, as can be seen in Table 8.1 – even the sign of the ozone effect is not ascertained. None of the estimates, however, come anywhere near the

Table 8.1. A Summary of published estimate of solar radiative forcing. 1st column: reference; 2nd: nominal solar variability; 3rd and 4th: solar UV radiative forcing at the top of atmosphere and at the tropopause; 5th: solar-induced ozone change; 6th, 7th and 8th: impact of ozone change on shortwave and longwave components of radiative forcing and the net effect; 9th: percentage amplification of solar forcing due to change in ozone

author	solar change	ΔS RF (toa)	ΔS RF (tpse)	ΔO_3	O_3 SW effect	O_3 LW effect	net O_3 effect	RF amp (%)
Haigh 1994	11-year amp	0.13	0.11	+ve peak near 40 km	-0.03	+0.02	-0.01	-9
Hansen et al 1997	11-year amp	0.13	0.11	+ve 10-150 hPa			+0.05	+45
Myhre et al 1998	11-year amp	0.13	0.11	+ve	-0.08	+0.06	-0.02	-18
Wuebbles et al 1998	c1680-c1990	0.49<0.70	0.42<0.60	+ve 10-150 hPa			-0.13	-21<-30
Larkin et al 2000	11-year amp	0.13	0.11	+ve (as H94)	-0.06	+0.11	+0.05	+45
		0.13	0.11	+ve (SBUV/TOMS)	-0.03	+0.08	+0.05	+45
Shindell et al 2001	1680-1780	0.30<0.39	0.26<0.33	+ve (upper strat) +ve (lower strat)			+0.02	+6<+8

factor 2 or 3 amplification suggested in Sect. 8.2 to be necessary to explain observations. We now consider mechanisms for potentially amplifying the solar influence which involve dynamical links between the stratosphere and troposphere and not (directly) radiative forcing.

Wave-Mean Flow Interactions in the Middle Atmosphere

In lecture 7 we discussed the existence of a link between solar activity, the QBO and the spring breakdown of the winter polar stratospheric vortex. The final warming of the winter polar stratosphere is brought about by upward-propagating planetary scale waves. The propagation of these waves, which are continually produced in the troposphere through the effects of topography or land-sea temperature contrasts, is determined by the thermodynamic structure of the stratosphere and Kodera has argued that a modification of the state of the stratosphere, such as induced by the absorption of enhanced solar UV radiation, will modify the wave propagation. Furthermore, the effects of this wave-mean flow interaction can propagate downwards into the troposphere. Thus solar heating of the upper stratosphere might influence the state of the late winter troposphere. Independent studies of the QBO have shown that the temperature of the polar winter lower stratosphere is related to the phase of the QBO in the tropics. Although details of this mechanism need to be ascertained it does appear to be able to explain, at least qualitatively, the links between solar variability, the QBO and winter polar temperatures shown in Fig. 7.10.

Coupling Between the Tropical Lower Stratosphere and Tropospheric Circulation

The first GCM studies of the impact of solar UV-induced variations in stratospheric temperature and ozone on the dynamical structure of the troposphere were presented by Haigh, (1996). In that paper a number of experiments were carried out using a variety of assumptions concerning ozone changes. All the experiments showed the same pattern of response in tropospheric temperatures and winds but with different magnitudes depending on the specifications of the UV and ozone changes. An example is shown in Fig. 8.8. At higher solar activity the sub-tropical jets are weaker and move slightly poleward. This theoretical result is very similar to that determined as the solar signal in observational data (Fig. 3.11).

Figure 8.9 shows results from a similar model experiment for the tropospheric mean meridional circulation. At solar maximum the winter Hadley cell is clearly weaker and broader than it is at solar minimum. Again these results are similar to those deduced from observational studies (see Fig. 3.12). The apparent success of the GCM simulations is reassuring but further study is required to provide a detailed understanding of the mechanisms whereby the effects take place, and thus an ability to predict future responses.

In order to investigate these mechanisms a GCM with highly simplified representations of radiative and cloud properties has been used in a series of experiments designed to analyse the processes taking place. The advantage of this approach over using a full GCM is that many runs can be carried out without straining computing resources. The experiments discussed below were designed to investigate how radiative perturbations to the lower stratosphere affect tropospheric dynamics. Two types of perturbation were applied: in the first a uniform perturbation to stratospheric heating was imposed while in the second the perturbation was largest at the equator and smoothed to zero at the poles. These experiments were designed based on the analysis of temperature fields shown in Fig. 3.11; the first representing the stratospheric response to volcanic (stratospheric) aerosol and the second to solar variability.

Figure 8.10 shows some of the simplified GCM results; it should be noted first that, although the perturbations were applied only in the stratosphere, a response is clearly seen in the troposphere. The uniform stratospheric perturbation causes the sub-tropical jets to weaken and move equatorward and the Hadley cells to weaken and shrink. In response to the perturbation with the latitudinal gradient both the jets and the Hadley cells again weaken but this time the patterns move poleward.

The magnitudes of the perturbations applied in these experiments was much larger than might be expected from volcanic or solar influences so that no direct comparison with observations is appropriate. It is clear, however, that the dynamical responses are qualitatively very similar to those shown in Fig. 3.11 for the observed zonal wind responses to volcanic and solar forc-

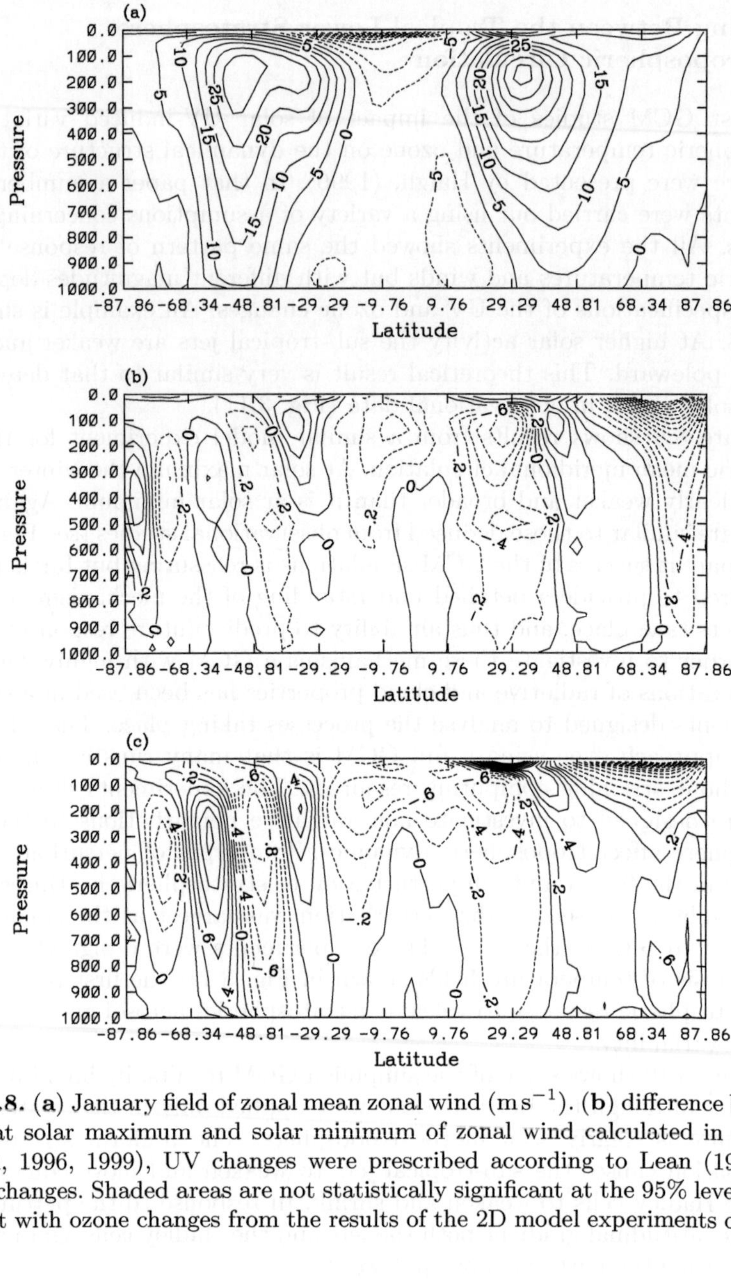

Fig. 8.8. (**a**) January field of zonal mean zonal wind ($m\,s^{-1}$). (**b**) difference between fields at solar maximum and solar minimum of zonal wind calculated in a GCM (Haigh, 1996, 1999), UV changes were prescribed according to Lean (1989), no ozone changes. Shaded areas are not statistically significant at the 95% level (**c**) As (**b**) but with ozone changes from the results of the 2D model experiments of Haigh (1994)

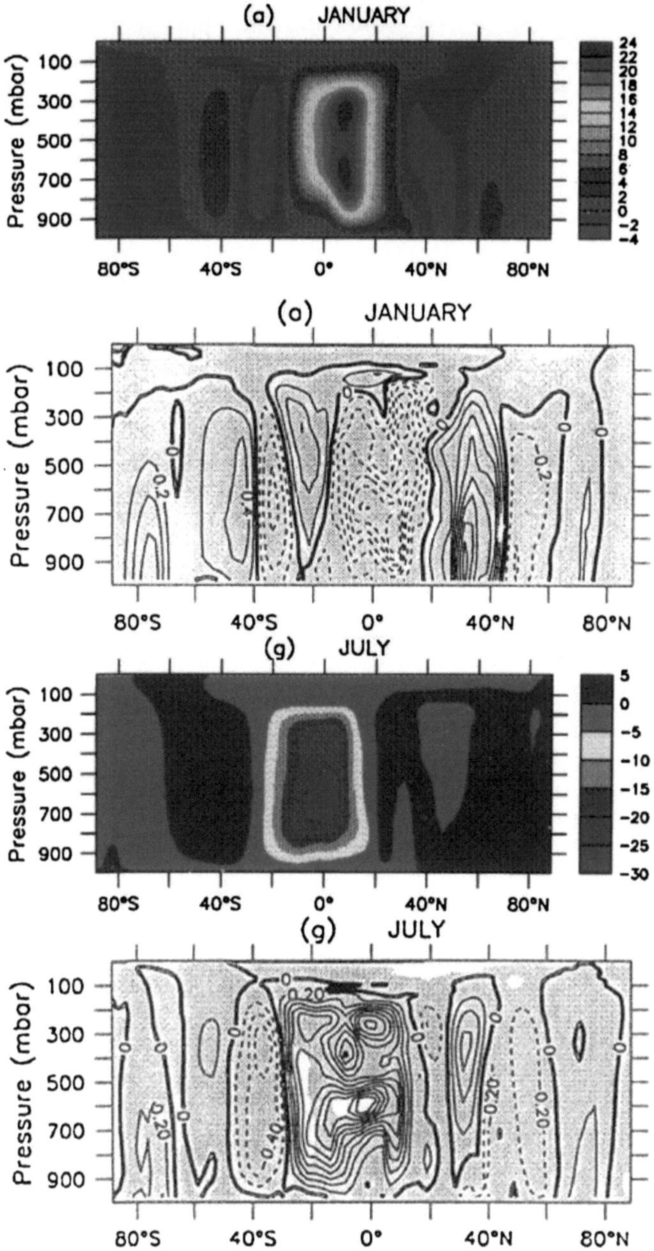

Fig. 8.9. Results from the model experiments outlined in the caption to Fig. 8.8 for mean meridional circulation. From *top*: mean field for January; difference between solar max and solar min in January; mean field for July; difference between solar max and solar min in July. Units: 10^{10}kg s^{-1}

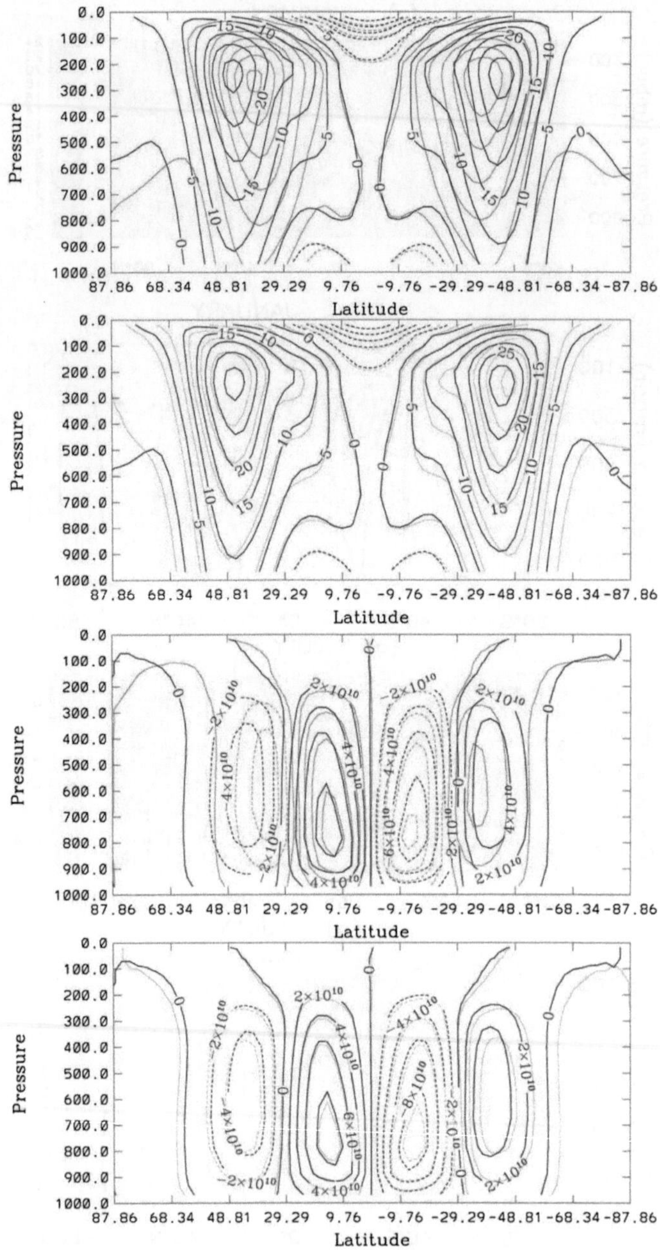

Fig. 8.10. Results from experiments with a simplified GCM. *Top panel*: zonal mean zonal wind, dark: control run, light: uniform stratospheric perturbation; *Second panel*: as top panel but for latitudinally varying perturbation; *Third panel*: as top panel but for mean meridional circulation; *Fourth panel*: as third panel but for latitudinally varying perturbation. From Haigh et al. (2004)

ing and in Fig. 3.12 for the solar response in vertical velocity. This work is beginning to provide us with an understanding of how, through the spectral composition of solar irradiance, apparently small changes in TSI may significantly impact the circulation of the lower atmosphere.

9 The Earth's Electric Field and Ionisation of the Atmosphere

It has been proposed, originally by Dickinson (1975) who acknowledged that his idea was entirely speculative at the time, that variations in cosmic rays could provide a mechanism whereby solar activity would produce a direct impact on cloud cover by modulating atmospheric ionization, resulting in the electrification of aerosol and increasing the effectiveness of this aerosol to act as condensation nuclei. Other processes whereby changes in the Earth's electric field might modify cloud cover, or cloud properties, have also been proposed (see e.g. Tinsley, 2000). The processes involved are complex but if they do take place then there is scope for considerable amendment to the value for solar radiative forcing of climate based on incident irradiance alone. In this lecture we will look at how the Earth's electric field is produced, how ionisation of aerosol takes place and how changes in these factors might affect cloud properties. The material is borrowed extensively from a comprehensive review article by Harrison & Carslaw 2003.

9.1 Atmospheric Electrical System

The electric currents flowing within the atmosphere are largely due to the charge separation that takes place within thunderstorms causing positive charges to flow upwards, to a conducting layer of the atmosphere at around 80 km altitude known as the ionosphere, and negative charges to pass to the Earth's surface by lightning discharge.

Outside of thunderstorm regions it is generally assumed that there are no sources of charge separation and a small "fair weather " ion drift current flows in the opposite sense, with negative ions flowing upwards and positive ones downward, driven through the electrical resistance of the air by the surface-ionosphere voltage difference. In fair weather regions the potential gradient has a value of about $150\,\mathrm{V\,m^{-1}}$ but it can reach a magnitude of $\sim 10^5\,\mathrm{V\,m^{-1}}$ in thunderstorms.

These currents, along with a summary of the processes which involve ions and electrification, are illustrated in Fig. 9.1. The processes include ionisation of the atmosphere, the involvement of the ions in aerosol charging and nucleation, and the enhancement of the local electric field in non-thunderstorm cloud. In the following sections we will look at some of these processes and how they might be affected by variations in solar activity.

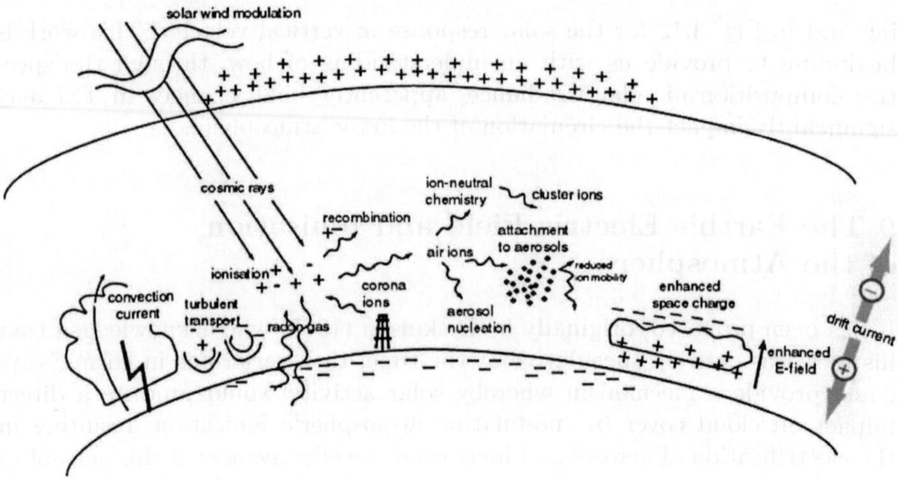

Fig. 9.1. Schematic of the atmospheric processes involving ions and electrification (from Harrison and Carslaw, 2003)

9.2 Atmospheric Ionisation

Ionisation of molecules in the lower atmosphere is brought about by cosmic rays and by naturally-occurring radioactivity. The latter consists of airborne alpha-particle emitters (such as radon gas) and direct gamma radiation from the soil. Cosmic radiation consists of extremely high energy (>GeV) particles, mostly protons and helium nuclei. Both sources cause an electron to separate from a molecule of nitrogen or oxygen; the electron then being captured by a neutral molecule on a very short timescale. Thus equal numbers of positive and negative ions are produced. Other processes can introduce a net charge into the atmosphere; these include combustion, rainfall and breaking ocean waves.

Cosmic rays are responsible for ∼20% of the ionisation over land surfaces and is the principle source of ionisation over the oceans. The ionisation rate increases with altitude reaching a peak near 15 km (see Fig. 9.2(a)). The geographical distribution of cosmic ray ionisation is strongly modulated by geomagnetism with the rays tending to follow the magnetic field lines down to the magnetic poles. Thus only very high energy cosmic radiation reaches the Earth's surface in low latitudes while much lower energy rays penetrate at high latitudes.

Solar activity modulates the heliospheric magnetic field which acts as a shield to cosmic rays. Thus, during periods of higher solar activity fewer cosmic rays reach the Earth, although the modulation primarily affects lower energy cosmic radiation. At the Earth's surface cosmic rays are monitored by neutron monitors which detect the disintegrated particles (e.g. pions, muons) produced when cosmic radiation impacts atmospheric particles. Figure 9.2(b)

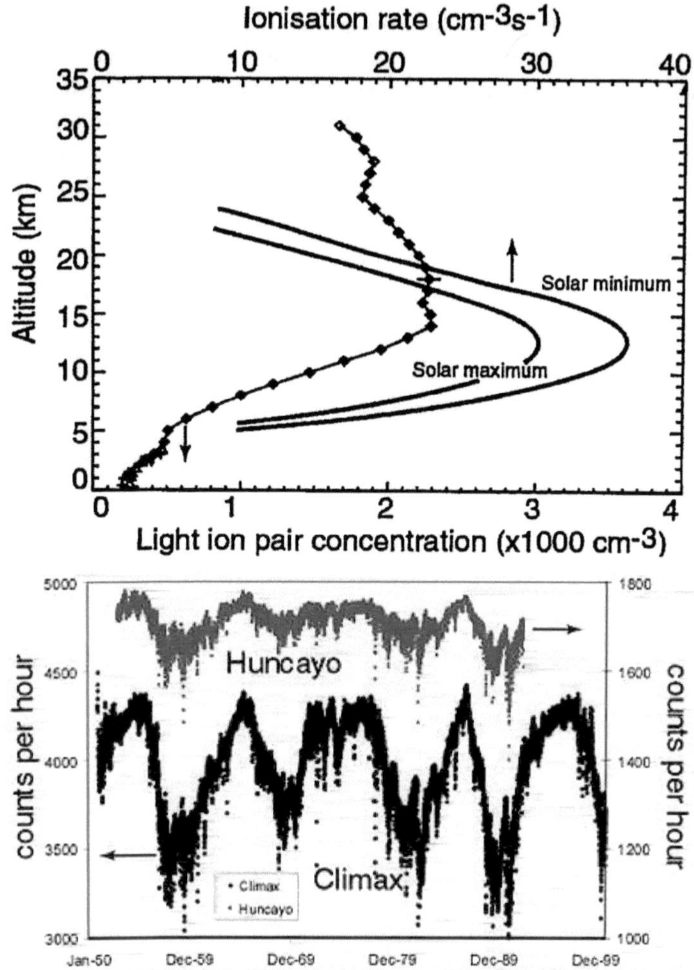

a

b

Fig. 9.2. (**a**) Variation of ionisation rate, and typical ion concentration, with height. (**b**) Variation with time of the neutron count rate at two surface stations: Climax, latitude 39.4°N and Huancayo, latitude 12.0°S (Harrison and Carslaw, 2003)

presents a time series of the neutron count rate at two surface stations. The lower latitude station clearly shows lower counts and weaker solar cycle modulation.

The density of ions (a typical profile is shown in Fig. 9.2(a)) is determined by a balance between the ionisation rate and the recombination of positive and negative ions. Collision between a neutrally-charged aerosol and an ion results in charge transfer (the electrification of the aerosol and the removal of the ion). Although ionisation results in equal numbers of positive and negative ions variations in atmospheric electrical parameters can result in local regions of net space charge which can be important in aerosol processes. Regions of one polarity are created when a gradient in air conductivity is produced in the presence of non-zero current density. Conductivity is reduced where air ions are scavenged by aerosol or cloud drops thus unipolar charge regions may occur on the upper and lower surface of cloud and aerosol layers (see Fig. 9.3). Above and below the cloud the electric fields has its fair weather value but the vertical drift of current ions in the cloud enhances the field within.

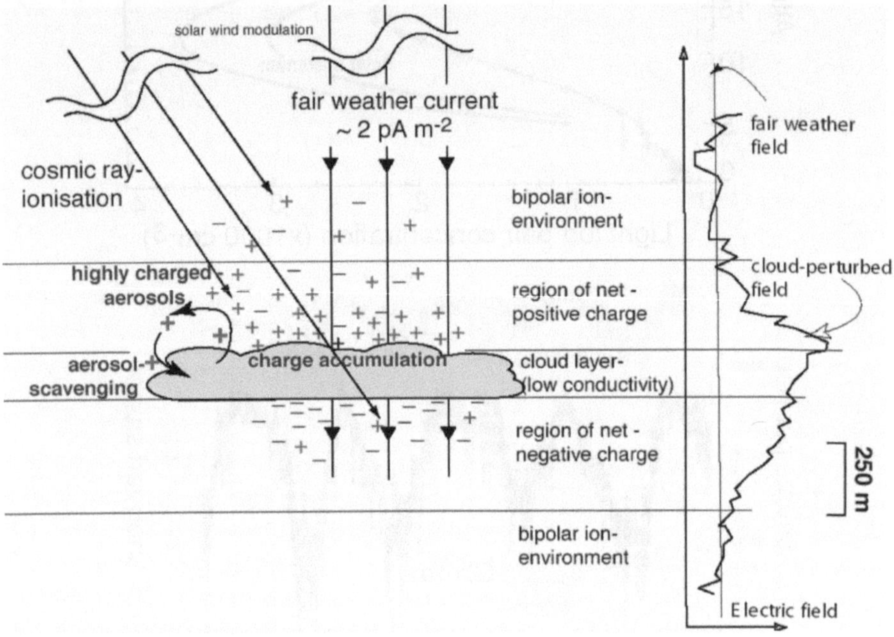

Fig. 9.3. Charge structure around an isolated layer of cloud in the presence of an electric field (Harrison and Carslaw, 2003)

9.3 Modulation of Cloud Condensation Nucleus Concentrations by Cosmic Rays

The mechanism proposed by Dickinson (1975), and adopted by Marsh & Svensmark (2000) to account for the correlation of tropical marine low cloud cover and cosmic radiation (Fig. 3.13), requires that modulation by solar variability of cosmic rays causes a response in the concentration of cloud condensation nuclei. Several consecutive processes (summarised in Fig. 9.4) need to take place in order for this to come about. Firstly air ions are produced by the action of cosmic rays; this is not controversial, as discussed above, but

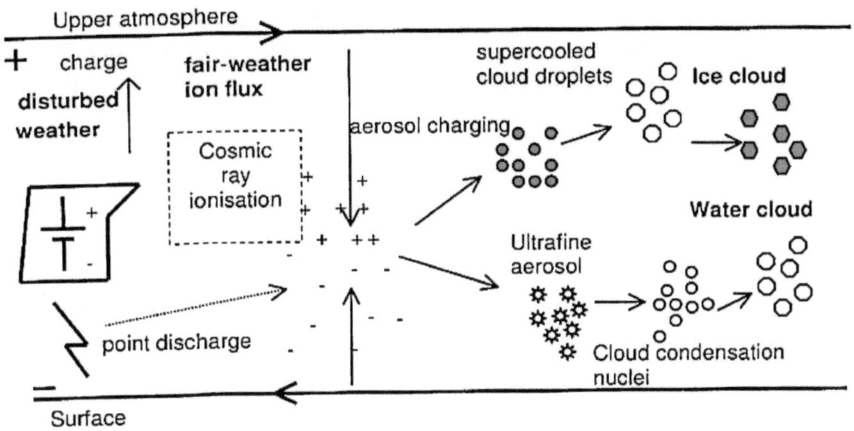

Fig. 9.4. Stages in process whereby atmospheric ionisation results in the formation of cloud droplets/crystals (Harrison and Carslaw, 2003)

it should be noted that these ions cannot act directly as cloud condensation nuclei as their small size would require the surrounding air to have a very high level of supersaturation which does not occur in the atmosphere.

The air ions produced by the cosmic rays may act as sites for the nucleation of new ultrafine aerosol (or condensation nuclei, CN). The mechanism then hinges on the extent to which these CN may grow into particles large enough (>80 nm) to become cloud condensation nuclei (CCN) and whether this process is enhanced by the particles being charged. Growth may occur through condensation of water vapour or other soluble gases or through coagulation among neutral and charged particles. Some observational evidence (Yu and Turco 2000) suggests that charged molecular clusters grow faster than neutral clusters and chemical box models (Yu and Turco 2001) have been able to simulate this effect. However, to reach CCN size would take several days and whether the growth can be maintained depends on the supply of vapour and competitive sources of new aerosol and CCN all of which vary with location, altitude and time of year. Yu (2002) suggests that conditions in the lower troposphere may be more favourable than at higher altitudes.

Even if by such a mechanism it proves feasible to produce a measurable effect on cloud cover or properties, the magnitude, and even the sign, of the impact on radiative forcing remains uncertain as it will depend on the cloud location, altitude and physical properties, as discussed in lecture 6.

9.4 Effects of Varying Electric Field on Ice Cloud Processes

At temperatures between –40°C and 0°C liquid water cannot freeze spontaneously by homogeneous nucleation but requires nuclei on which to start freezing. However, only a small proportion of atmospheric aerosol form suitable sites. It has been observed in laboratory experiments that imposition of a large electric field can cause the freezing of supercooled water droplets. Other work has shown that particles which are usually poor ice nuclei became much more efficient when electrically charged. Such processes are collectively known as electrofreezing.

In a series of papers Brian Tinsley and co-workers have proposed that solar modulation of cosmic ray ionisation may affect electrofreezing processes. The latent heat released during freezing is available for modifying the weather systems of which the ice cloud is a part. This hypothesis awaits confirmation that electrofreezing processes really do take place in the atmosphere and also that the latent heat released is sufficient to produce the apparent effects on cyclone development.

10 Conclusions

Radiation from the Sun ultimately provides the only energy source for the Earth's atmosphere and changes in solar activity clearly have the potential to affect climate. There is statistical evidence for solar influence on various meteorological parameters on all timescales, although extracting the signal from the noise in a naturally highly variable system remains a key problem. Changes in solar irradiance undoubtedly impact the Earth's energy balance, thermal structure and composition but in a complex and non-linear fashion and questions remain concerning the detailed mechanisms which determine to what extent, where and when these impacts are felt. Variations in cosmic radiation, modulated by solar activity, are manifest in changes in atmospheric ionisation but it is not yet clear whether these have the potential to significantly affect the atmosphere in a way that will impact climate. It is only by further investigation of the complex interactions between radiative, chemical and dynamical processes in the atmosphere that these questions will be answered.

References

References in Text

Baldwin M.P., Dunkerton T.J.: Stratospheric harbingers of anomalous weather regimes. *Science*, **294**, 581–584 (2001)

Bond G. et al.: Persistent solar influence on north Atlantic climate during the Holocene. *Science*, **294**,, 2130–2136 (2001)

Crowley T.J.: Causes of climate change over the past 1000 years. *Science*, **289** 270–277 (2000)

Dickinson R.E.: Solar variability and the lower atmosphere. *Bull. Am. Meteorol. Soc.*, **56**, 1240–1248 (1975)

Eddy J.A.: The Maunder Minimum. *Science*, **192**, 1189–1202 (1976)

Friis-Christensen E., Lassen K.: Length Of The Solar-Cycle – An Indicator Of Solar-Activity Closely Associated With Climate. *Science,* **254,** 698–700 (1991)

Gleisner H., Thejll P.: Patterns of tropospheric response to solar variability. *Geophys Res Lett,* **30,** art. no. 1711 (2003)

Haigh J.D.: The role of stratospheric ozone in modulating the solar radiative forcing of climate. *Nature*, **370**, 544–546 (1994)

Haigh J.D.: The impact of solar variability on climate. *Science*, **272**, 981–984 (1996)

Haigh J.D.: A GCM study of climate change in response to the 11-year solar cycle. *Quart.J.Roy.Meteorol.Soc.*, **125**, 871–892 (1999)

Haigh J.D.: The effects of solar variability on the Earth's climate. *Phil. Trans. Roy. Soc A.*, **361**, 95–111 (2003)

Haigh J.D., Blackburn M., Day R.: The response of tropospheric circulation to perturbations in lower stratospheric temperature. Submitted to *J. Clim.* (2004)

Harrison R.G., Carslaw K.S.: Ion-Aerosol-Cloud Processes in the Lower Atmosphere. *Rev. Geophys.*, **41**, 3 (2003)

IPCC: Climate Change 2001: The Scientific Basis. (CUP 2001)

Jackman C.H. et al.: Northern Hemisphere atmospheric effects due to the July 2000 solar proton event. *Geophys Res Lett*, **28**, 2883–2886 (2001)

Keeling C.D., Whorf T.P.: Atmospheric CO_2 records from sites in the SIO air sampling network. In Trends: A Compendium of Data on Global Change. Carbon Dioxide Information Analysis Center, Oak Ridge National Laboratory, U.S. Department of Energy, Oak Ridge, Tenn., U.S.A. (2004)

Kiehl J.T., Trenberth K.E.: Earth's annual global mean energy budget *Bull. Am. Meteorol. Soc.*, **78**, 197–208 (1997)

Kodera K.: On The Origin And Nature Of The Interannual Variability Of The Winter Stratospheric Circulation In The Northern-Hemisphere. *J Geophys Res-Atmos,* **100**, 14077–14087 (1995)

Labitzke K., van Loon H.: On The Association Between The QBO And The Extratropical Stratosphere. *J Atmos Terr Phys* **54** 1453–1463 (1992)

Labitzke K., van Loon H.: Connection Between The Troposphere And Stratosphere On A Decadal Scale. *Tellus A* **47**, 275–286 (1995)

Larkin A.: An investigation of the effects of solar variability on climate using atmospheric modesl of the troposphere and stratosphere. Ph.D. thesis, University of London (2000)

Larkin A., Haigh J.D., Djavidnia S.: The effect of solar UV irradiance variations on the Earth's atmosphere. *Space Science Reviews*, **94**, 199–214 (2000)

Lassen K., Friis-Christensen E.: Solar cycle lengths and climate: A reference revisited – Reply. *J. Geophys. Res.*, **105**, 27493–27495 (2000)

Laut P., Gundermann J.: Solar cycle lengths and climate: A reference revisited. *J. Geophys. Res.*, **105**, 27489–27492 (2000)

Lean J.: Contribution Of Ultraviolet Irradiance Variations To Changes In The Sun's Total Irradiance. *Science*, **244**, 197–200 (1989)

Lean J.: Variations In The Sun's Radiative Output. *Rev Geophys,* **29,** 505–535 (1991)

Ludlam F. H.: Clouds and storms: the behavior and effect of water in the atmosphere (Pennsylvania State University Press 1980)

Mann M.E., Bradley R.S., Hughes M.K.: Northern hemisphere temperatures during the past millennium: Inferences, uncertainties, and limitations. *Geophys Res Lett* **26**, 759–762 (1999)

Marsh N.D., Svensmark H.: Low cloud properties influenced by cosmic rays. *Phys Rev Lett* **85**, 5004–5007 (2000)

Matthes K. et al.: GRIPS solar experiments intercomparison project: initial results. *Papers in Meteorology and Geophysics*, **54**, 71–90 (2003)

Nakajima T., King M.D.: Determination Of The Optical-Thickness And Effective Particle Radius Of Clouds From Reflected Solar-Radiation Measurements. 1. Theory *J Atmos Sci*, **47** , 1878–1893 (1990)

Petit J.R. et al.: Climate and atmospheric history of the past 420,000 years from the Vostok ice core, Antarctica. *Nature*, **399**, 429–436 (1999)

Ramanathan V., Vogelmann A.M.: Greenhouse effect, at-mospheric solar absorption and the Earth's radiation budget: from the Arrhenius-Langley era to the 1990s. *Ambio* **26**, 38–46 (1997)

Ringer M.A., Shine K.P.: Sensitivity of the Earth radiation budget to interannual variations in cloud amount. *Clim. Dynam.*, **13**, 213–222 (1997)

Robock A.: Stratospheric forcing needed for dynamical seasonal prediction. *B Am Meteorol Soc* **82**, 2189–2192 (2001)

Shindell D.T. et al.: Solar forcing of regional climate change during the Maunder Minimum. *Science*, **294**, 2149–2152 (2001)

SOLICE: Final Report of European Community Framework Programme 5 project Solar Influences on Climate and the Environment (2004)

Stauffer B.: Long term climate records from polar ice. *Space Sci Rev*, **94**, 321–336 (2000)

Strobel D.F.: Parameterization of the atmospheric heating rate from 15 to 120 km due to O_2 and O_3 absorption of solar radiation. *J. Geophys. Res.*, **83**, 6225–6230 (1978)

Tinsley B.A.: Influence of the solar wind on the global electric circuit and inferred effects on cloud forcing, temperature and dynamics in the troposphere. *Space Sci. Rev.* **94**, 231–258 (2000)

Van Loon H., Shea D.J.: A probable signal of the 11-year solar cycle in the troposphere of the northern hemisphere. *Geophys. Res. Lett.* **26**, 2893–2896 (1999)

Vlachogiannis D. and Haigh J.D.: The impact of solar proton events on lower stratospheric ozone. pp. 275–278 of Atmospheric Ozone, Proc QOS Acquila 1996, ed R D Bojkov and G Visconti (1998)

White W.B. et al.: Response of global upper ocean temperature to changing solar irradiance. *J Geophys Res*, **102**, 3255–3266 (1997)

Williams V., Austin J. and Haigh J.D.: Model simulations of the impact of the 27-day solar rotation period on stratospheric ozone and temperature. *Adv. Space Res.*, **27**, 1933–1942 (2001)

Yu F.: Altitude variations of cosmic ray induced production of aerosols: Implications for global cloudiness and climate. *J. Geophys. Res.*, **107**, 1118 (2002)

Yu F., Turco R.P.: Ultrafine aerosol formation via ion-mediated nucleation. *Geophys. Res. Lett.*, **27**, 883–886 (2000)

Yu F., Turco R.P.: From molecular clusters to nanoparticles: Role of ambient ionisation in tropical aerosol formation. *J. Geophys. Res.*, **106**, 4797–4814 (2001)

Background and Further Reading

Andrews D.G.: An Introduction to Atmospheric Physics (CUP 2000)
Burroughs W.J.: Weather Cycles: Real or Imaginary (CUP 1992)
Friis-Christensen E. et al. (Eds.): Solar Variability and Climate (Kluwer 2000)
Hobbs P.V. (Ed.): Aerosol-Cloud-Climate Interactions (AP 1993)
Holton J.R.: An Introduction to Dynamic Meteorology (AP 1992)
Houghton J.T.: The Physics of Atmospheres (CUP 1977)
Hoyt D.V., Schatten K.H.: The Role of the Sun in Climate Change (OUP 1997)
Nesme-Ribes N. (Ed.): The Solar Engine and its Influence on Terrestrial Atmosphere and Climate. NATO ASI series (Springer-Verlag 1994)
Salby M.L.: Fundamentals of Atmospheric Physics (AP 1996)
Wallace J.M., Hobbs P.V.: Atmospheric Science: an Introductory Survey (AP 1977)

Solar Outputs, Their Variations and Their Effects on Earth

M. Lockwood

[1] Space Science and Technology Department, Rutherford Appleton Laboratory, Chilton, UK
[2] School of Physics and Astronomy, University of Southampton, Southampton, UK
M.Lockwood@rl.ac.uk

1 Introduction to the Sun and the Solar Activity Cycle

The Sun is the source of the energy that powers our climate and allows life on Earth. It also provides particles (and the energy with which to accelerate them) which bombard the Earth: these have a variety of "space-weather" effects on both natural phenomena and man-made systems. At the same time, the Sun generates the heliosphere, which isolates our solar system from interstellar space and shields Earth from energetic particles generated, for example, by supernova explosions.

There is one factor which plays a key role in the variations of all of these solar outputs – the Sun's magnetic field. The following sections look at the origin, evolution and effects of the solar magnetic field, starting from Table 1, which lists some of the basic characteristics of the Sun.

The visible solar surface is called the *photosphere* and lies at an average heliocentric distance $r = R_S = 6.96 \times 10^8$ m (see Table 1). The regions below the photosphere are not directly observable and our knowledge of them comes from application of the helioseismology technique, from numerical models and, now that we understand more about their mass and oscillations, from neutrinos which can escape the interior without interacting [Bahcall, 2001].

The models must be constrained by one key output of the Sun, our best estimate of the total power output which is dominated by the total electromagnetic power or *luminosity*, L (see Table 1). The electromagnetic power falling on unit area at the mean Earth-Sun distance ($r = R = 1\,\text{AU}$), is the *total solar irradiance*, I_{TS}, and has been measured from space with high accuracy since 1978. If the Sun emitted isotropically, L would be equal to $4\pi R^2 I_{TS}$. We can average out longitudinal structure in the solar emission by averaging over solar rotation intervals (close to 27 days as seen from Earth or from a satellite in orbit around the L1 point where the gravitational pull of the Earth and the Sun are equal, and for which $r \approx 0.99R$). However, we have never measured the latitudinal variation and the irradiance emitted over the solar poles.

Table 1. The characteristics of the Sun (S.I. units)

Solar radius, R_S (radius of the visible disc, the photosphere)	6.9599×10^8 m = 109.3 R_E (an Earth radius, 1 $R_E = 6.37 \times 10^6$ m)
Solar mass, m_s	1.989×10^{30} kg = 3.33×10^5 m_E (an Earth mass, $m_E = 5.97 \times 10^{24}$ kg)
Surface area	6.087×10^{18} m^2 ($1.19 \times 10^4 \times$ that of Earth)
Volume	1.412×10^{27} m^3 ($1.304 \times 10^6 \times$ that of Earth)
Age	4.57×10^9 yr
Luminosity	3.846×10^{26} W
Power emitted in solar wind	3.55×10^{14} W
Surface temperature	5770 K
Surface density	2.07×10^{-5} kg m^{-3} ($1.6 \times 10^{-4} \times$ the density of air at Earth's surface)
Surface composition (by mass)	70% H, 28% He, 2% (C, N, O, ...)
Central temperature	1.56×10^7 K
Central density	1.50×10^5 kg m^{-3} (8× the density of gold)
Central composition by mass	35% H, 63% He, 2% (C, N, O, ...)
Mean density	1.40×10^3 kg m^{-3} (0.25× the mean density of Earth)
Mean distance from Earth, d_{ES}	1.50×10^{11} m = 1 AU = 215 R_S
Mean angle subtended by a solar diameter at Earth = $2\tan^{-1}(R_S/d_{ES})$	0.532°
Mean solid angle subtended by a solar disc at Earth = $\pi(R_S/d_{ES})^2$	6.7635×10^{-5} sr
Surface gravity	274 m s^{-2} (27 × the gravity at Earth's surface)
Escape velocity at surface	6.18×10^5 m s^{-2}
Equatorial rotation period (w. r. t. fixed stars)	24.6 days (frequency, f = 470 nHz)
Equatorial rotation period (w. r. t. Earth)	27 days
Polar rotation period (w. r. t. fixed stars)	38.6 days (frequency, f = 300 nHz)
Solar wind mass loss rate	1.5×10^9 kg s^{-1}
Inclination of equator (w. r. t. ecliptic)	7°

At the centre of the Sun lies the *core* ($0 < r < 0.25R_S$) where the high pressure and temperature cause the thermonuclear reactions which power the Sun. The energy is then passed, mainly by the diffusion of gamma rays and X-rays, through the *radiative zone* ($0.25R_S < r < 0.7R_S$). Were it not to interact, a photon would cross the radiative zone in 2 s; however photons are scattered, absorbed and re-radiated so many times that this journey takes 10 million years. Above the transition region at around $r = 0.7R_S$, the energy is brought to the surface by large scale circulation across the *convection zone*, driven by buoyancy forces. The upflows and downflows are seen in the surface by the pattern of small *granules* (of order 1 Mm across), with hotter, rising material appearing brighter than cooling, falling material in dark lanes [Hirzberger et al., 2001] that are about 400° cooler. These ubiquitous cellular features cover the entire Sun except for those areas covered by sunspots. They are the tops of small and shallow convection cells. Individual granules last for only 18 minutes on average. The granulation pattern is continually evolving as old granules are pushed aside by newly emerging ones. The circulation flow speeds within the granules are typically $1\,\mathrm{km\,s^{-1}}$ but can reach supersonic speeds exceeding $7\,\mathrm{km\,s^{-1}}$. The granulation of the quiet photosphere can be seen outside the sunspots in Figs. 11 and 13. In fact, simulations and observations show that granules are a shallow, surface effect [Steiner et al., 1998]. The surface upflows and downflows are also organised into larger-scale circulation cells: *mesogranulation*, with typical cell sizes between 3 and 10 Mm and a lifetime of around 1 hour; *supergranulation*, with average cell size is 20–30 Mm and the average lifetime is about one day, and *giant cells* extending 40–50° in longitude and less than 10° in latitude, with a lifetime of about 4 months [Ploner et al., 2000, Beck et al., 1998]. Supergranules are associated with the *network* pattern of emission intensities in the overlying chromosphere.

The *chromosphere* is the lower part of the solar atmosphere and 2.5×10^6 m thick (so it covers $1R_S < r < 1.004R_S$). The temperature of the atmosphere increases dramatically at the transition region and is very high (of order 2×10^6 K) throughout the main part of the solar atmosphere, the *corona.* As can be seen during eclipses, the corona has no clear outer edge; instead it evolves into the heliosphere, the region of space dominated by the solar wind outflow of ionised gas (plasma) and the weak magnetic field, also of solar origin, that is carried with it. One convenient threshold that can be thought of a separating the corona from the heliosphere is $r = 2.5R_S$, beyond which the magnetic flux pulled out of the Sun is approximately constant [Suess, 1998].

The journey to the Earth takes solar photons 500 s but the thermal charged particles of the solar wind take anything between about 2.5 and 6 days. More energetic particles travel more rapidly, for example a 100 MeV *solar proton event (SPE)* would take only 20 min to reach Earth. The solar wind is slowed at a termination shock which theory predicts could be anywhere between r of 10 AU and 120 AU, but is generally well beyond all the planets. Towards the end of 2003 there was considerable debate in the literature that the termination shock may have been observed for the first time as it moved in, and then out again, over the Voyager-1 spacecraft at $r = 85$ AU and 87 AU, respectively [Krimigis et al., 2003, McDonald et al., 2003]. The failure of the thermal plasma instrument on this 26-year old craft makes the data ambiguous, but the detected energetic particles, predicted to be accelerated at the shock, mean that it was relatively nearby even if it didn't actually pass over the craft. Beyond this shock, the slowed solar wind continues to flow out to the outer boundary of the Sun's sphere of influence, the *heliopause*, the location of which could, in principle, vary between about $r = 50$ AU and 150 AU, depending on the pressure of the interstellar wind which meets the solar wind at this boundary.

1.1 The Solar Interior

The high temperatures in the solar interior mean that it is almost fully ionised. The primary ions are protons and in the core, the high pressure of the overlying layers, along with the high temperature, overcomes Coulomb electrostatic repulsion, and so can press two protons together

$$^{1}\mathrm{H}^{+} + {}^{1}\mathrm{H}^{+} \rightarrow {}^{2}\mathrm{D}^{+} + \beta^{+} + \nu^{e} \quad (+1.44\,\mathrm{MeV}) \tag{1}$$

where the superscripted number gives the atomic mass number of each reactant. The products are a deuterium nucleus, a positron and an electron neutrino and 1.44 MeV of energy. This reaction proceeds only relatively slowly, but is followed by two further reactions

$$^{2}\mathrm{D}^{+} + {}^{1}\mathrm{H}^{+} \rightarrow {}^{3}\mathrm{He}^{2+} + \gamma \qquad (+5.5\,\mathrm{MeV}) \tag{2}$$

where γ is a gamma-ray photon

$$^{3}\mathrm{He}^{2+} + {}^{3}\mathrm{He}^{2+} \rightarrow {}^{4}\mathrm{He}^{2+} + {}^{1}\mathrm{H}^{+} + {}^{1}\mathrm{H}^{+} \quad (+12.9\,\mathrm{MeV}) \tag{3}$$

The net effect of this chain is to convert 4 protons into a helium ion, with a mass loss of $\delta m = (4m_H - m_{He}) = 0.029$ amu. Thus the energy released is

$$\delta\mathrm{E} = \delta\mathrm{m}c^{2} = 27\,\mathrm{MeV} = 4.3259 \times 10^{-12}\mathrm{J} \tag{4}$$

where c is the velocity of light.

The above chain of nuclear reactions is called the proton–proton chain and is the most important of several that are active. To supply the present-day luminosity of the Sun ($L = 3.846 \times 10^{26}$ W, see Table 1) requires the reaction chain to be completed $N = (L/\delta E) = 9 \times 10^{37}$ times per second, for which protons in the core are used up at a rate of $3.6 \times 10^{38}\ \mathrm{s}^{-1}$. Thus proton mass is used up at the rate of $6 \times 10^{11}\ \mathrm{kg\,s}^{-1}$ (considerably greater than the current loss rate of protons in the solar wind outflow which is about $1.5 \times 10^9\,\mathrm{kg\,s^{-1}}$). The model of the solar interior used below yields a total mass of protons in the core ($r < 0.2R_S$) of 4×10^{29} kg which, at the present consumption rate, would be all used up in 2×10^{10} yr. In fact, models predict that the depletion of hydrogen would take effect after about 5×10^7 yr, causing the first major solar expansion (into a red giant star) that marks the end of the Sun's *hydrogen-burning phase* [Schröder et al., 2001].

The density distribution $\rho(r)$ within the solar interior is principally determined by the balance between the (inward-directed) gravity, and the (outward-directed) gas pressure gradient. In *hydrostatic equilibrium* the pressure gradient is (neglecting any magnetic pressure and for spherical symmetry)

$$\nabla P = -\frac{\delta P}{\delta r} = \rho \boldsymbol{g} \tag{5}$$

where the pressure $P = N k_B T$, N is the number density ($= \rho/\mu$, where μ is the mean mass), k_B is Boltzmann's constant ($= 1.3806 \times 10^{-23} JK^{-1}$) and T is the temperature. The gravitational acceleration is

$$g(r) = \frac{GM(r)}{r^2} \tag{6}$$

where $M(r)$ is the mass contained within the sphere of radius r

$$M(r) = 4\pi \int_0^r r^2 \rho(r) \mathrm{d}r \tag{7}$$

The many collisions ensure that the protons, helium ions, heavier ions and electrons share the same temperature T. The high value of that temperature ensures that almost all ions are fully ionised (H^+, He^{2+}, O^{8+}, etc.) at all layers except the cooler photosphere where the change of ionisation state (with protons de-ionising exothermically) is a factor in the formation of the granular convection near the surface. Neglecting ions heavier than Helium, the total pressure is

$$P(r) \sim (n_e + n_H + n_{He} + \ldots) k_B T \tag{8}$$

where the number densities of hydrogen, helium and electrons are n_H, n_{He} and n_e, respectively. Because the plasma is electrically neutral and throughout the Sun $(n_{He}/n_H) \sim 0.08$

$$n_e = n_H + 2n_{He} + 8n_O + \ldots \sim n_H + 2n_{He} \sim 1.16 n_H \tag{9}$$

and the mean mass is

$$\mu = \{m_H(1 + 4 \times 0.08) + 1.16 m_e\}/\{1 + 0.08 + 1.16\} \sim 0.6 m_H \tag{10}$$

Equation (8) becomes

$$P(r) \sim (\rho(r)/\mu)k_B T(r) \tag{11}$$

the solution to which is

$$P(r) = P_o \exp\left[-\int_0^r \mathrm{d}r/H(r)\right] \tag{12}$$

where P_o is the pressure at the centre of the Sun and the scale height is

$$H = k_B T(r)/\{\mu g(r)\} \tag{13}$$

The temperature profile $T(r)$ must be calculated from conservation of energy, allowing for the heat input profile by nuclear reactions and the heat transport by radiative and convective processes. Once this is done (11), (12) and (13) give us the associated pressure and density profiles.

Figures 1–3 give radial profiles of some key parameters from the model of the solar interior used by the GONG helioseismology project, and as derived by Christensen-Dalsgaard et al. [1996]. Figure 2 shows that at r greater than about $0.25\,R_S$, the heat flux $F(r)/L$ is equal to its maximum value of unity because there are no nuclear reactions outside the core to add to the energy flux. This is also shown by the profile of R_H, the rate of change of hydrogen mass due to nuclear reactions. The fraction of the mass made up by hydrogen, X_H, falls in the core because the sedimentation of the heavier products of the fusion reactions. This effect is also mirrored in the mean mass, μ, (given in Fig. 1 in units of a hydrogen atom mass m_H, see 10) which depends on both the ion composition and the charge state: μ is 0.5 m_H for a fully ionized gas of pure hydrogen. The zero-age Sun was made of 70% Hydrogen, 28% Helium, with the remaining 2% accounting for all heavier chemical elements; the corresponding μ for such a mixture is 0.605 m_H which has hardly been influenced by hydrogen burning and which still applies throughout most of the Sun. Near the photosphere μ rises because the proton gas recombines to give hydrogen atoms due to the steep fall in the temperature in the surface layer (see in Fig. 3).

Figure 4 shows how a major change takes place at the boundary between the radiative zone (RZ) and convective zone (CZ). The plot shows the rotation rate ($f = \Omega/2\pi = 1/T$, where Ω is the angular velocity and T is the rotation period) as a function of r/R_S for different heliographic latitudes, λ. These data are from interpretations of helioseismic oscillations observed by SoHO and the ground-based GONG network. Inside the RZ, the rotation rate is approximately independent of λ and r ($f \approx 430\,$nHz and thus the core and radiative zone rotate with a period T = 26.9 days, which is T' = 28.9 days when viewed from Earth). However, in the CZ the equator is seen to rotate faster than the poles. On average, f is near 300 nHz at the poles and 470 nHz at the equator (T of 38.6 and 24.6 days, respectively). The boundary between

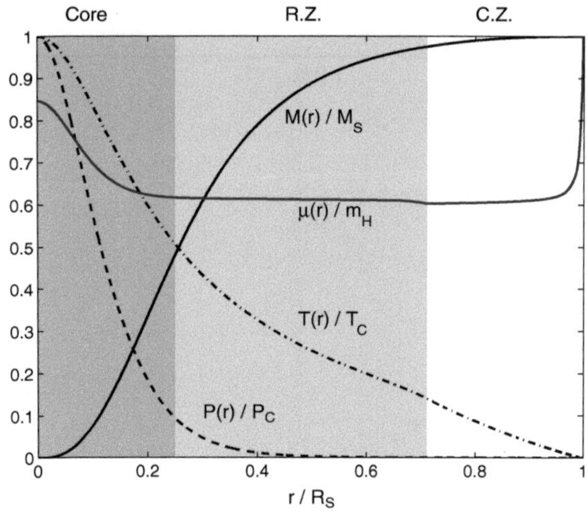

Fig. 1. The variation of various parameters in the solar interior from the model by Christensen-Dalsgaard et al. [1996]. Parameters are shown as a function of heliocentric distance r, as a ratio of the mean photospheric radius, R_S, with the *shading* giving the approximate limits of the three major regions of the interior: the core ($r < 0.25R_S$), the radiation zone (RZ, $0.25R_S \leq r < 0.75R_S$), and the convection zone (CZ, $0.75R_S \leq r < R_S$). The variations shown are the mass inside r, M, as a ratio of the solar mass M_S (*black solid line*); the temperature T as a ratio of T_C, its value at $r = 0$ (*dot-dash line*); the pressure P as a ratio of P_C, its value at $r = 0$ (*dashed line*); and the mean mass μ in units of the hydrogen atom mass, m_H (*thinner solid line*)

this co-rotating inner region and the differentially rotating convection zone is called the *tachocline* [Spiegel and Zahn, 1992]. Some flow shear is also seen near the photosphere at the top of the CZ. The flow in the CZ is further complicated by a meridional circulation which is from the equator to the poles higher in the CZ, with return flow in the opposite direction lower in CZ, near the tachocline [Giles et al., 1997]. This circulation is expected as a consequence of the differential rotation which, via the coriolis force, it acts to reduce [Gilman and Miller, 1986]. At r greater than about $0.8\,R_S$, the poleward flow is of order $20\,\mathrm{m\,s^{-1}}$, which calls for equatorward return flow at $r/R_S < 0.8$ of about $3\,\mathrm{m\,s^{-1}}$. The circulation is shown schematically in Fig. 5. The flows are further complicated by torsional oscillations of super- and under-rotation (see Fig. 15) which extend deep into the CZ [Howe et al., 2000b]. In addition, surface features show that rotation is slightly different in the two solar hemispheres [Antonucci et al., 1990].

In order to understand the RZ–CZ transition, it is useful to look at the mean free path λ_{mfp} of a photon in the radiative zone, which is related to the particle number density n, the mass density $\rho = n\mu$ and the photon–particle interaction cross section σ_{ph} by

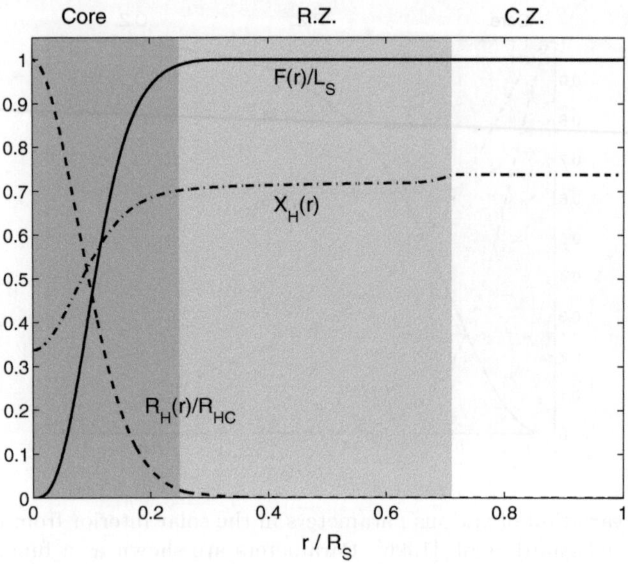

Fig. 2. Same as Fig. 1 for: the energy flux F, as a ratio of the surface luminosity L (*solid line*); the rate of change of hydrogen mass, R_H, as a ratio of R_{HC}, its value at $r = 0$ (*dashed line*); the hydrogen abundance by mass, $X_H(r)$ (*dot-dashed line*)

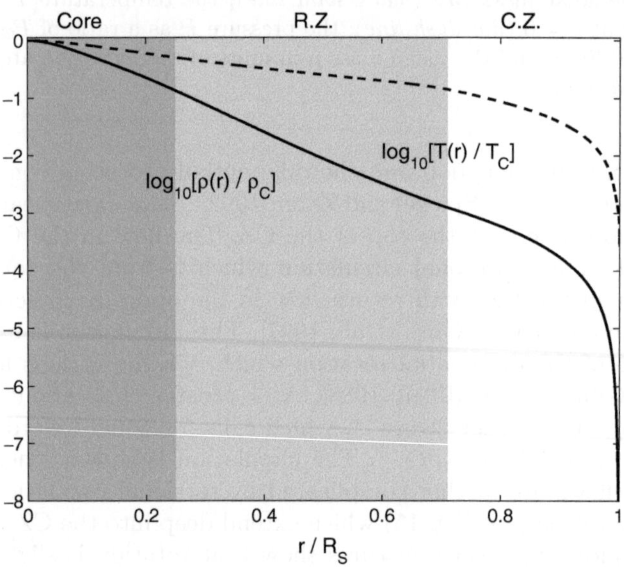

Fig. 3. Same as Fig. 1 for: the logarithm of T/T_C, where T is the temperature and T_C is its value at $r = 0$ (*dashed line*); the logarithm of ρ/ρ_C, where ρ is the mass density and ρ_C is its value at $r = 0$ (*solid line*)

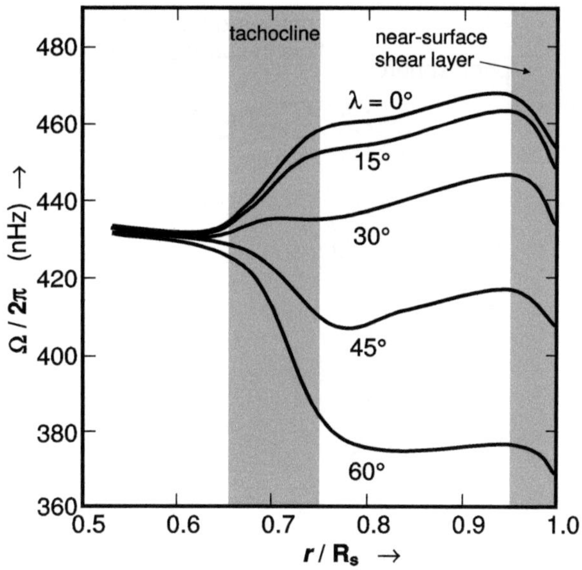

Fig. 4. Average rotation rates inferred from the helioseismic inversion of over 4 years of GONG data. Shear layers are *shaded* and occur at the base of the convection zone as well as near the surface. Contours of angular velocity Ω are shown as a function of r/R_S for various heliographic latitudes. The slower/faster rotation rate in, respectively, the polar/equatorial convection zone and photosphere can be seen. The lower shear layer is called the tachocline, and marks the boundary of differential rotation (and of the convection zone) below which the Sun approximately rotates as a solid body (Adapted from Howe et al. 2000a)

$$\lambda_{mfp} = (n\sigma_{ph})^{-1} = (\kappa\rho)^{-1} \tag{14}$$

where κ is the opacity. At r near $0.2\,R_S$, in the outer part of the core, ρ is about $10^4\,\mathrm{kg\,m^{-3}}$ and, allowing for all scattering processes, κ can be estimated to be of order $0.4\,\mathrm{m^2\,kg^{-1}}$, which yields λ_{mfp} of $2\times10^{-4}\,\mathrm{m}$ or $2.9\times10^{-13}R_S$. Thus to travel just 1% of R_S, the photon must be scattered or absorbed/re-radiated 3.5×10^{12} times. The energy transport by these photons obeys a diffusion equation and for spherical symmetry this yields a radial energy transport by photons of

$$F_{ph} = -D_{ph}\frac{\delta\epsilon_{ph}}{\delta r} \tag{15}$$

where ϵ_{ph} is the total photon energy density and the diffusion coefficient is given approximately by

$$D_{ph} \approx \lambda_{mfp}\left(\frac{c}{3}\right) = \frac{c}{3\kappa\rho} \tag{16}$$

Because the spectrum of photons is that of a blackbody radiator, the Stefan–Boltzmann law applies, which means that the energy flux from a

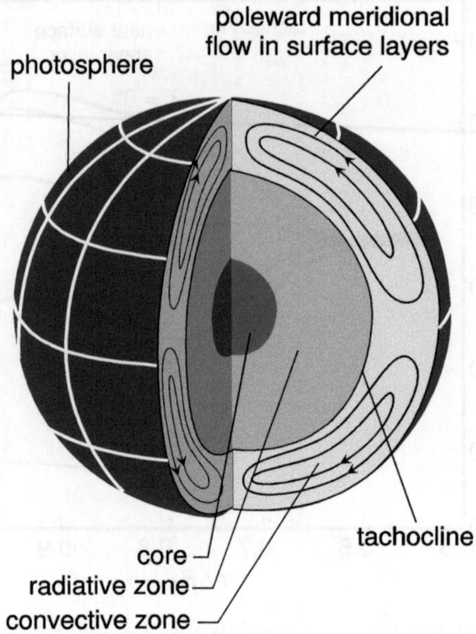

Fig. 5. Cutaway schematic of the solar interior, showing the core, the RZ and the CZ. Above the tachocline (and perhaps in an "overshoot" layer at the top of the RZ), flows in the CZ show a meridional circulation

surface is

$$F = \sigma_{SB} T^4 \tag{17}$$

where the Stefan–Boltzmann constant, $\sigma_{SB} = 5.6696 \times 10^{-8}\,\mathrm{W\,m^{-2}\,K^{-4}}$, and the photon energy density is

$$\epsilon_{ph} = \frac{4F}{c} = \frac{4\sigma_{SB} T^4}{c} \tag{18}$$

From (15), (16) and (18)

$$F_{ph} = -\frac{4\sigma_{SB}}{3\kappa\rho} \frac{\delta T^4}{\delta r} \tag{19}$$

At $r > 0.25R_S$, the heat flux is constant because there are no nuclear reactions and in steady state this equals the total luminosity L radiated by the Sun divided by the surface area (see Fig. 2). Thus from (19) this defines the temperature profile associated with the radiative processes

$$\left[\frac{\delta T}{\delta r}\right]_r = -\frac{3\kappa\rho}{(16\sigma_{SB} T^3)} \frac{L}{4\pi R_S^2} \tag{20}$$

Equations (20) and (8) form a coupled pair (both contain T and ρ) which can be solved to give the profiles like those shown in Figs. 1–3.

In the transition zone, the rate of energy transfer by these radiative processes becomes too small, and bulk motion – i.e. convection – takes over the upward heat transport. At the base of the convection zone the temperature is about 2×10^6 K. This is cool enough for the heavier ions (such as carbon, nitrogen, oxygen, calcium, and iron) to hold onto some of their electrons. This makes the material more opaque (increased κ). Figure 6 gives the variations of the key parameters in (20), using the same model used to derive Figs. 1–3 [Christensen-Dalsgaard et al., 1996]. It can be seen that as r increases, the T^{-3} term increases rapidly, as does the opacity κ, such that, even though the density ρ falls, the net effect is that the magnitude of the temperature gradient $[\delta T/\delta r]_r$ (also called the "*lapse rate*") increases (i.e. the gradient becomes more negative), as shown in Fig. 7.

The convective instability occurs where the magnitude of the lapse rate due to radiative processes becomes too large. To understand this instability better, Bernouille's relation, for steady flow without heat sources or sinks and applied here for low-speed flow and neglecting magnetic pressure, shows

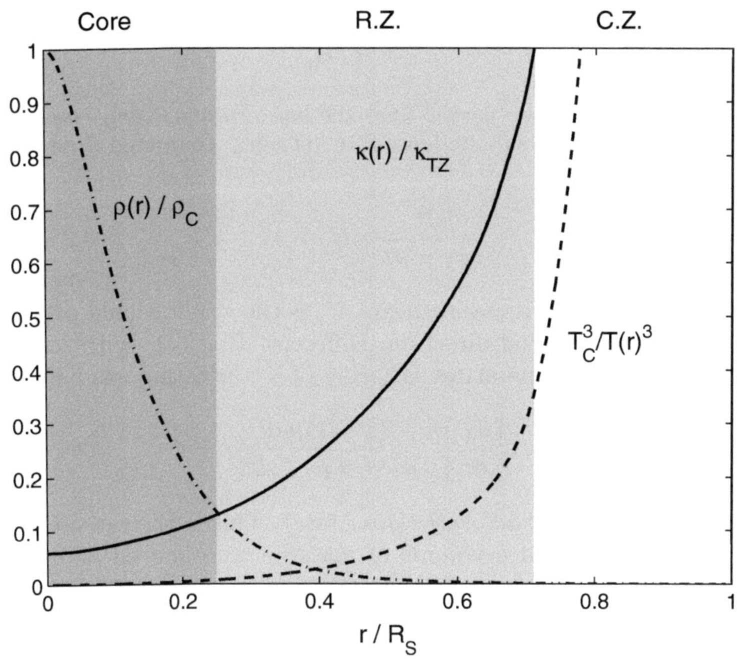

Fig. 6. Same as Fig. 1 for: the opacity, κ (normalised to κ_{TZ}, its value at the RZ–CZ transition zone); $(T_C/T)^3$, where T is the temperature and T_C is its value at $r = 0$ (*dashed line*); ρ/ρ_C, where ρ is the density and ρ_C is its value at $r = 0$ (*dot-dashed line*)

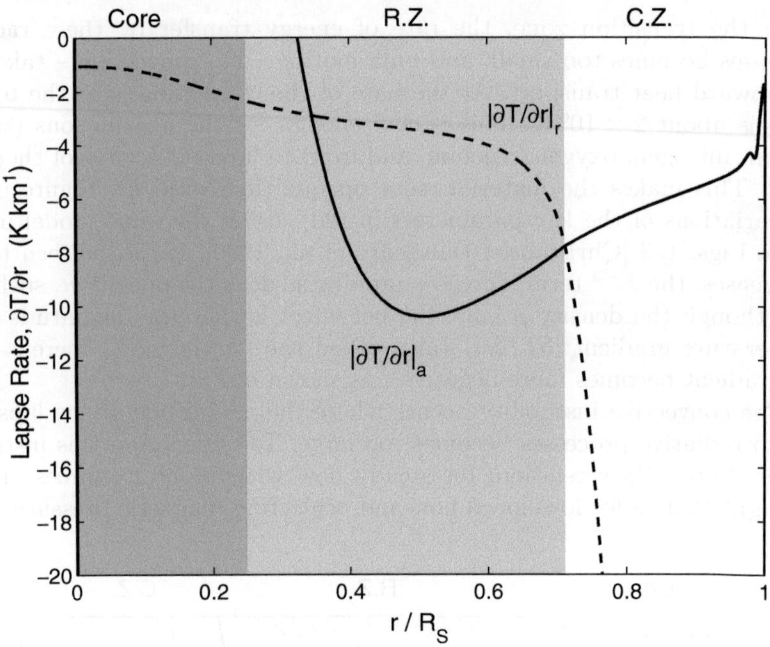

Fig. 7. Same as Fig. 1 for: (*dashed line*) the lapse rate $[\mathrm{d}T/\mathrm{d}r]_r$ computed from (20) and (*solid line*) the adiabatic lapse rate $[\mathrm{d}T/\mathrm{d}r]_a$, computed from (22)

$$\frac{\gamma P}{(\rho\gamma - \rho)} + U_g = k_r \tag{21}$$

where γ is the ratio of the specific heats, U_g is the gravitational potential and k_r is a constant in a radial direction. Differentiating (21) with respect to r, yields the adiabatic temperature variation of a convecting gas parcel

$$\left[\frac{\mathrm{d}T}{\mathrm{d}r}\right]_a = \frac{(\gamma - 1)\mu}{(\gamma k_B)} \frac{\mathrm{d}U_g}{\mathrm{d}r} \tag{22}$$

which is also called the "*adiabatic lapse rate*". This is the rate at which the temperature would fall if a volume of material were moved higher without adding heat. If the lapse rate given by (20) (i.e. the temperature gradient associated with radiation, $[\mathrm{d}T/\mathrm{d}r]_r$) is larger in magnitude than the magnitude of the adiabatic lapse rate given by (22) (i.e. the temperature gradient associated with convective motion) then if a parcel of plasma is moved upward by a small amount the plasma within it cools at $[\mathrm{d}T/\mathrm{d}r]_a$ compared with the cooling at lapse rate $[\mathrm{d}T/\mathrm{d}r]_r$ associated with radiation of the surrounding plasma. Thus this parcel becomes warmer and less dense than the surrounding cooler (denser) plasma and moves further upward under buoyancy forces. The transition zone is where $[\mathrm{d}T/\mathrm{d}r]_a \approx [\mathrm{d}T/\mathrm{d}r]_r$.

This can be seen to be the case in Fig. 7 which plots $[\mathrm{d}T/\mathrm{d}r]_a$ and $[\mathrm{d}T/\mathrm{d}r]_r$, as given by (22) and (20) respectively, for the same model of the interior presented in Figs. 1–3 and 6. The two can be seen to be equal at the boundary between the RZ and the CZ. Below this transition $[\mathrm{d}T/\mathrm{d}r]_a < [\mathrm{d}T/\mathrm{d}r]_r$ (i.e. the adiabatic lapse rate is more negative than the radiative lapse rate), which means that buoyancy forces act to suppress any motions of a plasma parcel with respect to its surroundings and the plasma is stable to the convective instability. (In Earth's troposphere an analogous situation leads to a stable atmospheric inversion). In the convective zone $[\mathrm{d}T/\mathrm{d}r]_a > [\mathrm{d}T/\mathrm{d}r]_r$ (i.e. the adiabatic lapse rate is less negative than the radiative lapse rate) and the convective instability sets in. This means plasma parcels which move up are forced further up, whereas those that move down are forced further down and circulation cells with up and down flows are established. (In the analogous situation in Earth's troposphere, strong convection and thunderstorms can result. Note that in the Sun, ionisation state plays the role that water vapour plays in the atmospheric case). The stability condition is called the Schwarzschild condition [Schwarzschild, 1906].

Convective motions carry heat rapidly to the surface. The fluid expands and cools as it rises. At the visible surface the temperature has dropped to 5770 K and the density is only $2 \times 10^{-4}\,\mathrm{kg\,m^{-3}}$, as shown in Fig. 3.

1.2 The Solar Dynamo

The magnetic field of the Sun is generated by currents in the Sun's interior, in accordance with Ampère's law. Section 2 will outline the derivation of the magnetic induction equation and show how, for large-scale plasma, this leads to the concept of "frozen-in flux" which means that plasma and field move together. A consequence of this is that fluid motions can amplify a small pre-existing field. In the solar dynamo, the most important plasma motions are differential rotation with angular velocity that is a function of both solar latitude and radial distance, $\Omega(r, \lambda)$, and the meridional circulation, both of which are found in the CZ, predominantly above the tachocline. Full dynamo theory is too complex to consider here [see reviews by Weiss, 1994, Schmitt, 1993, Schüssler et al., 1997] and has been greatly constrained in recent years by the revolution in our knowledge of the solar interior's structure and dynamics brought about by the helioseismology technique [Nandy, 2003]. Figure 8 gives a schematic illustration of two key effects, based on the original concepts introduced by Babcock [1961] and Parker [1955] who considered the effects of a prescribed pattern of flow on the field; full dynamo models also need to incorporate the feedback effect that the field has on the pattern of flow.

If we have a small "seed" field in the north–south direction, the differential rotation will generate large east–west fields by winding up the field, as shown for the southward-pointing poloidal seed field in parts (a)–(c) of Fig. 8. As discussed above, the polar convection zone rotates every 34 days, whereas

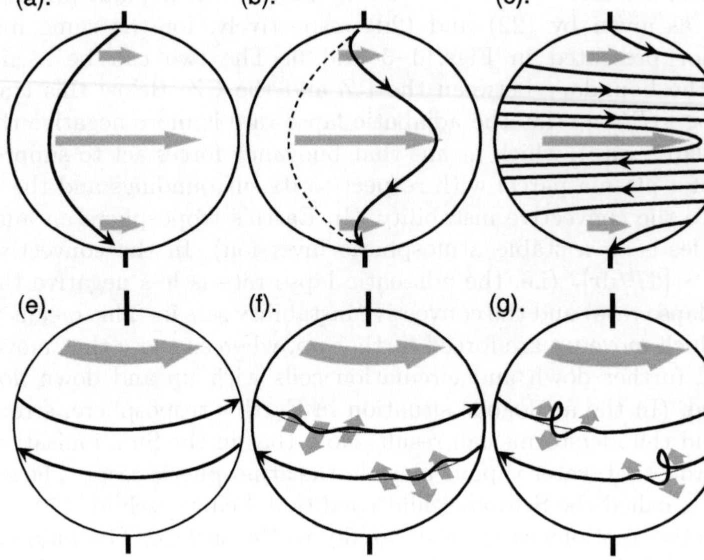

Fig. 8. Schematic illustration of the solar magnetic dynamo effects. (**a**)–(**c**) The "omega" effect: a weak "seed" pre-existing magnetic field line (*with small arrow*) is wound up into a strong toroidal component by the differential rotation of convection zone plasma (*thick grey arrows*) into which the field is frozen. (**e**)–(**f**) The "alpha" effect: radial motions cause a twisting of the toroidal field under the coriolis force, generating a poloidal field component

the equatorial convection zone rotates every 25.5 days. Thus after 34 days the polar regions have rotated by 360° but the equatorial region has rotated by 480°, 120° further than the poles, as shown in Fig. 8b. After 527 days the field would be wound up as in Fig. 8c. It can be seen that this effect generates strong east–west or "toroidal" field out of a weak seed field, this is called the *omega* effect.

The toroidal field generated has opposite senses in the two hemispheres. Thus where field rises up through the photosphere (emerges) in loops connecting *bipolar magnetic regions*, BMRs [Harvey and Zwaan, 1993, Harvey, 1992], the leading associated sunspots (and active region faculae) will have opposite field polarities in the two hemispheres, as is observed. These polarities reverse with each new solar cycle, telling us that the toroidal field has swapped polarity, and thus so has the initial seed field from which it grew.

A second effect (the *alpha* effect) arises from the convection cells and eddies which cause radial movements of plasma and the frozen-in magnetic field in the convection zone. As it rises, a plasma parcel and its frozen-in toroidal field are twisted by the coriolis force, generating a north–south or "poloidal" field from the toroidal field, as shown in parts (c)–(e) of Fig. 8. The twist is such that the following spot(s) of a BMR are at higher latitudes than the leading spots giving a "tilt" to the BMR in both hemispheres, as is observed (Joy's Law). Note that the poloidal field generated in Fig. 7f is southward in both hemispheres for this southward seed field, but this would reverse polarity with the seed field polarity.

The α-effect was first introduced by Parker [1955] and, because it regenerates poloidal field, is a fundamental part of the solar dynamo. A major difference between the wide range of dynamo models proposed is where the α-effect takes place. Buoyancy considerations for magnetic flux tubes mean that they only take about one month to rise up through the entire CZ . This means that most of the magnetic flux in most of the CZ is concentrated in small-scale intermittent features, as we see in the photosphere, and this is why strong (10^4–10^5 G), long-lived (~10 years) toroidal field is thought to be stored in the flow shear layer at the base of the CZ. This is called the overshoot layer which is slightly sub-adiabatic but into which convection penetrates. Models must predict a dynamo wave which propagates equatorward once every solar cycle to reproduce the *butterfly diagram* (see below) and in the most recent models meridional circulation is a vital part of this. This circulation is thought to draw down poloidal seed field at high latitudes (of a polarity which reverses every 11 years) through the CZ and ensures that the strongest toroidal fields are generated at low latitudes, where the field becomes strong enough to erupt and rise through the CZ to give the active regions (see Fig. 10). It has been postulated that there are two forms of emergence through the solar surface, with a turbulent *weak-field dynamo* in addition to the *strong-field dynamo* [Cattaneo and Hughes, 2001]. These two are coupled, but the former gives irregular fields while the latter gives the strong ordered fields of active regions and is predicted only at latitudes below about 35°. After they have risen rapidly through the CZ , the loops predicted by the α-effect emerge through the photosphere into the solar atmosphere, as shown by Fig. 9. The magnetic fields in sunspots and BMRs that penetrate the photosphere remain rooted in the overshoot layer at the base of the CZ, as shown in Fig. 10b.

In order to explain the fact that the polarity of emerged field associated with leading/trailing spots migrates equatorward/poleward, respectively, Leighton [1969] introduced the concept of turbulent diffusion of field under the effect of supergranules as they form and dissipate. In addition to the differential rotation and diffusion effects introduced by Babcock and Leighton, we now know we must also allow for meridional flow, as revealed by helioseismology observations of the pattern of flow in the convection zone, as discussed above.

For total solar irradiance variations, a key element is the small magnetic flux tubes outside of active regions. Some of this is remnant, dispersed flux left over from active regions produced by the strong dynamo and predicted by "mean field theory". However, there is growing awareness of the role of the weak, turbulent dynamo action of granular and supergranular flows which

Fig. 9. Image of coronal loops forming a bipolar magnetic region (BMR), imaged using the 17.1 nm FeIX line (corresponding to about 10^6K) by the TRACE (Transition Region And Coronal Explorer) satellite [Aschwanden and Title, 2004]

causes *ephemeral flux* to appear preferentially in the centres of supergranules and then be swept to the dark lanes between the supergranules [Schrijver et al., 1997, Cattaneo and Hughes, 2001] where it forms the network.

1.3 The Photosphere

The photosphere is the visible surface of the Sun and a layer which is about 100 km thick. When we look at the limb of the solar disc, as opposed to the centre, we see light that has taken a slanting path through this layer and this gives "*limb darkening*" as we only see the upper, cooler and dimmer regions of the photosphere.

As discussed earlier, the photosphere bears the signature of convection in the underlying CZ on a range of temporal and spatial scales with granules, mesogranules, supergranules and giant cells. However, the most well-known symptoms of the Sun's magnetic cycle are sunspots which have been studied since the work in the early 17th century by Galileo Galilei and Christoph Scheiner. Sunspots are darker because they are cooler patches of the surface where large magnetic fields inhibit the convective upflow of energy from below. Temperatures in the central *umbra* of spots are around 3800 K (i.e. 2000 K cooler than the surrounding undisturbed photosphere) and the

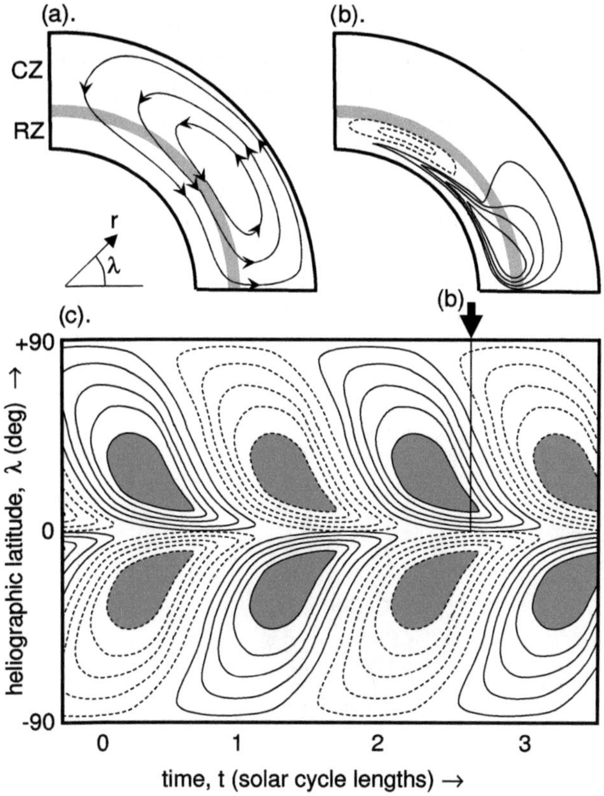

Fig. 10. Schematic of solar dynamo based on a the model simulations by Nandy and Choudhuri [2002]. In this model, the meridional CZ circulation (shown in a) penetrates below the tachocline (*the grey band*) to form an overshoot layer where toroidal field is generated by the omega effect and can be stably stored; when this field exceeds 10^5 G it is made to erupt and generates poloidal field by the alpha effect which is only active at the top of the CZ. Part (**c**) shows the stored toroidal field as a function of latitude and time: eastward field is given by *solid* contours, westward by *dashed* contours. The areas *shaded grey* are where field exceeds 10^5 G. The model predicts an equatorward-propagating dynamo wave, which yields major erupting flux in a butterfly pattern at latitudes below about 40°. The poloidal field emerging in the active region bands migrates poleward, under the meridional circulation, acting in concert with supergranular diffusion and differential rotation. At high latitudes it sinks through the CZ with the meridional circulation and becomes the new, reversed polarity, seed field. This grows under the omega effect and spreads equatorward to replace the old-cycle polarity field in the overshoot layer. Part (**b**) shows a latitudinal distribution of toroidal field at the time labelled "b" in (**c**). At this time the new-cycle polarity toroidal field is growing at high latitudes while the old-cycle polarity field is still present and erupting at low latitudes

magnetic field there is typically 0.1 T (roughly 1000 times greater than the average photospheric field). The magnetic field is weaker and more horizontal in the surrounding, less dark, more structured *penumbra* (see Fig. 11). Spots generally last for several days, although very large ones may survive for several months. They usually form in groups and may be unipolar but are usually paired with a neighbouring spot or spots in a BMR. Some spots are more complex than this in their magnetic topology. Spot sizes vary greatly, but typical umbral and penumbral diameters are 20×10^6 m and 40×10^6 m, respectively. The lower temperature in spots causes the surface to be a little lower than for the quiet photosphere (the "*Wilson depression*").

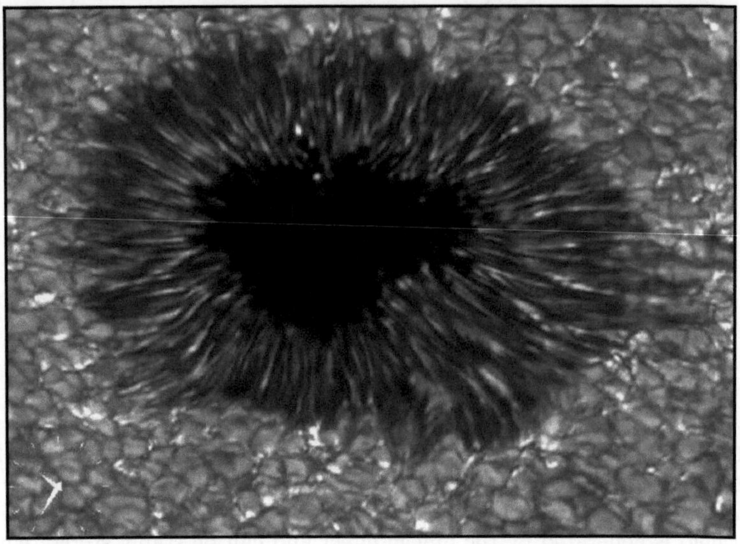

Fig. 11. High-resolution image of a sunspot showing the *dark* central umbra, filamentary penumbra and the granulation of the photosphere surrounding the spot

Sunspots occur in relatively narrow latitudinal bands near the solar equator (mainly below heliographic latitude of 30°) as shown in Fig. 12. Heinrich Schwabe first noted in 1843 that their occurrence shows a strong modulation on decadal timescales (for historical review, see Cliver, 1994). At each minimum of this cycle the Sun is almost, but not completely, free of spots and the amplitude of the maxima evolves on century timescales called Gleissberg cycles. Figure 12a shows that the first spots of each new cycle appear at the highest latitudes and that spot occurrence migrates equatorward in both hemispheres during each cycle, giving the famous butterfly diagram (Spörer's law). For some cycles, the high-latitude spots of the new cycle appear before the low-latitude spots of the old cycle have faded away, for other cycles there is no such overlap.

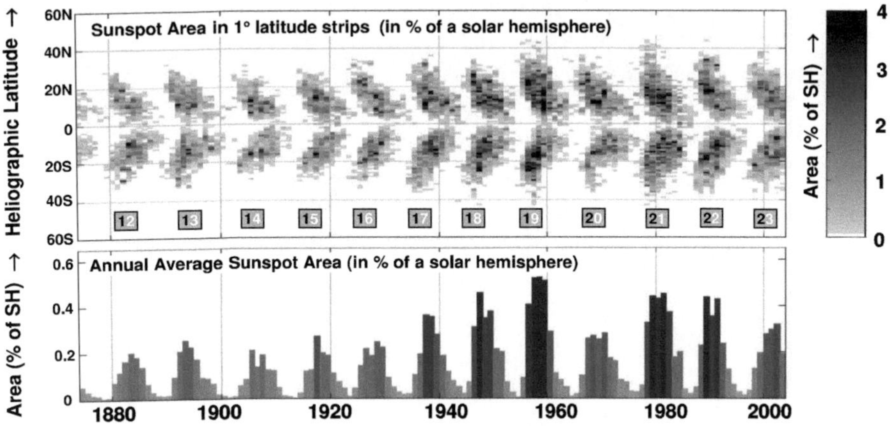

Fig. 12. Daily sunspot data that has been averaged over annual intervals. (*Bottom*) The total area covered by sunspots for solar cycles (11–23) (A_S – given in % of the visible solar hemisphere). The data are from Greenwich (1874–1976) and Mount Wilson (1982–present) observations. Data for (1977–1981) come from the former Soviet Union and are also used to intercalibrate the other two datasets over the interval for which it is available (1968–1992). (*Top*) The distribution of that area as a function of heliographic latitude and time – the famous butterfly diagram (after Foster, 2004)

The leading group of spots of the two that are paired in a BMR have consistently the same magnetic field polarity (inward or outward) in one hemisphere during any one solar cycle. It is also at a lower latitude giving the tilt angle of the BMR (see Fig. 13). The polarity of the leading spot is opposite in the two hemispheres and changes with each new solar cycle. This reveals that the full magnetic cycle is not the 11 years of the sunspot cycle but 22 years (the "*Hale cycle*").

The association of the sunspot cycle and the full magnetic cycle of the Sun is made clear by Fig. 14. The top panel shows the solar cycle in sunspot area and the lower panel shows longitudinally averaged radial magnetic field measured by solar magnetographs. These instruments use the Zeeman splitting of spectral lines to measure the line-of-sight component of the magnetic field, B. This is converted to radial field (B/μ) with the assumption that the field is radial, where the position parameter, $\mu = \cos\theta$ (θ is the heliocentric angle, so $\mu = 1$ at the disc centre and $\mu = 0$ at the limb). Note that there is no information from the limb and although features on the equatorial limb are later seen when they rotate through the disc centre, no information is available from the solar poles and the uncertainty in the radial field is larger at higher heliographic latitudes.

Figure 14 shows that the regions where sunspots occur (the butterfly "wings") are regions where intense field, of both polarities, emerges through the photosphere. This field migrates poleward at a rate which would take it

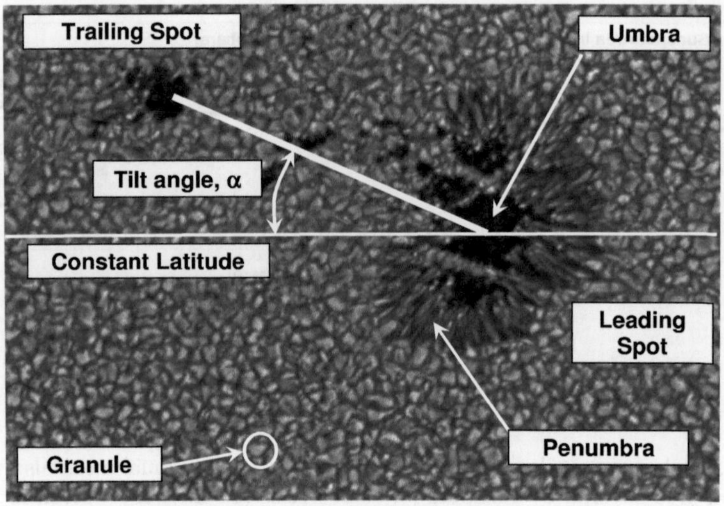

Fig. 13. A pair of spots showing the tilt angle with the leading spot at lower latitudes

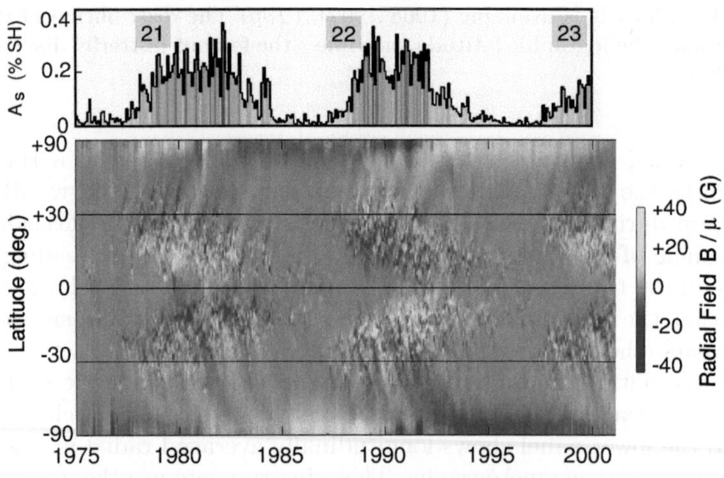

Fig. 14. The association of sunspots and magnetic field seen in magnetogram data. (*Top*) The total area covered by sunspots (A_S – given in % of the visible solar hemisphere) for 1975 to 2000, covering solar cycles (21, 22) and the start of (23). (*Bottom*) Longitudinal averages of the radial field (B/μ) as a function of latitude (positive northward) and time, where B is the line-of-sight field observed by magnetographs (positive outwards) and $\mu = \cos\theta$, where θ is the heliocentric angle

from 30° to the pole in about 1 year. In fact, progressively, the polarity of the trailing (more poleward) spots comes to dominate as it migrates poleward. Field of the other polarity tends to drift equatorward and fade away. This behaviour is reproduced by numerical models which follow the evolution of a BMR under the combined effects of differential rotation, meridional flow and diffusion discussed in Sect. 1.2 [Wang et al., 2000a,b, Mackay et al., 2002, Mackay and Lockwood, 2002, Schrijver et al., 2002]. For cycle 21, like all odd-numbered cycles, the polarity of the trailing spots is inward (negative field, shaded blue) in the northern hemisphere and outward (positive, shaded yellow) in the southern hemisphere. These polarities are reversed in the next cycle (cycle number 22), but the start of cycle 23 (beginning about 1997) shows a return to the same behaviour as cycle 21. Note that early in each sunspot cycle, the polar field in each hemisphere has the opposite polarity to the dominant polarity which is simultaneously emerging and migrating poleward from the sunspot belt. The arrival of the new polarity field from lower latitudes reduces the flux in the polar corona until, roughly one year after each sunspot maximum, the polar field polarity reverses. This new polarity then persists until about 1 year after the next solar maximum when it is flipped back again. Thus the polar field also shows a 22 year cycle, flipping in sense every 11 years. This means that all features of the magnetic cycle show opposite polarities during even and odd cycles, and this must be predicted by any successful dynamo model.

Dynamo models, like that illustrated in Fig. 10, predict that although sunspots only start to form (at the start of each new solar cycle) when the dynamo wave reaches a latitude λ below about 40°, the wave itself formed earlier than this at higher latitudes. In these 2-dimensional simulations, the alpha effect occurs in a thin layer at the top of the CZ, whereas the omega effect is mainly in the overshoot layer, just beneath the CZ. Collectively, features of the dynamo wave seen before the onset of the spots themselves (or perhaps after they have ceased) are called the "*extended solar cycle*", beginning before the previous solar maximum and lasting for between 18 and 22 years. If the average, background, differential rotation is removed from helioseismology observations of solar rotation, a pattern of torsional oscillations is revealed, as shown in Fig. 15. These oscillations clearly follow the dynamo cycle, but their role in field generation and eruption is not yet understood. They are shown here as one illustration of the extended solar cycle. Wilson et al. [1988] have linked the early appearance of these torsional oscillations with other symptoms of the extended solar cycle, including *ephemeral flux emergence* [Zwaan, 1987], chromospheric plages [Harvey, 1994] and coronal emissions (most clearly seen in the FeXIV emission patterns presented by Altrock, 1997).

Ephemeral flux is the small-scale end of a continuous (and approximately linear) distribution of BMR [Zwaan, 1985, Schrijver and Title, 1999, Schrijver and Zwaan, 2000] and may emerge as part of the turbulent weak-field

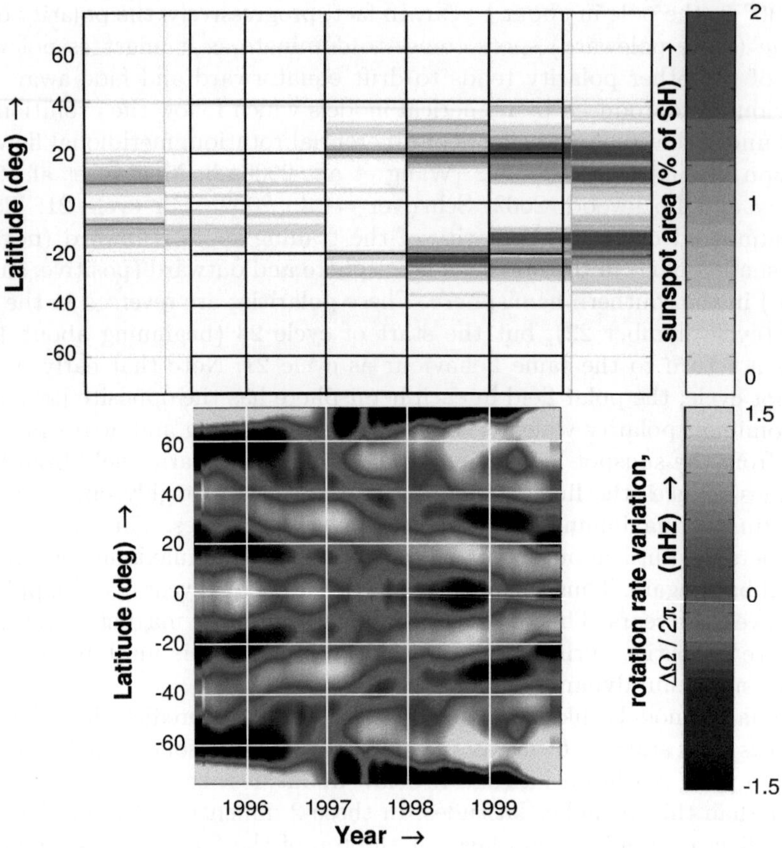

Fig. 15. Observations of torsional CZ oscillations around the solar minimum separating cycles (22) and (23). The sunspot areas (*top*) show little overlap between these two cycles but the pattern of super- and under-rotation (in *yellow* and *dark blue*, respectively, in the *bottom* panel) show an extended solar cycles (Adapted from Howe et al., 2000b)

dynamo, as opposed to the strong-field dynamo thought to be responsible for active regions. Simulations suggest that the observed distribution of BMR scales may result from emergence, of small-scale ephemeral flux and large-scale active regions BMRs, both of which give rise to intermediate-scale BMRs [Schrijver et al., 1997]. If the emerged flux in an active region is strong enough, it resists dispersion by supergranular flows for a duration of between a few days and a few weeks. Eventually all flux becomes subject to diffusive random-walk dispersion under granular and supergranular flows. When opposite polarity flux tubes collide they partially cancel; when same polarity tubes collide, they temporally coalesce. Ephemeral flux emergence can replace surface flux on a timescale of about 2 days, compared to the 6 months that differential rotation takes to spread emerged flux and the 1–2 years for flux to evolve towards the pole. Thus flux is constantly replaced as it migrates in the large scale-motions shown in Fig. 14: although emerged flux migrates poleward over long distances, individual flux tubes do not. Flux topologies are changed by *magnetic reconnection* (see Sect. 2.5) taking place in the CZ and in the solar atmosphere. Much of the flux is lost, by *flux cancelling* [Schrijver and Title, 1999, Close et al., 2003]. In fact, flux is often seen to disappear from the overlying corona and chromosphere before it vanishes in the photosphere, implying that much, or maybe even all, the cancelled flux is, in fact, subducted below the surface rather than cancelled, [Harvey et al., 1999].

The presence of weak emerged field outside of active regions is stressed in Fig. 16, which shows the variations of total (unsigned) magnetic flux in pixels where the mean field strength exceeds 25 G (roughly equivalent to active regions) and in pixels where it is less than 25 G (roughly equivalent to ephemeral regions, decayed active regions, intranetwork flux and network flux, collectively termed the *magnetic carpet*, Title and Schrijver, 1998). The average photospheric magnetospheric field is higher at sunspot maximum than at minimum by a factor of about 3. Figure 16 appears to show that the flux outside active regions is greater than the active region flux at sunspot minimum (as one would expect) but is only about half of it at sunspot maximum. However, one must bear in mind that within the $1'' \times 1''$ resolution pixels of the Kitt Peak magnetograms, used to generate Fig. 16, small regions of

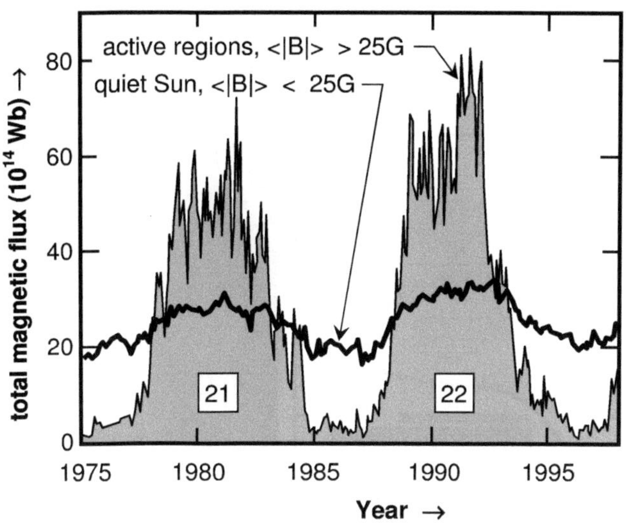

Fig. 16. The variation of the total, unsigned, magnetic flux observed using Kitt Peak magnetograms in active regions and in the quiet Sun, which can broadly be separated into regions where mean field strength per pixel is, respectively, greater than or less than 25 G (after Harvey, 1997)

oppositely-directed field tend to give smaller $|B|$ than would be observed with higher spatial resolution; the importance of this effect being higher in lower resolution magnetograms. Krivova et al. [2002b] and Krivova and Solanki [2004] have used high-resolution SoHO MDI magnetograms ($0.045'' \times 0.045''$ resolution) to show that pixels that are smaller in area by a factor of 500 than for the Kitt Peak data give $|B|$ that is larger by a factor of 2.5 for the quiet Sun. In comparison, the equivalent factor for the larger-scale magnetic fields of active regions is only 1.1.

A *facula* ("torch") is a small but bright spot on the photospheric surface. Faculae are considerably smaller than sunspots but are much more numerous. They are most easily observed near the limb of the solar disc where they have more *contrast* with respect to the quiet Sun. Like sunspots, they are regions where the magnetic field threads the photosphere, the main difference being that they are considerably smaller in diameter. The cross-sectional area of these tubes increases with height above the surface and form bright regions in the chromosphere called plages. Faculae can be observed in white light and at various wavelengths in the solar continuum emission; however, they are often most readily seen in chromospheric emissions. Figure 17 shows the Sun in Calcium K spectral line emissions using a filter with a 1 nm bandpass centered at 393.4 nm: this allows significant contributions from heights in the upper photosphere as well as from the low chromosphere. The image reveals a few isolated dark sunspots, surrounded by bright faculae in the active region bands, as well as *network faculae* which are found all over the solar disc (see Fig. 19).

Faculae cluster around sunspots and sunspot groups in active regions and their occurrence rises and falls with the sunspot cycle. Figure 18 shows

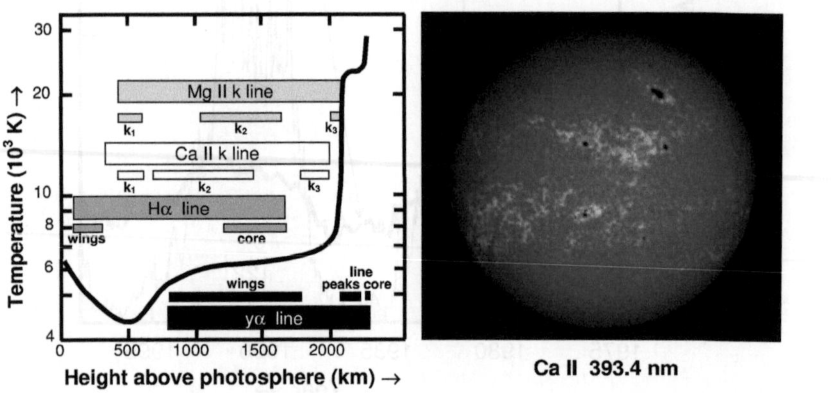

Fig. 17. (*Left*) The profile of chromospheric and transition region temperature, as a function of the height above the photosphere. The heights at which various emissions are generated are marked. (*Right*) An image in the 393.4 nm Ca II line emissions arising in the photosphere and lower chromosphere, showing sunspots and active-region and network faculae

that they cover an area which is roughly 10 times the sunspot area at all phases of the solar cycle [Chapman et al., 1997]. The correlation coefficient of facular area and sunspot area is 0.917 (significant at the 99.97% level, see Wilks, 1995, Lockwood, 2002a): the slope of the linear regression fit shown in Fig. 18 is $\mathrm{d}A_f/\mathrm{d}A_G = 10.2 \pm 0.8$ and the intercept means that there is an area of faculae at sunspot minimum of $A_f = 1357\,\mathrm{ppm}$, when the Sun is completely free of detectable spots ($A_G = 0$).

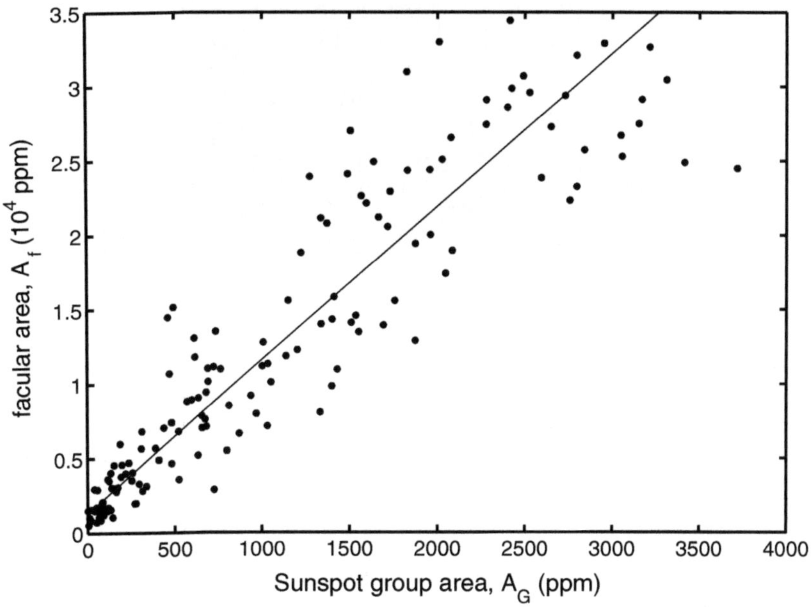

Fig. 18. Scatter plot of facular area A_f (in parts per million of a solar hemisphere, as measured by San Fernando observatory) against the total area of sunspot groups A_G (as measured at Mt. Wilson) for a whole solar cycle (1988–2000). It can be seen that at all phases of the solar cycle, faculae cover roughly 10 times the area covered by sunspots

Figure 19 demonstrates that not all faculae are found in the active region bands. Network faculae are found at all latitudes [Walton et al., 2003]. They sit in the lanes of the chromospheric network, where supergranulation flows cause magnetic flux to collect. Table 2 contrasts the properties of spots and faculae.

1.4 The Solar Atmosphere

The solar atmosphere is most easily seen during a total eclipse: when the moon blocks out direct light from the visible disc, Thompson scatter of the

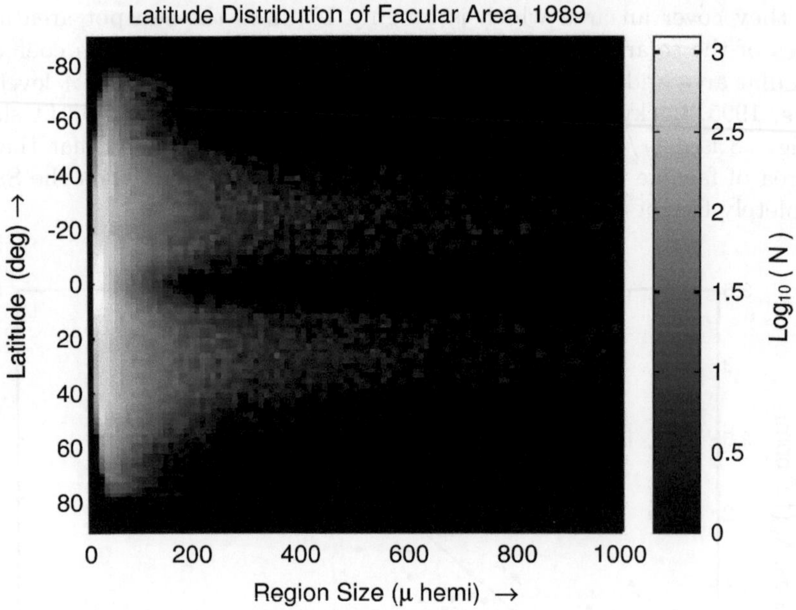

Fig. 19. Distribution of the number N of bright facular pixels in Ca II K images as a function of size and latitude, as observed during 1989 by San Fernando observatory. The scale bar indicates the logarithm of the number of such features in each size (in a millionth of a solar hemisphere) and latitude bin. The largest areas are covered by features in the two active region bands defined, but network faculae are seen at all latitudes (from Walton et al., 2003)

Table 2. Comparison of sunspots (average of umbrae and penumbrae) and faculae

	Spots	Faculae
Surface temperature at optical depth $\tau = 2/3, T_S$	≈4100 K	≈5920 K
% of solar hemisphere, $< f >$ at solar maximum	∼0.3%	∼3%
Magnetic field, B	≈ 0.1–0.3 T	≈0.1 T
Contrast at $\mu = 0.2$ (near limb)	∼0.3	∼1.1
Contrast at $\mu = 1$ (disc centre)	∼0.3	∼0.999–1.01
Radius, r	∼10000 km	< 100 km
Lifetime, t	<100 days	∼1 hour
Wilson depression at $\tau = 2/3\, d$	∼600 km	∼200 km

photospheric light allows us to see the thin, hot plasma of the *solar corona*. The corona can also be viewed using a *coronagraph*, such as the LASCO instruments on board SoHO, which use an occulting disc to obscure the photosphere. The striations and loops seen in the corona (see Fig. 9) reflect the presence of magnetic field which has emerged through the solar surface and which dominates the behaviour of the solar atmosphere. The corona is exceptionally hot and the processes which elevate the temperature of 5770 K in the photosphere to of order 10^6 K in the corona (see Fig. 20) are still a matter of great debate and the focus of much research; however, there is general agreement that coronal heating involves the magnetic field and the twisting up of that field by the complex motions and evolution of the surface field (see reviews by Narain and Ulmschneider, 1990, 1996, Gomez et al., 2000).

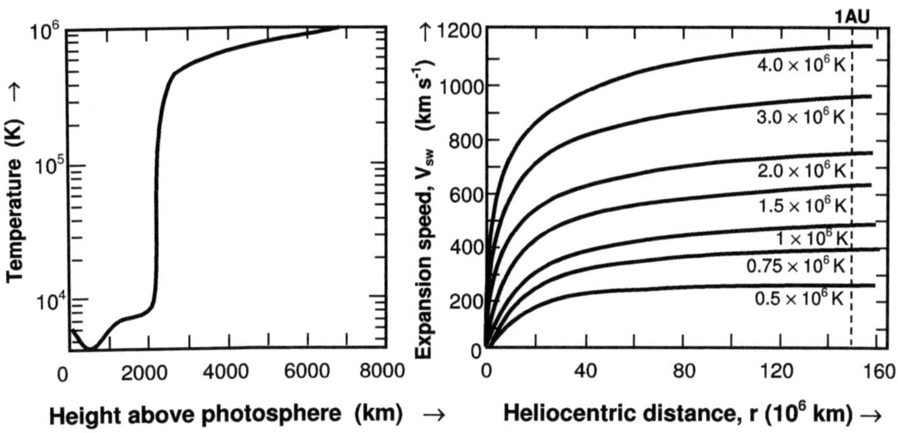

Fig. 20. (*Left*) A typical profile of the temperature T in the solar atmosphere. The effect of coronal heating is seen as the rise from 10^4 K to over 0.5×10^6 K across the transition region between the chromosphere and the corona (adapted from Noyes, 1982). (*Right*) Theoretically-derived speeds of the solar wind, V_{SW}, as a function of heliospheric distance, r, for a variety of coronal temperatures, T, between 0.5×10^6 K and 4×10^6 K (adapted from Parker, 1958, 1963)

The high temperatures mean that the coronal plasma is fully ionised and are also responsible for driving the supersonic and super-Alfvénic solar wind that blows close to radially away from the Sun (as illustrated by the simple model results presented in Fig. 20). The high temperatures are required because the solar wind must escape the gravitational potential well of the Sun. The escape velocity from the surface of the Sun is $v_e = 625\,\mathrm{km\,s^{-1}}$ which means a proton requires an energy $1/2 m_p v_e^2 > 2\,\mathrm{keV}$ to escape to infinity from the surface of the Sun and about 0.5 keV to escape from $r = 5R_S$. Note that ion velocity v and energy E are related by

$$[v \text{ in km s}^{-1}] = 13.861 \left\{ \frac{[E \text{ in eV}]}{a} \right\}^{1/2} \tag{23}$$

where a is the ion mass in atomic mass units (amu) and $1\,\mathrm{eV} = 1.602 \times 10^{-19}$ J. In a plasma, the effect of the magnetic field on charged particle motions means that temperatures are generally different in the field parallel and field-perpendicular directions. The mean energies in these directions are $1/2k_BT_\parallel$ and $k_BT_\perp$ (because they have 1 and 2 degrees of freedom respectively): for gyrotropic plasma (distribution function symmetric around the magnetic field direction) the average 3-dimensional temperature is $T = (T_\parallel + 2T_\perp)/3$ which therefore corresponds to a mean total energy (the sum of the parallel and perpendicular thermal energies) of $(3/2)k_BT$. The energy of thermal motion can thus be calculated from

$$[E \text{ in eV}] = \frac{[T \text{ in K}]}{(1.1605 \times 10^4)} \tag{24}$$

Thus the thermal energy of a 10^6 K proton in the corona is about 100 eV. Note that this is considerably lower than the escape velocities discussed above.

We do not have a full understanding of how the solar wind is driven and evolves, partly because satellite observations have not been possible in the region where it is accelerated and partly because fully self-consistent theoretical modelling has also not been possible. A number of different approaches have been tried, each with different approximations, none of which can be tested against in-situ data [Cranmer, 2002]. The fluid approximation investigates only the moments of the plasma (density, temperature, velocity and heat flux. However, a specific particle *distribution function* (see Sect. 3.2) must be assumed and it is not clear if the solar wind is best treated as a Maxwellian population and if the components of the plasma (ions and electrons, different ion species, different energy populations of the same species) require separate analysis. Kinetic treatments of the solar wind avoid some of these difficulties because they compute the distribution function rather than assuming its form; however, solutions are only possible if many simplifications are made. Equation (25) gives Parker's original solution for an isothermal magnetohydrodynamic plasma which reveals that the solar wind speed increases with increasing temperature, as shown in Fig. 20. A derivation of this equation is given by Hundhausen [1995].

$$\begin{aligned} v^2 - (2k_BTm)\left(1 + \ln\left(\frac{mv^2}{2k_BT}\right)\right) \\ = \left(\frac{8k_BT}{m}\right)\ln\left(\frac{r}{r_C}\right) + 2GM_s\left(\frac{1}{r} - \frac{1}{r_C}\right) \end{aligned} \tag{25}$$

where v is the plasma velocity, T the plasma temperature (the sum of the electron and ion temperatures), m is the mean ion mass, G is the gravitational constant, M_S is the solar mass, r is the heliocentric distance and r_C is the "critical radius" and is equal to $GM_sm/(4k_BT)$. Generalisation for, e.g., a realistic temperature profile derived from heat conduction yields similar solutions to (25) provided $T(r)$ falls off less rapidly than $(1/r)$. The isothermal solution applies to zero heat flux, and varying the heat flux to give $T = 0$ at

$r = \infty$ gives $T(r) \propto r^{-2/7}$. In this case, the acceleration of the solar wind takes place mainly at $r < 10R_S$ and, unlike the isothermal profiles shown in Fig. 20, flow speed remains approximately constant at $r > 10R_S$.

Coronal holes were first recognised by Waldmeier [1957, 1975] who noted long-lived regions of very low intensity in the coronal green line emissions (530.3 nm). Subsequently they were observed as dark patches in UV and X-ray images and associated with largely unipolar regions of open magnetic flux and fast solar wind flow [Krieger et al., 1973]. The definition of open flux will be discussed later, but for now we just note that it is magnetic field which extends out into the heliosphere, rather than looping back to the solar surface within a few solar radii. The field-free solutions to the solar wind acceleration, such as (25) apply most readily to these regions of open flux in which we observe the fast solar wind with typical velocities V_{SW} of 700 km s^{-1} at $r > 10R_S$. This fast solar wind outflow depresses the coronal plasma densities [Wang et al., 1996].

Figure 21 is a combination of images which show a dark coronal hole in the northern hemisphere. The solar atmosphere is seen in visible light during eclipses or in coronagraph data because it scatters light generated in the photosphere. It can also be seen in the EUV or X-ray wavelengths at which the hot coronal plasma emits. The lower coronal densities in coronal holes means

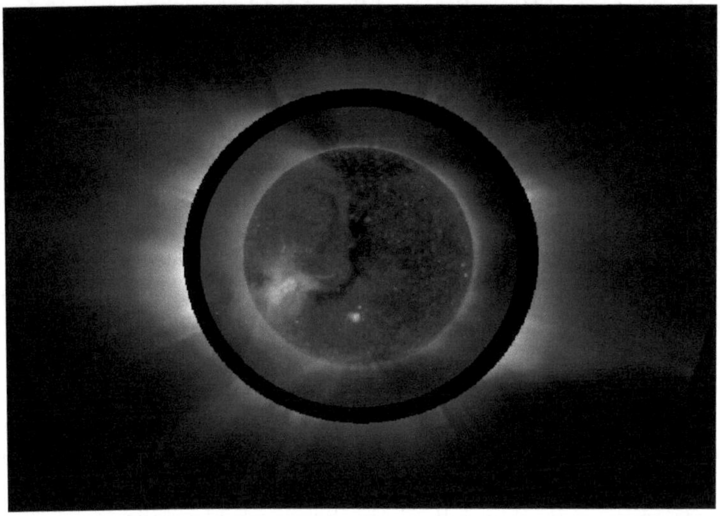

Fig. 21. An EIT image overlaid on the occultation disc of the LASCO C2 coronagraph, taken during the declining phase of the solar cycle. The tilt of the Sun allows us to see the northern polar coronal hole. In addition, a J-shaped extension to this coronal hole can be seen reaching down to low latitudes and into the southern hemisphere, ending at a bright active region. Both the EIT and LASCO instruments are on the SoHO spacecraft (This coronal hole has been analysed by Zhao et al., 1997)

that within them the intensities of scattered and emitted light are suppressed. The combined EIT/LASCO image was recorded during the declining phase of the solar cycle and at a time when the tilt of the Sun's axis with respect to Earth makes a northern hemisphere coronal hole clearly visible: in fact, coronal hole morphology is found to change radically during the solar cycle. At sunspot minimum the coronal holes are two large, contiguous regions, of opposite magnetic polarity, around the solar poles, but as solar maximum is approached these break up into smaller, more transient patches with both field polarities occurring at all latitudes in both hemispheres [Maravilla et al., 2001]. In the declining phase of the solar cycle, the polar coronal holes begin to regroup (but with reversed polarities), but with coherent extensions down to lower latitudes (as can also be seen in Fig. 21). Coronal holes and their low latitude extensions in the declining phase do not show the differential rotation of the underlying photosphere and CZ, rather they co-rotate "rigidly" with a period of about 27 days, when viewed from Earth [Wang et al., 1996].

The solar wind speed depends not only on the coronal temperature, but also on the flux tube area expansion between the base and the top of the corona [Wang and Sheeley, 1990, Wang, 1995, Wang et al., 1996]. Outside the coronal holes, where the magnetic field is in closed loops with smaller area expansion factors, the *slow solar wind* (V_{SW} of typically $350\,\mathrm{km\,s^{-1}}$ at $r > 10R_S$) is found (see review by Poletto, 2004). Figure 22 shows the latitudinal variation of the solar wind flow speed at sunspot minimum as seen by the Ulysses satellite, the first mission to study the heliosphere at latitudes away from the ecliptic plane. The flow is seen to be fast at all latitudes above about 30° but slow or mixed at lower latitudes where the superposed coronagraph image reveals higher density plasma in the streamer belt. Where the flow is fast, low coronal densities are seen in two large coronal holes. The magnetic field is almost exclusively inward in the northern hemisphere and outward in the south, as seen in the photosphere at the same time (seen for 1993–1995 in Fig. 14). This clear-cut field topology is a feature of the sunspot-minimum Sun. Because the radial field changes polarity across the streamer belt, it must contain a disc-like *heliospheric current sheet* (HCS) which separates the two magnetic hemispheres.

The lower part of Fig. 23 shows the variations of the fast and slow solar wind observed by the Ulysses spacecraft [McComas et al., 2002a,b, 2003]. The interpretation of this plot is complicated by the fact that the sunspot number changes on a comparable timescale to the change in the latitude of the Ulysses spacecraft. The left of the lower panel applies to near sunspot minimum, when Ulysses remained continuously in the northern polar coronal hole down to a latitude λ of about 30°, observing (fast) solar wind speeds V_{SW} of near $750\,\mathrm{km\,s^{-1}}$. It subsequently moved in and out of the slow solar wind in the streamer belt every 27 days because of the inclination of the solar equator, before becoming continuously within the streamer belt between about +15° and −5°, where $V_{SW} \approx 350\,\mathrm{km\,s^{-1}}$. The evolution of the coronal

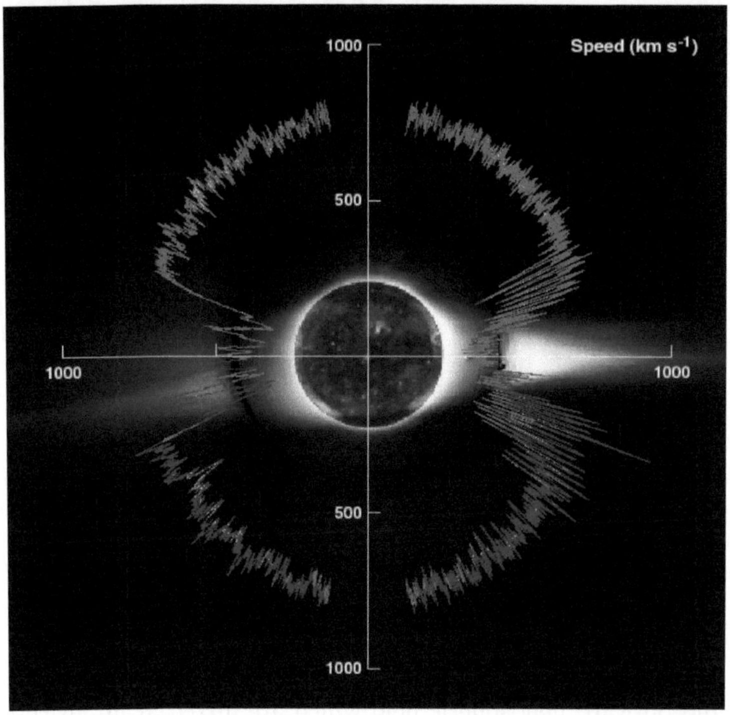

Fig. 22. Dial plot of the solar wind velocity as a function of heliographic latitude as seen near sunspot minimum by the SWOPS thermal plasma instrument on the Ulysses satellite during 1993–1995. Where the magnetic field seen by the FGM instrument on the same craft is inward (toward the Sun) the plot is coloured *blue* (true throughout almost all of the southern hemisphere) and where outward it is coloured *red* (as throughout almost all of the northern hemisphere). The plot is superposed on a sunspot-minimum image of the solar disc, as seen in Extreme Ultraviolet (EUV) by the EIT instrument on the SoHO satellite, and an image of the corona made by the LASCO C2 coronagraph on SoHO: these data show the polar coronal holes and the streamer belt [from McComas et al., 1998]

holes and the streamer belt during this interval have been studied by Wang et al. [2000c] and is shown by the upper panel of Fig. 23. The first coronagraph image in the upper panel shows this clear, single, equatorial streamer belt at sunspot minimum. The other images show that this progressively broke up into smaller streamers at all latitudes as the solar activity increased. While between latitudes $-5°$ and $-45°$ (the end of the plot), Ulysses observed flow which oscillated between about 400 km s^{-1} and 600 km s^{-1} as it moved in and out of the increasing number of streamer belts. At sunspot maximum, both inward and outward magnetic field is seen at all latitudes, as are the streamers which are mixed with smaller coronal holes. The HCS becomes increasingly

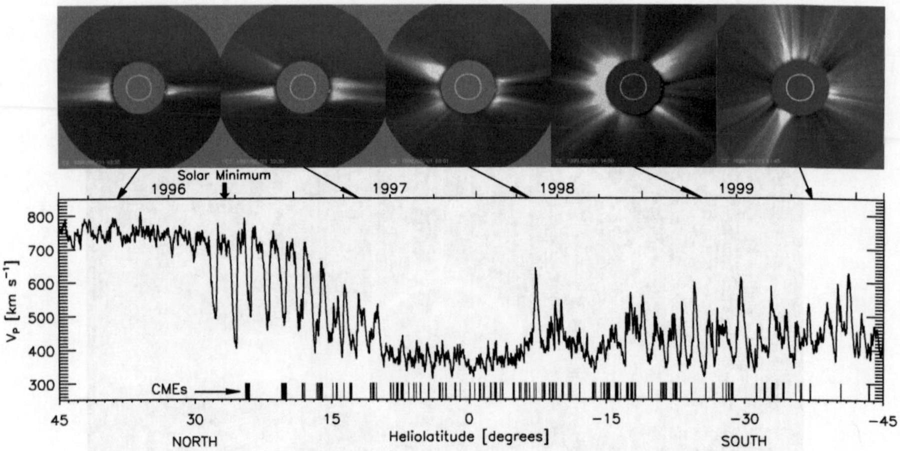

Fig. 23. (*Top*) A series of SoHO/LASCO coronagraph images, taken as solar activity increases (taken at times when Ulysses was at the latitudes given by the arrows pointing to the lower panel). (*Bottom*) The solar wind flow speed seen by the Ulysses spacecraft as a function of its latitude λ (Courtesy of the SoHO/LASCO and Ulysses/SWOOPS instrument teams)

warped and may even develop into multiple current sheets separating the inward and outward heliospheric field which are mixed at all latitudes.

The low-latitude extensions to coronal holes in the declining phase of each solar cycle (such as that seen in Fig. 21) are an important feature for the Earth. Like the polar coronal holes, fast solar wind emanates from these features and, because it flows almost radially, the fast flow arising from the equatorial part of a coronal hole extension will intersect the Earth. The fast solar wind meets slow solar wind ahead of it and forms a *corotating interaction region* (CIR) and as they sweep over the Earth, these cause disturbances to the geomagnetic field and to near-Earth space that repeat every 27 days. These are called *recurrent geomagnetic storms*. Not all geomagnetic disturbances are recurrent. A second class of geomagnetic storm, the occurrence of which peaks as sunspot maximum, is random in its timing and these occur because the solar wind is not steady but shows large enhancements called *Coronal Mass Ejections* (CMEs), as illustrated by the sequence of images, roughly 40 min. apart, shown in Fig. 24. On average, a CME contains about 10^{13} kg of material, moving at about $350\,\mathrm{km\,s^{-1}}$, and so constitutes a total energy of about 10^{24} J. By way of comparison, the Sun loses of order 10^{14} kg per day in the total solar wind. Thus CMEs form a significant contribution to the solar wind. On average, 1 CME occurs every 4 days at sunspot minimum, but this rate rises 2 CMEs per day at sunspot maximum. The directionality of these events, with respect to the ecliptic plane, changes over the solar cycle and 1 CME hits Earth every 2 weeks at sunspot minimum, but this rises to 4 per week at maximum. The event shown in Fig. 24 is clearly visible,

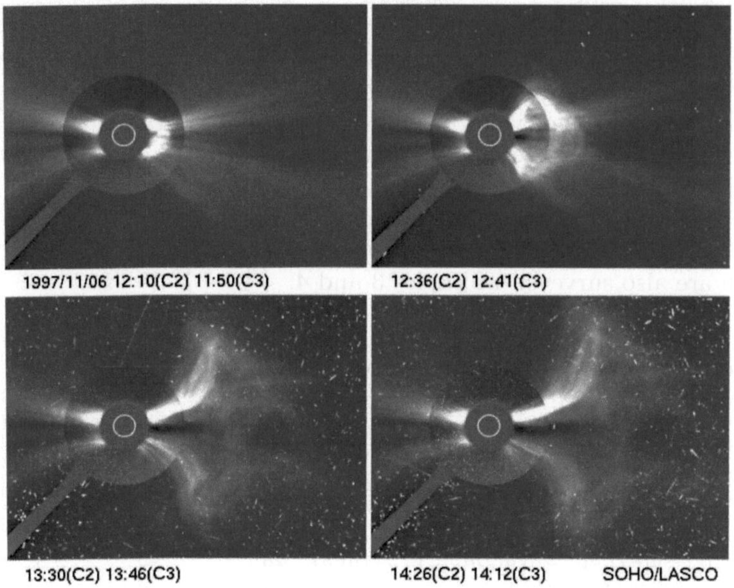

Fig. 24. A sequence of images that are combinations of data from the LASCO C2 and C3 coronagraphs on the SoHO spacecraft, showing a particularly large coronal mass ejection (CME). Note the increasing number of energetic particle strikes on the imager CCDs

but is moving roughly perpendicular to the Sun–Earth line and will not hit Earth. CME events that are travelling toward the Earth form a "halo" in coronagraph images and these are much harder to detect. Those that do hit Earth can drive large (non-recurrent) geomagnetic storms, depending on the direction of the magnetic field within them. Some CME's contain high-density, low-temperature plasma which, from the charge state abundance, can be identified as coming from the photosphere: in such cases the CME has dragged a prominence feature after it and these events are thought to be particularly effective in driving storms in near-Earth space.

Note in Fig. 24 that there are an increasing number of spots on the images, caused by energetic particles striking the CCDs of the LASCO imagers. These particles are accelerated to very high energies at the shock front at the leading edge of the CME and/or by the associated flare. The particles have travelled rapidly along magnetic field lines to the SoHO craft and also impinge on Earth's magnetosphere.

Direct information about the solar wind in the acceleration region is restricted to data from long-baseline observations of *interplanetary scintillations* of radio galaxies caused by density variations in the propagating solar wind. This effect is analogous to the twinkling of visible stars caused by variations and turbulence in Earth's atmosphere. However, models and theories of the solar wind must also match Ulysses observations of the out-of-ecliptic heliosphere in addition to the long series of observations of the near-Earth solar wind in the ecliptic plane. Tables 3 and 4 summarise the results of a survey covering 2 solar cycles of hourly averages of data on the solar wind impinging on the near-Earth space environment [Hapgood et al., 1991]. The solar wind drags with it a weak magnetic field of solar origin called the *heliospheric field* which, when measured near Earth in the ecliptic plane, is referred to as the *Interplanetary Magnetic Field* (IMF), the characteristics of which are also surveyed in Tables 3 and 4.

Table 3. Distributions of hourly averages of solar wind parameters at Earth

	Largest	Smallest	Mode Value
Density, N_{SW} (m^{-3})	8.3×10^7	$\sim$0	6×10^6
Velocity, V_{SW} ($km\,s^{-1}$)	950	250	370
Plasma temperature, T_{SW} (K)	3.2×10^5	0.2×10^5	1.3×10^5
Dynamic pressure, $P_{SW} = N_{SW} m_{SW} V_{SW}^2$ (nPa)	28	$\sim$0	3
IMF field strength, B_{IMF} (nT)	85	$\sim$0	6
Northward IMF component, B_z (GSM) (nT)	27	−31	0
Radial IMF component, $\|B_r\| = -\|B_x\|$ (nT)	70	$\sim$0	5

1.5 Solar Output Power at Earth

To end this brief survey of the Sun and its atmosphere, it is instructive to compare the powers received by Earth in the form of the solar wind and electromagnetic radiations. Table 4 shows that the incident solar wind energy density is dominated by the bulk flow kinetic energy, $W_d \sim 10^{-9}\,J\,m^{-3}$ and thus the incident solar wind power density, $p_d = W_d V_{SW} \sim 5 \times 10^{-2}\,W\,m^{-2}$. This impinges on the magnetosphere, the region of near-Earth space that is dominated by the geomagnetic field, which presents a cross-sectional area of $A_m \sim \pi(15R_E)^2 \sim 3 \times 10^{16}\,m^2$ (the mean Earth radius, $1\,R_E = 6370\,km$). Thus the solar wind power incident on geomagnetic field is $A_m p_d \sim 1.5 \times 10^{15}\,W$ (equivalent to 600,000 large modern power stations of 2.5 GW each). When the IMF points southward (optimum conditions) the geomagnetic field extracts about 2% of incident energy, i.e. $\sim 3 \times 10^{13}\,W$ and of this about one third deposited in upper atmosphere and inner magnetosphere, i.e. about $1 \times 10^{13}\,W$ (4000 major power stations). The other two thirds are returned to the solar wind. By way of comparison, mankind currently uses (from all sources) about $10^{13}\,W$.

We can carry out equivalent calculations for the energy brought to Earth's coupled atmosphere/ocean/land system by electromagnetic radiations. The total solar irradiance, I_{TS}, is 1367 (±7) $W\,m^{-2}$ which is 27,000 times larger

Table 4. Typical values of other solar wind parameters at 1 AU

	Typical Value
Proton composition (% of ion gas by number density)	84%
He^{2+} ion composition (% of ion gas by number density)	15%
Heavier ion (mean 16 amu) composition (% of ion gas by number density)	1%
Mean ion mass, $< m_i >$ $\approx (0.84 \times 1 + 0.15 \times 4 + 0.01 \times 16)$	1.6 amu ($\equiv 2.67 \times 10^{-27}$ kg)
Bulk flow kinetic energy density, W_d $= (N_{SW} m_{SW} V_{SW}^2)/2$	$10^{-9}\,\mathrm{J\,m^{-3}}$
Magnetic energy density, $W_B = \mathrm{B}_{IMF}^2/2\mu_0$	$10^{-11}\,\mathrm{J\,m^{-3}}$
Thermal energy density, $W_{th} = N_{SW} k_B T_{SW}$	$10^{-12}\,\mathrm{J\,m^{-3}}$
Plasma beta, $\beta = W_{th}/W_B$ $= 2\mu_0 N_{SW} k_B T_{SW}/\mathrm{B}_{IMF}^2$	0.1
Alfvén speed, $V_A = \mathrm{B}_{IMF}/(\mu_0 N_{SW} < m >)^{1/2}$	$42\,\mathrm{km\,s^{-1}}$
Alfvén Mach number, $M_A = V_{SW}/V_A$	9
Sound speed, C_S	$60\,\mathrm{km\,s^{-1}}$
Proton temperature, $T_{\mathrm{H+}}$	1.2×10^5 K
Electron temperature, T_e	1.4×10^5 K
Proton–proton collision time	4×10^6 s
Electron–electron collision time	3×10^5 s

than the solar wind power density. However, its target, Earth's atmosphere, presents as smaller cross-sectional area of $A_E \approx \pi R_E^2 \approx 1.3 \times 10^{14}\,\mathrm{m}^2$. Thus the total power incident is $A_E I_{TS} \sim 1.8 \times 10^{17}$ W. Of this, close to one third is reflected back into space (Earth's "albedo") gives this input 1.2×10^{17} W (equivalent to 48×10^6 power stations and more than 10^4 times larger than received from solar wind).

Because of this great disparity in powers, the solar wind, and associated phenomena, have often been discounted as factors in studies of Earth's climate system. Whilst it is true that arguments in favour of such associations have often been based of inadequate statistical analysis, arguments based on comparison of magnitudes, have in the past often also been proved wrong by the discovery of previously unimagined mechanisms. An excellent example of this is the early debate about the influence of the Sun on geomagnetic activity. In 1863, William Thomson (later to become Lord Kelvin) calculated the strength of the Sun's apparently dipolar field at the Earth's surface using the expected $1/r^3$ dependence and concluded that it's effect was entirely negligible in magnitude, compared to the Earth's own field. So convinced was he by this superposition argument about relative magnitudes, that he dismissed the growing evidence for correlations between solar magnetic effects and geomagnetic activity, famously stating in his presidential address to the Royal Society in November 1892 [Thompson, 1893]:

"During eight hours of a not very severe magnetic storm, as much work must be done by the Sun in sending magnetic waves out in all directions through space as he actually does in four months of his regular heat and light. This result is absolutely conclusive against the supposition that terrestrial magnetic storms are due to magnetic action of the Sun; or to any kind of dynamical action taking place within the Sun, or in any connexion with hurricanes in his atmosphere ... The supposed connexion between magnetic storms and sunspots is unreal, and the seeming agreement between the periods has been a mere coincidence."

Of course Lord Kelvin knew nothing of the solar wind and its ability to drag frozen-in solar magnetic field with it, nor of magnetic reconnection and the complex interplay of plasma energy and magnetic field in Earth's magnetosphere nor how this results in the currents in Earth's upper atmosphere that generate geomagnetic activity. The first suggestion of what became called the "corpuscular hypothesis", and which grew into our modern-day understanding of the solar wind and its effects, was made in the same year by FitzGerald [1892]:

"... a sunspot is a source from which some emanation like a comet's tail is projected from the Sun ... Is it possible, then, that matter starting from the Sun with the explosive velocities we know possible there, and subject to an acceleration of several times solar gravitation, could reach the Earth in a couple of days?"

The lesson to be learned from such history is clear. Whilst it is true that apparent connections and correlations, by themselves, prove nothing, their investigation can sometimes lead the way to the discovery of un-envisaged physical mechanisms.

2 Fundamental Plasma Physics of the Sun and Heliosphere

Maxwell's equations can be reduced in complexity for plasmas because the free charges mean that they are exceptionally good electrical conductors. Specifically for phenomena of frequency less than about 10^{14} Hz, the displacement current $\delta D/\delta t$ is negligible (the "Q" of the medium is much less than unity). In addition, charges are free to move under Coulomb attraction/repulsion to null any net space charge. Thus the plasma is electrically neutral to a very good approximation (the space charge ρ_t is very close to zero). Equations (26)–(29) give the simplified relationships between the magnetic field $\boldsymbol{B}$ and the electric field $\boldsymbol{E}$ in the now standard differential and integral forms first introduced by Oliver Heaviside.

$$(\nabla \times \boldsymbol{B})/\mu_o = \boldsymbol{J} \quad ; \quad \oint_c \boldsymbol{B} \cdot \mathrm{d}l = \int_A \mu_o \boldsymbol{J} \cdot \mathrm{d}\boldsymbol{A} \quad \text{Ampère's Law} \tag{26}$$

$$\nabla \times \boldsymbol{E} = -\frac{\partial \boldsymbol{B}}{\partial t} \quad ; \quad \oint_c \boldsymbol{E} \cdot \mathrm{d}l = -\frac{\partial}{\partial t}\left\{\int_A \boldsymbol{B} \cdot \mathrm{d}\boldsymbol{A}\right\} \quad \text{Faraday's Law} \tag{27}$$

$$\nabla \cdot \boldsymbol{E} = \frac{\rho_t}{\epsilon_o} \approx 0 \quad ; \quad \int_A \boldsymbol{E} \cdot \mathrm{d}\boldsymbol{A} = \int_V \left(\frac{\rho_t}{\epsilon_o}\right) \mathrm{d}V \approx 0 \quad \text{Gauss' Law} \tag{28}$$

$$\nabla \cdot \boldsymbol{B} = 0 \quad ; \quad \int_A \boldsymbol{B} \cdot \mathrm{d}\boldsymbol{A} = 0 \quad \text{(magnetic monopoles do not exist)} \tag{29}$$

If we combine these equations with those of fluid dynamics we can derive a system of fluid equations to describe the motion of the plasma, called *magnetohydrodynamics* or MHD.

2.1 Ohm's Law for a Plasma

The momentum balance equation for species k (of number density N_k, pressure P_k, bulk flow velocity v_k, mass m_k and charge q_k) is

$$N_k m_k \left(\frac{\delta \boldsymbol{v}_k}{\delta t}\right) = \nabla P_k + N_k m_k \boldsymbol{g} + N_k q_k m_k [\boldsymbol{E} + \boldsymbol{v}_k \times \boldsymbol{B}] - \sum_{j \neq k} \boldsymbol{F}_{kj} \tag{30}$$

where the inertial term is on the left-hand side (LHS) and, from left to right, terms on the right-hand side (RHS) are due to pressure gradients, gravity, the Lorentz force on a charged particle and the frictional drag caused by collisions with other species. The force on the electrons due to the ions is $\boldsymbol{F}_{ei}$

$$\boldsymbol{F}_{ei} = \nu_{ei} m_e (\boldsymbol{v}_e - \boldsymbol{v}_i) = -\boldsymbol{F}_{ie} \tag{31}$$

where ν_{ei} is the collision frequency for momentum transfer. A plasma is quasi-neutral and so for a single-ion plasma $N_i = N_e$ and the current density is

$$\boldsymbol{J} = N_e e (\boldsymbol{v}_i - \boldsymbol{v}_e) = -\frac{N_e e \boldsymbol{F}_{ei}}{\nu_{ei} m_e} \tag{32}$$

We define the plasma velocity to be the average velocity of ions and electrons, weighted by their mass

$$\boldsymbol{V} = \frac{(m_e \boldsymbol{v}_e + m_i \boldsymbol{v}_i)}{(m_e + m_i)} \tag{33}$$

and use the vector relation

$$(m_e + m_i)[\boldsymbol{V} \times \boldsymbol{B}] = m_e[\boldsymbol{V}_e \times \boldsymbol{B}] + m_i[\boldsymbol{V}_i \times \boldsymbol{B}] \tag{34}$$

If we take (30) for electrons (multiplied by m_i and neglecting small electron pressure gradient, gravity and electron-neutral collision terms), subtract (30) for ions (multiplied by m_e and neglecting small ion pressure gradient and ion-neutral collision terms), assume steady state (all time derivatives are zero) and substitute using (32) (33) and (34) we derive *Ohm's law for a plasma*

$$\boldsymbol{J} = \sigma[\boldsymbol{E} + \boldsymbol{V} \times \boldsymbol{B}] \tag{35}$$

where the (electrical) conductivity, $\sigma = \{N_e e^2/(\nu_{ei} m_e)\}$. Applying the Lorentz transformations shows that the term in square brackets in (35) is the electric field in the rest frame of the plasma.

2.2 The Induction Equation

If we substitute Ampére's law (26) and Faraday's law in differential form (27) into Ohm's Law (35) and use (29) and the vector relation

$$\nabla \times \boldsymbol{B} = \nabla(\nabla \cdot \boldsymbol{B}) - \nabla^2 \boldsymbol{B} \tag{36}$$

we derive the *induction equation*

$$\frac{\partial \boldsymbol{B}}{\partial t} = \nabla \times (\boldsymbol{V} \times \boldsymbol{B}) + \frac{\nabla^2 \boldsymbol{B}}{(\mu_o \sigma)} \tag{37}$$

The first and second terms on the RHS are called the "convective" and "diffusive" terms, respectively. We define the *magnetic Reynolds number* R_m to be the ratio of these two terms

$$R_m = \frac{\{\nabla \times (\boldsymbol{V} \times \boldsymbol{B})\}}{\{\nabla^2 \boldsymbol{B}/(\mu_o \sigma)\}} \tag{38}$$

Taking orders of magnitude, the convective term $\{\nabla \times (\boldsymbol{V} \times \boldsymbol{B})\} \sim V_c B_c / L_c$ and the diffusive term $\{\nabla^2 \boldsymbol{B}/(\mu_o \sigma)\} \sim \{B_c/L_c^2\}\{1/(\mu_o \sigma)\}$, where V_c, B_c and L_c are the characteristic speed, field and scale length of the plasma in question. Thus

$$R_m \sim \mu_o \sigma V_c L_c \tag{39}$$

Table 5 gives typical values of the terms in (39) for various regions of the Sun–Earth system. Note that $R_m \gg 1$ in all these regions.

2.3 The Convective Limit: The Frozen-In Flux Theorem

If $R_m \gg 1$ we can neglect the diffusive term and the induction (37) becomes

$$\frac{\partial \boldsymbol{B}}{\partial t} = \nabla \times (\boldsymbol{V} \times \boldsymbol{B}) \tag{40}$$

We call this the convective limit. Satellite observations of $\boldsymbol{E}$, $\boldsymbol{V}$ and $\boldsymbol{B}$ show that (40) applies to a very high degree of accuracy, even in the F-region ionosphere where R_m is not as big as in other regions [Hanson et al., 1994].

Table 5. Calculations of magnetic Reynold's number

Region of space	σ (mhos m^{-1})	V_c (m s^{-1})	L_c (m)	R_m
Solar Convective Zone	10^2	10^5	10^6	10^7
Base of solar corona	10^3	10^5	10^6	10^8
Solar wind at r = 1 AU	10^4	10^5	10^9	10^{12}
Earth's Magnetosphere	10^8	10^5	10^8	10^{15}
Earth's Ionospheric F-region	10^2	10^3	10^5	10^4

If we compare (40) to Faraday's law (in differential form, 27) we derive that in the convective limit

$$\boldsymbol{E} = -\boldsymbol{V} \times \boldsymbol{B} \tag{41}$$

Note that this is often called the "infinite conductivity limit" but has arisen because L_c is large, as much as because σ is large (see Table 5). If we take the component parallel to $\boldsymbol{B}$, (41) shows $E_{\parallel} = 0$ and the field-perpendicular component $E_{\perp} = |\boldsymbol{E}|$. From (41), $\boldsymbol{E} \times \boldsymbol{B} = -(\boldsymbol{V} \times \boldsymbol{B}) \times \boldsymbol{B} = \boldsymbol{V} B^2$ and thus

$$\boldsymbol{V} = \frac{\boldsymbol{E} \times \boldsymbol{B}}{B^2} \tag{42}$$

The equations for the convective limit ($R_m \gg 1$, also called *ideal MHD*) are approximate, but work exceptionally well throughout almost all of the heliosphere. Note, however, the places where they do break down are very important in explaining the overall behaviour.

Consider a fixed loop in space C, threaded by a magnetic flux F and in a plasma with $R_m \gg 1$, as illustrated in the left hand side of Fig. 25. Faraday's law in integral form becomes, for ideal MHD

$$\frac{\partial F}{\partial t} = -\oint_c \boldsymbol{E} \cdot \mathrm{d}\boldsymbol{l} = \oint_c [\boldsymbol{V} \times \boldsymbol{B}] \cdot \mathrm{d}\boldsymbol{l} \tag{43}$$

dl is a segment of the loop C. If we define coordinates such that the loop element d$\boldsymbol{l}$ lies in the $+x$ direction (see right hand side of Fig. 25), and the y direction is towards the inside of the loop: dl = dx, and dy = dz = 0 then

$$[\boldsymbol{V} \times \boldsymbol{B}] \cdot \mathrm{d}\boldsymbol{l} = (V_y B_z - V_z B_y)\mathrm{d}l \tag{44}$$

Let us make an assumption in order to test if it is true. If $\boldsymbol{B}$ moves with the plasma velocity, then the rate of flux transport across dl is $\mathrm{d}f/\mathrm{d}t = (aB_{zy}\mathrm{d}l)$, where $\boldsymbol{V}_{yz}$ and $\boldsymbol{B}_{zy}$ are the components of $\boldsymbol{V}$ and $\boldsymbol{B}$, respectively, in the yz plane and a is the field perpendicular component of the velocity $\boldsymbol{V}_{yz}$ (see RHS of Fig. 25),

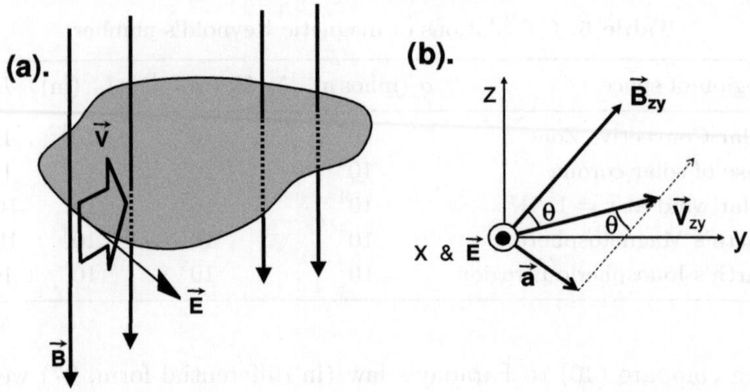

Fig. 25. Derivation of the frozen-in flux theorem. The x-direction in (**b**) is parallel to a segment of the loop in (**a**) which bounds the shaded area. The y-direction is normal to x and points to the center of the loop and z makes up the right-hand coordinate set

$$\frac{\mathrm{d}f}{\mathrm{d}t} = aB_{zy}\mathrm{d}l = (V_{zy}\sin\theta)B_{zy}\mathrm{d}l = |\boldsymbol{V}_{zy} \times \boldsymbol{B}_{zy}|\mathrm{d}l = (V_y B_z - V_z B_y)\mathrm{d}l \quad (45)$$

by (44)

$$\frac{\mathrm{d}f}{\mathrm{d}t} = [\boldsymbol{V} \times \boldsymbol{B}] \cdot \mathrm{d}\boldsymbol{l} \quad (46)$$

Integrating around loop C

$$\frac{\mathrm{d}F}{\mathrm{d}t} = \oint_c [\boldsymbol{V} \times \boldsymbol{B}] \cdot \mathrm{d}\boldsymbol{l} \quad (47)$$

Equation (47), derived by assuming that the magnetic field $\boldsymbol{B}$ moves with the plasma velocity, is the same as (43), derived by applying Faraday's law to the convective limit. Therefore the assumption must indeed be true, i.e. magnetic field does move with the plasma velocity. This is called the *frozen-in flux theorem*. It means that if a magnetic field line threads a series of plasma parcels at a certain time, when those parcels then move with the plasma velocity, as defined by (33), the field line will continue to thread the parcels, as illustrated by Fig. 26, for regions where the magnetic Reynolds number is large. As this applies in most regions of the heliosphere, this is a very powerful theorem in space plasma physics (and was invoked already in Sect. 1 in discussing the alpha and omega effects of the solar dynamo).

The consequences of frozen-in depend on the energy densities. For example, the energy density of the bulk flow of the solar wind W_d, dominates over both the thermal and magnetic energy densities W_{th}, and W_B. This means that frozen-in results in the field being dragged out and away from the Sun by the solar wind flow. In other regions (for example Earth's magnetosphere), W_B dominates over both W_d, and W_{th}. In these cases the frozen-in theorem means that the field constrains the plasma.

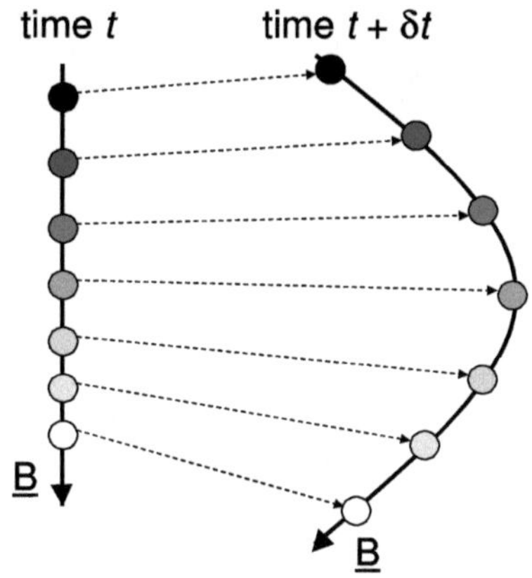

Fig. 26. The frozen-in flux theorem. Plasma elements which move between times t and $(t+\delta t)$, in the sense of their plasma velocity defined by (33), remain connected by the same magnetic field line

2.4 The Parker Spiral

Parker spiral theory is an example of the frozen-in theorem at work. The solar wind always blows almost radially away from the Sun and throughout the corona and the heliosphere $R_m \gg 1$ (see Table 5), so frozen-in applies (an important exception being at some current sheets, as we will see in the next section). The flow energy density W_d greatly exceeds the magnetic energy density W_B, so the solar wind flow drags the IMF with it. In this section, we discuss how the combination of radial flow and solar rotation winds the IMF into the Parker spiral.

As seen from Earth, the corona rotates with a period $\tau' = 27\,\text{days}$ (rotation rate $1/\tau' = 429\,\text{nHz}$). But in this time, Earth has moved along its orbit through an angle $\delta = 2\pi[\tau' \text{indays}]/365.25 = 0.464\,\text{radians}\ (26.6°)$. Hence in the time τ', the Sun has actually rotated through $(2\pi + \delta)$ and the period with respect to the fixed stars is $\tau = \tau' \times 2\pi/(2\pi+\delta) = 25.1\text{days}$ (an angular velocity $\omega = 2.90 \times 10^{-6}\,\text{rad}\,\text{s}^{-1}$, or a rotation rate of $1/\tau = 461\,\text{nHz}$).

Figure 27 shows schematically how the field, frozen into plasma parcels that move radially away from the Sun, are wound up into a spiral by the rotation of their footprints, that are rooted in the photosphere. Consider two plasma parcels that are connected by the same field line but which left the solar corona at times dt apart. Parcel 1 left first and will, at all times, have moved radially further away from the Sun by $V_{sw}dt$ where V_{sw} is the (radial) solar wind flow speed. Parcel 2 will be on a flow streamline that makes an angle ωdt with respect to that of parcel 1 because the Sun rotated through this angle in the interval between parcels 1 and 2 leaving the corona. At a radial distance r parcel 2 will be $\omega r dt$ from parcel 1 in the tangential direction. Thus the frozen-in field line will make an azimuthal angle

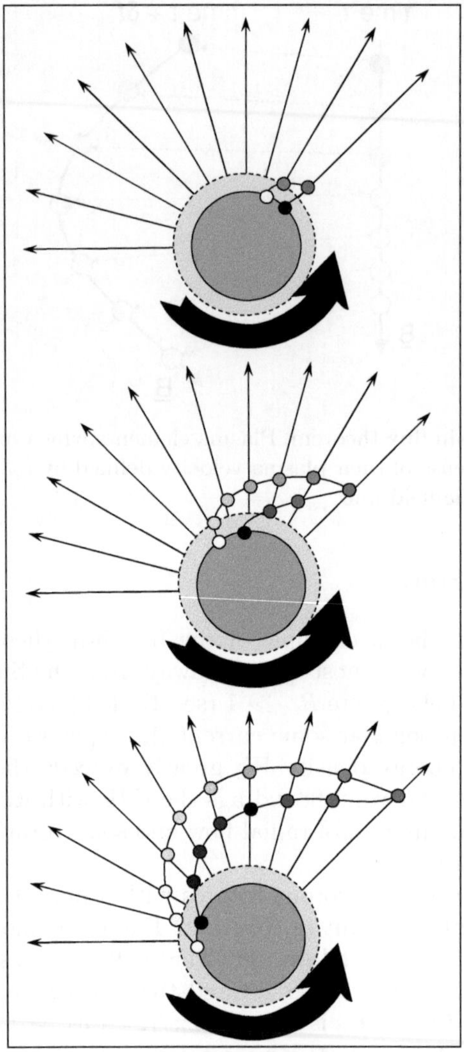

Fig. 27. Schematic illustration of the development of the Parker spiral in the heliosphere

$$\theta = \tan^{-1}\left\{-\frac{[B_Y]_{\mathrm{GSE}}}{[B_X]_{\mathrm{GSE}}}\right\} = \frac{r\omega}{V_{SW}} \tag{48}$$

with respect to the Sun–Earth line (the GSE X-axis). If the solar wind speed V_{SW} increases, (48) shows that the angle θ will decrease and the spiral will unwind.

On the other hand, if the speed V_{SW} decreases, (48) shows that θ will increase and the spiral will become more wound up. On average, the heliospheric field is well aligned with these "*gardenhose*" spiral directions, as predicted by (48). However, there are distortions caused by transient phenomena such as CMEs and CIRs and sometimes the field is even perpendicular to the orientation predicted by (48) – this is called an "*ortho-gardenhose*" orientation and, although rarer, it does exist. The most common orientation occurs at a garden hose angle that decreases as the solar wind speed increases, in very good agreement with the theory.

Equation (48) also predicts that the gardenhose angle will increase with increasing radial distance r, until θ becomes near 90°. For a relatively low V_{SW} of 350 km s^{-1}, θ will be 85° at $r = 9$ AU but $\theta = 89°$ is not achieved until $r = 46$ AU. For fast solar wind of 700 km s^{-1}, these θ values are achieved at r of 19 and 92 AU. As the angle increases, the magnetic field strength and pressure increase (and would even become infinite for $\theta = 90°$), this does not occur because the *termination shock* forms first.

The spiral angle of the heliospheric field has been monitored in the ecliptic plane by near-Earth craft and the average behaviour is very well described by Parker spiral theory (see Fig. 28) [Gazis, 1996, Stamper et al., 1999]. Forsyth et al. [1995] have shown that the average field seen by Ulysses also matches the predicted Parker spiral out of the ecliptic plane. In addition to these statistical studies of in-situ data, instantaneous spiral configuration has been monitored by remote sensing techniques. The spiral can be seen using the interplanetary scintillations technique. In addition, the vantage point of the Ulysses spacecraft has given a unique opportunity to observe the instantaneous spiral configuration when it was sited over one of the solar poles. *Flares* are explosive events on the solar surface which release bursts of energetic electrons which, because they travel at such high, superthermal velocities follow trajectories that are very close to field-aligned (in their flight time, the field line down which they travel is not moved much by the combination of corotation and radial flow of the solar wind). By tracking the radio emissions generated by these bursts the spiral configuration of the field has been mapped out and shown to be very close to the spiral predictions [Reiner et al., 1998].

2.5 The Diffusive Limit: Magnetic Reconnection

In current sheets, particle and magnetic pressures act to confine the current to a very narrow sheet. If the spatial scale becomes small enough, such that the magnetic Reynold's number, R_m becomes very small ($R_m \ll 1, L_c \ll \mu_o \sigma V_c$, 39), then the convective term in the induction equation becomes negligible and (37) reduces to

$$\frac{\partial \boldsymbol{B}}{\partial t} = \nabla^2 \boldsymbol{B}/(\mu_o \sigma) \tag{49}$$

If we have a thin, infinite planar current sheet 2L thick and we define z to be the sheet normal and the current density vector to be in the y direction (see Fig. 29), then this reduces to

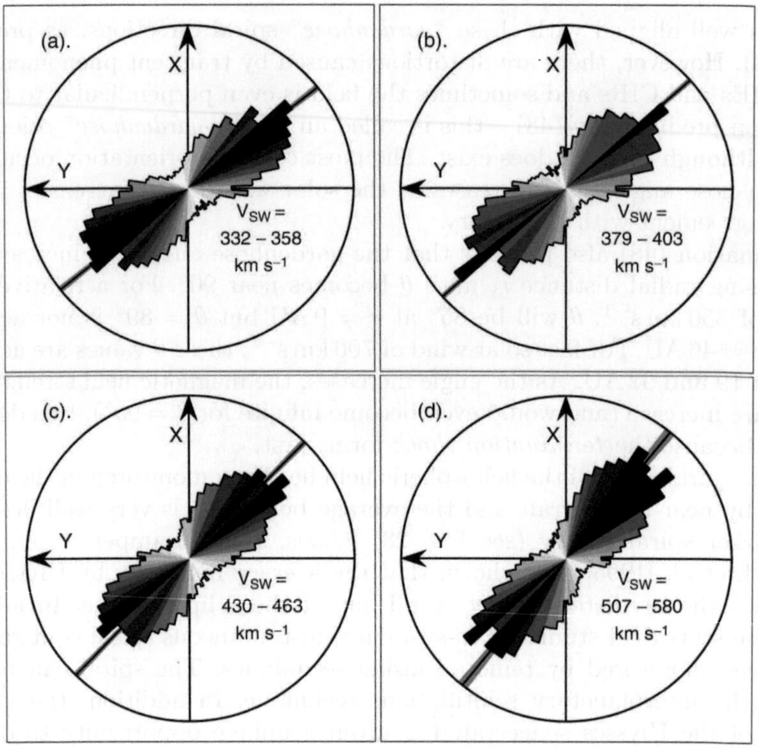

Fig. 28. Results of a survey of 142,186 hourly solar wind and interplanetary magnetic field (IMF) observations for 1963–2000 (the "Omnitape" data set, see Couzens and King, 1986). The data have been subdivided into 9 ranges of the solar wind velocity V_{SW} that give the same number of samples in each range. Results are shown here for: (**a**). 332–358 km s^{-1}; (**b**). 379–403 km s^{-1}; (**c**). 430–463 km s^{-1}; and (**d**). 507–580 km s^{-1}. The IMF garden hose angles θ, calculated from Parker spiral theory using (48), corresponding to the limits of the range in each case delineate the *gray* band plotted in each panel. Overlaid on this is a *black* line for the angle, again calculated using (48), corresponding to the mean velocity in each range. The grey-scale polar histograms give the number of observed IMF gardenhose angle observations that fall in 5° bins for the V_{SW} range in question. Both the length and *shading* of the bars are scaled according to the fraction of the total number of samples: the circle in each case marks the 4% contour. Orientations with $X > 0$ are "*toward*" solar magnetic sectors (the IMF field points toward the Sun), $X < 0$ are "*away*" sectors. Cases which line up well with the expected orientation are much more common and these are called "*gardenhose*" orientation, but there is spread and the white bars show a significant number orthogonal to the expected orientation and these are called "*ortho-gardenhose*" cases. These arise form local perturbations to the heliosphere due to the distorting of the Parker spiral field by phenomena like corona mass ejections (CMEs) and co-rotating interaction regions (CIRs). The predicted angle θ decreases as V_{SW} increases and this rotation is also seen in the most common orientations. Thus the spiral can be seen, on average, to unwind as the solar wind speed increases, as predicted by the theory

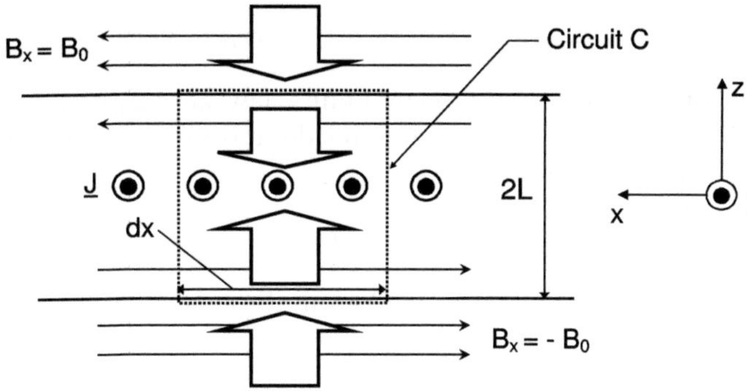

Fig. 29. An infinite, thin, planar current sheet

$$\frac{\partial B_x}{\partial t} = \left(\frac{\partial^2 B_x}{\partial z^2}\right)\left(\frac{1}{\mu_o \sigma}\right) \tag{50}$$

which is a diffusion equation (the term $1/(\mu_o\sigma)$ is sometimes called the "magnetic diffusivity", η)

Equation (50) predicts that the magnetic field $\boldsymbol{B}$ diffuses from high to low values. This means it diffuses toward the centre of the current sheet where there is a minimum in B. This is a breakdown of the frozen-in and ideal MHD.

In steady state, Faraday's law (in differential form) gives us

$$\nabla \times \boldsymbol{E} = -\frac{\partial B}{\partial t} = 0 \tag{51}$$

thus in steady state the electric field is curl-free, which means

$$\nabla \times \boldsymbol{E} = \left(\frac{\partial E_z}{\partial y} - \frac{\partial E_y}{\partial z}\right)\boldsymbol{i} + \left(\frac{\partial E_x}{\partial z} - \frac{\partial E_z}{\partial x}\right)\boldsymbol{j} + \left(\frac{\partial E_y}{\partial x} - \frac{\partial E_x}{\partial y}\right)\boldsymbol{k} = 0 \tag{52}$$

All three components must be zero, in particular

$$\left(\frac{\partial E_z}{\partial y} - \frac{\partial E_y}{\partial z}\right) = \left(\frac{\partial E_y}{\partial x} - \frac{\partial E_x}{\partial y}\right) = 0 \tag{53}$$

If we assume that there is no structure in the y direction (an infinite, flat current sheet) $\partial E_z/\partial y = \partial E_x/\partial y = 0$ and thus

$$\frac{\partial E_y}{\partial z} = \frac{\partial E_y}{\partial x} = 0 \tag{54}$$

Therefore steady state means that E_y is uniform around the current sheet. Well away from the current sheet (where the field is $B_o = B_x$ on one side and $B_o = -B_x$ on the other) frozen-in applies so $\boldsymbol{E} = -\boldsymbol{V} \times \boldsymbol{B}$. If there is no flow along the current sheet, $V = V_z$ and

$$E = E_y = -V_z B_o \qquad \& \qquad |E_y| = V_z B_o \tag{55}$$

At the centre of the sheet $B = 0$ so by Ohm's law $\boldsymbol{E} = \boldsymbol{J}/\sigma$. But $J = J_y$, so $E = E_y = J_y/\sigma$, where E_y is the same inside and outside the current sheet because of steady state. Thus

$$J_y = |\sigma V_z B_o| \tag{56}$$

Applying Ampére's law to the circuit C around the current sheet in Fig. (29)

$$\oint_C \boldsymbol{B} \cdot \mathrm{d}\boldsymbol{l} = \int_A \mu_o \boldsymbol{J} \cdot \mathrm{d}\boldsymbol{A}$$
$$2B_o \mathrm{d}x = \mu_o J_y 2L \mathrm{d}x$$
$$J_y = \frac{B_o}{\mu_o L} \tag{57}$$

Equating (56) and (57)

$$L = \frac{1}{\mu_o \sigma V_z}$$
$$R_m = \mu_o \sigma L V_z = 1 \tag{58}$$

In other words, the thickness of the current sheet adjusts to balance convective and diffusive terms such that the magnetic Reynolds number is unity. This equilibrium sheet is called a *Harris current sheet.* Magnetic field lines diffuse into the current sheet from both sides. Equation (58) shows that the inflow speed is $|V_z| = 1/(\mu_o \sigma L)$ and well away from the sheet (where frozen-in applies) the field is moving toward the centre of the current sheet. We have considered a steady state situation and so the antiparallel fields must be annihilating when they meet at the centre of the current sheet at a rate to match the inflow.

It was originally thought that this could give a way of destroying magnetic field – thereby quickly releasing the energy stored in the field and giving it to the particles (this would heat or accelerate the particles quickly, which may be important in explaining solar flares, for example). However, there is a problem which means that this process cannot take place for long. The problem is that outside of the sheet, the frozen-in theorem still applies so field lines moving into the sheet to replace annihilated ones bring frozen-in plasma with them. Thus there is an inflow of plasma from both sides. By continuity, this means the plasma concentration N within the current sheet rises. In addition, the energy released from the field raises the plasma temperature, T, and so the plasma pressure in the sheet, Nk_BT rises. The plasma pressure gradient increases until it applies enough force to the plasma and frozen-in field to choke-off the inflow and the process stops soon after it started.

However, if the breakdown of frozen-in does not take place everywhere in the current sheet, but just in a localised part (where the sheet is thinner

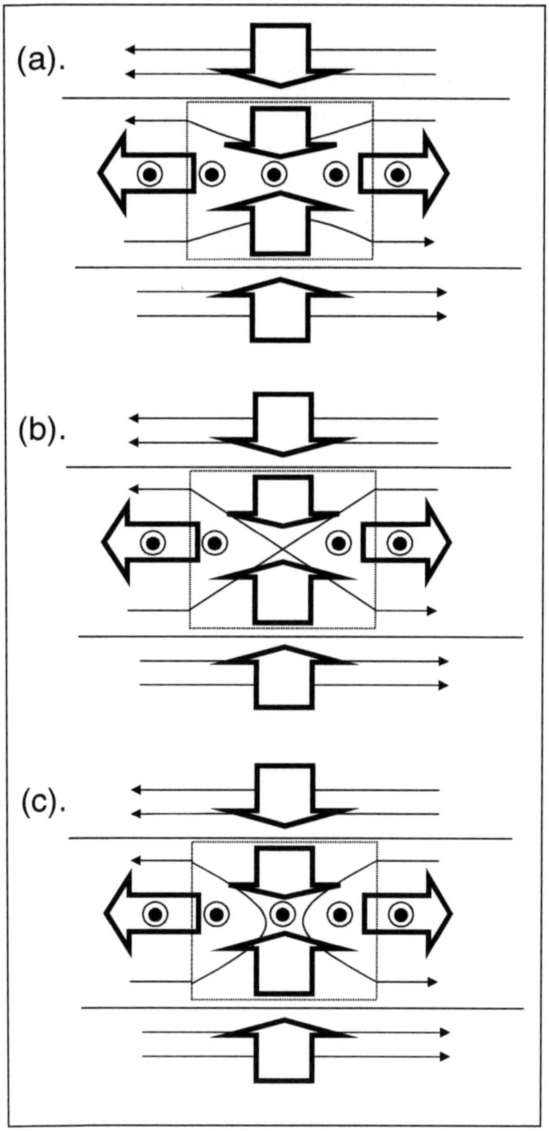

Fig. 30. Schematic of the process of magnetic reconnection

and/or there is anomalous resistivity) then the build-up of plasma can be prevented by letting it escape along the current sheet, as shown by the arrows in Fig. 30a). But what happens to the field lines at the singularity in the centre of this localised *diffusion region* where they meet? They cannot just annihilate there (that would violate Maxwell's equation $\nabla \cdot \boldsymbol{B} = 0$ and form magnetic monopoles in the diffusion region). In Fig. 30b two field lines of opposite directions come into contact and have simultaneously both the original topology (along the current sheet) and also a new one that threads the boundary. In Fig. 30c these field lines have evolved further and only have the new topology, threading the current sheet. Note that both the boundary normal field and the boundary tangential flow reverse across the diffusion region.

Away from the diffusion region, frozen-in applies and so field lines move with the plasma along the current sheet away from the singularity. This process is called *magnetic reconnection* and is arguably the most important in space plasma physics. It was first suggested by Dungey in 1953, but its significance for the space physics was not published until 1961. The term *reconnection* was adopted by Dungey because he originally thought that field lines did break and then join up again in the new topology. Then it was realised that this breakdown of Maxwell's laws was not necessary and the term *merging* was adopted by many scientists. The field lines that, for an instant, simultaneously have both the topologies and connect to the singularity at the centre of the diffusion region, are called the *separatrices* as they divide plasma and field lines that are moving towards the reconnection site (in the inflow region) from those that are moving away from it along the current sheet (in the outflow region).

The reconfiguration can proceed in a steady-state manner with inflow and outflow (making E_y the same everywhere for an infinite planar current sheet). We can analyse the region around the diffusion region, where frozen-in applies, and set boundary limits on the processes inside the diffusion region. Parker and Sweet analysed the simplest case, that is the symmetric case, where field and plasma conditions are the same on the two sides of the current sheet with reconnection taking place over a length D in the x direction. The geometry is shown in Fig. 31. If we give all parameters in the inflow region a suffix "in" and those on the other side of the separatrices an "out" label, in steady state,

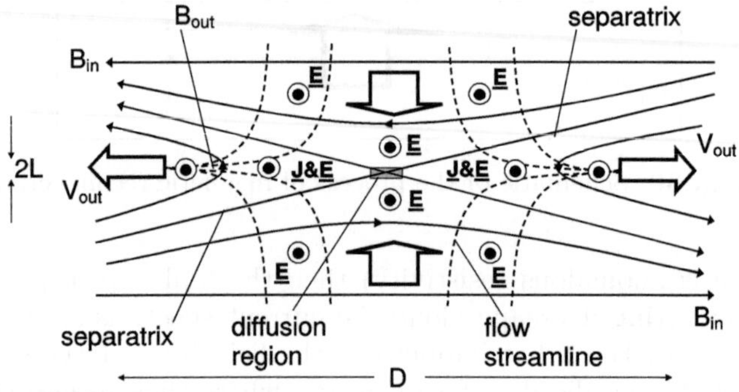

Fig. 31. Parker–Sweet reconnection geometry

$$E_y = V_{in}B_{in} = V_{out}B_{out} \tag{59}$$

The Poynting flux in a plasma is $\boldsymbol{S} = (\boldsymbol{E} \times \boldsymbol{B})/\mu_o$ and so the total power input from both sides, per unit length in y dimension is

$$P_{in} = 2DS = \frac{2DE_yB_{in}}{\mu_o} \tag{60}$$

By conservation of energy, this is equal to the rate at which energy is given to the outflowing plasma

$$P_{out} = \frac{2DE_yB_{in}}{\mu_o} = \frac{1}{2}\frac{\delta m}{\delta t}V_{out}^2 \tag{61}$$

where $(\delta m/\delta t)$ is the rate at which mass is transported into the outflow region. By conservation of mass this must equal the rate of mass inflow into the current sheet (from both sides of the boundary)

$$\frac{\delta m}{\delta t} = 2mN_{in}V_{in}D \tag{62}$$

substituting for $(\delta m/\delta t)$ in (61) and equating P_{in} and P_{out} (steady state)

$$\frac{2DE_yB_{in}}{\mu_o} = mN_{in}V_{in}DV_{out}^2 \tag{63}$$

from (59)

$$\frac{2DE_yB_{in}^2}{\mu_o} = mN_{in}E_yDV_{out}^2$$

$$V_{out} = \frac{(2)^{1/2}B_{in}}{(\mu_o mN_{in})^{1/2}} \tag{64}$$

$$V_{out} = (2)^{1/2}V_{Ain} \tag{65}$$

Where V_{Ain} is the *Alfvén speed* in the inflow region. Note that the outflow speed is independent of the electric field E_y. The motion of magnetic flux into the current sheet is associated with the electric field E_y, as is the transport of magnetic flux along the current sheet. An electric field is the same as a flux transfer rate per unit length (dimensionally, the unit of Volts is the same as $\mathrm{Wb\,s^{-1}}$) and so E_y is called the *reconnection rate*. Integrated along the length of the singularity in the y direction (the *X-line*, or *neutral line*), the electric field E_y gives a reconnection voltage (the total flux transfer rate).

Consider conservation of mass again

$$2DV_{in}mN_{in} = 2LV_{out}mN_{out} \tag{66}$$

where N_{in} and N_{out} are the plasma concentrations in the inflow and outflow regions respectively by (64) and (66)

$$V_{in} = \frac{(2)^{1/2} V_{Ain} (L N_{out})}{(D N_{in})} \tag{67}$$

Given that the current sheet has a width $L \approx 1/(\mu_o \sigma V_{in})$ (58)

$$V_{in} = \left\{ \frac{(2)^{1/2} (N_{out}/N_{in})}{R_{MA}} \right\}^{1/2} V_{Ain} \tag{68}$$

where R_{MA} is the inflow region Reynolds number ($= \mu_o \sigma V_{Ain} D$). We know R_{MA} is very large (frozen-in applies away from the current sheet) and that N_{out} and N_{in} are roughly the same, so by (64) V_{in} is very small. (This can also be seen from (67) because $L \ll D$). By (59), if V_{in} is small then E_y must be also. Thus the conclusion from Parker and Sweet's work was that reconnection takes place, but is too slow, i.e. E_y is too small to explain the phenomena observed. For example, let us quantify Parker–Sweet reconnection at Earth's magnetopause, a current sheet where the solar wind, and frozen-in interplanetary magnetic field, meet Earth's magnetospheric field (see Sect. 5.2). This is useful because we have good in-situ satellite data from this boundary and a variety of satellite and radar observations that show that reconnection in this current sheet yields voltages that can exceed 150 kV. At this current sheet, $\sigma \sim 10^8$ mhos m^{-1}, $B \sim 20$ nT, $N_{out} \sim N_{in} \sim 2 \times 10^7$ m^{-3} and $m \sim 1$ amu (proton plasma). These values yields $V_{Ain} = B_{in}/(\mu_o m N_{in})^{1/2} \sim 100$ km s^{-1} and so by (68) $V_{in} \sim 3 \times 10^{-3}$ m s^{-1}. By (59), this yields a reconnection rate of $E_y = V_{in} B_{in} \sim 6 \times 10^{-11}$ V m^{-1}. Thus even if we extend the reconnection singularity in the y direction into an X-line that is as much as $Y = 30 R_E$ long (i.e over the entire dayside of the magnetospheric surface), we have a voltage of $Y E_y \sim 12$ mV. This shows that reconnection, in this Parker–Sweet form at least, is wholly inadequate.

Thus after the work of Parker and Sweet it appeared that reconnection was a real phenomenon, but far too slow to do anything significant. This situation was changed by the work of Petschek. He postulated that MHD shocks could stand in the inflow to the current sheet, as shown in Fig. 32. These have several effects: (a) they deflect the flow so as to decrease the flow normal to the shock and increase the flow tangential to it; (b) they carry current so that they deflect and decrease the magnetic field; (c) they accelerate plasma (via the $\boldsymbol{J} \times \boldsymbol{B}$ force); (d) they compress the plasma so $N_{out} > N_{in}$; (e) they convert magnetic energy to particle kinetic energy; and, most importantly, they do all these things over a region of much greater extent than the diffusion region itself. We now know that they need not necessarily be shocks, Alfvén waves having the same effects.

We can gain some idea as to why this is so much more effective if we return to magnetic annihilation. Remember, we need to remove the outflow plasma to prevent the inflow being choked off. In Parker–Sweet, (Fig. 31) the outflow is restricted to the current layer, which we have shown is thin (only at the centre of the current sheet is B_{out} normal to the current sheet so

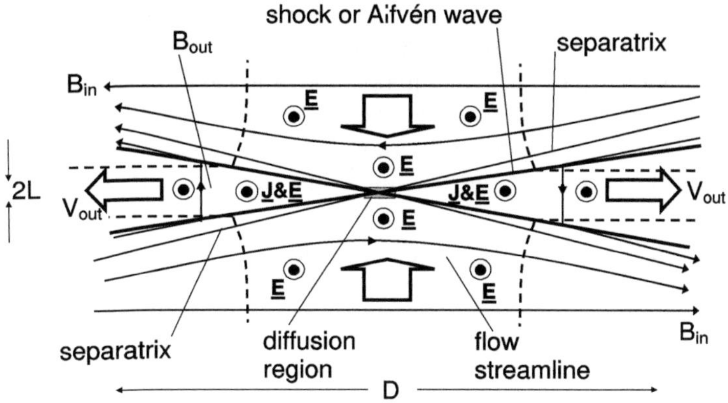

Fig. 32. Petschek reconnection geometry

that $\boldsymbol{V} = \boldsymbol{E} \times \boldsymbol{B}/B^2$ is fully along it). In Petschek reconnection, the outflow is everywhere between the two shocks (where B_{out} is normal to the current sheet) and so the outflow region is much broader. This allows the outflow rate to be much, much greater and so this does not limit the reconnection rate to anything like the same extent as in Parker–Sweet reconnection.

Analysis of Petschek reconnection is largely a matter of geometry. It is intricate in detail but similar in principle to the Parker-Sweet analysis. Two additional physical considerations that are needed are that, because $\nabla \cdot \boldsymbol{B} = 0$, the field component perpendicular to the shocks, $B_\perp$ must be the same on both sides of the shock and that mass conservation applies to the shock (so, in steady state, mass flux into shock equals mass flux out of it). The analysis concludes

$$V_{out} = V_{Ain} \cos(\chi) \tag{69}$$

Where χ is the angle between the current sheet and the shocks. This is almost the same equation as for Parker–Sweet ($\cos\chi$ has replaced $2^{1/2}$). It can be shown that

$$\frac{1}{2}\sin(\chi) \leq \left\{\frac{V_{in}}{V_{Ain}}\right\} \leq \sin(\chi) \tag{70}$$

so the inflow speed is a large fraction of the inflow Alfvén speed (remember for Parker–Sweet reconnection (P–SR), $V_{in}/V_{Ain} = \{(2)^{1/2}(N_{out}/N_{in})/R_{MA}\}^{1/2}$ which was very small because R_{MA} is very large). As a result, $E_y = V_{in}B_{in}$ is much larger than for P–SR.

It turns out that the optimum $V_{Ain}\sin(\chi)$ (giving the largest V_{in}) is at $\chi \leq 6°$ (i.e. small perturbation of the inflow). This means $\sin(\chi) \leq 0.1$ and by (70), $0.05 \leq V_{in}/V_{Ain} \leq 0.1$. If we return to the magnetopause current sheet conditions that we considered for P–SR, with $V_{Ain} = 100\,\text{km}\,\text{s}^{-1}$ Petschek reconnection can give us V_{in} up to $10\,\text{km}\,\text{s}^{-1}$ (this is very large compared to the V_{in} of $3 \times 10^{-3}\,\text{m}\,\text{s}^{-1}$ for P–SR) and for $B_{in} = 20\,\text{nT}$ gives E_y of

$1\,\mathrm{mV\,m^{-1}}$, which when applied to the reconnection X-line $Y = 20R_E$ long this gives a voltage of $V = 10^{-3} \times 20 \times 6370 \times 10^3 = 125\,\mathrm{kV}$. This is the sort of voltage that we see across the magnetosphere and thus Petschek reconnection, unlike P–SR, is fast enough to explain what we observe. There are other Alfvénic disturbances which can stand in the inflow and outflow regions and so modulate the reconnection rate and cause structure in the outflow layer.

2.6 The Role of Magnetic Reconnection in the Solar Corona and Inner Heliosphere

Observations show frozen-in magnetic field being continuously dragged over the Earth. At heliocentric distances of 1 AU, the flux transport is always outward because the solar wind is always away from the Sun. But this field is, at least initially, rooted in the base of the convection zone of the Sun. It is instructive to look at the rate of (unsigned) poloidal field transport in the ecliptic plane. Using (41), i.e. assuming frozen-in flux, the total poloidal flux transport rate Φ_{PE} across $r = R = 1\,\mathrm{AU}$ in the ecliptic plane is given by $2\pi R V_r |B_N|$, where V_r is the radial solar wind velocity and B_N is the poloidal field (here perpendicular to the ecliptic, i.e. $B_N = [B_Z]_{GSE}$). Using mode values of $V_r = 370\,\mathrm{km\,s^{-1}}$ and $|B_N| = 2\,\mathrm{nT}$, yields $\Phi_{PE} = 7 \times 10^8\,\mathrm{Wb\,s^{-1}}$. This is not the total transport of open flux to beyond $r = 1\,\mathrm{AU}$ because we have not considered toroidal field and some poloidal field will emerge only at higher heliographic latitudes and not be seen in the ecliptic plane. Nevertheless, the flux transfer rate Φ_{PE} would alone be able to replenish a typical total open flux of $F_s = 4 \times 10^{14}\,\mathrm{Wb}$ in a time (F_s/Φ_{PE}) of just 7 days and open flux would grow at a rate of at least $2 \times 10^{16}\,\mathrm{Wb\,yr^{-1}}$ if this were the only active process. Clearly open solar flux does not remain rooted in the base of the CZ and must become disconnected from the Sun. In steady state, the total connected open flux passing through $r = 1\,\mathrm{AU}$ would be balanced by the same amount of disconnected flux and so half of all flux transported by the solar wind would be disconnected. Imbalances between connected and disconnected flux transport will cause the open flux to grow and decay.

Reconnection has a key role in the disconnection of open flux. Figure 33 illustrates one way in which reconnection can reduce the amount of connected flux. It shows two BMRs that have emerged through the photosphere, and where some of this flux has risen through the corona to become open. (The *coronal source surface* is a convenient concept that is discussed further in the next section; it enables us to define any flux that threads it as "open"). Where open field lines (O) come into close proximity and have opposite polarity, reconnection can take place at an X-line X_E in the current sheet between them. If the reconnection voltage along such an X-line is V_O, then in a time Δt a flux $\Delta F_O = V_O \Delta t$ is reconfigured and the initial (unsigned – i.e. of either inward or outward polarity) open flux involved, $2\Delta F_O$ (with topology O), is halved to ΔF_O (with topology RO). The U-shaped field moving away above X_E is often called disconnected flux, but is, in fact, topologically still connected to

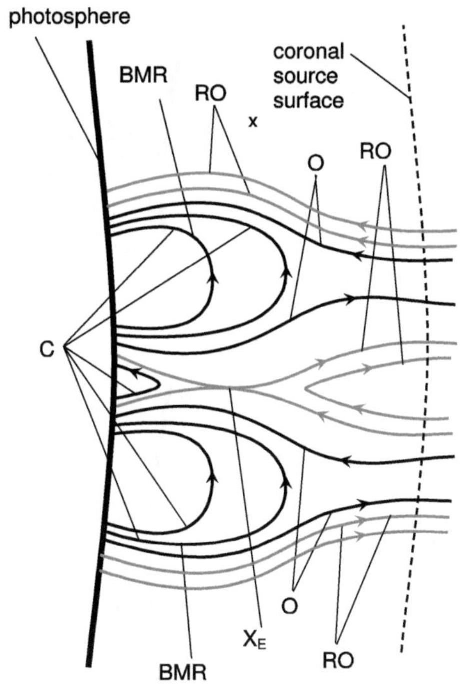

Fig. 33. Reconnection of open flux which has emerged in two different BMRs at a reconnection X-line X_E, which converts open flux like O (*shown in black*) into reconfigured open flux RO (*shown in grey*) and additional closed flux (C)

the Sun as it is part of the extended loop that makes up the reconfigured open flux RO (see the wider scale view given in Fig. 36). Such U-shaped structures have been observed by in-situ heliospheric observation, using the heat flux to deduce the field topology [McComas et al., 1991]. In addition, the loop of closed reconnected flux has been seen in falling downward through the corona following streamer disconnection events [Wang et al., 1999b] and flowing CME release events [Webb and Cliver, 1995, Simmnet, 1997, Wang et al., 1999a].

Reconnection can also help us understand how low-latitude coronal hole extensions, as seen in the declining phase of the solar cycle, can rigidly corotate whereas the photosphere beneath them shows differential rotation [Wang et al., 1996]. The concept is illustrated in Fig. 34. By reconnecting with a closed loop of the "magnetic carpet", C, the footprint of the open field line O moves from A to B and so the open flux can have a different motion to the photospheric rotation of the closed field lines. Wang et al. [1996] show how field lines must be opened at the leading edge of a coronal hole extension by reconnection of this type and then closed again at its trailing edge. In this way, the coronal hole extension can rotate faster than the closed photospheric

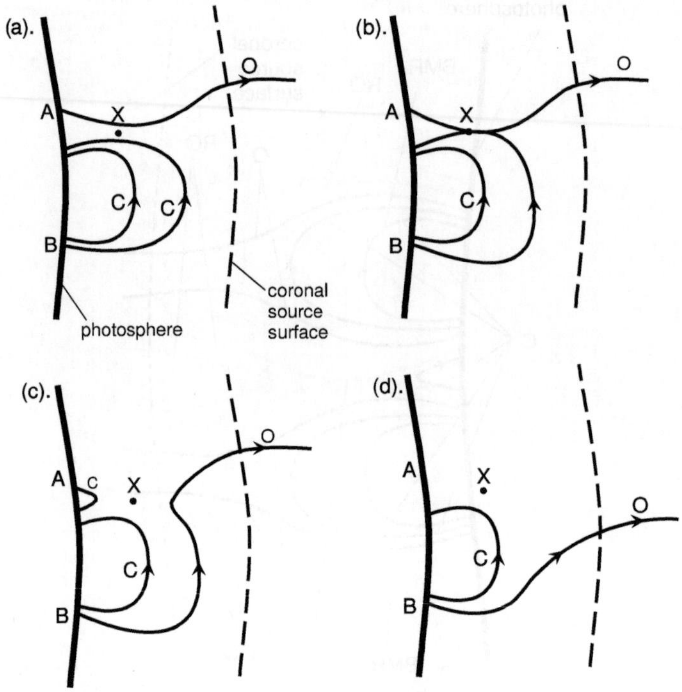

Fig. 34. The motion of open field line footprints caused by reconnection with closed flux in the corona. The footprint of open field line O moves from A to B due to reconnection at X with a closed loop (C). The closed loops are broken up into smaller structures and one is shown here descending back through the solar surface

flux. Note that at A, the original footpoint of the open field line, a closed flux loop is generated which is shown in Fig. 34 as falling back through the solar surface. This disappearance of flux from the photospheric surface was often termed *flux cancellation*, but Fig. 34 offers an explanation in terms of reconnection and flux being subsumed below the surface. The same effect would occur underneath the reconnection site X_E between the two BMRs in Fig. 33. Harvey and Hudson [2000] provide evidence that this submergence is the cause of the majority, perhaps all, flux cancellation sites by showing that the magnetic field disappears from the chromosphere first and from the photosphere soon after.

In Fig. 34 open flux is conserved and the mechanism shown in Fig. 33 cannot be the only way that open flux is lost because, although it helps open flux migrate towards the poles, it does not readily explain the reversal of the open field in polar coronal holes shortly after sunspot maximum, as seen in Fig. 14. Figure 35 shows another proposed mechanism which removes open flux of the old (previous cycle) polarity from the polar coronal hole (from Schrijver et al., 2002). Here successive reconnections draw the open field line

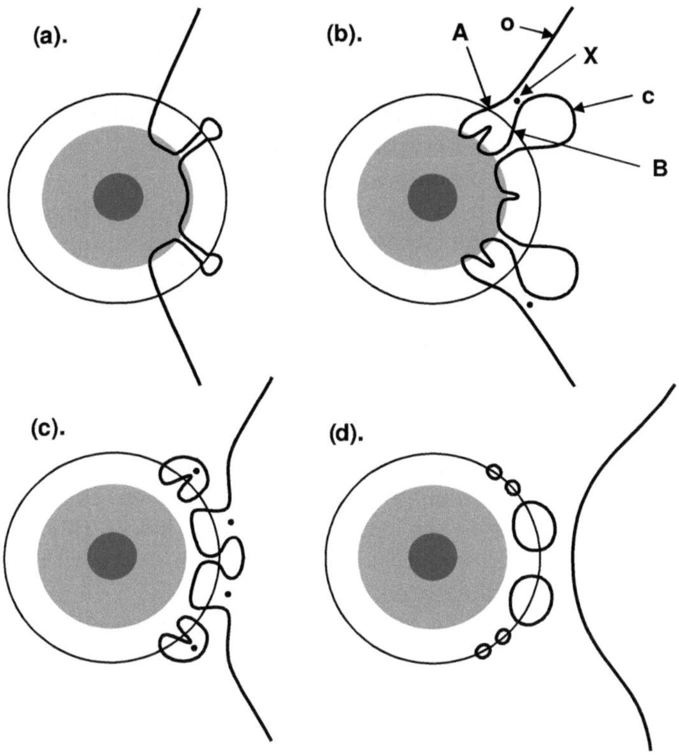

Fig. 35. The hemisphere-symmetric reconnection sequence which destroys polar open flux proposed by Schrijver et al. [2002]. (**a**) Shows field lines erupting from the overshoot layer at the base of the CZ. The same magnetic field lines enter the overshoot layer as polar, poloidal, open field. In part (**b**) the coronal loops associated with this emergence reconnect with the open segment of the same field line, making the open flux footprints migrate equatorward. The open flux is then disconnected by reconnection at low latitudes (**c** and **d**). Other reconnections reduce the large-scale surface flux to small-scale patches that can readily be dispersed by granular and supergranular motions

footprints to lower latitudes, where the flux is finally disconnected. This is a variant of the proposal by Fisk and Schwadron [2001], in which the open flux continues to migrate into the opposing hemisphere. The equatorward migration in the open flux proposed in Fig. 34 can help generate the open flux that appears at lower latitudes as the cycle progresses (the alternative mechanism being emergence of new open flux in active regions with poleward migration).

The open flux can be estimated from photospheric magnetogram data using the *potential field source surface* (PFSS) procedure [Schatten et al., 1969] by adopting a number of assumptions. The surface field is assumed to be radial, so that the component normal to the surface can be computed from the observed line-of-sight component (and, even then, no information is available from near the poles). The field is also assumed to be radial at a coronal source surface which may only be a hypothetical surface, but which is usually assumed to be spherical, heliocentric and at $r = 2.5R_s$. The corona is assumed to be current-free between the photosphere and the coronal source surface ($\nabla \times \boldsymbol{B} = 0$, an assumption that is, in fact, inconsistent with the occurrence of reconnection in the corona) and Laplace's equation is solved for Carrington maps of the photospheric field by assuming that all fields are constant over each Carrington rotation interval. Field lines which reach the coronal source surface are defined as open and the flux they constitute quantified. Wang et al. [2000b] used the PFSS method to study open flux evolution. Large concentrations of open flux are deduced in the active region belts, and in the polar coronal holes, similar to the surface flux shown in Fig. 14. However, the open flux is much more concentrated into patches than the surface flux and, although some poleward migration of the trailing spot polarity from the active regions towards the poles is seen, this is not as clear as for the poleward surges in the surface flux. Nevertheless, this implies much of the open flux seen at lower latitudes emerged there, rather than migrating equatorward in manner illustrated in Fig. 35.

Figure 36 illustrates some other reconnection scenarios relevant to the growth and decay of open flux. An X-line of the type discussed in Fig. 33 is shown at equatorial latitudes, X_E, and this converts the loops of open flux

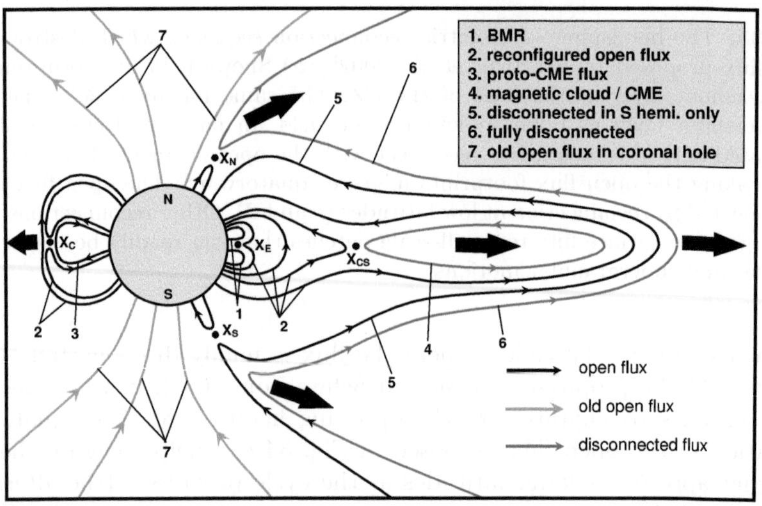

Fig. 36. Potential magnetic reconnections in the solar corona/inner heliosphere and open field line topologies (see text for details). *Light grey* field lines are "old" open flux which collected in the polar coronal hole during the previous cycle; *darker grey* field lines have been completely disconnected from the Sun and black field lines are open flux that has emerged during the current solar cycle

from BMR in opposite hemispheres (labelled 1 in Fig. 36) into reconfigured open flux with photospheric footpoints in opposing hemispheres (labelled 2). reconnection in the central current sheet separating the two magnetic hemisphere (at X_{CS}) can generate magnetic islands and plasmoids (labelled 4); however, it must be remembered that Fig. 36 is only a 2-dimensional slice and such field lines are likely to be helical flux that is still connected at both ends, out of the plane of the diagram [Gosling et al., 1995]. Complete disconnection requires reconnection with open flux, such as shown at reconnection sites X_N and X_S. Figure 36 has been drawn such that some field lines are disconnected in the southern hemisphere first: field line 5 has been reconnected at X_S, and so has been disconnected from the southern hemisphere but not the north. Only when it is subsequently also reconnected at X_N is the flux fully disconnected from the Sun (field line 6). Topologies 2, 5 and 6 can sometimes be recognised in the heliosphere from observations of superthermal electron flows called strahl [McComas et al., 1991, Larson et al., 1997]. These electrons are generated in the solar corona and flow away from the Sun. Bidirectional strahl reveals a field line directly connected to the sun at both ends (topology 2), whereas unidirectional strahl implies direct connection at one end only (topology 5). Complete absence of strahl may indicate topology 6, but care must be taken because strahl electrons are readily scattered into less easily observed and more isotropic distribution functions (*halo* electrons) by structures such as corotation interaction regions (CIRs) where fast and slow solar wind flows meet. Thus the absence of strahl does not uniquely label an open field line as disconnected [Larson et al., 1997].

The reconnection at X_S and X_N reduces the flux of old open flux which accumulated in the polar cap during the previous solar cycle (topology 7) and so can help explain why the polar coronal hole flux decays in the rising phase of each solar cycle, prior to the reversal of the polar field shortly after solar maximum. After all the old-polarity open flux in the polar caps has been removed, the poleward motion of the more open flux allows the accumulation of a polar coronal hole of the new polarity during the declining phase of each cycle. Fisk and Schwadron [2001] suggest that this mechanism is inadequate because it can only take place on the edges of the coronal holes and so mechanisms like that shown in Fig. 35 are also required.

In addition to its role in the large-scale evolution of the magnetic field in the corona and heliosphere, magnetic reconnection plays a key role in transient events. The explosive release of energy in flares is caused by the release of magnetic energy made possible by the reconfiguration of the field. In addition, CME release models generally invoke reconnection, although buoyancy and other factors are also important. To the left of Fig. 36 is an example of such a model in which "tether" field lines (of topology 2) are eaten away by reconnection at the top of the emerging CME bubble (field line 3). This is called the "breakout" CME release model and involves a quadrupolar field configuration in which the inner part of the central field line arcade are sheared by antiparallel footpoint motions near the equator causing the proto-CME field lines to bulge upward [Antiochos et al., 1999]. Clearly, evolution of flux topology in each such CME release also has implications for the overall cycle of open magnetic flux. However, this is certainly not the only model of CME release and there is currently much debate in the literature, evaluating each against observations. A recent review of CME release models has been given by Klimchuk [2003].

Figures 33–36 illustrate how magnetic reconnection is a vital part of the observed magnetic cycle and even in the rotation of the solar corona. The relative roles of the different types of reconnection in the evolution of the open solar flux is still a matter of debate.

2.7 The Ulysses Result and the Coronal Source Flux

There is no clear distinction between closed field lines, like those labelled C in Figs. 33 and 34, and more distended loops that extend out into the heliosphere, which we call open. However, there is an important difference because field lines that reach radial distances, r, large enough to be frozen into the solar wind outflow will be dragged out to the outer heliosphere, whereas closed loops in the lower corona do not necessarily evolve the same way. As mentioned in the previous section, a convenient concept used to separate these two classes of magnetic flux loop is the *coronal source surface.* This can be defined as where the magnetic field becomes approximately radial. As such, it is quite possible that there are times and places where such a surface does not exist. In practice, we usually take the coronal source surface to be spherically symmetric at $r = 2.5R_s$. This is very valuable as it allows us to quantify the total open magnetic flux of the Sun, which we call the coronal source flux, F_S.

Because of the high solar wind velocity, the magnetic flux crossing the heliospheric current sheet(s) between the coronal source surface and Earth ($r = R = 1\,\mathrm{AU}$) is a small fraction of F_S [Lockwood, 2002c], which means by conservation of magnetic flux, we can use the equation

$$F(r) = 4\pi r^2 \frac{\langle |B_r(r)| \rangle}{2} \approx F_s = 4\pi (2.5R_S)^2 \frac{\langle |B_r(r = 2.5R_S)| \rangle}{2} \qquad (71)$$

where $\langle |B_r(r)| \rangle$ is the mean of the absolute value of the radial field, averaged over the heliocentric sphere of radius r. This definition quantifies the *signed* open flux (i.e. the total of one polarity): assuming that there are no magnetic monopoles in the Sun (or, more precisely, that there is no imbalance in the numbers of opposite polarity monopoles), half the field through any surface around the Sun will be *toward* and half *away*, hence the inclusion of the factor 2 in (71). The *unsigned* flux is simply $2F_S$.

A useful parameter for evaluating the interplay between the particles and magnetic field of a plasma is its beta, the ratio of the thermal particle pressure to the magnetic pressure

$$\beta = \frac{2\mu_o N k_B T}{B^2} \tag{72}$$

The magnetic pressure acts perpendicular to the field and so if we are concerned with the heliospheric tangential pressures, we require the tangential plasma temperature but the radial field.

From (71) the average B_r will fall with a $1/r^2$ dependence with increasing r. Similarly, given the solar wind velocity V_{SW} is approximately constant at r greater than about $10R_s$, and the total flux of particles must be conserved, the solar wind density N_{SW} must also fall with a $1/r^2$ dependence. The solar wind solutions require the plasma temperature fall of with less than a $1/r$ dependence and $T_{SW}(r) \propto r^{-2/7}$ is a useful approximation. Using these dependences on r, with typical values near Earth ($r = 1\,\mathrm{AU}$) of $5 \times 10^6\,\mathrm{m}^{-3}$, $5\,\mathrm{nT}$ and $10^5\,\mathrm{K}$ for N_{SW}, B_r and T_{SW}, respectively, (72) gives the variation of β shown in Fig. 37.

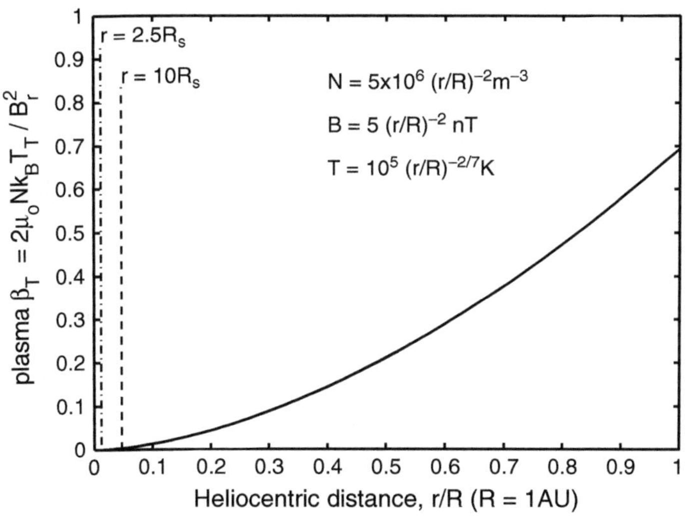

Fig. 37. The variation of the tangential plasma beta with radial distance r

It can be seen that although β approaches unity near Earth, it is very small in the region where the solar wind is accelerated. In reality, N_{SW} will fall off less rapidly than assumed because of the rise in V_{SW} with r and so β will fall to even lower values between the coronal source surface at about $10R_S$. These low β values mean that the tangential magnetic pressure will be much greater than the tangential thermal plasma pressure. The solar wind is flowing approximately radially and so the dynamic pressure does not contribute significantly to the tangential stress balance. As a result, the flow in the low-β region will become slightly non-radial such that by about $r = 10R_S$ any tangential magnetic pressure differences have been ironed out.

When this has been achieved, the radial field B_r is approximately independent of latitude, as has been observed by the Ulysses spacecraft, the first craft to view the heliosphere outside the ecliptic plane. The latitudinal uniformity of the radial field B_r was first found to apply as the satellite passed from the ecliptic plane to over the southern solar pole [Smith and Balogh, 1995, Balogh et al., 1995]. Suess and Smith [1996a] and Suess et al. [1996b] then provided the above explanation in terms of the pressure transverse to the flow in the expanding solar wind. The result is consistent with the heliosphere containing thin current sheets, but not "volume currents" spread over a larger cross-sectional area.

Subsequently, this result has been confirmed during the pole-to-pole "fast" latitude scan during the first perihelion pass and during the second ascent of Ulysses to the southern polar region (Lockwood et al. [1999b] and Smith et al. [2001], respectively). Recently, the second perihelion pass has also underlined the generality of the result [Smith and Balogh, 2003].

The first perihelion pass took place during the interval September 1994–July 1995 when solar activity was low (the average sunspot number during the pass was $\langle R \rangle = 23.5$). On the other hand, the second perihelion pass (December 2000–October 2001) was near sunspot maximum ($\langle R \rangle$ was 106.5). Thus the result appears to apply at all phases of the solar cycle. Figures 38 and 39 (from Lockwood et al., 2004) shows the results for the two perihelion passes. The difference between the solar minimum and solar maximum heliosphere is immediately apparent in the radial field. As discussed before, the field at sunspot minimum is separated into two clear hemispheres of toward and away field, with only a relatively flat HCS between the two arising in the equatorial streamer belt. However, at sunspot maximum there are several regions of toward and away flux with current sheets between them at all latitudes. It is not clear if there are indeed multiple current sheets or if this is a single HCS that has been severely warped so that it intersects any one meridian at several different latitudes. If we average over 27-day solar rotation periods, the modulus of the radial field is very similar to that seen simultaneously at Earth in both the sunspot-minimum and -maximum cases [Lockwood et al., 2004].

Smith and Balogh [1995] noted that the uniformity of the radial field allowed computation of the total open solar flux from radial field values, wherever they are measured, because the mean value $\langle |Br| \rangle$ equals the observed value. From (71)

$$F_S = \frac{4\pi r^2 |B_r(r)|}{2} \tag{73}$$

Using the data summarised by Figs. 38 and (39), Lockwood et al. [2004] have shown that the error in the F_S estimates made using (73) are less than 5% for averaging timescale >27 days, on which longitudinal structure is averaged out. This applies at both sunspot minimum and maximum.

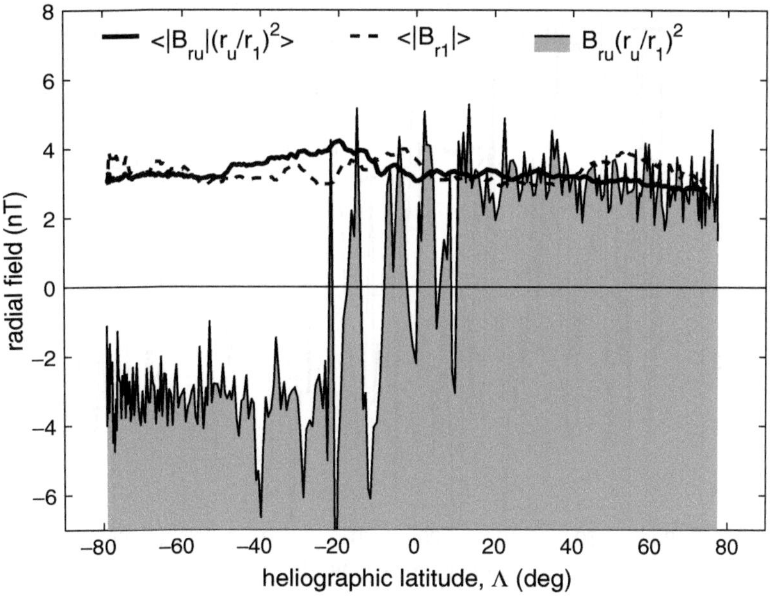

Fig. 38. Data from the first perihelion pass by the Ulysses spacecraft (the first "fast latitude scan" which took place between September 1994 and July 1995, near sunspot minimum), as a function of the craft's heliographic latitude. The *thin* line bounding the *grey* area shows daily means of the radial field observed by Ulysses, B_{ru}, range-corrected for the radial distance of Ulysses, $r = r_u$ to $r = r_1 = 1\,\mathrm{AU}$ using a $1/r^2$ dependence. The *thick black* line gives the 27-day means of the modulus of this value. The *dashed* line shows the corresponding means of the values seen simultaneously near Earth by the IMP-8 spacecraft

As discussed in the previous section, the PFSS method allows us to compute the open flux from surface magnetogram data. Despite the assumptions involved, and complications that can be introduced by magnetogram saturation effects and the lack of information over the solar poles, the results from the PFSS method and from near-Earth methods using (73) are rather similar [Wang and Sheeley Jr., 1995, 2004, Lockwood, 2003, Lockwood et al., 2004]. This has been true for almost 3 solar cycles now, which demonstrates that the Ulysses result that the field is independant of latitude is a general one.

Using (73) we can use all the data on the radial field B_r that has been obtained since the start of the space age to study the open solar flux. Intercalibration of the various magnetometer datasets is an issue here, particularly for the earliest data. However, by comparisons of all available data Couzens and King [1986] have arrived at the most reliable composite dataset of the early data, which has subsequently been continued (the "Omnitape" set). The results are shown in Fig. 40. It can be seen that the flux has been of order $5 \times 10^{14}\,\mathrm{Wb}$ and showed strong solar cycle variations (amplitude of order

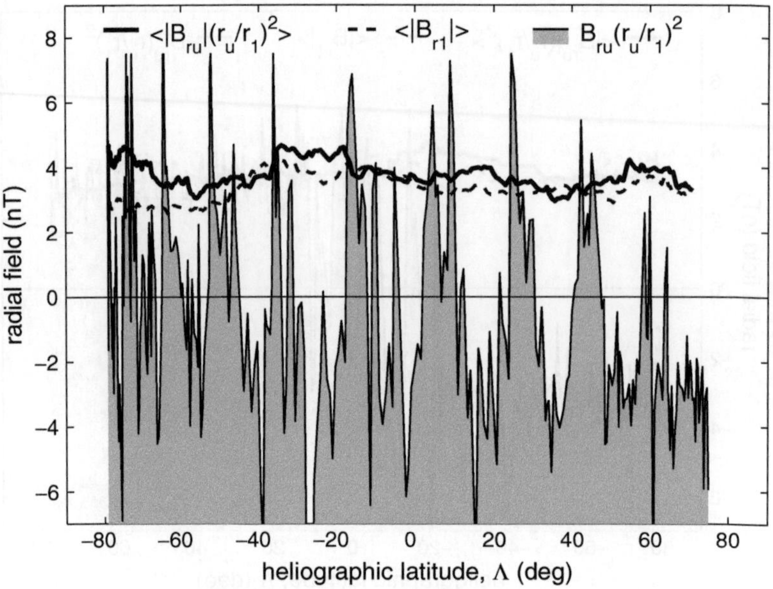

Fig. 39. Same as Fig. 38 for the second perihelion pass by the Ulysses spacecraft (the second fast latitude scan which took place between December 2000 and October 2001, near sunspot maximum). Simultaneous near-Earth data in this case comes from the ACE spacecraft

3.5×10^{14} Wb) during cycles 21 and 22, but a variation of smaller amplitude during cycles 23 and 20. In fact, cycle 20 showed very little variation at all (although calibration of these earliest data may be a factor here). The largest values are seen early in the declining phase of the sunspot cycle, just after the polar field has changed polarity.

3 The Heliosphere, Cosmic Rays and Cosmogenic Isotopes

3.1 Cosmic Rays

The previous sections have described how the Sun ejects a continuous, but highly variable stream of solar wind plasma into the heliosphere, which carries with it a magnetic field, the total flux of which is F_S which we can estimate from near-Earth measurements. The heliospheric field dominates the behaviour of the plasma out to the boundary where the heliosphere meets interstellar space, the *heliopause*. The field in the outer heliosphere, beyond the termination shock, is weaker than in the inner heliosphere where it follows a Parker spiral configuration, perturbed by the warped HCS and features such as CIRs and CMEs . The heliosphere acts as a shield for Earth because it

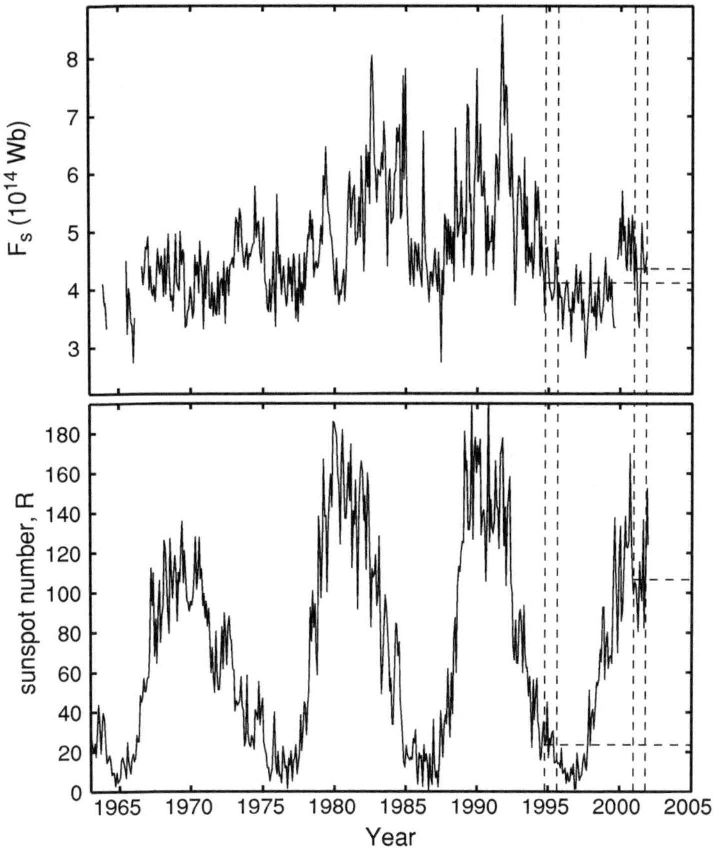

Fig. 40. (*Top*) The open solar flux, F_S, derived from near-Earth observations of the radial component of the IMF and using the Ulysses result. (*Bottom*) The sunspot number, R. The vertical *dashed* lines give the times of the two Ulysses fast latitude (perihelion) passes and the *dashed horizontal* lines in the upper panel are obtained by integrating the Ulysses data over all latitudes. The corresponding *horizontal dashed* lines in the lower panel are the mean values of R over the duration of the perihelion passes

reduces the flux of energetic particles called cosmic rays reaching the inner solar system. There are three classes of cosmic rays:

Galactic Cosmic Rays (GCRs). These are accelerated at the shock fronts of explosive galactic events such as supernovae. The flux of GCRs incident on our heliosphere is expected to vary on very long timescales as our Sun passes through the spiral arms of our galaxy. There is some evidence from meteorites for this long timescale variation (Shaviv, 2002). These particles mainly have energies between about one and a few tens of GeV, with a low flux in a high energy tail to the spectrum. The flux of GCRs has been continuously and systematically monitored now for over 50 years by a network of ground observatories which measure the neutrons or muons they produce in the atmosphere. Earth's magnetic field adds to the shielding caused by the heliosphere and limits the ranges of particle that can be seen at any one site. The key parameter is the *ridgidity* of the GCRs which is a measure of the tendency of the particle to keep moving in a straight line. Ridgidity is measured in GV, but because the particles move at close to the speed of light, their energy is close to the ridgidity value expressed in GeV. The geomagnetic field places a cut-off threshold on the particles that can be seen at any one site. Close to the geomagnetic equator this is close to 15 GV, whereas at mid latitudes particles of a few GV and above can be seen. At the polar latitudes the geomagnetic cut-off falls below a 1–2 GV cut-off set by the atmosphere. Secular changes in the geomagnetic field will cause changes in the cut-off rigidity of a given site to change [Bhattacharyya and Mitra, 1997].

Anomalous Cosmic Rays . These originate from neutral particles in the local interstellar wind that therefore drift across the heliopause before they are ionised within it. They are accelerated at the heliopause and/or termination shock.

Solar Cosmic Rays. These are generated at the shock fronts of explosive events on the Sun, for example the leading edge of CMEs. Because they are somewhat lower energies (up to several hundred MeV) and come from the Sun, these are now generally referred to as *solar energetic particles* (SEPs).

The energy and composition spectra of GCRs provide unique information on astrophysical processes, but interpretation is complicated by the effects of magnetic fields which influence the particle's trajectory, particularly within the heliosphere [e.g. Ginzburg, 1996]. At Earth, GCRs (and the secondary products generated when they hit the atmosphere) can deposit significant charge in small volumes of semiconductor to cause malfunctions in the avionics of spacecraft and aeroplanes [e.g. Dyer and Truscott, 1999]. In addition, the implications for human health of prolonged exposure to cosmic rays in high-altitude aircraft has been the focus of recent study [Shea and Smart, 2000]. GCRs also generate conductivity in the sub-ionospheric gap, allowing current to flow in the global electric thunderstorm circuit [e.g. Bering et al., 1998, Harrison, 2003] and it has been suggested in recent years that they influence the production of certain types of cloud with considerable implications for Earth's climate [Marsh and Svensmark, 2000b]. The spallation products of GCRs hitting atomic oxygen, nitrogen and argon in Earth's atmosphere (cosmogenic isotopes, stored in reservoirs such as tree trunks and ice sheets) are often used as indicators of solar variability in paleoclimate studies [e.g. Bond et al., 2001, Neff et al., 2001], although the implied links between total solar irradiance variations and cosmic ray shielding by the heliosphere are not yet understood [Lockwood, 2002a,b]. In all these studies, understanding how the heliosphere influences GCR fluxes and spectra, of both hadrons and electrons [Heber et al., 1999], is of key importance.

3.2 Cosmic Ray Modulation by the Heliosphere

The modulation of GCRs is described by Parker's transport equation [Parker, 1965, Potgieter, 1998] an expression giving the GCR phase space density, $f(\boldsymbol{r}, \boldsymbol{v}, t)$, where $\boldsymbol{r}$ is heliocentric position vector, $\boldsymbol{v}$ is the GCRs' velocity and t is time

$$\frac{\partial f}{\partial t} = \frac{\partial}{\partial x_i}\left[\kappa_{ij}^S \frac{\partial}{\partial x_j}\right] - \boldsymbol{V}_{SW} \cdot \nabla f - \boldsymbol{V}_d \cdot \nabla f + \frac{1}{3}\nabla \cdot \boldsymbol{V}_{SW}\left[\frac{\partial f}{\partial \ln p}\right] + Q \quad (74)$$

Phase space density $f(\boldsymbol{r}, \boldsymbol{v}, t)$ is also called the particle distribution function and is the number of particles per unit volume of ordinary space that also fall into unit volume of velocity space. Phase space is thus a 6-dimensional space with three spatial dimensions (x, y, and z, where $r^2 = x^2 + y^2 + z^2$) and the three corresponding velocity dimensions (v_x, v_y and v_z). If N is the number of particles in a 3-dimensional volume $\mathrm{d}^3\boldsymbol{r} = \mathrm{d}x\mathrm{d}y\mathrm{d}z$ and also within a velocity space volume $\mathrm{d}^3\boldsymbol{v} = \mathrm{d}v_x\mathrm{d}v_y\mathrm{d}v_z$ (i.e. with x between x and $(x + \mathrm{d}x)$, v_x between v_x and $(v_x + \mathrm{d}v_x)$, ...etc.), the phase space density is $f(\boldsymbol{r},\ \boldsymbol{v},\ t) = N/(\mathrm{d}^3r\mathrm{d}^3v)$. Phase space density therefore has units of $\mathrm{m}^{-6}\,\mathrm{s}^3$. For azimuthal symmetry, as generally imposed on charged particles by the presence of a magnetic field, $E = m\boldsymbol{v}^2/2$ yields $\boldsymbol{v}\mathrm{d}^3\boldsymbol{v} = (2/m^2)E\mathrm{d}E\mathrm{d}\Omega$ where E is energy and Ω is solid angle. The total flux is the number of particles passing through unit area (normal to unit vector $\boldsymbol{n}$) in unit time

$$F(\boldsymbol{r}, t) = \int_V f(\boldsymbol{r}, \boldsymbol{v}, t)\boldsymbol{n} \cdot \boldsymbol{v}\mathrm{d}^3\boldsymbol{v} = \int_E \int_\Omega f(\boldsymbol{r}, E, t)\left(\frac{2}{m^2}\right) E\mathrm{d}E\mathrm{d}\Omega \quad (75)$$

F is measured in $\mathrm{m}^{-2}\,\mathrm{s}^{-1}$. We define the *differential number flux*, $\mathrm{j}(\boldsymbol{r},\ \mathrm{t})$ (also called the particle *intensity*) to be the total flux per unit energy and per unit solid angle and so from (75)

$$j(\boldsymbol{r}, t) = \frac{\mathrm{d}^2F(\boldsymbol{r}, t)}{\mathrm{d}E\mathrm{d}\Omega} = f(\boldsymbol{r}, E, t)\left(\frac{2}{m^2}\right) E \quad (76)$$

GCR differential number flux is usually measured in $\mathrm{m}^{-2}\,\mathrm{s}^{-1}\,\mathrm{sr}^{-1}\,\mathrm{GeV}^{-1}$.

The terms on the right-hand side of (74) allow for (in order): diffusion (due to scattering from irregularities), convection (due to bulk solar wind flow), particle drifts – gradient and curvature drifts due to changes of the heliospheric field within a particle gyroradius [Jokipii et al., 1977, Jokipi, 1991], adiabatic cooling (or heating) and any local source Q (for example the addition of anomalous cosmic rays). The factor κ_{ij}^s is the symmetric diffusion coefficient and $\boldsymbol{V}_{SW}$ is the outward solar wind velocity. The theory of cosmic ray transport in the heliosphere [Fisk, 1999, Moraal et al., 1999] is mature and work in recent years has been mainly to evaluate the magnitude, spatial and energy dependence of the different terms in the Parker equation. Our present understanding has been obtained through theoretical estimations of the different modelled parameters and their comparison to the limited data available. Figure 41 shows a typical GCR spectra (differential number flux, as given by (76), as a function of energy) simulated using (74), and as obtained from balloon flights above Earth's atmosphere. The figure demonstrates that the energy-dependent shielding effect [Dorman et al., 1997] of the heliosphere, as predicted using the Parker equation, is greater at lower energies. A good fit to observations can be obtained at all phases of the solar cycle [e.g. Bonino et al., 2001] but there are key free parameters. In particular, the initial interstellar GCR spectrum (dashed line) is assumed and is not independently measured.

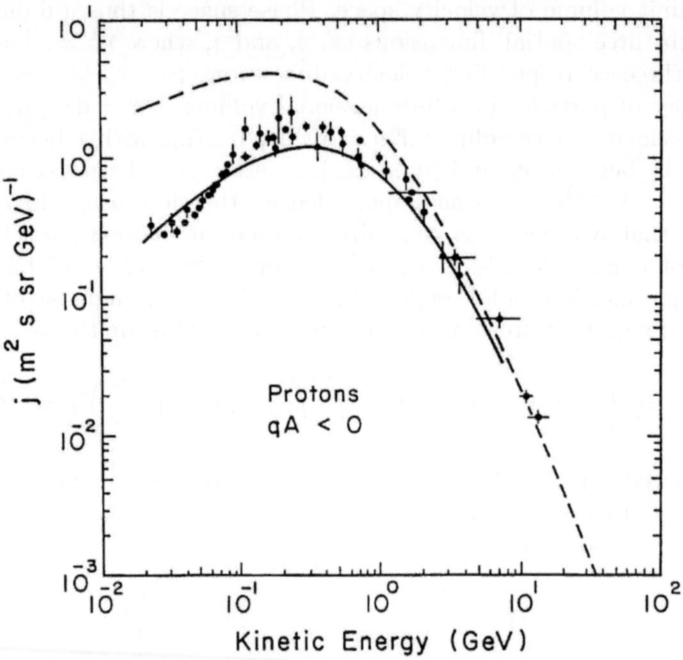

Fig. 41. An example of a proton GCR spectrum (differential number flux j as a function of energy) near Earth. The *solid* line is a prediction using Parker's transport equation at a given phase of the solar cycle and with negative polarity ($A < 0$) polar solar field. This assumes the interstellar spectrum of GCRs outside the heliosphere shown by the *dashed* line. The points are the corresponding observed spectrum from ionisation chambers carried on balloon flights

More information is available from satellites at various distances from Earth in the ecliptic plane and recently the Ulysses spacecraft has monitored the spectra out of the ecliptic. As a result of these satellite observations, the modulation effects of outward convection and adiabatic energy losses by the solar wind speed are relatively well understood [Goldstein, 1994]. Similarly, drift effects in the smooth background field are also well understood (even near discontinuities such as the heliospheric current sheet, the termination shock and the heliopause). The problems arise mainly from the great uncertainties remaining about the effect of irregularities in the magnetic field and our lack of understanding of the scattering of charged particles that they cause parallel and perpendicular to the heliospheric field [Moraal et al., 1999]. To help this analysis we only have theoretical estimates of the diffusion coefficient κ_{ij}^{s}, derived from first principles by looking at charged particle scattering caused by complex space-time dependent plasma turbulences [Parhi et al., 2001]. The effect of irregularities produces a good correlation between the charged GCR propagation and heliospheric magnetic field variations, which has recently been observed [Droge, 2003].

Because the amplitude of magnetic irregularities in the heliosphere increases with the magnitude of the field, we would expect a strong anti-correlation between heliospheric field strength and GCR fluxes, if the diffusion term dominates. Initial studies of the cosmic ray flux and the interplanetary magnetic field did not find a strong relationship between the two [e.g. Hedgecock, 1975]. This may have been because of poor calibration of the dataset, or because the first cycle observed (sunspot cycle 20) was a very unusual one, following the strongest solar cycle since reliable sunspot data began (which is also inferred to be the strongest solar cycle in the last 1.3 millenia [Usoskin et al., 2003c, 2004]). Subsequently, cycles 21, 22 and 23 have shown a very strong and highly significant anticorrelation between GCR fluxes and the magnitude of the local heliospheric field at Earth (the interplanetary magnetic field or IMF) [Cane et al., 1999, Belov, 2000, Lockwood, 2003].

These strong anti-correlations between GCR fluxes and the heliospheric field have recently led researchers to investigate the effect of solar modulation of GCRs using much simpler concepts than the full Parker equation. For example Wibberenz and Cane [2000] and Wibberenz et al. [2002], assumed the radial diffusion coefficient scales as some power of the magnitude of the IMF [Burlaga, 1987] and also assumed the presence of continuous recovery processes (related to particle entry into depleted regions of the heliosphere by drift and diffusion processes). From this, these authors have developed a simple model which reproduces the cosmic ray intensity variations very well in the last four solar cycles. This model assumes steady-state and spherical symmetry and the cosmic ray intensity is perturbed by increases in the IMF that propagate away from the Sun and cause a reduction in the GCR radial diffusion coefficient. The assumed inverse coupling of the magnetic field with cosmic ray spatial diffusion coefficients leads to the concept of propagating diffusive barriers first introduced by Perko and Fisk [1983]. The decrease in GCR fluxes associated with these barriers is followed by a recovery process determined by both diffusion mechanisms and the large-scale influence of drifts. Longer recovery times are therefore expected for periods of solar field polarity $A < 0$ when particle inflows are along the heliospheric current sheet than for $A > 0$ where inflows are expected from over the poles. The recent work by Ferreira et al. [2003] to include the interplay between these diffusive barriers and large scale drifts in a full time dependent model has shown very promising results concerning the charge sign-dependent modulation effects predicted by drift theory.

Because the open solar flux quantifies the total field in the heliosphere, good anticorrelation is also expected between it and the GCR fluxes, as shown by Fig. 42. This plot shows the time-variation of the total open solar magnetic flux estimate, $[F_S]_{aa}$, derived from the aa geomagnetic index using the procedure of Lockwood et al. [1999a,b]. These open flux estimates agree very closely with those from near-Earth IMF observations (as shown in Fig. 40), after the latter commence in 1965 and will be discussed in more detail later. The black

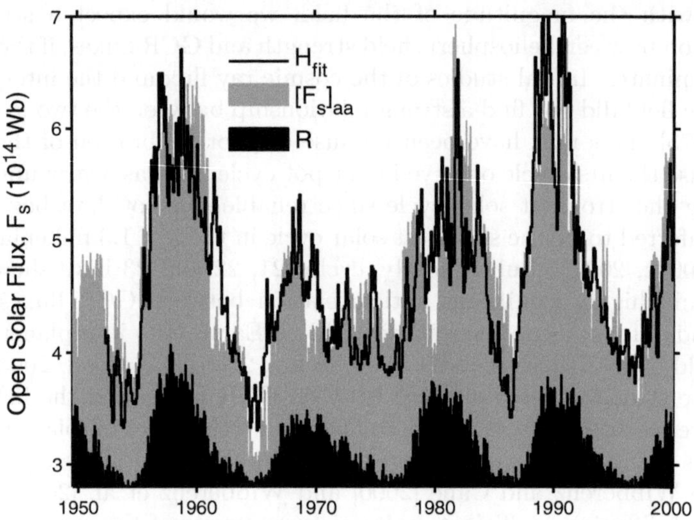

Fig. 42. Variations of monthly means of solar, heliospheric and cosmic ray data since 1950. The *grey* histogram gives the open solar magnetic flux $[F_S]_{aa}$, deduced from the aa geomagnetic index using the method of Lockwood et al. [1999a,b]. The *black* line is H_{FIT}, the best fit of the anti-correlated cosmic ray counts H, observed by the equatorial Huancayo and Hawaii neutron monitors (which form a homogeneous data sequence of positively-charged hadron GCRs with rigidities exceeding 13 GV). The *black* histogram gives the sunspot number, R, for comparison. The shield presented by the open solar flux is strongest (peak $[F_S]_{aa}$) shortly after each sunspot maximum, giving minima in H and peaks in H_{FIT}. The correlation coefficient between H and $[F_S]_{aa}$ for the full interval of coincident data (1953–2001) is $c = -0.87$, which means that $c^2 = 0.75$ of the variation in the cosmic ray flux is explained (in a statistical sense) by the open solar flux. Allowing for the persistence in both the H and the $[F_S]_{aa}$ data series, the significance of this correlation, S, exceeds 99.999%, i.e. there is less than a 0.001% probability that this result was obtained by chance [after Lockwood, 2003]

line shows H_{FIT}, the best linear regression fit to $[F_S]_{aa}$ of the GCR count rate H, as observed by neutron monitors at Hawaii and Huancayo. These two stations provide a long, continuous and homogeneous data series on GCRs of rigidity exceeding 13 GV. As well as matching the short-period variations in H_{FIT}, the alternately rounded and the V-shaped minima in $[F_S]_{aa}$ match those in H_{FIT}. This has often been cited as evidence for polarity-dependent drifts of GCRs at sunspot minimum; however, Fig. 42 suggests an alternative explanation, in that this appears to be a feature of open flux emergence.

Figure 43 stresses the role of individual diffusive barriers in shielding Earth from GCRs. These data were taken around the time of a major geomagnetic storm on 14 July 2000 (as this was the anniversary of the storming of the prison in Paris, it was termed the "Bastille day storm"). During this event, a CME passed close by the Earth, sufficient to cause shielding of GCRs in what is termed a *Forbush decrease* [Cane, 2000]. The second CME hit the Earth and its arrival was heralded by energetic protons accelerated at the shock front on the leading edge of the CME. This lead to the *ground level enhancement* (GLE) where large fluxes of solar protons reach ground level, most readily in the polar caps. At sunspot maximum GCR shielding is thought to be due to the combination of many such diffusive barriers (not only from CMEs but also from co-rotating interaction regions, CIRs) which merge together in the outer heliosphere into a *global merged interaction region* that provides an effective shield for GCRs [McDonald et al., 1993].

The location where the bulk of the GCR shielding takes place will depend on their energy. The correlations between GCR fluxes and the open solar flux, such as that shown in Fig. 42, are strongest at short lags. This is demonstrated by Fig. 44, from the work of Rouillard and Lockwood [2004]. This figure

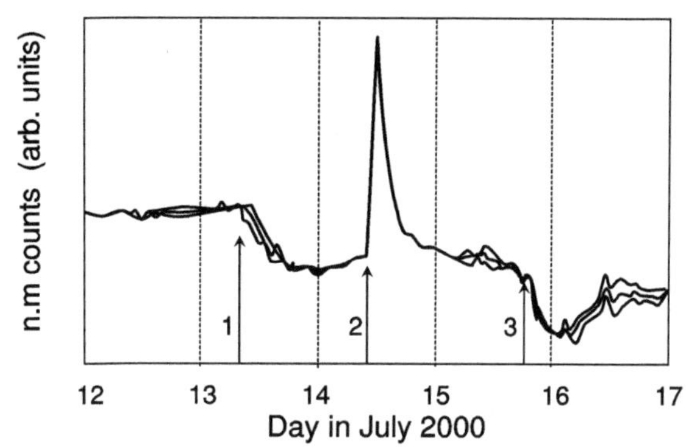

Fig. 43. Cosmic rays observed by high-latitude neutron monitors (at Thule in the north and McMurdo in the south) during the "Bastille Day" storm of 14 July 2000. The vertical arrows 1 and 3 show the onset of Forbush decreases caused by the passage nearby of large CME events. The arrow 2 marks the start of a "ground-level enhancement" of solar protons (i.e. an SPE) generated at the shock on the leading edge of the second CME which impinged on the Earth

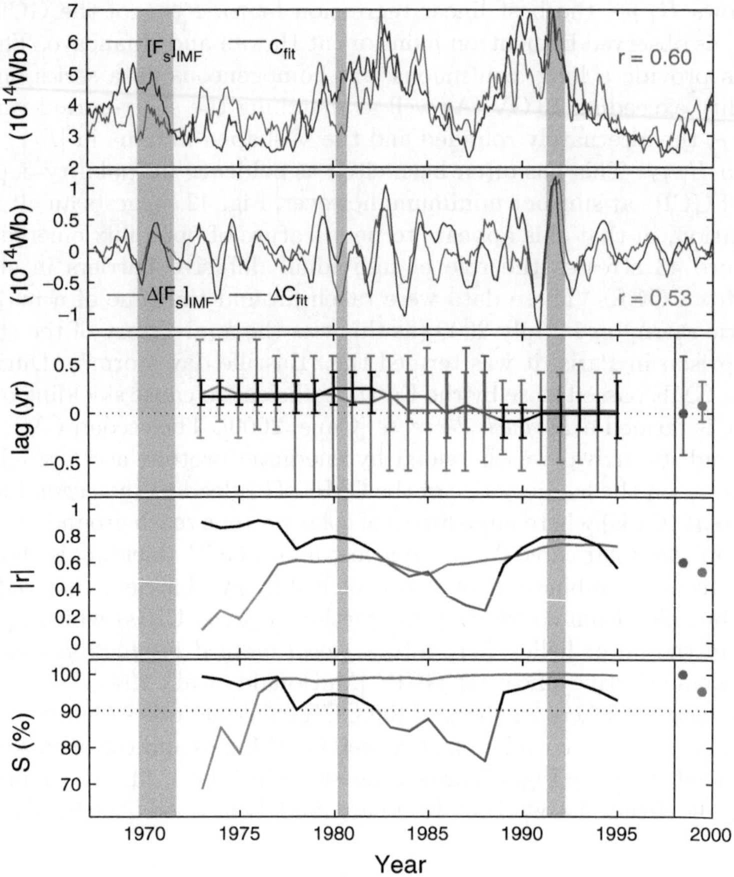

Fig. 44. Analysis of the correlation between GCR fluxes, C, observed at Climax (rigidity $> 3\,\mathrm{GV}$) and the open solar flux $[F_S]_{IMF}$ derived from the radial component of the IMF observed by near-Earth spacecraft. The *top* panel shows $[F_S]_{IMF}$ (*in red*) and the best linear regression fit of C to $[F_S]_{IMF}$ (C_{FIT}, *in blue*). The correlation coefficient is 0.60 and is significant at greater than the 99.99% level. The second panel shows $\Delta[F_S]_{IMF}$ and ΔC_{FIT} obtained by passing $[F_S]_{IMF}$ and C_{FIT} through a high-pass filter which supresses variations of period longer than 5 years and reveals the second largest periodicity in the spectra, a variation with period 1.68 years. The correlation coefficient is 0.53 and is significant at greater than the 95% level. The third, fourth and fifth panels show the results of correlations on 11-year sliding windows of the data giving, respectively, the best-fit lag δt, the correlation coefficient C and the significance S. In these three panels, *red* and *black* data points and curves are for the unfiltered data $[F_S]_{IMF}$ and C_{FIT} (*black* is used where $S > 90\%$ and *red* where $S \leq 90\%$), *blue* and *green* are for the filtered data $\Delta[F_S]_{IMF}$ and ΔC_{FIT} (*green* is used where $S > 90\%$ and *blue* where $S \leq 90\%$). The mauve and *grey* points to the right of the lower three panels are the results for the full data sequences of, respectively, unfiltered and filtered data [from Rouillard and Lockwood, 2004]

analyses the relationship between the open solar flux derived from near-Earth measurements of the heliospheric field (using (73) and as shown in Fig. 40) and cosmic rays of rigidity exceeding 3 GV observed by the Climax neutron monitor. After the solar cycle period of about 11 years, the second strongest peak in the power spectrum is a persistent variation at 1.68 years which is revealed by passing the data through a high-pass filter. This frequency has been noted before in GCR data and connected with several features on the Sun showing the same periodicity [Valdés-Galicia et al., 1996, Valdés-Galicia and Mendoza, 1998, Wang and Sheeley Jr., 2003b]. It is clearest at sunspot maximum where it is probably related to the *Gnevyshev gap* [Gnevyshev, 1967, 1977, Wang and Sheeley Jr., 2003b] in solar activity. Both filtered and unfiltered data are analysed in Fig. 44, using 11-year sliding windows which reveal that the anticorrelation is usually significant for both the 11-year and the 1.68-year variations. The correlation can be seen to fail for the unfiltered data for solar cycle 20, but not for the filtered data. This suggests that the early IMF data may indeed have suffered from calibration drifts (which would influence the unfiltered data but be sufficiently slow to be removed by the high-pass filter) and that the anticorrelation was really present throughout the interval, as found for the open flux derived from the aa geomagnetic index (Fig. 42). The feature to note is that the peak correlations are all found for relatively short lags δt ($\sim$ 1 month). For a fast solar wind flow speed of $V_{SW} = 700\,\mathrm{km\,s^{-1}}$, this places the bulk of the shielding at distances of $(V_{SW}\delta t) \approx 12\,\mathrm{AU}$, which is considerably closer than where MIRs are thought to form.

Lockwood [2001a] has used annual means to demonstrate that the anticorrelation between GCRs and the open solar flux holds for the earliest GCR data taken using ionisation chambers (see Fig. 45). The Fredricksberg and Yakutsk detectors are well intercalibrated and the dashed line shows the scaled variation $[F_S]_F$, from the best-fit linear regression of $[F_s]_{aa}$ with Forbush's original data, as presented by McCracken and McDonald [2001]. These data were taken by a network of 5 "Carnegie Type C" ionisation chambers established in 1936–7 which were monitored closely and corrected for sensitivity changes [Forbush, 1958]. McCracken and McDonald point out that Forbush's data show a downward drift in average cosmic ray fluxes between 1936 and 1958, consistent with the downward drift in ^{10}Be isotope abundances at this time (see Sect. 5.1). This drift is sometimes suppressed by re-calibrations of the data which implicitly, or explicitly, assume that it is instrumental in origin [Ahluwalia, 1997]: it appears as an upward drift in $[F_S]_F$ in Fig. 45, is consistent with the open flux variation deduced from aa. In addition, Neher made observations from high altitude ionisation chambers from 1933 to 1965. The intercalibration of the instruments was quoted as being better than 1% [Neher et al., 1953]. These data, scaled using a linear regression to give $[F_S]_N$, are shown by the stars in Fig. 45. Both of these early cosmic ray data sets are, like the later observations from both neutron

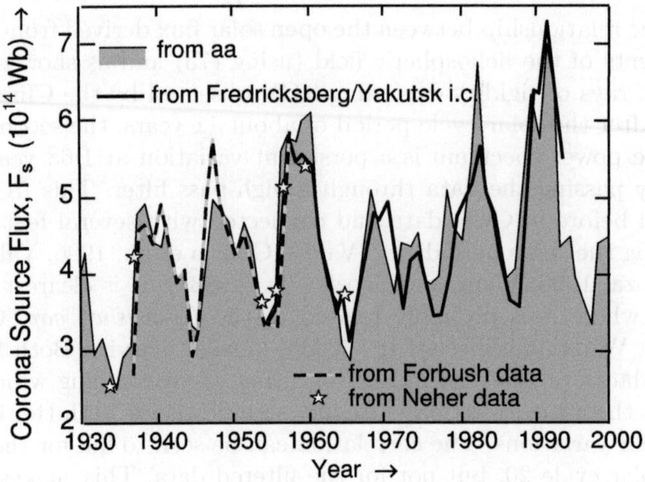

Fig. 45. Annual means of the coronal source flux derived from the aa index $[F_S]_{aa}$ (*grey* area bounded by *thin black* line) and best-fit linear regression of the cosmic ray counts from various ionisation chambers. The *thick black* line is the variation deduced from the data from the Fredricksberg and Yakutsk instruments, $[F_S]_{ic}$. The *dashed* line shows the variation scaled from Forbush's original data, $[F_S]_F$, and the stars are from Neher's data, $[F_S]_N$

monitors and ionisation chambers, entirely consistent with the open solar flux variation deduced from the aa index. The early data indicate a fall in cosmic ray fluxes during the 20th century towards present-day levels, consistent with a rise in the open solar flux.

3.3 Cosmogenic Isotopes

The observations of cosmic ray fluxes shown in Fig. 45 are the oldest "as-it-happened" observations of cosmic rays available. To extend the data sequence further back in time requires us to look at some products of GCR precipitation that have been stored in terrestrial reservoirs. In particular, the ^{14}C and ^{10}Be isotopes are produced as spallation products when GCRs interact with O, N & Ar in Earth's atmosphere. Figure 46 illustrates the processes by which these isotopes are deposited in their respective reservoirs. A key point is that the deposition of these two isotopes is radically different [O'Brien, 1979, Stuiver and Quay, 1980, Bard et al., 1997, Beer et al., 1990, Beer, 2000]. In both cases about 2/3 of the production is in the stratosphere, 1/3 in the troposphere. The ^{10}Be takes about 1 week to be deposited in an ice sheet from the troposphere, but of order a year from the stratosphere: it becomes attached to aerosols before precipitating into ice sheet. The upper layers of the ice sheet can be dated by counting layers of enhanced abundance of photosensitive molecules (produced much more rapidly in summer) and from

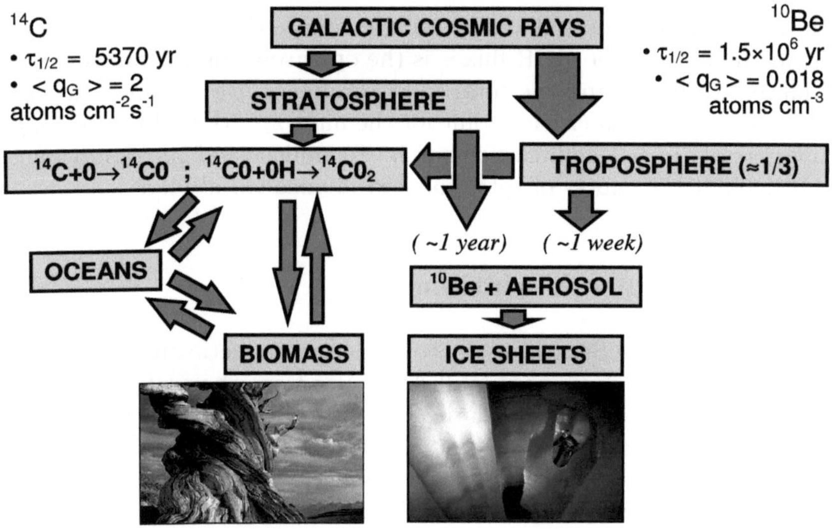

Fig. 46. Schematic illustration of the deposition of cosmogenic isotopes in terrestrial reservoirs

identifiable volcanic dust layers from known eruptions. However, for deeper layers (further back in time), dating requires modelling of the flow of the ice sheet. The ^{10}Be abundance can also be monitored using cores taken from ocean sediments. Although differences due to local climate changes can be found in ^{10}Be records from different sites, the agreement is generally very good, especially in the long-term trends [McCracken, 2004].

The ^{14}C isotope, on the other hand, is exchanged as part of the carbon cycle with two major reservoirs, the oceans and the biomass. Understanding the abundance in, for example ancient trees like bristlecone pines, in terms of the production rate requires modelling to allow for the exchange and time constants of these reservoirs. The time constants smooth out the solar cycle variation in ^{14}C but century-scale variations can be seen and compared to those in ^{10}Be. The one common denominator between the deposition chains for ^{14}C and ^{10}Be is the production by the flux of incident GCRs. Thus when a phenomenon correlates well with the inferred production rate of both these isotopes we can be sure that it is the production, and not the deposition, that is causing the variations – i.e. the phenomenon is correlating with the incident flux of GCRs.

In paleoclimate studies, it is often assumed that the cosmogenic isotope abundances are an index of solar variability, in the sense of the total solar irradiance variability discussed in the next section (see for example, Bond et al., 2001). This may be valid but it would rely on a connection that we do not yet understand and cannot, as yet, verify. Strictly speaking, cosmogenic isotopes tell us about the flux of cosmic rays bombarding the Earth. The previous section has laid out the growing evidence that the dominant factor in the modulation of the GCR fluxes is the open solar flux. However, if there is a link between this and the total solar irradiance, we certainly do not yet understand it. This is a crucial point for the interpretation of the cosmogenic isotope records in paleoclimate research. To understand its significance we need to look at the causes of total solar irradiance variability.

4 Solar Irradiance Variations

The luminosity of the Sun, L, is the total electromagnetic energy, integrated over all wavelengths, emerging in all directions from the surface of the Sun. Because we have never observed the Sun from over its poles, we have no observations of L, but theory suggests a value near 3.845×10^{26} W. The total solar irradiance I_{TS} is the total power received (again, integrated over all wavelengths) per unit normal area in the ecliptic plane at $r = R_1$ from the Sun, where R_1 is the mean Earth–Sun distance, 1 AU $= 1.496 \times 10^{11}$ m. If the Sun were to radiate isotropically, I_{TS} would be $L/(4\pi R_1^2)$ which for the above luminosity gives a value of $1367.2\,\mathrm{W\,m^{-2}}$. Given this is close to the observed values of I_{TS} (see Fig. 47), the above estimate of L suggests the Sun is indeed close to being an isotropic radiator. The intensity of a point on the Sun I is the power radiated by unit area of the solar surface into unit solid angle. If the Sun is featureless, I everywhere equals $\langle I \rangle_D$, the disc-averaged intensity, which in turn is related to the total solar irradiance by

$$I_{TS} = \langle I \rangle_D \left(\frac{\pi R_S^2}{R_1^2} \right) \tag{77}$$

giving $\langle I \rangle_D = 2.011 \times 10^7\,\mathrm{W\,m^{-2}\,sr^{-1}}$ for the above value for I_{TS} of $1367.2\,\mathrm{W\,m^{-2}}$.

Because of the variability and uncertainty of atmospheric absorption, accurate measurements of the total solar irradiance (hereafter referred to as TSI) require space-based observations and absolute radiometry from space is very demanding. To allow for instrument degradation caused by exposure, self-calibrating instruments usually have two identical channels, one used all the time and the other only rarely. Nevertheless, different instruments give different absolute values for TSI and degrade at different rates. Figure 4.1 shows the PMOD composite derived from a variety of instruments with best allowance for their degradation and inter-calibration [Fröhlich and Lean, 1998a,b, Fröhlich, 2000, 2003]. Others, for example the ACRIM composite [Willson, 1997, Willson and Mordvinov, 2003], show similar features but differ in subtle, yet important, ways.

The most apparent feature in Fig. 47 is the solar cycle variation. The large downward spikes in TSI that are common at sunspot maximum are caused by the passage of individual sunspots or sunspot groups across the visible

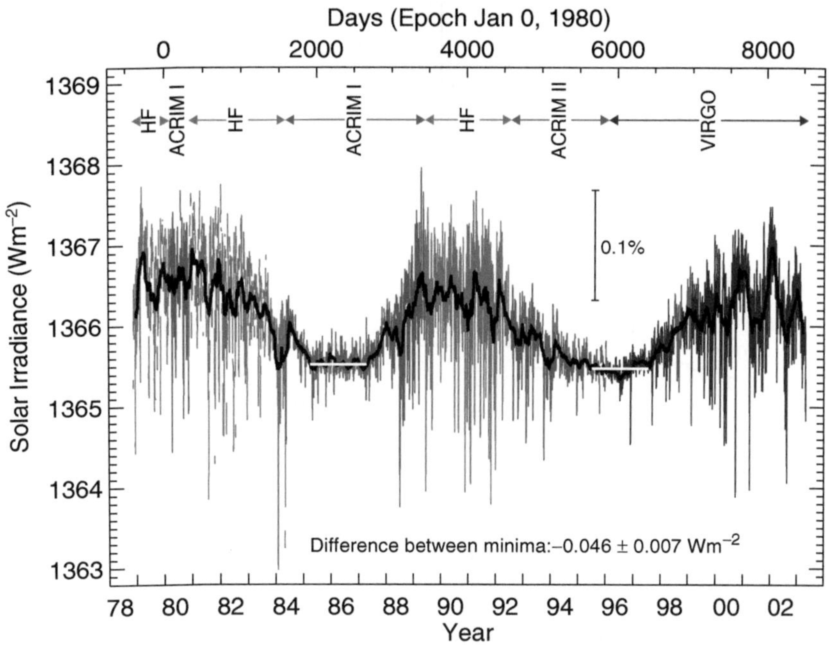

Fig. 47. Composite of several datasets from different spacecraft showing the total solar irradiance variation since 1979. Data are from the HF instrument on Nimbus 7, the ACRIM-1 radiometer on SMM, ACRIM-2 on UARS and VIRGO on SoHO. Daily values are shown in *red*, *blue* and *green* (for HF, ACRIM 1–2, and VIRGO, respectively), monthly means in black [from Fröhlich, 2003]

disc. However at sunspot maximum, TSI is enhanced despite the presence of more dark spots, the reason being the increase in small, bright faculae.

The effect of individual sunspot or facular groups is most readily seen at sunspot minimum when it is possible to have just one such feature on the visible disc at any one time, as for the data shown in Figs. 48 and 49 which were both recorded during 1996. Figure 48 shows the effect of a sunspot group passing over the visible disc. The three panels show the surface area covered by sunspots and faculae (A_S and A_F, respectively) and the percentage change in TSI relative to the value in the absence of the isolated feature. The top panel shows the group grew as it approached the central meridian (on day 330), and the bottom panel shows that the TSI decayed in response. Although the group subsequently decayed somewhat as it moved across the visible disc, there was a significant drop in A_S when it rotated through the eastern limb on day 336. The rise in TSI was more gradual because it depends on the area of the spots on the disc (the filling factor) which is lower for a constant surface area of spots if they are closer to the limb. The middle panel shows that facular occurrence was sporadic at this time.

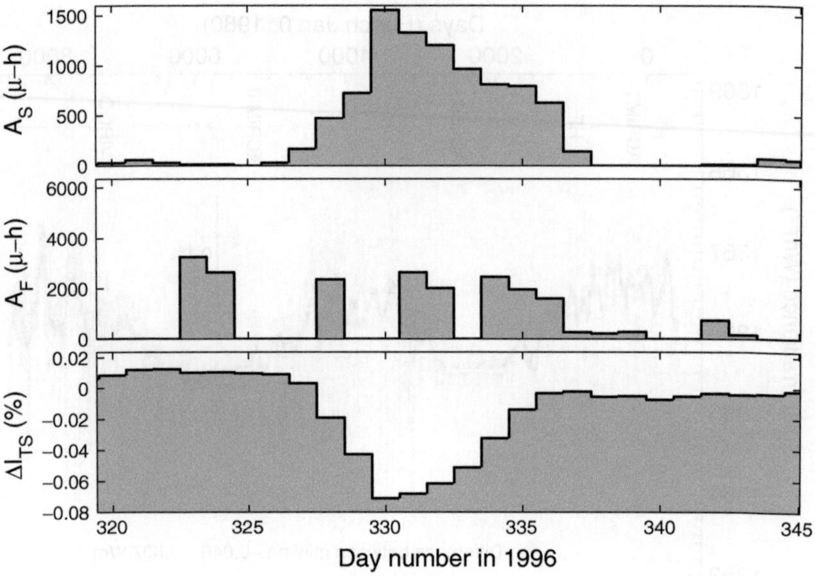

Fig. 48. The effect of an isolated sunspot group observed near solar minimum in daily means. From *top* to *bottom* the panels show: the disc-integrated surface area of sunspot groups, A_S (in millionths of a solar hemisphere); the disc-integrated surface area of faculae, A_F (also in millionths of a solar hemisphere); and the percentage change in TSI relative to that in the absence of the spot group

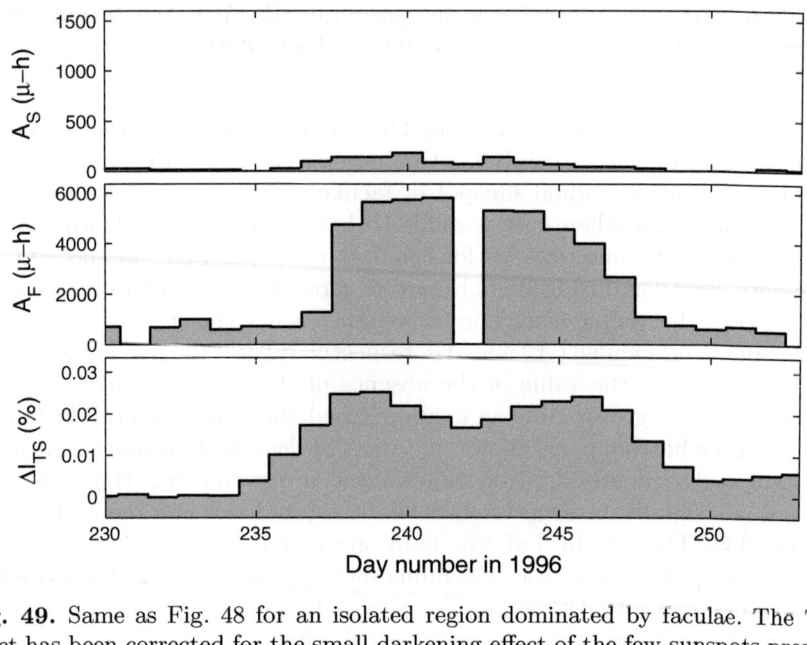

Fig. 49. Same as Fig. 48 for an isolated region dominated by faculae. The TSI effect has been corrected for the small darkening effect of the few sunspots present

Figure 49 is an equivalent plot for the passage of an area dominated by faculae which passed through the central meridian on day 242. In this region only a few small sunspots were present, as shown by the top panel. The bottom panel shows the change in TSI due to the faculae, which has been corrected for the small darkening effect of the sunspots present using the *photometric sunspot index*, PSI (see Sect. 4.12). The bottom panel shows that faculae have more effect on TSI when they are nearer the limb than when near the disc centre. An important point to notice is the difference between the facular area A_F and the TSI perturbation. As the region passed through the central meridian on day 242, A_F fell to almost zero, although the faculae were still having considerable effect on the TSI. This shows that the faculae were brighter than the quiet photosphere at the disc centre, but their contrast was not great enough for them to be classed as faculae. This threshold effect is also seen in the sharp rise in A_F as the region rotated away from the west limb of the Sun, even though the rise in TSI was more gradual. Thus care must be taken when using facular observations to deduce TSI behaviour as considerable brightening can be present even if features with sufficient contrast to be called faculae are absent.

In addition to these changes in the TSI, there are changes in the spectrum of received radiation. This is shown in Fig. 50 [adapted from Lean, 1991]. The upper panel shows a typical solar spectrum, along with that for a 5770 K blackbody radiator and the spectrum that penetrates Earth's atmosphere to the surface. The bottom panel shows the variability of the spectral irradiance, defined as the difference between the solar minimum and maximum values, as a ratio of the solar minimum value. It can be seen that variability is greater at the shorter wavelengths, with variations well above average in the EUV and UV. However, the upper panel shows that the power in this part of the spectrum is lower. Near the peak of the spectrum, at visible wavelengths, variability is close to the average value for all wavelengths (about 0.1%). In the near IR the variability is lower than the average.

From (77), variations in the Sun's total solar irradiance I_{TS} at constant R_1 arise through changes in the disc-average surface intensity $\langle I \rangle_D$ and/or the radius, R_S. On decadal timescales and less, changes are caused by the magnetic fields in the photospheric surface and in the underlying convection zone. In addition, we expect variations on much longer timescales of 10^6–10^8 years as a consequence of stellar evolution and the burning of hydrogen in the solar core [Schröder et al., 2001]. Our knowledge of the solar cycle variations comes from the observations made by high-resolution radiometers in space over the past 25 years, as shown in Fig. 47. Our understanding of the secular change, on the other hand, comes from surveys of astronomical data on other stars. For Earth's climate, intermediate variations on timescales of 10–10^3 years are of particular importance. Because these timescales are considerably shorter than the time constant for energy transfer from the Sun's core to the surface or for any warming or cooling of the convection zone, the relevant

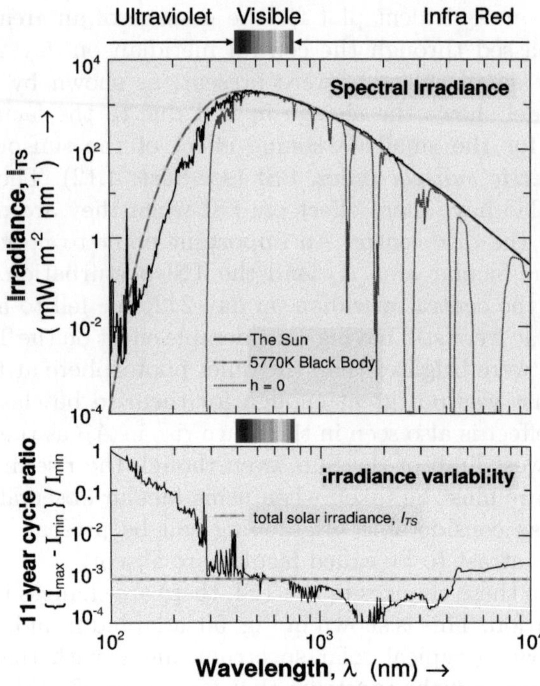

Fig. 50. (**a**) The spectrum of total solar irradiance, compared with that of a 5770 K black body radiator. The *blue dotted* line shows the spectrum of radiation reaching the surface of the Earth. (**b**) The spectral variability of the irradiance defined as the fractional difference between the solar maximum and minimum values. The *horizontal dashed* line gives the corresponding value for the total solar irradiance that is the integral over all wavelengths [after Lean, 1991]

changes are most likely to be magnetic in origin, as they are over the solar cycle. The variations in luminosity and radius may be caused by magnetic effects taking place either within the convection zone or in the photospheric surface.

4.1 Surface Effects

Magnetic fields threading the photosphere influence the solar output by modulating the emissivity of the surface. The larger of the photospheric flux tubes (above a radius threshold of about 250 km) cause sunspots to appear on the solar surface. The blocking of upward heat flux by the magnetic field in sunspots was originally suggested by Bierman [1941] and the mathematical treatment supplied by Spruit [1981, 1991, 2000] is discussed in the following sections. This blocking reduces the surface temperature from the normal value of near 5770 K to near 4000 K. Thus spots radiate less than the surrounding photosphere and appear dark. On the other hand, smaller-scale photospheric magnetic flux tubes (radius below about 250 km) cause bright faculae on the solar surface. The most widely-cited theory of these faculae invokes magnetic flux tubes threading the surface, the main difference between them and spots being that their smaller radii allows radiation from the surrounding walls to maintain the temperature near the ambient 5770 K. The enhanced magnetic pressure in the tube ($B^2/2\mu_o$) requires a lower particle pressure (Nk_BT) in equilibrium and because T is constant, the particle concentration N must be reduced. This increases the optical depth, allowing the observer to see deeper into the Sun, where the temperature is higher. As a result, faculae have a reverse effect to sunspots, being brighter than the surrounding photosphere, and giving an excess emission. The walls of the facular flux tubes are most visible, especially nearer the solar limb. This effect is often referred to the as the "*bright wall effect*" and explains why faculae are brighter closer to the limb [Spruit, 1976, Deinzer et al., 1984a,b, Knölker et al., 1988, Steiner et al., 1996]. However, there are other theories of faculae, for example, the "*hillock*" model is claimed to have several advantages in explaining the brightening very close to the limb [Schatten et al., 1986]. Individual faculae have a much smaller effect than that of individual spots, but there are many more of them such that their combined brightening effect on average exceeds the darkening effect of spots by factor of about 2.

4.2 Subsurface Effects

These can be split into two separate phenomena which affect the convection zone:

1. *Shadows* (The *"alpha effect"*). Magnetic fields in the convection zone can interfere with convection, causing a reduction in the efficiency of heat transport towards the surface.
2. *Sources and sinks* (The *"beta effect"*). The creation of a magnetic field involves the conversion of energy of motion into magnetic energy. Since the motions in the solar envelope are thermally driven, this ultimately means conversion of thermal energy into magnetic energy. Where field decays the opposite will happen, and magnetic energy will be converted back into heat.

4.3 Timescales

The analysis pioneered by Spruit [1976, 1981, 1991, 2000] shows how thermal disturbances in the convection zone evolve on two different timescales. The longest of these timescales is the thermal timescale (τ_T) of the convection zone as a whole (which is also called the Kelvin–Helmholtz timescale). This is the timescale for warming or cooling the entire convection zone and is of the order 10^5 years because the thermal capacity of the convection zone is very large. Even if the central heat source of the Sun were to be switched off completely, the internal thermal structure and surface luminosity would start to change only on this extremely long timescale.

The thermal time scale can be defined as a function of depth z, by considering the time taken for a heat flux through the depth z, $F(z)$, to take away the internal energy stored in a scale height $H(z)$ which equals $H(z)U(z)$

$$\tau_T(z) = H(z)U(z)/F(z) \tag{78}$$

where U is the thermal energy per unit volume at a depth z. Because the heat flow into the surface must equal the heat flux radiated by the surface in steady state, the luminosity, L, equals the average upward heat flux per unit area at the surface, $\langle F(z=0)\rangle$, multiplied by the surface area, $4\pi R_S^2$.

$$L = 4\pi R_S^2 \langle F(z=0)\rangle \tag{79}$$

$\tau_T(z)$ is the timescale on which the heat flux profile, and the observed surface luminosity, would start changing if the heat flux in the Sun were to be interrupted at a depth z. The thermal timescale is a strong function of depth, due to the rapidly increasing temperature and density (see Fig. 3). Some rough values for $\tau_T(z)$ are 10^5 yr at the base of the convection zone ($z = 2\times10^5$ km), 10 years at a depth $z = 16{,}000$ km, 10 hours at $z = 2000$ km and 1 hour at $z = 1000$ km. Thus the speed of the thermal response of the Sun depends critically upon the location of the magnetic disturbance.

The second timescale involved in thermal changes is the diffusive time scale $\tau_D(z)$. This is the timescale on which differences in entropy between different parts of the convection zone are equalled out.

$$\tau_D(z) = \frac{d^2}{K_t} \tag{80}$$

where d^2 is the volume considered, and K_t is the turbulent diffusivity. From the "mixing length" theory it can be estimated that K_t is of order $10^9\ \mathrm{m^2\,s^{-1}}$ at all z. For the base of the convection zone ($z = 2\times10^5$ km) τ_D is about 1 yr (so $\tau_T/\tau_D \approx 10^5$), for above the depth of $z = 2000$ km, it is about 1 hr ($\tau_T/\tau_D \approx 10$) and for $z > 1000$ km it is about 15 min ($\tau_T/\tau_D \approx 4$). Thus at the surface τ_T and τ_D are of a similar magnitude, but as we move into deeper layers, the thermal time scale increases rapidly and at the bottom of the convection zone, the thermal timescale is longer than the diffusive timescale by up to 10^5 years. Since the thermal timescale is so much larger than the diffusive timescale (even at $z = 2000$ km, $\tau_D \approx \tau_T/10$), then it can be considered that the changes below the surface do not occur in thermal equilibrium. This means that upward heat flux blocked, for example under sunspots, is stored in the convection zone.

4.4 The Heat Flow Equation

The thermal adjustment of the convection zone can be described by the energy equation from the first law of thermodynamics

$$\rho T \frac{\mathrm{d}S}{\mathrm{d}t} = -\nabla \cdot F + G \tag{81}$$

where ρ is the density, T the temperature, S the entropy per unit mass and F is the energy flux (convection plus radiation). G includes sources and sinks of heat. In the mixing length-approximation F can be written as

$$F = -K_t \rho T \nabla S \tag{82}$$

giving F as a function of the local entropy gradient ∇S. Using (82), we can write (81) as

$$\rho T \frac{\mathrm{d}S}{\mathrm{d}t} = K_t \nabla \cdot (\rho T \nabla S) + G \tag{83}$$

where K_t is assumed to be constant, since it is approximately independent of depth.

A quasi-hydrostatic approximation is introduced to describe the change in pressure due to the local acceleration due to gravity

$$\frac{\mathrm{d}P}{\mathrm{d}z} = g\rho \tag{84}$$

where P is the gas pressure. (Note that using this equation means we can only look at processes on timescales longer than the hydrodynamic adjustment time which is about 1 hour for the Sun as a whole). We can assume that the convection zone is thin enough to make g approximately constant, so (83) has a solution of form

$$P = P_O e^{z/H} = P_O e^{\mu} \tag{85}$$

where μ is a Lagrangian depth coordinate

$$\mu = \ln\left(\frac{P}{P_O}\right) \tag{86}$$

where P_O is the reference gas pressure at the surface ($z = 0$, $\mu = 0$). Equation (83) can now be rewritten as

$$H^2 \frac{\mathrm{d}S}{\mathrm{d}t} = K_t \frac{\partial^2 S}{\partial \mu^2} + K_t(1 - \nabla)\frac{\partial S}{\partial \mu} + \frac{H^2}{\rho T} G \tag{87}$$

where $H = \mathrm{d}z/\mathrm{d}\mu$ is the pressure scale height; $\nabla = \partial \ln T/\partial \mu$ is the logarithmic temperature gradient; $\partial S/\partial \mu = c_p(\nabla - \nabla_a)$ where $\nabla_a = (1 - \gamma)$ is the adiabatic gradient and c_v and c_p are the specific heats at constant pressure and volume, respectively, the ratio of which is γ. Because pressure is a Langrangian variable its perturbation $P' = 0$ and thus $S' = c_p(T'/T)$.

If we return to (87) and we neglect sources and sinks ($G = 0$) and reduce to vertical variations, (so the operator $\nabla \cdot$ becomes $\partial/\partial z$) we can see mathematically where the two timescales come from

$$\frac{\mathrm{d}S}{\mathrm{d}t} = K_t \frac{\partial^2 S}{\partial z^2} + \frac{K_t}{H}\frac{\mathrm{d}S}{\mathrm{d}z} \tag{88}$$

where $H = [\partial \ln(\rho T)/\partial z]^{-1}$ is the pressure scale height. If the first term on the right hand side dominates then

$$\frac{\mathrm{d}S}{\mathrm{d}t} = K_t \frac{\partial^2 S}{\partial z} \tag{89}$$

which is a diffusion equation, so the entropy and all dependent parameters will evolve on the diffusive time scale which, from the form of (89), is (z^2/K_t) as expected from (80).

If the second term in (88) dominates then

$$\frac{\mathrm{d}S}{\mathrm{d}t} = \frac{K_t}{H}\frac{\partial S}{\partial z} \tag{90}$$

writing S as $c_v(\ln P - \gamma \ln \rho)$, where c_v and c_p are the specific heats at constant pressure and volume, respectively, the ratio of which is the polytropic index, γ. Using (83) for $G = 0$, (90) becomes

$$\frac{\mathrm{d}}{\mathrm{d}t}(\ln P - \gamma \ln \rho) = -\frac{F}{\rho T c_v H} = -\frac{F}{UH} \tag{91}$$

where the internal energy per unit volume, U, is $\rho T c_v$. This gives the thermal time constant of UH/F, as given in (78).

4.5 Polytropic Model

As discussed above, magnetic fields can either introduce a new energy source/sink by magnetic flux being destroyed/created in the convection zone (a beta perturbation), or they can change the energy transport coefficient (an alpha perturbation). To understand these two separate effects one needs to model the variation of key parameters with depth in the convection zone. In order to solve the heat transport equations, Spruit [1976] introduced a "pseudo-polytropic" model of the convection zone, which is a linear variation of temperature with depth. The depth dependence can be described using a convenient depth parameter ζ

$$\zeta = 1 + \frac{z}{(n+1)H_O} \tag{92}$$

where n is a model index. The pressure, density and scale height are then given by

$$P = P_O \zeta^{n+1} \tag{93}$$

$$\rho = \frac{P_O}{(gH_O)\zeta^n} \tag{94}$$

$$H = H_O \zeta \tag{95}$$

At the photospheric surface ($z = 0$), $\zeta = 1$. A good fit to the Sun's convection zone inferred from helioseismology observations is $n = 2$, $H_O = 1.5 \times 10^7$ cm and $P_O = 4 \times 10^4$ Pa. Because the surface gravity is $g_O = 274\,\mathrm{m\,s^{-2}}$, this gives a surface density $\rho_O = P_O/(g_O H_O)$ of $10^{-3}\,\mathrm{kg\,m^{-3}}$. The logarithmic temperature gradient is fixed by the value of n

$$\nabla = \frac{\partial \ln T}{\partial \mu} = \frac{\partial \ln T}{\partial \ln P} = \frac{1}{n+1} \tag{96}$$

Spruit shows that the solution of (91) using the polytropic model is that a small fractional temperature perturbation (T'/T) gives a heat flux perturbation F' of

$$\frac{F'}{F} = \frac{-T}{T_O} + \frac{\zeta^{n+1}}{\delta_O(n+1)} \frac{\partial\left(\frac{T'}{T}\right)}{\partial \zeta} \tag{97}$$

where

$$\delta_O = \frac{F_O H_O}{T_O K_t \rho c_p} \tag{98}$$

Equation (97) applies up to the base of a surface layer. Spruit showed that the so-called "*superadiabaticity*" $\delta = (\nabla - \nabla_a) = \delta_O \zeta^{-n}$ is small everywhere but increases rapidly with decreasing depth close to the surface layer. This means that the solution cannot apply to this thin "superadiabatic" surface layer as well as to the remainder of the convection zone. Thus the convection zone model given by (92–94) must be used in conjunction with a thin emitting surface layer model.

4.6 The Surface Boundary Layer

At the surface, the temperature is T_S and the heat flux is F_S. If we assume a blackbody radiation $F_S = \sigma T_S^4$ (the Stefan–Boltzmann law, where σ is the Stefan–Boltzmann constant) and differentiate

$$\frac{\mathrm{d}F_S}{\mathrm{d}T_S} = 4\sigma T_S^3 = 4\frac{F_S}{T_S} \tag{99}$$

using the perturbation notation $F_S' = \mathrm{d}F_S$ and $T_S' = \mathrm{d}T_S$

$$\left(\frac{F_S'}{F_S}\right) = 4\left(\frac{T_S'}{T_S}\right) \tag{100}$$

The surface temperature T_S will, in general, depend on the solar radius, R_S, the surface heat flux F_S at the surface, and S_O, the entropy at the base of the surface layer. For small perturbations we can write

$$T_S' = \left.\frac{\partial T_S}{\partial R}\right|_{F,S_O} R' + \left.\frac{\partial T_S}{\partial F}\right|_{R,S_O} F' + \left.\frac{\partial T_S}{\partial S_O}\right|_{R,F} S_O' \tag{101}$$

To calculate the first term, we investigate the dependence of the boundary layer on surface gravity. This is determined mostly by the dependence of opacity at the surface on temperature and density. Since $g \propto R_S^{-2}$ we find

$$\left.\frac{\partial T_S}{\partial R}\right|_{F,S_O} \approx \frac{0.6T_S}{R_S} \tag{102}$$

This term can be seen to depend upon the solar radius. The last two terms can be calculated from the polytropic model solution, at the surface where $\zeta = 1$

$$\frac{\partial T_S}{\partial F_S} = \frac{T_S}{F_S}\left[\exp\left(-\frac{2}{3}\delta_O\right) - 1\right] = \frac{T_S'}{F_S'} \tag{103}$$

for constant pressure $S_O' = c_p(T_O'/T_O)$, thus

$$\frac{\partial T_S}{\partial S_O} S_O' = T_S\left(\frac{T'}{T_O}\right) \tag{104}$$

which is the temperature perturbation at the top of the envelope. From (99) we get

$$\frac{F_S'}{F_S} = -0.6\eta\frac{R_S'}{R_S} + \eta\left(\frac{T_O'}{T_O}\right) \tag{105}$$

where

$$\eta = \left[\frac{5}{4} - \exp\left(-\frac{2}{3}\delta_O\right)\right]^{-1} \tag{106}$$

Because $n = 2$, $\delta_O \approx 0.25$, this gives $\eta \approx 1.8$. Equation (105) gives the surface flux change if changes in radius and temperature below the surface layer are known. Since luminosity, $L = 4\pi R_S^2 F$ then

$$\frac{\mathrm{d}L}{\mathrm{d}t} = 8\pi R_S\frac{\partial R_S}{\partial t}F_S + 4\pi R_S^2\frac{\partial F_S}{\partial t} \tag{107}$$

$$L' = 4\pi R_S^2 F_S\left(\frac{2}{R_S}R_S' + \frac{F_S'}{F_S}\right) = L\left(\frac{2R_S'}{R_S} + \frac{F_S'}{F_S}\right) \tag{108}$$

substituting (105) yields

$$\frac{L'}{L} = (2 - 0.6\eta)\left(\frac{R_S'}{R_S}\right) + \eta\left(\frac{T_O'}{T_O}\right) \tag{109}$$

so from this we can look at the effects on the luminosity of radius changes and temperature changes at the base of the surface layer. For $\eta \approx 1.8$, the two coefficients in (109) are ≈ 0.9 and ≈ 1.8. We will later use this equation to evaluate the relative effects of surface temperature and radius on luminosity.

4.7 Effect of Blocked Heat Flux

Not all heat that is blocked under a sunspot is stored in the convection zone, and a fraction α will re-appear at the surface, depending on the depth of the blockage. Convective flows and the transport of heat by turbulent diffusion may result in more of the blocked heat emerging in bright rings around spots than would otherwise be expected from diffusion considerations alone. Bright rings were first reported by Waldmeier [1957, 1975] and have recently been studied in detail by Rast et al. [1999, 2001] (See Fig. 51).

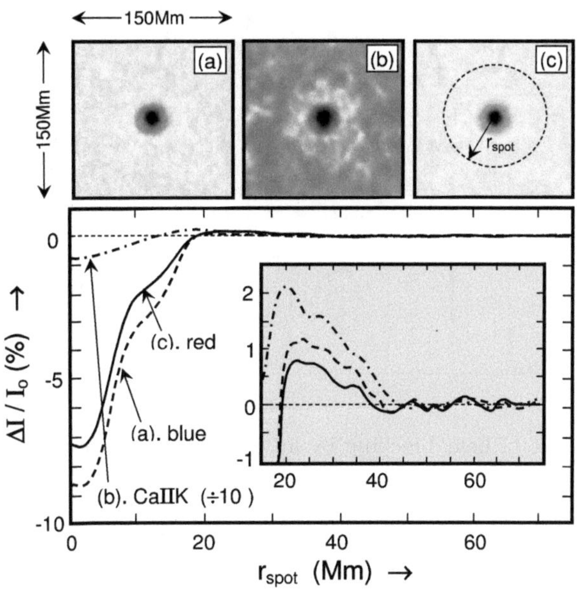

Fig. 51. Observations of the intensity in and around a sunspot by Rast et al. [1999, 2001]. Images (**a**), (**b**) and (**c**) show a 150 Mm × 150 Mm area around sunspot NOAA 8263 as observed through, respectively, the *blue*, Ca II K, and *red* filters of the PSPT (Precision Solar Photometric Telescope) on 6 July 1998. (**d**) Azimuthal averages of the residual intensity (given as $\delta I/I_O$, where $\delta I = I_{(r_{spot})} - I_O$ and $I_{(r_{spot})}$ and I_O are, respectively, the intensities at r_{spot} and of the undisturbed photosphere) as a function of distance r_{spot} from the spot centre for all three wavelengths: *dashed*, *solid* and *dot-dash* curves correspond to blue continuum, *red* continuum and Ca II K intensities, respectively. The Ca II K-line intensities are reduced by a factor of 10 to fit on the same scale, and the spot centre is defined as the intensity centroid of the spot umbra and penumbra. Intensities are enhanced in a region surrounding the spot and extending about one sunspot radius outward from the outer penumbral boundary. The inset shows detail of the bright-ring region. The results show that an intensity increase of about 0.5–1% in the continuum emissions within the ring, consistent with a temperature rise of 10 K over the quiet photosphere

On a timescale t which exceeds the thermal timescale for the depth d of the heat block, $t > \tau_T(d) > \tau_D(d)$ (remember that for $d = 1000\,\text{km}$, $\tau_D(d) \approx 0.25\,\text{hr}$ and $\tau_T(d) \approx 1\,\text{hr}$), then both the heat flux profile and the temperature profile above the base of the spots ($z < d$) will have adjusted to the presence of the spot. Figure 52 illustrates schematically the heat flow in the vicinity of a spot.

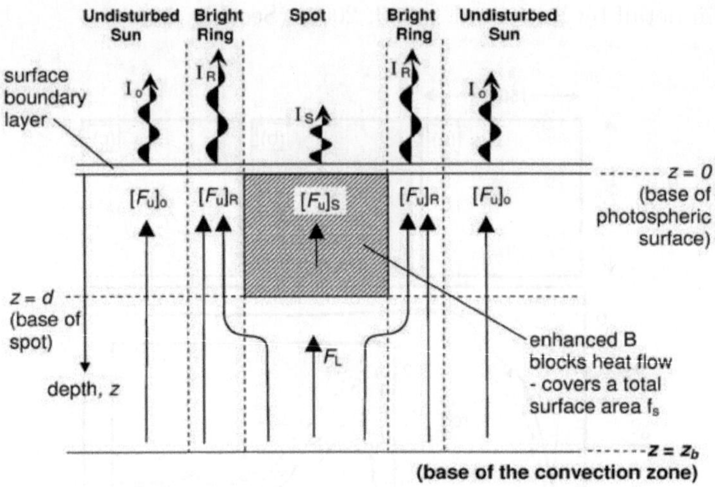

Fig. 52. Schematic of heat blocking by an enhanced field region below a sunspot in the surface layer

Just below the sunspots ($z = d$) the average upward heat flux is F_L and in the surface layer effected by sunspots ($z < d$) the average heat flux is F_U. In the absence of spots, $F_L = [F_U]_O$ everywhere (where the subscript O denotes the undisturbed flux) and the luminosity, $L_O = 4\pi R_S^2 F_L$ from (79). In Fig. 52, we divide this upper layer into three classes when sunspots are present: (1) undisturbed sun (covering a fraction f_O of the surface and through which the heat flux is $[F_U]_O$); (2) dark spots (taken to be here an average of umbra and penumbra and covering a fraction f_S of the surface and through which the heat flux is $[F_U]_S$); (3) bright rings around spots (covering a fraction f_R of the surface and through which the heat flux is $[F_U]_R$). The average flux through the upper layer is then

$$F_U = f_O[F_U]_O + f_S[F_U]_S + f_R[F_U]_R \tag{110}$$

where $f_O + f_S + f_R = 1$. From (79)

$$\frac{L'}{L_O} = \frac{F'_U}{[F_U]_O} = \frac{(F_U - [F_U]_O)}{[F_U]_O} \tag{111}$$

We define α as the fraction of the blocked heat flux that still reaches the surface. The blocked flux is $f_S([F_U]_O - [F_U]_S)$ and the flux returned in bright rings is $f_R([F_U]_R - [F_U]_O)$. Thus

$$\alpha = \frac{f_R}{f_S} \frac{\left(1 - \frac{[F_U]_R}{[F_U]_O}\right)}{\left(1 - \left(\frac{[F_U]_S}{[F_U]_O}\right)\right)} \tag{112}$$

From (112), (111) and (110)

$$\frac{L'}{L_O} = f_S \left(1 - \frac{[F_U]_S}{[F_U]_O}\right)(1-\alpha) \tag{113}$$

Note that it is often assumed (indeed it is in Fig. 54 of the present text) that $[F_U]_S$ is vanishingly small (i.e. all upward heat flux is blocked in sunspots), in which case (113) reduces to $L'/L_O = f_S(1-\alpha)$ and from (112) the corresponding α is $(f_R/f_S)(1 - [F_U]_R/[F_U]_O)$.

From Fig. 51 we have that $(f_R/f_S) \approx (\pi(2r_S)2 - \pi r_S^2)/\pi r_S^2 = 3$ (where r_S is the spot radius), and that the average intensity (by area, rather than radius) over the bright ring is $[F_U]_R \approx 1.03[F_U]_O$. From (112) this gives $\alpha = 0.09$, i.e. roughly 10% of the blocked flux still reaches the surface. Note that the fraction $(1-\alpha)$ (i.e 90% of the blocked flux) that does not reach the surface is stored in the deeper layers of the convection zone.

Figure 51 is an example of a very bright ring and the value of $\alpha = 0.09$ is a relatively large value. Nevertheless it is instructive to compare with the solution for the polytropic model. Spruit [1976] derives an expression

$$\alpha = 1 + \left[\left\{\exp\left(\frac{-(n+1)}{n}\delta_O \zeta_d^{-n}\right) - 1\right\} \left\{\frac{5}{4}\exp\left(\frac{n+1}{n}\delta_O\left(1-\zeta_d^n\right)\right) - 1\right\}^{-1}\right]^{-1} \tag{114}$$

where, from the definition of ζ (92)

$$\zeta_d = 1 + \frac{d}{(H_O)(n+1)} \tag{115}$$

If we consider spots deep enough so that $d \gg (n+1)H_O$, then (114) reduces to

$$\alpha = \zeta_d^{-n}\left(\frac{n+1}{n}\right)\delta_O\left\{\frac{5}{4}\exp\left(\frac{n+1}{n}\delta_O\right) - 1\right\}^{-1} \tag{116}$$

Taking a value for n of 2 (gives $\delta_O = 0.25$), typical of the convection zone

$$\alpha \approx 0.5\zeta_d^{-2} = 0.5\left(1 + \frac{d}{3H_O}\right)^{-2} \tag{117}$$

For $d \geq 3500$ km (the magnitude of d implied by helioseismology data) and $H_O = 1.5 \times 10^5$ m, (117) yields $\alpha \leq 0.6\%$, which is much smaller than the 9% inferred above: by (117), α of 9%, requires a depth d of just 612 km. Thus either spots are much shallower than we have inferred from helioseismology data or, more likely, that the combination of turbulent diffusion and convective flows may be more effective in bringing blocked heat flux to the surface than the above diffusion theory predicts.

4.8 Effect of Radius Changes

Due to the appearance of spots at the surface of the Sun, the temperature outside the spots is slightly increased. Hydrostatic equilibrium requires that the stellar radius, as would be measured outside the spots, is then slightly greater. Inside the spots the temperature is lower, and so the local stellar radius is reduced (the Wilson depression), so we have to distinguish between the radius change inside and outside the spots. We can calculate the radius changes outside the spots from hydrostatic equilibrium applied to the full CZ

$$R'_S = \int_O^D \frac{T'}{T} \mathrm{d}z \tag{118}$$

for small T'/T, and for $\delta_O < 1$ we find

$$R' = \frac{9}{2} \delta_O H_O (1 - \alpha) f_S \frac{3 d H_O}{(d + 3H_O)^2} \tag{119}$$

For very deep and very shallow spots, R' is negligible, d has a maximum at $3H_O$, such that:

$$R_S < \frac{9}{8} \delta_O H_O (1 - \alpha) f_S \tag{120}$$

Equation (117) gives a theoretical value for this condition $d = 3H_O$ of $\alpha = 0.125$ (in fact comparable with the value deduced from observed bright rings). Figure 10 shows that f_S can have a range of values up to about 0.003, and for this upper limit the radius change (R_S) outside the spot is 1.2×10^3 m, which is extremely small, when compared to the radius of the Sun as a whole, and so will not easily be detected. The ratio R'_S/R_S is of order 2×10^{-8}. On the other hand, the mean temperature change associated with spots is $T' = f_S(T_O - T_S) \times 0.003(5770 - 4100)$, giving $T'/T \sim 9 \times 10^{-4}$. By (109), this means that radius changes due to sunspots have negligible effect compared to the surface temperature change they cause. Radius changes and their effects have been reviewed by Noël [2004].

Taking the average surface temperature inside spots to be $T_S = 4100$ K, and the surface temperature outside the spots to be $T_O = 5770$ K, then the ratio of heat flux inside and outside the spots is

$$\frac{[F_U]_S}{[F_U]_O} = \left(\frac{\sigma T_S^4}{\sigma T_O^4} \right) = 0.25 \tag{121}$$

The heat flux blocked $[F_U]_O - [F_U]_S = 0.75[F_U]_O$ and of that, up to about $\alpha = 0.1$ still makes it to surface, i.e. $0.075[F_U]_O$. Therefore the heat flux making it to the surface is $(0.25+0.075)[F_U]_O = 0.325[F_U]_O$ and the blocked flux is $0.675[F_U]_O$ If we look at the luminosity of the Sun when clear of spots and when a fraction f_S of the surface is covered with spots, by (78)

$$\frac{L'}{L_O} = \frac{4\pi R_S^2[F_U]_O(0.675 f_S)}{4\pi R_S^2[F_U]_O} = 0.675 f_S \tag{122}$$

Which for a large f_S of 0.3% at sunspot maximum (Fig. 12) yields L'/L_O of 0.2%. For isotropic effects this will equal I'_{TS}/I_{TS}. Thus we can explain the large spikes reducing TSI at sunspot maximum in Fig. (47) which are roughly of this magnitude. An average value of f_S at sunspot maximum is nearer 0.15%, compared with an average near zero at sunspot minimum (see top panel of Fig. 7). Thus we would expect I'_{TS}/I_{TS} due to sunspots to be of order 0.1% over the solar cycle, as is observed (Fig. 18). In Sect. 4.12 we repeat this calculation allowing for umbrae and penumbrae separately in the derivation of the photometric sunspot index which quantifies the sunspot darkening effect.

4.9 Effects of Magnetic Field: The β Effect

As mentioned in Sect. 4.2, magnetic fields below the surface of photosphere have two effects. (1) The creation/destruction of magnetic fields at a depth z will cause a sink/source of energy. This is called the β effect. (2) Magnetic interference with convection motions, which is called the α effect. From the equations of heat flow, solved using the polytropic model with a superadiabatic surface layer, we can look at the development of the luminosity and the height profiles for both these effects. In this section we study the β effect. (The α effect will be addressed in the next section).

If we assume that magnetic flux is created or destroyed at a depth z_G, in a layer dz thick then the rate of growth of magnetic energy is

$$\frac{\mathrm{d}}{\mathrm{d}t}\left\{\left|\frac{B^2}{2\mu_O}\right|\mathrm{d}z\right\} = G \tag{123}$$

where G is the sink term in (81). If we solve the heat balance equation above and below z_b

$$F_{(z>z_b)} - F_{(z>z_b)} = \frac{\mathrm{d}\left\{\mathrm{d}z\frac{B^2}{2\mu_O}\right\}}{dt} \tag{124}$$

since the timescales are long, the solution to (124) will involve the thermal mode. Computed variations are shown in Fig. 53. Upon a sudden increase of B ($G > 0$, i.e. an energy sink), L initially rises due to expansion of R_S under the enhanced magnetic pressure (this effect is also seen at large times t). At intermediate t, L is reduced because of the cooling at $z = z_G$ which spreads to surface on a timescale $t_D(z_G)$. At large t, as exemplified by t_f, the sink is supplied almost entirely from the larger heat reservoir below the increasing field ($z > z_G + \mathrm{d}z$) and not from above it. At $t \geq t_f$, the temperature perturbation T'/T is constant with depth and the temperature must be continuous across z_b. Figure 53 also shows the solution for the heat flux and it can be seen that the heat flux which is destroyed in the sink, is almost entirely supplied from below the level z_b. A consequence of this is that only a very weak signal (if any) reaches the surface. A signal of any magnitude will only reach the surface if the sink varies quickly enough (shorter than the diffusive timescale), so there is very little luminosity change associated with such sinks.

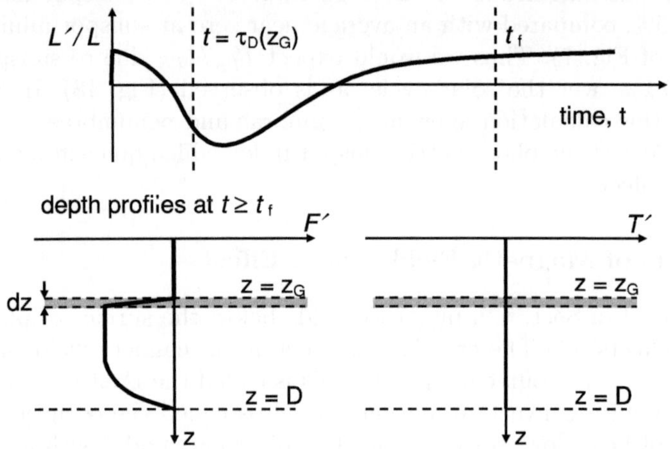

Fig. 53. Predictions of the β effect where the field B changes in a slab, dz thick at depth z_G. (*Top*) The time variation in the fractional luminosity perturbation L'/L. (*Bottom left*) The profile of the perturbation in heat flux, F' at $t = t_f$. (*Bottom right*) The profile of the perturbation in temperature, T' at $t = t_f$

4.10 Effects of Magnetic Field: The α Effect

By its influence on convective motions, a magnetic field can locally increase the entropy gradient required to transport a given energy flux. In the α effect, the balance between magnetic energy density and the kinetic energy density in convective turbulence is important. When these two energy densities are comparable, the magnetic field has what is called the *equipartition strength*, B_e such that

$$\frac{1}{\delta} \approx \beta_e = \frac{2\mu_O P}{B_e^2} \tag{125}$$

where the superadiabaticity δ is $(\nabla - \nabla a)$. The condition that $\beta_e \delta \approx 1$ allows B to start reducing the degrees of freedom of convective flow which means that a larger entropy gradient is needed to transport the same energy flux. To allow for this effect, we change the mixing-length expression for the convective energy flux by introducing an additional factor. Substituting $H = \partial z/\partial\mu$ and $\partial S/\partial\mu = c_p(\nabla - \nabla_a) = c_p\delta$ (see Sect. 4.4) into (82), for radial stratification $(-\nabla S = \partial S/\partial z = (1/H)\partial S/\partial\mu = c_p\delta/H)$, yields that without this factor

$$F = K_t \rho c_p (T/H)\delta \tag{126}$$

In order to allow for the loss of the degrees of freedom this becomes

$$F = K_t \rho c_p \left(\frac{T}{H}\right)\left(\delta - \frac{q}{\beta}\right) = K_t \rho c_p \left(\frac{T}{H}\right)\left(\delta - \frac{qB^2}{2\mu_O P}\right) \tag{127}$$

where q is a factor near unity. We can then introduce the α-effect by making q unity in a layer between $(z_b - d)$ and z_b at a time $t = 0$. This reduces F in the layer according to (127) so that the temperature T above the layer $z < z_b$ starts to fall, but at $z \geq z_b$ (in and below the layer) T stays essentially constant because the heating effect is negligible (due to the large thermal time constant of the convection zone).

Figure 55 gives three sets of profiles of the perturbations to the energy flux and temperature (as was given in the lower panel of Fig. 53 for the β-effect) and Fig. 54 gives the fractional luminosity variation (as given in the top panel of Fig. 53 for the β-effect).

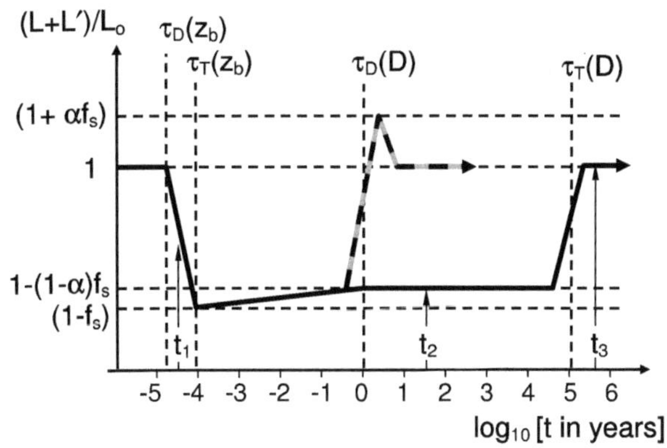

Fig. 54. Predictions of the α effect: the time variation (on a log scale) of the ratio of the perturbed and undisturbed luminosity $(L + L')/L$. The profiles at times t_1, t_2 and t_3 are given in Fig. 55. The luminosities given assume that all flux is blocked in sunspots ((112) with $[F_U]_S = 0$ so $L'/L = f_S(1 - \alpha)$ once the bright rings are established and $L'/L = -f_S$ while α is zero). f_S is the fraction of the solar surface affected by sunspots. The *dashed* line shows the variation if the sunspot blocking is switched off after 100 days

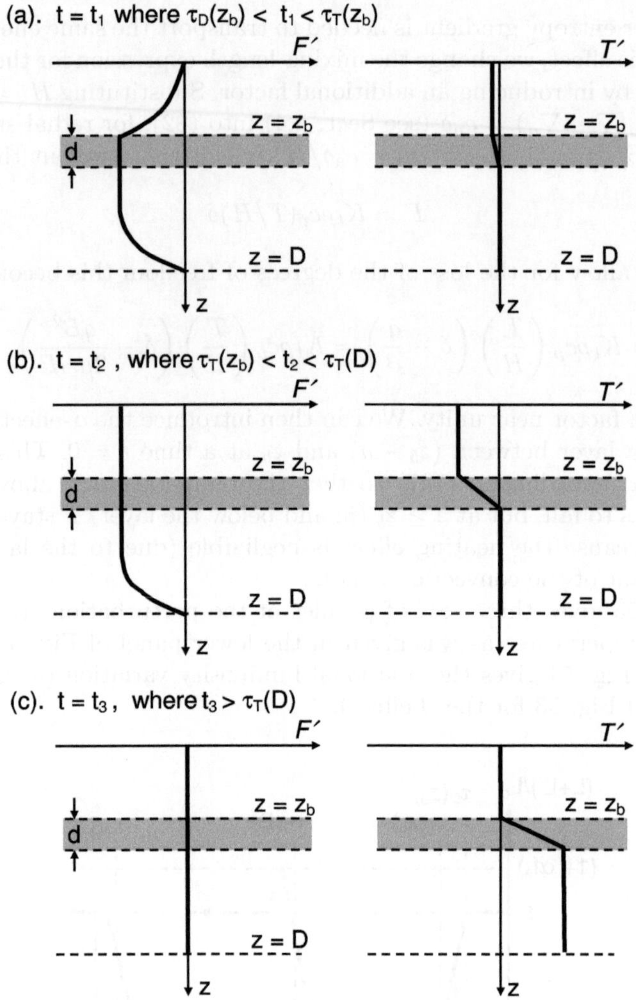

Fig. 55. Predictions of the α effect. The magnetic field influences the entropy gradient needed in a layer d deep below a depth $z = z_b$. (*Left*) The profile of the perturbation in heat flux, F' at various t marked in Fig. 54. (*Right*) The profile of the perturbation in temperature, T' at the same t

At $t = t_1$, where $\tau_D(z_b) < t_1 < \tau_T(z_b)$ (top row of Fig. 55), adjustments have occurred on the diffusive timescale in all but the deepest convection zone, but the layers above z_b have not yet returned to thermal equilibrium. The temperature below z_b has not changed, but the temperature above z_b is slightly reduced due to the mismatch between the heat fluxes at $z = 0$ and $z = z_b$.

At $t = t_2$, where $\tau_T(z_b) < t_2 < \tau_T(D)$, the layers above z_b have returned to thermal equilibrium, while the layers below have not and are heating up – but only very, very slowly. The surface flux is reduced during this period, and the temperature of the layers above z_b are reduced.

On very long timescales $t_3 > \tau_T(D)$, the whole convection zone has returned to equilibrium, the temperature below the layer has increased and the flux and temperature above the layer have returned to their $q = 0$ values.

We here consider a $z_b = 10^6$ m which gives us typical timescales. For this depth $\tau_D(z_b) = 10^3$ s ($\approx$ 15 minutes) and $\tau_T(z_b) = 3.34 \times 10^3$ s ($\approx$ 1 hour), and remember $\tau_D(D) \approx 1$ year and $\tau_T(D) \sim 10^5$ years. The time series of surface luminosity is displayed in Fig. 54.

The effect of the dark spots resulting from the α-effect blocking begin to appear after $\tau_D(z_b) \approx 15$ min when the flux and temperature begin to fall at the surface. The full darkening is achieved by $\tau_T(z_b) \approx 1$ hour. The return of some of the blocked heat flux first appears at this time and the bright rings reach full luminosity at $\tau_D(D) \approx 1$ year, when the disturbance has propagated through the entire convection zone. This marks the start of the "quasi-static phase" where the entire convection zone is heating up on the thermal timescale of $\tau_T(D) \sim 10^5$ years.

If the spot is switched off, the heat stored in the deepest layers is released. If the spot is switched off during the quasi-static phase, the luminosity is enhanced to $L_0(1 + \alpha f_s)$, where L_0 is the undisturbed luminosity, and then decays to L_0: both these changes take place on the diffusive timescale $\tau_D(D) \approx 1$ year. The typical lifetime of spots on the surface of the Sun is 8.6×10^6 s ($\sim$ 100 days) which is between $\tau_T(z_b)$ and $\tau_D(D)$ and these changes, shown by the dashed line in Fig. 54, are likely to occur. For completeness, we note that if the blocking is switched off after a time $\tau_T(D) \sim 10^5$ years, the luminosity is enhanced to $L_O(1 + f_s)$, and then returns to L_O, both change on timescales of $\tau_T(D)$.

4.11 Effects of Magnetic Fields: Quantifying Surface Effects

The intensity of a region of the Sun, I, is a function of the disk position parameter μ, defined by

$$\mu = \cos\theta \tag{128}$$

where θ is the angle that the region subtends with the Earth–Sun line at the centre of the Sun: $\mu = 1$ at the centre of the visible disk and $\mu = 0$ at the photospheric limb. If the surface area of the region is ds, it forms an area da = μds on the disc. The mean value of the intensity, averaged over the whole disc is:

$$\langle I \rangle_D = 2 \int_0^1 I(\mu)\mu \mathrm{d}\mu \tag{129}$$

The TSI, I_{TS}, is related to the disc-averaged intensity $\langle I \rangle_D$ by (77). The quiet-Sun intensity variation can be written as $I(\mu) = I_O L_D(\mu)$, where $L_D(\mu)$

is the *limb darkening* function and I_O is the intensity of the quiet Sun at the centre of the visible disc. The normal photosphere is darker at the limb because unit optical depth, $\tau = 1$, is reached at a greater height, where the photosphere is cooler. A multi-wavelength limb darkening function has been derived by Neckel and Labs [1994] who give the wavelength-dependent coefficients for a polynomial function of the disc parameter function μ. The SoHO MDI instrument measures the continuum emission at 676.8 nm and linearly interpolating the coefficients for 669.400 nm and 700.875 nm gives a limb-darkening function for 676.8 nm of

$$L_D = 0.3544 + 1.3472\mu - 1.9654\mu^2 + 2.5854\mu^3 - 1.8612\mu^4 + 0.54\mu^5 \quad (130)$$

The resulting limb-darkening function, $L_D(\mu)$ is shown in Fig. (56) which is similar to, but more precise than, the frequently-used Eddington function that is also shown in the figure. The latter is a useful approximation that allows some analytic solutions and is given by

$$L_D(\mu) = \frac{(3\mu + 2)}{5} \quad (131)$$

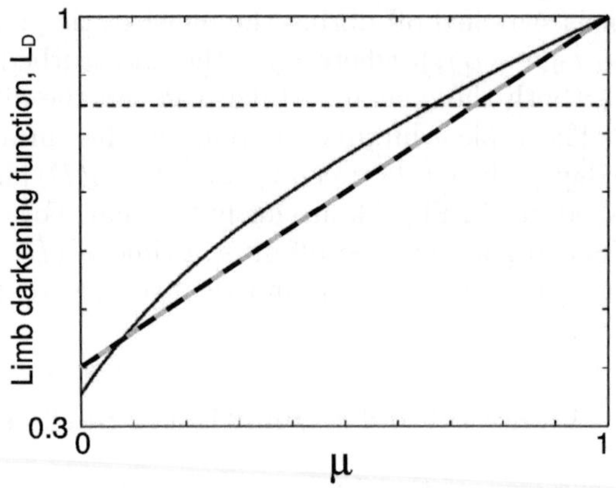

Fig. 56. (*Thick solid line*) The limb-darkening function, $L_D(\mu)$ for a wavelength of 676.8 nm from the polynomial expression by Neckel and Labs [1994], as given in (128). The disc-average value is shown by the horizontal *dashed* line $\langle L_D \rangle_D = 0.8478$. For comparison, the *dashed* line shows the Eddington limb darkening function used in the derivation of the photometric sunspot index

4.12 Sunspot Darkening

The darkening by sunspots is quantified by the photometric sunspot index, PSI [Willson et al., 1981, Hudson et al., 1982, Fröhlich et al., 1994]. In general, the intensity of the umbra of a spot varies with the spot's μ-value, μ_S, as

$$I_U = I_{U0} g_U(\mu_S) \tag{132}$$

and similarly for the spot penumbrae

$$I_P = I_{P0} g_P(\mu_S) \tag{133}$$

If the area of a spot umbra is A_U and of its penumbra is A_P, then the total area of the spot is $A_S = A_U + A_P$. Note that these surface areas give areas on the visible disk of $\mu_S A_S$, $\mu_S A_U$ and $\mu_S A_P$. From (77), the spot changes the irradiance by

$$\begin{aligned}\Delta I_S = & \left(\frac{\mu_S A_U}{R_1^2}\right)[I_0 L_D(\mu_S) - I_{U0} g_U(\mu_S)] \\ & + \left(\frac{\mu_S A_P}{R_1^2}\right)[I_0 L_D(\mu_S) - I_{P0} g_P(\mu_S)]\end{aligned} \tag{134}$$

where I_O is the quiet-Sun intensity at the disc centre.

To derive the PSI, it is assumed that all umbra have a common temperature, as do all penumbra and that all spots have the same ratio of their areas A_U/A_P. In addition, the limb darkening function is assumed to be the same for umbra, penumbra and the quiet sun, so $g_U(\mu) = g_P(\mu) = L_D(\mu)$. Equation (134) then reduces to

$$\Delta I_S = A_S \left(\frac{L_D(\mu_S)}{R_1^2}\right)[I_{S0} - I_0] \tag{135}$$

where

$$I_{S0} = \left(\frac{A_P}{A_S}\right) I_{P0} + \left(\frac{A_U}{A_S}\right) I_{U0} \tag{136}$$

and ΔI_S is defined as positive for an irradiance increase. From (77) and (129), the quiet-Sun irradiance, Q_O, is given by

$$Q_0 = 2\pi I_O \left(\frac{R_S}{R_1}\right)^2 \int_0^1 L_D(\mu)\mu \mathrm{d}\mu \tag{137}$$

and from this and (135)

$$\frac{\Delta I_S}{Q_0} = \left(\frac{\mu_S A_S}{\pi R_S^2}\right)\left[\frac{L_D(\mu_S)}{2\int_0^1 L_D(\mu_S)\mu \mathrm{d}\mu}\right]\left(\frac{I_{S0}}{I_0} - 1\right) \tag{138}$$

We here define the contrast to be

$$C_S = \frac{I_{S0}}{I_0} - 1 = \frac{(I_{S0} - I_0)}{I_0} \tag{139}$$

Note that with this definition, positive/negative contrast corresponds to a brightening/darkening, respectively. The *filling factor* is the fraction of the disk covered by the spot(s), $\alpha_S = (\mu_S A_S/\pi R_S^2)$. We here use the Eddington limb darkening profile which is plotted as a dashed line, in Fig. 56. Integration yields that the square term in brackets in (138) is equal to $(3\mu_S+2)/4$. Summing over all the spots present on the visible disk we get the total darkening (P_{SI} in $\mathrm{W\,m^{-2}}$):

$$\sum_S \Delta I_S = -P_{SI} = Q_0 \sum_S \frac{A_S}{\pi R_S^2} C_S \frac{(3\mu_S+2)}{4} \mu_S \tag{140}$$

This is a definition of the *photometric sunspot index* (PSI), P_{SI}, which quantifies the effect of sunspots on the total solar irradiance. Note that sunspot contrasts are negative, by the definition used, and so PSI is defined as positive if the increase ΔI_S is negative and the Sun is darkened. Because for monthly averages of PSI longitudinal effects are averaged out, the only influence of μ_S is through the latitudinal structure of sunspot occurrence. This influence is relatively small and so the variation of PSI is dominated by that in the total sunspot area $\Sigma_S A_S$ on these timescales (as demonstrated by Fig. 96).

We use estimates of the temperatures of the umbra, penumbra and quiet Sun of $T_U = 4240\,\mathrm{K}$, $T_P = 5680\,\mathrm{K}$, and $T_{QS} = 6050\,\mathrm{K}$ [Allen, 1973]. The fraction of the area of an average spot is 0.18 for umbra and 0.82 for penumbra ($A_U/A_S = 0.18$, $A_P/A_S = 0.82$). Using the Stefan–Boltzmann law for a blackbody radiator, so intensity I is proportional to T^4, the average contrast for a spot is

$$C_S = \left(\frac{A_U}{A_S}\right)\left\{\left(\frac{T_U}{T_{QS}}\right)^4 - 1\right\} + \left(\frac{A_P}{A_S}\right)\left\{\left(\frac{T_P}{T_{QS}}\right)^4 - 1\right\} = -0.32 \tag{141}$$

In fact, C_S shows some dependency on spot size and position, and to generate PSI, Fröhlich et al. [1994] employ

$$C_S = -0.2231 - 0.0244\log_{10}(\mu_S A_S) \tag{142}$$

We can get a rough estimate of the peak PSI by adopting the simple spot contrast given by (141). If spots are spread evenly over the surface, integration gives that the disc average of $\mu_S(3\mu_S+2)/4$ in (140) is 0.708. Monthly values of the total spot area $\mu_S A_S$ peaks at sunspot maximum at about $(3\times10^{-3})A_{SH}$, where A_{SH} is the area of a solar surface hemisphere. With typical sunspot maximum values of about $(1.5\times10^{-3})A_{SH}$. From (138), the sunspot darkening $\Delta I_S/Q_0 = 0.07\%$, which for $Q_0 = 1365\,\mathrm{W\,m^{-2}}$ yields $\Delta I_S \approx 1\,\mathrm{W\,m^{-2}}$.

There are a number of second-order corrections to the simple formulation of PSI given by (140) and (142) which are implemented by Fröhlich et al. [1994] and in the data used here in subsequent sections. These corrections allow the PSI to accurately reproduce the observed effect on total solar irradiance of sunspot groups and individual sunspots as they rotate across the solar disc.

4.13 Facular Brightening

As discussed earlier, if the magnetic flux tube is smaller in diameter than a sunspot it can emit more radiation than the surrounding photosphere, this is called a facula ("torch"). There are two main theories of faculae: the bright wall model Spruit [1976], Deinzer et al. [1984a,b], Knölker et al. [1988], Steiner et al. [1996] and the bright cloud model [Schatten et al., 1986].

In the bright wall model, faculae are very similar to sunspots, except that the radius of the flux tubes is smaller, allowing radiation from the tube walls to maintain the temperature: the increased optical depth inside the tube allows radiation from lower, hotter layers to escape, giving enhanced emission. The hot cloud model is dynamical in that it considers the effect of upflows which carry heat blocked in sunspots to the surface. A major difference between these models is the height of the surface in the faculae, compared to the surrounding photosphere – the hot wall predicting that the surface is depressed whereas the hot cloud model predicts that it is raised, so the latter is often referred to as the "hillock" model.

Figure 57 illustrates the bright wall model. Due to the pressure exerted by the magnetic field, the gas pressure inside the tube will be smaller than the surrounding photosphere at the same depth. This low gas pressure (and hence density) inside faculae will cause the opacity to be less than the surrounding area, so the optical depth unity will occur at a greater depth than for the surrounding photosphere.

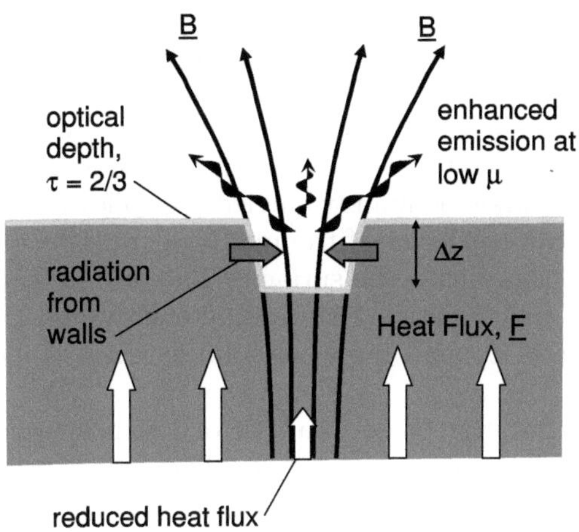

Fig. 57. The bright wall model of faculae. The enhanced magnetic pressure in the flux tube means it is evacuated of gas, but radiation from the walls maintains the temperature even though the upward heat flux is inhibited and reduced compared to its value outside the facula. As a result, the constant optical depth (the $\tau_0 = 2/3$ contour is shown here) is depressed by Δz

If we take a flux tube of $B_f = 1000\,\mathrm{G}$, which is in pressure equilibrium with it's surroundings, then

$$\left(\frac{B_f^2}{2\mu_0}\right) + N_f k T_f = N_{QS} k T_{QS} \tag{143}$$

where T_f is the gas temperature in the facula and T_{QS} is the temperature in the photospheric surface. If the tube is thin enough, then the horizontal exchange of radiation ensures that $T_{QS} = T_f$ so that the concentration difference between the quiet photosphere and the facular tube is

$$(N_{QS} - N_f) = \frac{B_f^2}{(2\mu_0 k T_{QS})} \approx 5 \times 10^2 \mathrm{m}^{-3} \tag{144}$$

The $\tau = 2/3$ level, for example, will occur at a greater depth inside the facular tube and thus the temperature will be higher at this contour and so more radiation is emitted. This will apply for all faculae which are smaller than the mean free path of photons at $\tau = 2/3$, which will be around 50 km. For tubes that are larger than this, the exchange of heat with the walls becomes less effective. Faculae of greater radius can still appear bright at the limb (small μ), but begin to behave in a similar fashion to spots at the disc centre where the centre of the tube will appear cooler than its surroundings so the tube will not appear bright at the disc centre. Such tubes are often called *micropores*. The bright wall model requires the blocked heat flux to cause the tube floor to be cooled so that the flux tube appears dark when viewed from above. However, this tube would still contribute enhanced emission near the solar limb where the bright walls become more visible.

The hot cloud model is illustrated by Fig. 58. In this model, the heat blocked by sunspots is conducted to the surface by magnetic flux tubes. At lower altitudes the upflow is mainly carried by upflowing protons which rise into the neutral hydrogen layer of the photosphere (which normally extends down to about 2000 km below the surface). The recombination of the ionised hydrogen is exothermic, releasing additional energy and the gas is lifted by buoyancy. This forms a small bump or hillock which is most visible on the limb of the sun.

Much evidence for the bright wall theory comes from the variation of contrast with the position parameter, μ. The hillock model predicts that faculae will bright right out to the solar limb ($\mu = 0$), whereas the hot wall theory predicts that the Wilson depression will cause faculae to vanish close to the limb. Much evidence of the contrast of faculae, as a function of μ, has been interpreted as favouring the hot wall model [Topka et al., 1997, Sánchez Cuberes et al., 2002] and this model has gained widespread acceptance. However, there are problems, for example satisfactory explanation of the cool floor (and thus lower contrasts at μ near unity) requires careful tuning of the model. Recent very high-resolution observations by Berger et al.

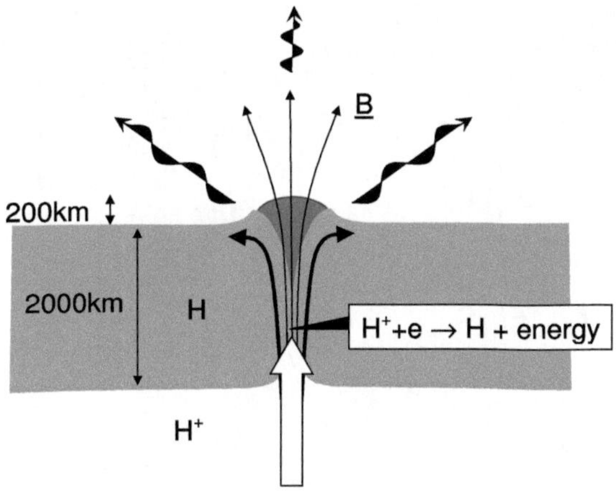

Fig. 58. The hot cloud model of faculae. Upflows of hydrogen ions are driven up into the neutral hydrogen layer. These ions recombine exothermically, releasing more energy and driving the flows up and apart and so a bright hillock appears [after Schatten et al., 1986]

[2003], shown in Fig. 59, offer the potential to distinguish between the hillock and hot wall models.

The Photometric Facular Index (PFI) quantifies the effect of faculae, in the same way that the PSI does for sunspots. The contrast of faculae and micropores is

$$C = \frac{I_f}{I_{QS}} - 1 = \frac{(I_f - I_{QS})}{I_{QS}} \tag{145}$$

giving, for the Eddington limb darkening function

$$\Delta I_f = Q_0 \sum_f \alpha_f C \frac{(3\mu_f + 2)}{4} \tag{146}$$

where α_f is the disk facular filling factor of faculae, and A_f is the surface area covered by faculae. At unit optical depth the surface temperature in faculae is about 150 K higher than the quiet Sun, and thus $T_f = 6200\,\mathrm{K}$. Again assuming a blackbody radiator, (145) yields a contrast c_f for faculae of

$$C_F = \left(\frac{T_f}{T_{QS}}\right)^4 - 1 \approx 0.103 \tag{147}$$

At sunspot maximum the total area of faculae is roughly 10 times that of sunspots, thus $\Sigma_f A_f$ peaks at sunspot maximum at about $(1.5 \times 10^{-3})A_{SH}$. From (147), (146) and the fact that the disc average of $\mu_f(3\mu_S+2)/4 = 0.708$, $\Delta I_f/Q_0 = 0.21\%$, which for $Q_0 = 1365\,\mathrm{W\,m^{-2}}$ yields a facular brightening of $\Delta I_f \approx 3\,\mathrm{W\,m^{-2}}$.

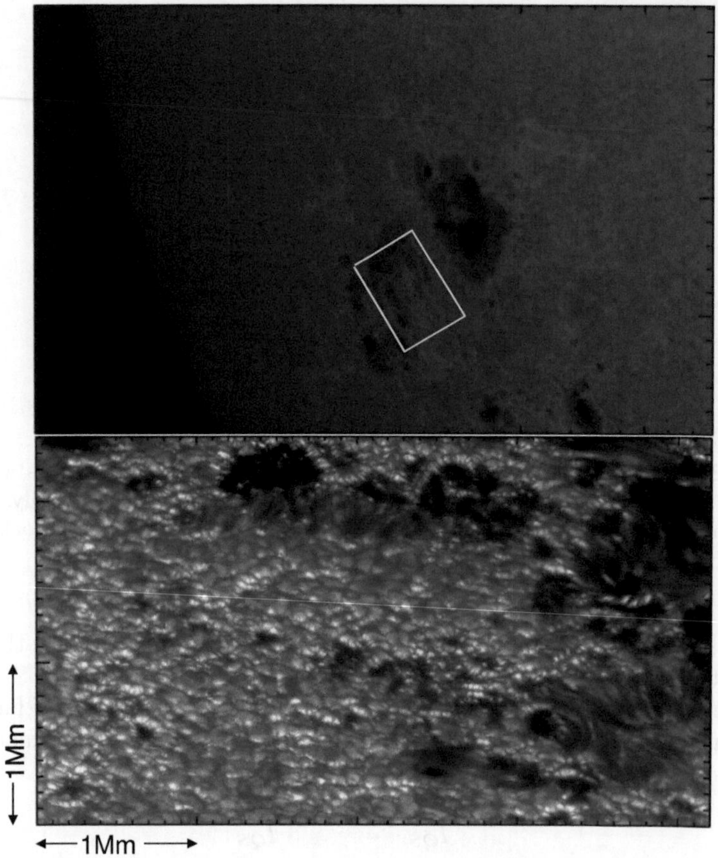

Fig. 59. Images of a solar active region taken on 24th July 2002, near the eastern limb of the Sun, as recorded by TRACE (*top*) and in a filtergram taken in 488 nm light by the Swedish 1-meter Solar Telescope (SST) on the island of La Palma (*bottom*). In the upper plot, tickmarks are 10,000 km apart and the *yellow* box outlines the approximate SST field-of-view in the image shown underneath. TRACE has 10 times lower spatial resolution than the SST and so faculae show only as vague bright patches surrounding the active regions in the upper image. Only when looking at active regions towards the solar limb with the 70 km spatial resolution of the SST do the three-dimensional aspects of the photosphere and faculae become apparent. In the *lower* image, tickmarks are 1000 km apart and the limb is towards the top of the right hand corner. The structures in the *dark* sunspots in the upper central area of the image show distinct elevation above the dark "floor" of the sunspot. There are numerous bright faculae visible on the edges of granulation that face towards the observer [Berger et al., 2003]

Thus these broad considerations predict that sunspots cause a darkening of about $1\,\mathrm{W\,m^{-2}}$, whereas faculae cause a brightening of about $3\,\mathrm{W\,m^{-2}}$ at sunspot maximum, relative to sunspot minimum. Together these cause a net solar cycle variation in total solar irradiance of amplitude of about $2\,\mathrm{W\,m^{-2}}$, as has been observed over recent solar cycles. Note that the facular contrast is roughly one third of that of spots, but that they cover roughly 10 times the area.

Ortiz et al. [2002] have provided a more precise algorithm for computing the contrast of small flux tubes (faculae and micropores), as a function of the field observed in a pixel of the Michelson Doppler Interferometer (MDI) on the SoHO satellite. Contrasts of MDI pixels were evaluated at 676.8 nm, using the definition given in (145), relative to a field-free quiet sun intensity, corrected for limb darkening. Ortiz et al. studied the contrasts as a function of (B_{MDI}/μ) and μ where B_{MDI} is the field detected by MDI in the line-of-sight direction. If we assume the field is radial, the field magnitude is (B_{MDI}/μ) and plots like Fig. 60 show that this yields very low scatter in the data. The lines in Fig. 60 are Ortiz et al.'s fit to the data.

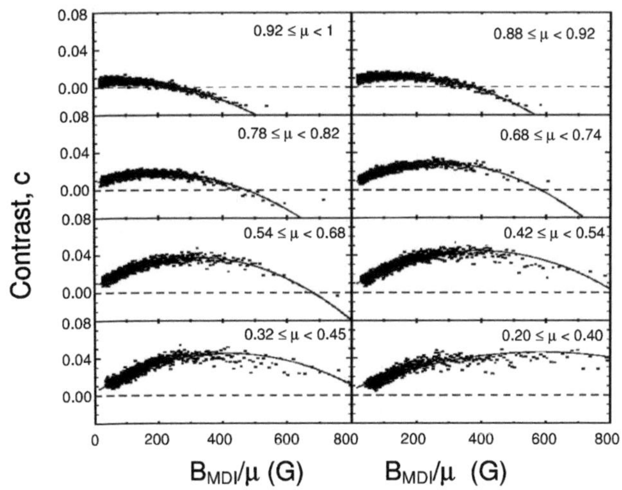

Fig. 60. Observations of the facular/micropore contrast C as observed at 676.8 nm by Ortiz et al. [2002] and sorted as a function of the disc position parameter μ and the radial field component (B_{MDI}/μ), where B_{MDI} is the line-of-sight field observed in an MDI pixel. The lines show the fits using the algorithm given in (147)

It should be noted that these data are specific to the pixel size of the MDI instrument ($2'' \times 2''$). This is because the facular flux tubes are not resolved and all faculae tend to have roughly the same order of field magnitude, $B_f \sim 1000\,\mathrm{G}$. This means that the B_{MDI} values are strongly dependent on the facular filling factor in a pixel, rather than the value of B_f. Ortiz et al. [2002] derive the best-fit polynomial

$$C\left(\left|\frac{B}{\mu}\right|,\mu\right) = (0.48 + 9.12\mu - 8.50\mu^2) \times 10^{-4} \times \left|\frac{B}{\mu}\right| + (0.06 - 2.00\mu - 1.23\mu^2) \times 10^{-6} \times \left|\frac{B}{\mu}\right|^2 + (0.63 + 3.90\mu + 2.82\mu^2) \times 10^{-10} \times \left|\frac{B}{\mu}\right|^3 \quad (148)$$

which is plotted in Figs. 60 and 61. Ortiz [2003] has demonstrated that this function is valid throughout the solar cycle.

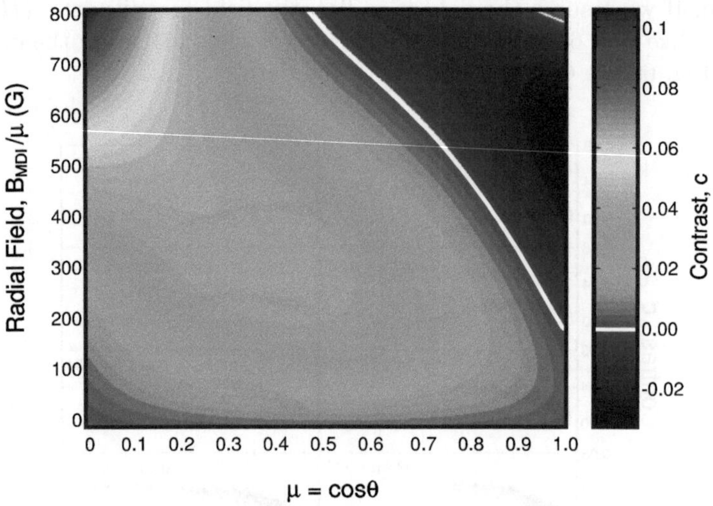

Fig. 61. Plot of the best-fit contrast at a wavelength of 676.8 nm of faculae and micropores, as a function of radial field strength $|B/\mu|$ and disc position μ, from the polynomial fit by Ortiz et al. [2002] and as given by (147). Note that *dark* micropores are observed near the disc centre (μ near unity) and larger field values. Both contrasts and field values relate to pixels of the size of the MDI instrument

4.14 Three- and Four-Component Models of TSI

In recent years, several studies have been able to explain almost all of the observed variations in the total solar irradiance by summing the effect of surface magnetic features (e.g. Solanki and Fligge 2002, Krivova et al. 2003, Solanki and Krivova 2003). The models have reproduced both the 27-day variations (as dark and bright features rotate over the visible disk) and the rising phase of the solar cycle (as the total magnetic flux threading the photosphere increases). The main assumption of these models is that changes are entirely caused by the magnetic field at the solar surface, as seen by high-resolution magnetograms. In the 3-component model, the entire photosphere is divided into just three components: quiet Sun, sunspots and faculae. The 4-component models are a refinement of this, making the distinction between umbra and penumbra, rather than using an average sunspot contrast. Magnetograms are used to determine the filling factor of each surface type at a given disc position (μ) and then a model of each of the three classes used to compute the intensity of each magnetogram pixel, these are then averaged to give the disc-averaged intensity which is converted into TSI using (77). The only free parameter is a pixel filling factor to allow for the fact that faculae are too small to be resolved in the magnetogram data.

In the 4-component model, the irradiance is computed from the disc-averaged intensity for a given wavelength λ and time t, given by (127)

$$\begin{aligned}\langle I\rangle_D(\lambda,t) = 2\int_0^1 & [\alpha_P(\mu,t)I_P(\mu,\lambda) + \alpha_U(\mu,t)I_U(\mu,\lambda) \\ & + \alpha_F(\mu,t)I_F(\mu,\lambda) + \alpha_Q(\mu,t)I_Q(\mu,\lambda)]\mu\mathrm{d}\mu \end{aligned} \tag{149}$$

where $\alpha(\mu,t)$ is the filling factor at position μ and time t. Because the entire disc is assumed to consist of only the 4 components (subscripts P, U, F and Q stand for, respectively, penumbra, umbra, facula and quiet Sun),

$$\alpha_P(\mu,t) + \alpha_U(\mu,t) + \alpha_F(\mu,t) + \alpha_Q(\mu,t) = 1 \tag{150}$$

The intensities of pixels of each type $I_P(\mu,\lambda)$, $I_U(\mu,\lambda)$, $I_F(\mu,\lambda)$ and $I_Q(\mu,\lambda)$ depend on position on the disc and wavelength, but are assumed to be independent of time, t.

If we adopt the convention that all positive contrasts C are brightenings, as in (145) (so for dark umbrae and (less dark) penumbrae $C_U < C_P < 0$ whereas for faculae $C_F > 0$), substituting (150) into (149),

$$\begin{aligned} I_{TS}(\lambda,t) = 2\left(\frac{\pi R_S^2}{R_1^2}\right) I_O \int_0^1 & L_D(\mu,\lambda)[\alpha_P(\mu,t)\{C_P(\mu,\lambda)+1\} \\ & + \alpha_U(\mu,t)\{C_U(\mu,\lambda)+1\} + \alpha_F(\mu,t)\{C_F(\mu,\lambda)+1\} \\ & + \{1-\alpha_P(\mu,t) - \alpha_U(\mu,t) - \alpha_F(\mu,t)\}]\mu\mathrm{d}\mu \end{aligned} \tag{151}$$

In order to compute the TSI from (151) we require a model of the contrasts C_U, C_P, and C_F as a function of position μ and wavelength λ (but note that these are independent of time t – the time dependence is entirely due to that in the filling factors α, which are functions of μ and t, but not λ). Every pixel in the magnetogram for time t that falls on the visible disc is then classified as either umbra, penumbra, facula or quiet Sun to derive the filling factors. We also need to adopt best fit values of the limb darkening function $L_D(\mu,\lambda)$

and of the quiet-Sun intensity of the disc centre, I_O, when free of all magnetic features.

The closeness with which the observed TSI can be reconstructed with a 3-, or better still, 4-component model is underlined by Figs. 62 and 63. Figure 62 compares the reconstructed and observed data for two intervals, one near solar minimum, the other near solar maximum. Agreement is very good and the reconstruction of TSI variations due to dark and bright solar features moving across the disc is well replicated. In addition the rise due to the solar cycle is well reproduced. The overall agreement for all daily data during the rising phase of solar cycle 23 is shown as a scatter plot in Fig. 63. The ideal slope is shown by the solid line, the best-fit the dashed line. It can be seen that the model reconstructs the observed TSI exceptionally well. Recently Wenzler et al. [2004] have used Kitt Peak magnetograms to carry out the same test using all the TSI observations (since 1978). Again the agreement is excellent.

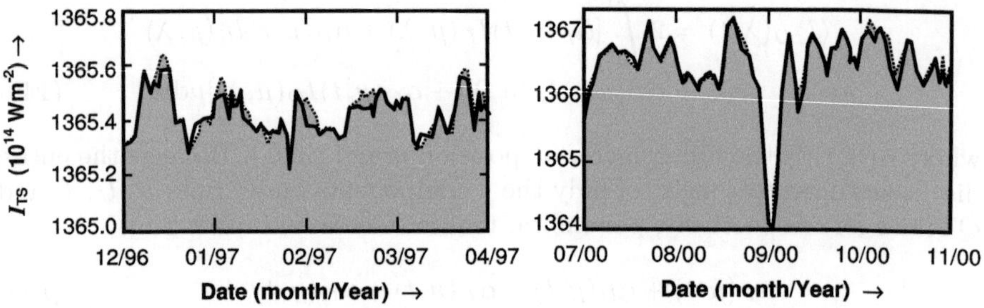

Fig. 62. Reconstruction of TSI from magnetogram data by a 4-component model [after Krivova et al., 2003]. The area *shaded grey*, bounded by a *dotted* line, gives the measured TSI as observed by the VIRGO instrument on SoHO, the *solid black* line is the reconstruction using the model and (151). The intervals are near sunspot minimum and maximum (*left* and *right* respectively)

The quality of the agreement between the observed and modelled TSI in these studies leave little room for β-effects or α-effects due to fields deep inside the convection zone, although variations in the intensity of the limb photosphere have been explained in terms of such effects in the past [Libbrecht and Kuhn, 1984, Kuhn et al., 1988, Kuhn and Libbrecht, 1991]. Thus shadow effects are not needed to explain recent solar cycle changes in the TSI, which are well explained by the effects of magnetic field in the solar surface.

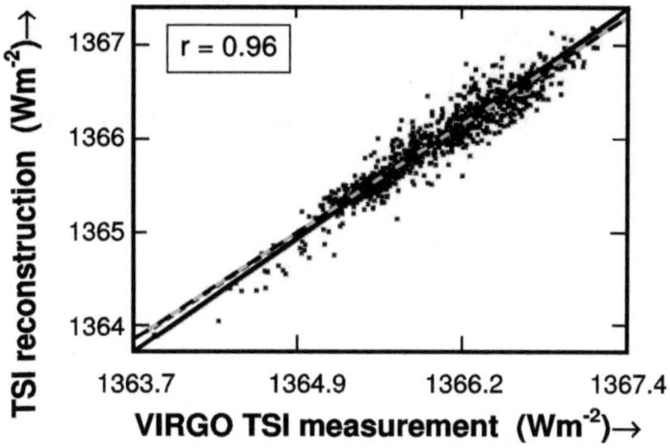

Fig. 63. Reconstruction of TSI from magnetogram data by a 4-component model. The scatter plot compares the measured TSI, as observed by the VIRGO instrument, to the reconstruction using the model and (148). The best fit linear regression (*dashed line*) matches the ideal (*solid*) line exceptionally well [after Krivova et al., 2003]

5 Variability on Century Timescales

The effects of variations in solar outputs, on timescales of decades and less, are expected, in large degree, to be smoothed out in Earth's surface temperatures. This is because of the long time-constants of Earth's coupled ocean–atmosphere system and, in particular, the large thermal capacity of the oceans [Wigley and Raper, 1990]. However, the surface temperature record inferred for the last few centuries shows variations on timescales of a few decades and greater which are readily detected above the inter-annual variability [Rind and Overpeck, 1993, Mann et al., 1999, Jones et al., 2001]. This places limits on the time constants for the terrestrial response to changes in the radiative climate forcing [Hansen et al., 1997]. Thus century-scale variations in solar outputs would not be smoothed out. It is important to characterise these properly when evaluating the relative effects of all other long-term influences on Earth's climate (e.g. Crowley, 2000). Modern-day studies of long-term solar change and its effect on climate was pioneered by Eddy [1976]. In this section we look at the evidence for variations in the solar outputs on 100-year timescales.

5.1 Long-Term Variations in Sunspots and Cosmogenic Isotopes

The longest sequence of measurements relevant to solar variability is the sunspot number. Regular observations began in Zurich in 1749 and the *Wolf sunspot number* was devised in 1848 by Johann Rudolf Wolf, director of the Zurich Observatory (and hence this is also called the *Zurich sunspot number*). Neither the number of individual spots, S, nor the number of spot groups, G, fully describe the level of activity and sunspot number is defined as

$R_Z = k(10G + S)$, where the factor k allows for differences between observers and their methods, sites and equipment. The *international sunspot number* R is compiled using the same basic algorithm as R_Z by the Sunspot Index Data Centre in Belgium and is a weighted mean (usually the weighting factor k is less than unity) for a global network of observatories (usually exceeding 6 in number). Another sunspot number is generated by NOAA in Boulder, USA and this is systematically about 25% higher than the international sunspot number (the differences arising from the observatories used and the weighting factors applied).

Observations of sunspots were made before 1749 and Wolf sunspot numbers can be generated back to 1700. The earliest values are generally reliable in annual means but the data is often too sparse for them to be reliable on a monthly basis. To give a sunspot index less susceptible to sparse data, Hoyt and Schatten [1998] devised the *group sunspot numbers*, $R_G = (12.08/N)\Sigma_{i=1}^{N} k_i G_i$, where G_i is the number of sunspot groups seen by the ith of N observers, for whom the weighting factor is k_i. The factor 12.08 is derived to make R_G equal the international sunspot number R for the interval between 1874 and 1976, when the Royal Greenwich Observatory generated a homogeneous and highly reliable sequence of sunspot group observations. Hoyt and Schatten were able to derive R_G data back to 1610 when sunspot observations became more common following the invention of the telescope. However, their error analysis reveals that it is highly desirable to have at least 4 widely-spaced observers and for the earliest data this causes errors in monthly values: annual means are more reliable because they average out such errors [Usoskin et al., 2003c].

The top panel of Fig. 64 shows the full sequence of annual means of the group sunspot numbers, R_G. The solar cycle can clearly be observed, as can the Maunder minimum, the extended period of very few detectable spots between about 1650 and 1700 [Eddy, 1980]. Reviews of how this minimum became to be accepted as real have been given by Letfus [2000] and Cliver [1994].

The lower panel of Fig. 64 shows annual values of the abundance of the ^{10}Be isotope, as measured and dated in the Dye-3 ice core taken from the Greenland ice sheet [Beer et al., 1990, 1998, Beer, 2000]. In general, the considerably greater precipitation rates into the Greenland ice sheet make the cosmogenic isotope data more reliable and easier to date that those from regions of relatively low precipitation, for example Antarctica [McCracken, 2004]. The solar cycle can also be seen in these cosmogenic isotope data.

The long-term drifts in these parameters are revealed by the 11-year running means, in which solar cycle variations are smoothed out. In Fig. 65, the ^{10}Be isotope abundance scale has been inverted so that direct comparison can be made with the corresponding means of R_G. As well as the Maunder minimum, the Dalton minimum can be seen at about 1790–1830 in both data series, and there is a weaker minimum around 1900. The quasi-periodic

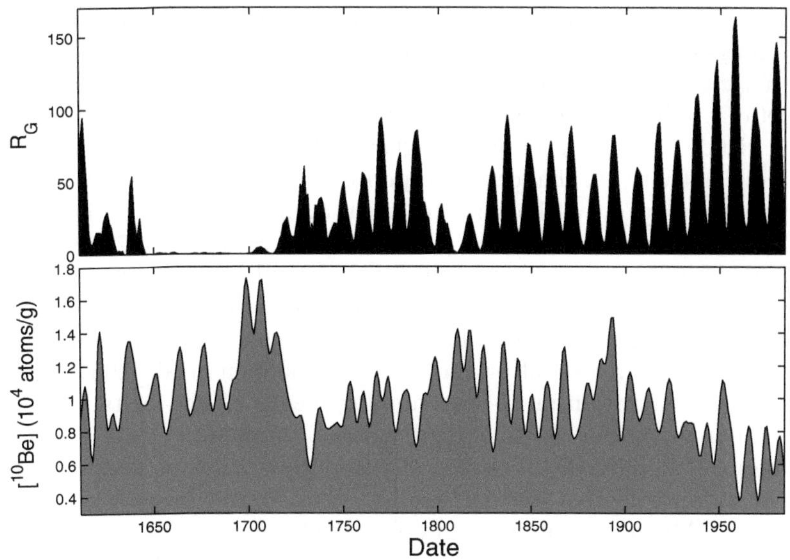

Fig. 64. (*Top*) Annual means of the group sunspot number, R_G, as compiled by Hoyt and Schatten [1998]. (*Bottom*) The abundance [^{10}Be] of the ^{10}Be isotope as measured and dated in the Dye-3 Greenland ice core by Beer et al. [1990]

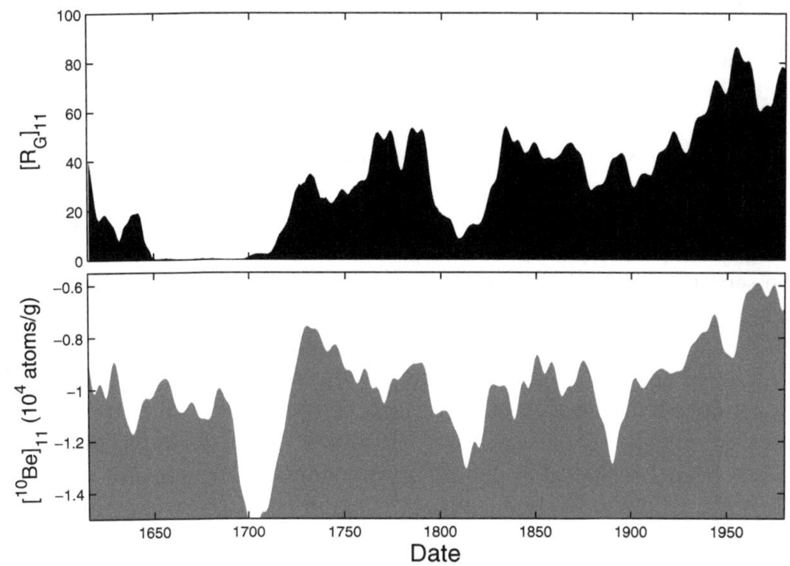

Fig. 65. (*Top*) Eleven-year running means of the group sunspot number, $[R_G]_{11}$. (*Bottom*) Eleven-year running means of the abundance of the ^{10}Be isotope as measured in the Dye-3 Greenland ice core, $[^{10}\mathrm{Be}]_{11}$. Note that the $[^{10}\mathrm{Be}]_{11}$ scale has been inverted to allow direct comparison with $[R_G]_{11}$

behaviour with period of 80–100 years was first noted by Gleissberg [1944]. These long-term changes are mirrored in data on the terrestrial response to solar activity – specifically geomagnetic activity and auroral activity. Geomagnetic activity will be discussed in the next section. Aurora is present, at some latitude and strength, on all nights, but moves to lower latitudes as geomagnetic activity is enhanced in response to the enhanced solar wind speed and, more importantly, the size and orientation of the heliospheric field at Earth. Legrand and Simon [1985], Simon and Legrand [1987], and Legrand and Simon [1991] have investigated the threshold latitude which makes the percentage of auroral nights at and below that latitude a good indicator of solar-terrestrial activity. Auroral occurrence has shown considerable changes over the past 500 years [Silverman, 1992]. Pulkkinen et al. [2001] show that the long-term variation in 11-year running means of low-latitude aurorae is very similar to that in the smoothed sunspot numbers.

One notable difference between the smoothed sunspot numbers and the ^{10}Be abundance is that the latter only rises to the largest values at the end of the Maunder minimum. Letfus [2000] shows the same is true of the occurrence of low-latitude aurora which falls to its lowest values only by the end of the Maunder minimum in sunspot data. On the other hand, the inferred production rate of the ^{14}C isotope [Kocharov et al., 1995] shows a longer-lived maximum. Thus both cosmogenic isotopes (shielded from Earth by the local heliospheric field) and auroral activity (enhanced when the local heliospheric field is enhanced) provide some evidence that the magnetic flux in the heliosphere decayed relatively slowly even when flux emergence through the solar surface was sufficiently reduced that almost no sunspots were seen.

Figure 66 shows the solar cycle variations in both the ^{10}Be abundance and the sunspot number R_G for 1830–1980. These data have been detrended by subtracting the simultaneous 11-year running means to give $\Delta[^{10}\mathrm{Be}](t) = [^{10}\mathrm{Be}](t) - [^{10}\mathrm{Be}]_{11}(t)$ and $\Delta R_G(t) = R_G(t) - [R_G]_{11}(t)$. It can be seen that the ^{10}Be abundance clearly reflects the solar cycle variation, although there are some phase differences in the earlier data shown. There are three factors which contribute to such phase discrepancies. The first is the dating of the ice core, the second is the delay in deposition of the isotopes produced in the stratosphere into the ice sheet, and the third is the sparcity of sunspot observers for early data. As an example of the latter, there has been considerable debate as the whether the sunspot observations failed to record a small, short sunspot cycle in the, otherwise, unusually long cycle number 4 which lasts from 1846 to 1880 in the group sunspot number and Wolf sunspot number data series [Usoskin et al., 2003b]. In this context, Krivova et al. [2002a] show that both ^{10}Be and ^{14}C cosmogenic isotope abundances, and the low-latitude auroral activity, all follow the Hoyt and Schatten and Wolf sunspot number data series, with an unusually long cycle number 4.

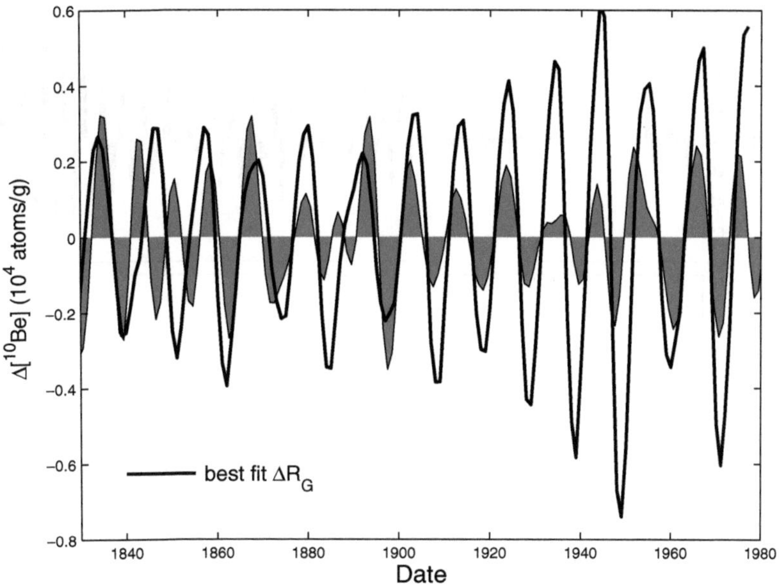

Fig. 66. Low pass filtered data on the group sunspot number and the ^{10}Be isotope abundance in the Dye-3 Greenland ice core. The *grey* area is $\Delta[^{10}\text{Be}]$, the deviation of annual values of the abundance $[^{10}\text{Be}]$ from the 11-year means $[^{10}\text{Be}]_{11}$; the *black* line is the best fit of ΔR_G, the corresponding deviation of annual values of R_G from $[R_G]_{11}$. The solar cycle variations of the two parameters can be seen to be in phase throughout most of the interval shown

Figure 67 shows the full data sequences of the detrended ^{10}Be data and the group sunspot number and a second major difference can be seen: whereas there are almost no spots in the Maunder minimum (and thus no cyclic behaviour), oscillations near 11 years are seen in the ^{10}Be data throughout the Maunder minimum [Beer et al., 1998]. This suggests that open flux may well have still emerged in the Maunder minimum, and what distinguishes this interval is not a complete lack of flux emergence but a lack of emergence of BMRs of sufficient strength and / or size to be seen as sunspots. Solar cycles in the terrestrial response during the Maunder minimum are also seen in geomagnetic activity [Feynman and Crooker, 1978, Cliver et al., 1998] and, to a lesser extent, in the occurrence of low-latitude aurorae [Letfus, 2000]. (Note, however, that the spectral analysis by Silverman [1992] and Silverman and Shapiro [1983], of auroral records for 1650–1725 did not detect any 11-year oscillation). The continued cyclic activity during the Maunder minimum in these indicators can be interpreted as showing that magnetic flux continued to emerge through the solar surface during the Maunder minimum, but that almost none of it was in the form of the strong, large BMR flux tubes that give sunspots. One possibility is that the strong solar dynamo ceased to operate,

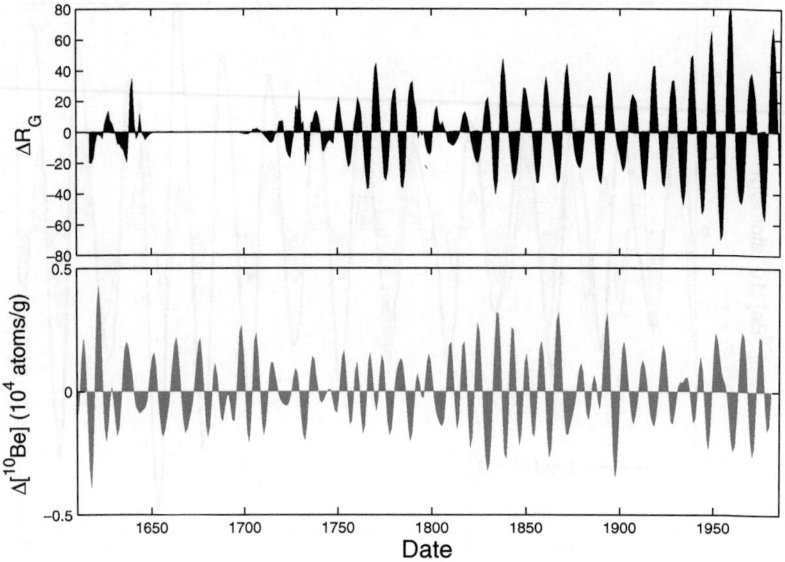

Fig. 67. The full sequences of the detrended ^{10}Be and sunspot data ($\Delta[^{10}\mathrm{Be}]$ and ΔR_G in the *top* and *bottom* panels, respectively) since 1600

but the weak, turbulent dynamo continued and produced flux which lines up to make a nett contribution to the open flux, possibly at a lower level or maybe even at the same level as during modern times. Searches that have been made for phase skips in the cosmogenic isotope abundance cycles as they may reveal the dynamo behaviour through the Maunder minimum have been inconclusive [Feynman and Gabriel, 1990].

5.2 Geomagnetic Variations

The previous section made a number of references to geomagnetic observations. To understand their implications properly it is important to understand these data and how they are influenced by solar magnetism. One of the longest available datasets is the *aa index*, compiled for 1868–1968 by Mayaud [1971, 1972] and subsequently continued to the present day. The *aa* index quantifies geomagnetic activity from the range of variations in the geomagnetic field during three-hourly intervals, recorded since 1868 by pairs of near-antipodal, mid-latitude magnetometers in England and Australia. Because the instruments used have been carefully cross-calibrated and because the data have been processed in a uniform way, this is a highly valuable and homogeneous data series. The *aa* index is defined as the average of the *aa* values from the two near-antipodal magnetometer stations. The exact location of the stations used has varied. Initially, Greenwich and Melbourne were employed. However, the Australian station was moved in 1919 to Toolangi and in 1980 to Canberra; the UK station was moved to Abinger in 1926 and to Hartland in 1957. For each change in location, a correlation analysis was carried out and calibration factors used to allow for changes in geomagnetic latitude and local effects. The procedure used to derive *aa* is demonstrated by Fig. 68.

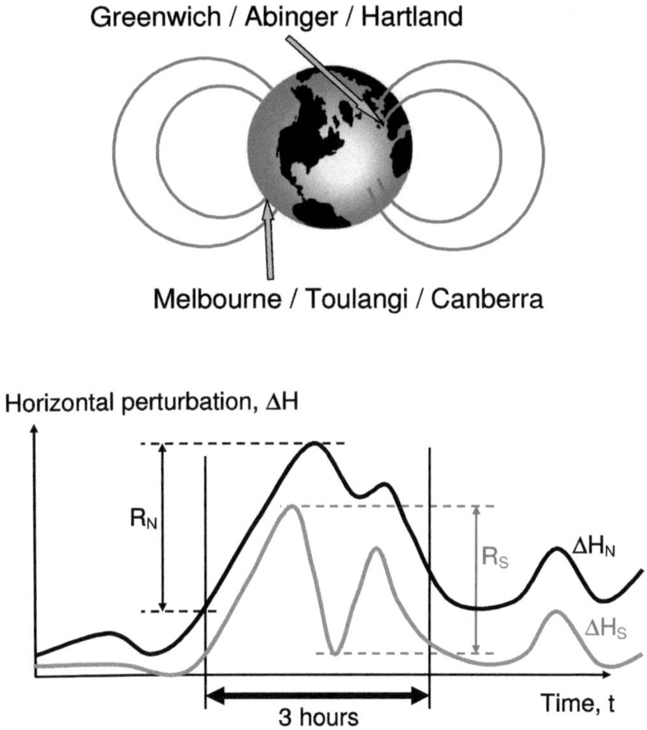

Fig. 68. The compilation of the *aa* geomagnetic index. The range (R_N and R_S for the northern and southern hemisphere observatories) of the variation in the observed horizontal component of the magnetic field (ΔH_N and ΔH_S) is scaled in each three-hour interval. These are then converted into K values, after the quiet diurnal variation has been subtracted and, using station-dependent scaling factors, *aa* indices are derived for the northern and southern hemisphere separately (aa_N and aa_S respectively). The *aa* index is the arithmetic mean, $aa = (aa_N + aa_S)/2$

Variations in the observed magnetic field are driven by many factors, ranging from thermal ionospheric tides and winds to the effect of transient solar disturbances in the solar wind. The size of the perturbation seen depends on a number of factors including the ionospheric conductivities (due to photoionisation by solar EUV and X-ray radiations and to particle impact ionization by auroral precipitation). One key phenomenon is the magnetospheric substorm, the occurrence and severity of which are controlled by the strength and orientation of the local heliospheric field (the IMF). The reasons can be understood from Poynting's theorem for energy flow applied to the magnetosphere [Cowley, 1991].

If we compress a magnetic field by ∂v in volume, we do work against the magnetic pressure of $\partial W_B = (B_2/2\mu_0)\partial v$. Thus the rate at which energy is stored in the field in a volume v is

$$\frac{\partial W_B}{\partial t} = \frac{\partial}{\partial t}\left(\int_v \left(\frac{B^2}{2\mu_0}\right) \mathrm{d}v\right) = \left(\frac{1}{\mu_0}\right) \int_v \left(\boldsymbol{B} \cdot \frac{\partial \boldsymbol{B}}{\partial t}\right) \mathrm{d}v \tag{152}$$

If we substitute from Faraday's law, $\nabla \times \boldsymbol{E} = -\partial \boldsymbol{B}/\partial t$, use the vector relation $\nabla \times (\boldsymbol{E} \times \boldsymbol{B}) = \boldsymbol{B} \cdot (\nabla \times \boldsymbol{E}) - \boldsymbol{E} \cdot (\nabla \times \boldsymbol{B})$, and apply Ampére's law, $(\nabla \times \boldsymbol{B}) = \mu_0 \boldsymbol{J}$, the definition of Poynting flux, $\boldsymbol{S} = (\boldsymbol{E} \times \boldsymbol{B})/\mu_0$, and the divergence theorem, $\int_v \nabla \cdot \boldsymbol{S} \mathrm{d}v = \int_A \boldsymbol{S} \cdot \mathrm{d}\boldsymbol{a}$ (where the surface A surrounds the volume v), we derive Poynting's theorem for a plasma:

$$\frac{\partial W_B}{\partial t} = -\int_A \boldsymbol{S} \cdot \mathrm{d}\boldsymbol{a} - \int_v \boldsymbol{E} \cdot \boldsymbol{J} \mathrm{d}v \tag{153}$$

The first term on the right is the divergence of the Poynting flux and the second is the ohmic heating term. Using Ohm's law, $\boldsymbol{J} = \sigma[\boldsymbol{E} + \boldsymbol{V} \times \boldsymbol{B}]$, it can be shown that the ohmic heating term has two parts, the resistive energy dissipation and the mechanical work done against the $\boldsymbol{J} \times \boldsymbol{B}$ force.

If we consider steady state, $\partial W_B/\partial t$ is zero and a region where $\boldsymbol{J} \cdot \boldsymbol{E} > 0$ is a sink of Poynting flux, i.e. energy goes from the electromagnetic field into the particles (giving acceleration and heating). Conversely regions of $\boldsymbol{J} \cdot \boldsymbol{E} < 0$ are sources of Poynting flux, i.e. energy goes from the particles into the electromagnetic field.

In current-free regions in non-steady situations, $\int_A \boldsymbol{S} \cdot \mathrm{d}\boldsymbol{a} = \partial W_B/\partial t$. In this case, the divergence in the Poynting flux is balanced by changes in the energy stored in the local magnetic field. A sink/source of Poynting flux is a region where the energy stored in the field is increasing/decaying.

A southward IMF (in the $-Z_{GSM}$ direction), frozen into the solar wind flow (in the $-X$ direction), gives a dawn-to-dusk electric field (in the $+Y_{GSM}$ direction) in the Earth's frame ($\boldsymbol{E} = -\boldsymbol{V}_{SW} \times \boldsymbol{B}$). The geomagnetic field forms an obstacle to the solar wind flow and generates a low-density cavity within it, the *magnetosphere*. Because the solar wind is super-Alfvénic (the Alfvén Mach number, $M_A = V_{SW}/V_A$ is of order 9, see Table 4), a bow shock forms upstream of the boundary between the solar wind and the magnetosphere, the *magnetopause*. The slowed and heated solar wind between the bow shock and the magnetopause is called the magnetosheath, as shown in Fig. 69. Poynting flux is radially away from the Sun in the solar wind and is enhanced at the bow shock (BS) and magnetosheath (MS) where $\boldsymbol{J} \cdot \boldsymbol{E} < 0$ (meaning that kinetic energy of the solar wind flow is converted here into Poynting flux) because of the currents associated with the draping of IMF field lines around the magnetosphere in the magnetosheath. The Chapman–Ferraro (C–F) currents flow in the magnetopause and are associated with

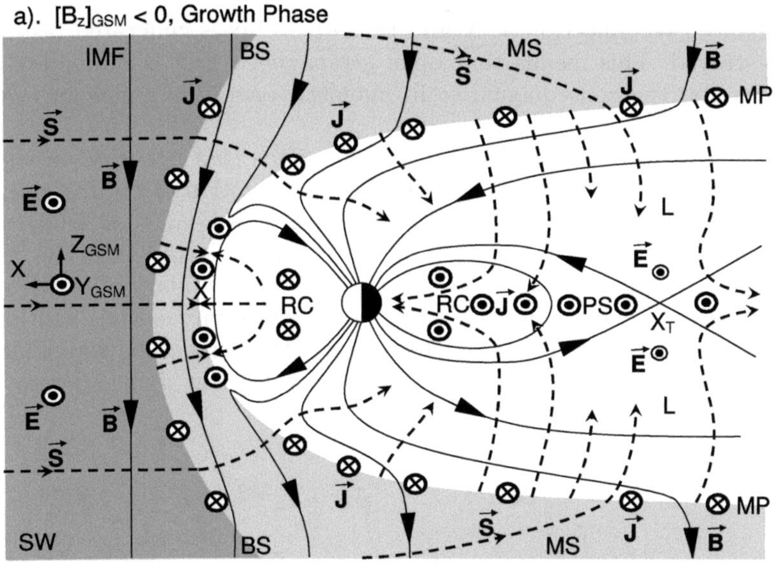

Fig. 69. Schematic of energy flow in the noon–midnight plane of Earth's magnetosphere, shortly after the interplanetary magnetic field (IMF) has turned southward in the GSM frame ($[B_Z]_{GSM} < 0$) – i.e. during the growth phase of a magnetospheric substorm. Other features labelled are the solar wind (SW, *shaded dark grey*), the bow shock (BS) which is the outer boundary of the magnetosheath (MS, *shaded light grey*), the magnetopause (MP) which bounds the magnetosphere (*shaded white*), of which the magnetospheric tail lobes (L), the plasma sheet (PS) and the ring current (RC) are labelled. Dashed arrows give the Poynting flux, $\boldsymbol{S}$, *solid* lines with arrows are the magnetic field, $\boldsymbol{B}$, and vectors in the dusk/dawn directions (out of the plane of the diagram) show the electric field, $\boldsymbol{E}$, and the current density, $\boldsymbol{J}$. (Note that the distortion of the figure introduced by the need to foreshorten the magnetospheric tail means that $\boldsymbol{S}$ does not appear orthogonal to $\boldsymbol{B}$ in some places.) Reconnection sites in the dayside magnetopause and cross-tail current sheet are labelled X and X_T, respectively

the difference between the high-field in the magnetosphere and the generally lower fields outside it. The C–F currents are dawnward on the long tail lobe boundaries, making $\boldsymbol{J} \cdot \boldsymbol{E} < 0$ and so these surfaces are sources of Poynting flux and energy is extracted here from the sheath plasma flow. Near the nose of the magnetosphere the currents are duskward (i.e. $\boldsymbol{J} \cdot \boldsymbol{E} > 0$). This part of the magnetopause is a sink of Poynting flux, consistent with the outflows away from the reconnection site, X. The oppositely-directed fields of the two tail lobes are separated by the cross-tail current where $\boldsymbol{J} \cdot \boldsymbol{E} > 0$. This sink of Poynting flux is consistent with the tail reconnection site X_T and where much of the energy extracted from the solar wind is deposited to generate the energetic plasma of the *plasma sheet* (PS) and the *ring current* (RC). In a *substorm growth phase*, the reconnection rate at X_T has yet to respond to the enhanced reconnection at X and thus open flux is generated faster than it is destroyed. This means that open geomagnetic flux accumulates in the *tail lobes* and the consequent rise in magnetic energy is a sink of Poynting flux.

Figures 69 and 70 consider the two situations that occur when the IMF points southward. For this IMF orientation, the motion of the frozen-in field causes an electric field in the Earth's frame $\boldsymbol{E} = -\boldsymbol{V}_{SW} \times \boldsymbol{B}$ which points from dawn to dusk. That electric field is communicated into the magnetosphere by magnetic reconnection at the dayside boundary of Earth's magnetosphere (at X, with a dawn-to-dusk reconnection rate electric field), which generates open geomagnetic flux (that threads the magnetopause). This open flux is moved into the tail lobe by the solar wind flow and is destroyed by

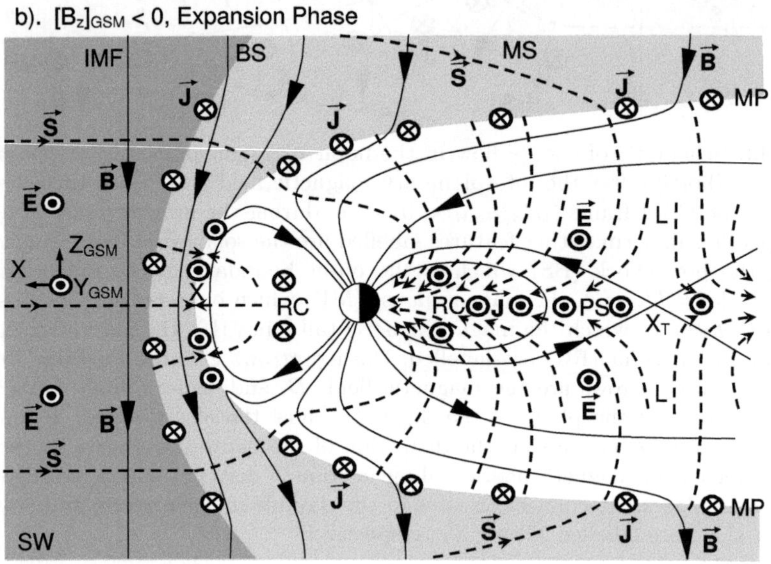

Fig. 70. Same as Fig. 69, for when the IMF has persisted in a southward orientation (for longer than a typical growth phase duration of 40 min.) and the onset of fast reconnection in the cross-tail current sheet (at X_T, which is usually a new reconnection site closer to Earth) means that the substorm has developed into an expansion phase. In this phase, the magnetic energy and open geomagnetic flux stored in the tail lobes during the growth phase is released by the rate of reconnection at X_T exceeding that at X. Now $\partial W_B/\partial t$ is negative in the tail lobes which become sources of Poynting flux. The field configuration changes in the near-Earth tail associated with this release of energy drive currents in the midnight sector ionosphere (the so called "current wedge") and ionospheric conductivity is enhanced by the precipitation of the particles energised in the plasma sheet. Thus energy is deposited in the upper atmosphere by both joule heating and energised particle precipitation. These currents give geomagnetic activity, detected by ground-based magnetometers and quantified by indices such as *aa*

reconnection (again associated with dawn-to-dusk electric field) in the cross-tail current sheet (at X_T). Figure 69 shows the situation shortly after a southward turning of the IMF, the reconnection voltage along X exceeds that along X_T, which has yet to respond to the IMF change. This applies throughout an interval termed the substorm growth phase. Along the long boundaries of the tail of the magnetosphere $\boldsymbol{J} \cdot \boldsymbol{E} < 0$, making these regions sources of Poynting flux, where energy is extracted from the flow of the shocked solar wind in the magnetosheath. This energy is stored in the tail lobes as magnetic energy ($\partial W_B / \partial t > 0$) as magnetic flux is appended to the tail lobes because open flux is generated and appended to the tail faster than it is being destroyed.

This accumulation of energy in the tail lobes during the growth phase cannot continue indefinitely and as the cross-tail current in the near-Earth tail increases it becomes unstable and fast reconnection is established at a tail reconnection site X_T. This destroys open flux more rapidly than it is produced giving, $\partial W_B / \partial t < 0$. Thus the energy stored in the tail lobes during the growth phase is released into the inner magnetosphere and nightside upper atmosphere. This is called the *substorm expansion phase* and is illustrated in Fig. 70. The reconfiguration of the field in the near-Earth tail means that the inner edge of the cross-tail current is diverted to flow through the ionosphere in the *substorm current wedge*, and the precipitation of energetic particles (produced in the plasma sheet sink of Poynting flux) enhances the aurora and ionospheric conductivities. Magnetometers at high and middle latitudes show deflections due to the auroral electrojet, the part of the current wedge in the ionosphere. Growth phases typically last about 45 minutes and for steady southward IMF, these substorm cycles of energy storage and deposition last of order 1.5 hours and set the observed range of variation in the three hour intervals of the *aa* index.

A very different situation prevails when the IMF points northward, as demonstrated by Fig. 71. In this case, the long boundaries of the geomagnetic tail are sinks of Poynting flux and energy deposition in the near-Earth magnetosphere and ionosphere is restricted to weak directly-driven deposition on the dayside and the weak remnants of prior periods of southward IMF on the nightside. Thus the energy deposition, and the geomagnetic activity associated with it, are strong functions of the IMF orientation [Arnoldy, 1971, Baker, 1986, Bargatze et al., 1986, Stamper et al., 1999].

Care must be taken when interpreting magnetometer data because of a number of complicating factors [Mayaud, 1976, Baumjohann, 1986]. The substorm auroral electrojet forms in the midnight sector in a geomagnetic frame of reference and thus the magnetic local time (MLT) of the station is important, as well as its geomagnetic latitude (which determines how close the station is to the auroral oval). MLT is defined from the hour angle of the Sun at the point where the field line in question cuts the ecliptic plane and depends on the Universal Time (UT), the time-of-year and the geomagnetic coordinates of the station. The ionospheric conductivity within the

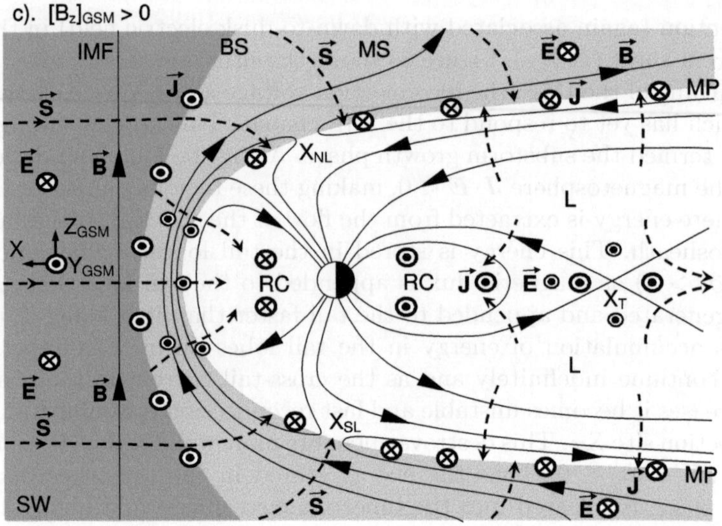

Fig. 71. Same as Figs. 69 and 70, for northward IMF in the GSM frame ($[B_Z]_{GSM} > 0$). The interplanetary electric field and the currents in the bow shock and magnetosphere (associated with the draping of the IMF round the magnetosphere) are all reversed compared to the $[B_Z]_{GSM} < 0$ case; however the strong magnetospheric field strengths mean that the C–F currents are not radically altered. Now the long boundaries of the tail lobes are sinks of Poynting flux, the energy going into accelerated outflow from reconnection sites which are on the sunward edges of the tail lobe magnetopause (X_{NL} and X_{SL} for the northern and southern hemisphere lobes, respectively). Because there is residual open flux in the tail lobes produced by prior periods of southward IMF, some reconnection continues at X_T, but at a much reduced rate (associated with weak dawn-to-dusk electric field). Note that the presence of dusk-to-dawn electric field nearer the magnetopause and dawn-to-dusk electric field in the centre of the tail means that $\nabla \times \boldsymbol{E}$ is non-zero and so, by Faraday's law, this is inherently a non-steady situation. The only part of the magnetopause that is a source of Poynting flux is the small region on the dayside. Energy deposition in the inner magnetosphere and ionosphere is restricted to small directly-driven effects on the dayside and weak remnant storage system release in the tail

auroral electrojet is largely set by the associated auroral particle precipitation; however, the currents detected at a magnetometer station away from the electrojet will also depend on the local conductivity which, in turn, depends on the solar local time (SLT, set by the hour angle of the Sun at the station and which therefore depends on the UT and the station's geographic coordinates), and also on the time-of-year.

However, these UT and seasonal effects (due to station position and ionospheric conductivities) are complicated by the effect of the tilt of Earth's magnetic dipole axis (see Fig. 72) on energy coupling between the solar wind and the magnetosphere. Earth's rotation axis makes an angle of 23° with

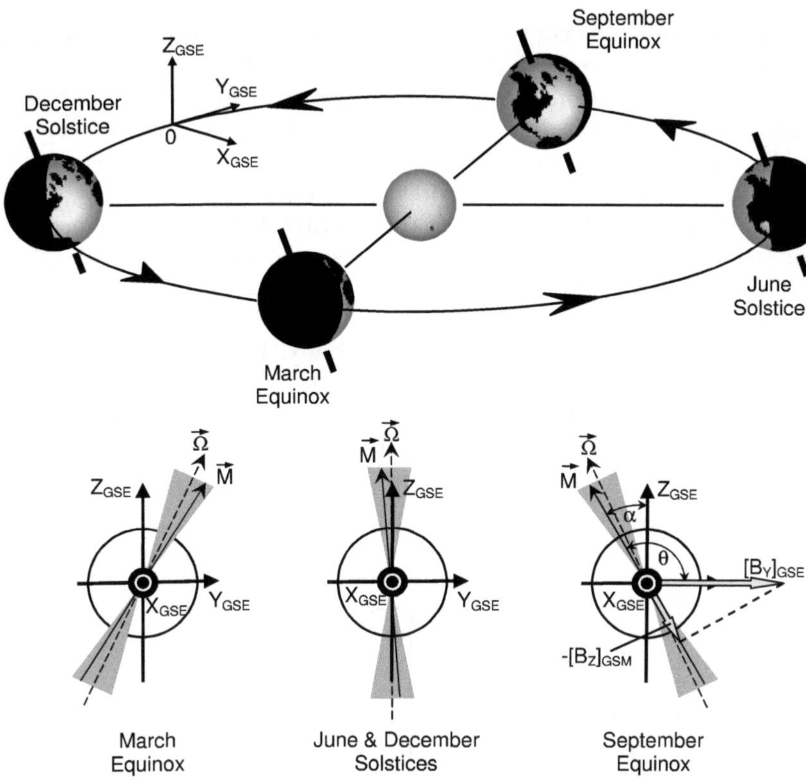

Fig. 72. (*Top*) The orientation of the Earth's rotation axis as a function of time and year. (*Bottom panel*) The Earth's rotation axis $\boldsymbol{\Omega}$ and magnetic dipole axis $\boldsymbol{M}$, as viewed from the Sun in the Geocentric Solar Ecliptic (GSE) frame, in which X_{GSE} points toward the Sun, Z_{GSE} is the northward normal to the ecliptic plane and Y_{GSE} makes up the right-hand coordinate set and is anti-parallel to the Earth's orbital motion). Every 24 hours the magnetic axis rotates around the rotation axis and so the $\boldsymbol{M}$ vector sweeps out the *grey* area in each case. The Geocentric Solar Magnetospheric frame shares the same X axis as the GSE frame, about which the Z axis is rotated through an angle α such that it lines up with the projection of $\boldsymbol{M}$ on the YZ plane. For the September equinox case in the right-hand figure, an IMF in the positive $+Y_{GSE}$ direction is shown, giving a southward IMF in the GSM frame ($[B_Z]_{GSM} < 0$). The IMF clock angle θ in the GSM frame is also shown

the Z_{GSE} axis and circles around once per year. The magnetic axis makes an angle of 11° with the rotation axis and circles around it once per day. Thus the rotation angle α between the GSE and GSM reference frames is a function of both UT and time of year. This is significant because the energy coupling characteristics, as discussed in Figs. 69, 70 and 71, depend on the northward IMF component in the GSM frame, $[Bz]_{GSM}$.

Figure 73 shows the occurrence of IMF orientations in the YZ plane of the GSE frame. It can be seen that $|B_Z|_{GSE}$ is generally smaller than $|B_Y|_{GSE}$ which makes the rotation angle α very important in generating the large negative $[B_Z]_{GSM}$ which drives geomagnetic activity. Figure 73 shows that some $[B_Z]_{GSM} < 0$ events can be caused by the occurrence of $[B_Z]_{GSE} < 0$ (with small α), but that a more frequent occurrence is large $|B_Y|_{GSE}$ which gives $[B_Z]_{GSM} < 0$ with a large α of the required polarity (as demonstrated in the last panel of Fig. 72). This effect is called the *Russell–McPherron effect* [Russell and McPherron, 1973] and the largest $|\alpha|$, giving the best solar wind–magnetosphere coupling is predicted to be at 22 UT at the March equinoxes and 10 UT at the September equinox. In fact, recent analysis of geomagnetic activity [Cliver et al., 2000] suggests an additional dependence on the sunward tilt of the Earth's magnetic axis (in the GSE ZX plane) that is not predicted by the theory of Russell and McPherron [1973]. The combined effect of the Russell–McPherron effect and the sunward tilt is called the "*equinoctial effect*".

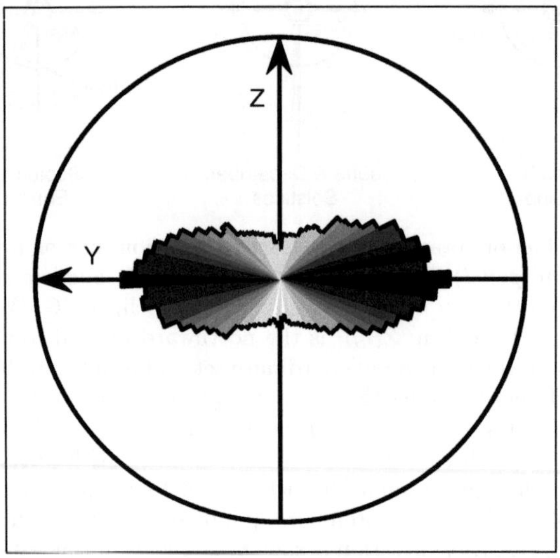

Fig. 73. The occurrence of orientations of the IMF in the GSE Y–Z plane for all hourly averages between 1964 and 2000 (see Fig. 28 for corresponding plots in the Y–X plane)

The role of the angle α means that data from any one magnetometer station, or meridional chain of magnetometers covering a limited range of longitudes, must be used with care to quantify geomagnetic activity because the MLT at which it is most sensitive to substorms occurs at certain UT (that depends on the longitude of the station) and thus there is an implicit selection of α values. For example, a station which approaches the auroral electrojet most closely at 16 UT or 04 UT relies more on large negative $|B_Z|_{GSE}$ events (CMEs, CIRs etc.) to give the $[B_Z]_{GSM} < 0$ which generates the activity it sees, whereas stations for which these times are 10 UT or 22 UT see more $[B_Z]_{GSM} < 0$ events because of the larger $|\alpha|$.

To investigate these effects on the *aa* index, the top panel of Fig. 74 shows the average values (over the full duration of the data sequence since 1868) of aa_N and aa_S as a function of UT and time of year. Both the English stations and the Australian stations show peak geomagnetic activity at the equinoxes, but the UT response is dominated by the station MLT effect and peaks near 21 UT for the English stations and 13 UT for the Australian stations. The lower panel compares the average of the two, the *aa* index, with the *Am* index which has been compiled since 1959 from 8 groups of 3 or 4 stations (including the *aa* stations) covering a full range of longitudes near 50° magnetic latitude in both hemispheres. The plots for the northern and southern hemisphere *Am* stations separately (called the *An* and *As* indices, respectively) show the same general features as Fig. 74(d). The limitation of deriving *aa* from just 2 of the 8 groups used by *Am* can be seen by comparing parts (c) and (d) of Fig. 74. The pattern for the *Am* data is consistent with the equinoctial effect, rather

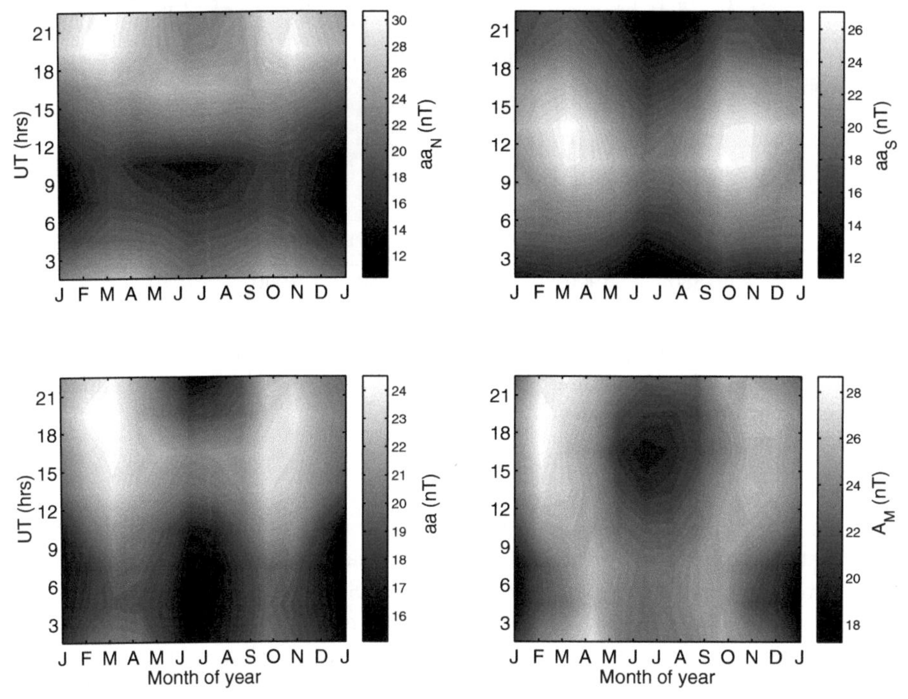

Fig. 74. The variations of various geomagnetic indices as a function of time of year and UT. (**a**) aa_S, (**b**) aa_N, (**c**) $aa = (aa_S + aa_N)/2$ and (**d**) A_m

than the Russell–McPherron effect (see Cliver et al., 2000). The pattern is not consistent with the third possible effect, the *axial effect*, caused by the Earth being furthest away from the heliographic equator near the equinoxes (which gives no UT dependence).

The limitations of a two-station index on timescales shorter than one year, as demonstrated by Fig. 74, were well understood by Mayaud who devised *aa* to reproduce annual means of geomagnetic activity. That this was successfully achieved is demonstrated by Fig. 75 which compares annual means of *aa* with those for the *Am* index and also for the *Ap* index, a range index which uses 12 stations at different longitudes, all in the northern hemisphere.

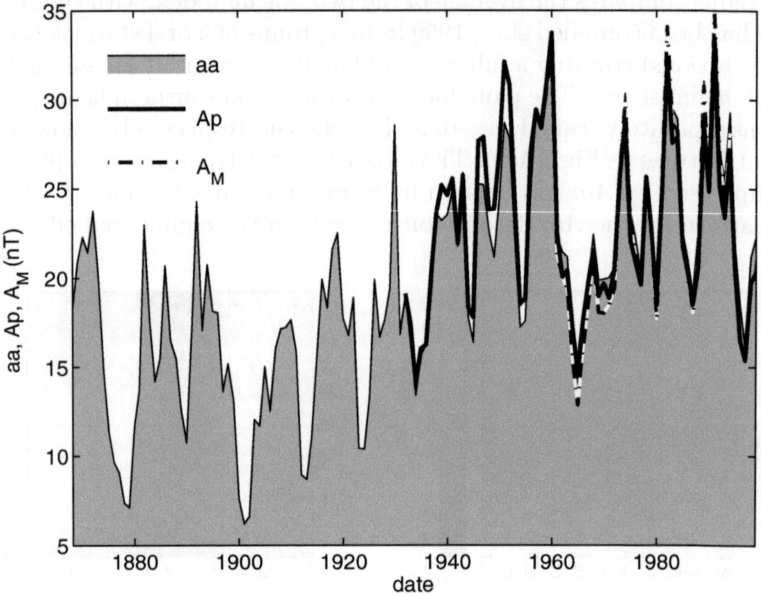

Fig. 75. Variations of annual means of the *aa* index (commencing 1868 and shown by the *thin black line* bounding the *grey* histogram), the A_m index (commencing 1959 and shown by the *dot–dash black* and *white line*) and the A_p index (*thick black line* commencing 1932)

The outstanding feature seen in the *aa* data is the long-term drift during the past 150 years. Application of these data by Lockwood and Stamper [1999] to estimate the open solar flux (see Sect. 5.3) has provoked some debate about the voracity of this long-term change (e.g. Svalgaard et al., 2004). However, there is considerable evidence from other sources that aa is correct: Nevanlinna and Kataja [1993] and Nevanlinna [2004] showed that the earliest *aa* values were consistent with a dataset for 1844–1899 from Helsinki; Cliver and Ling [2002] found similar trends to those in *aa* for other early geomagnetic indices and Pulkkinen et al. [2001] show that the occurrence of low-latitude aurorae follows a very similar long-term drift to *aa*. Figure 75 shows that the trend is also present in the *Ap* data series (starting 1932). Lockwood [2002c, 2003] has demonstrated that the drift is the same in form for aa_N and aa_S, which eliminates errors in the intercalibration of the aa magnetometers and site effects as a potential causes. In fact, the century-scale drift is found to be almost identical for aa_N and aa_S, if the amplitude of the solar cycle variations observed is used to re-calibrate the stations. Such checks are important because a number of factors can introduce drifts into the signal seen at any one station. The most obvious the changes are the locations of the *aa* station sites, but these must be put into the context of the changing geomagnetic field. Clilverd et al. [1998] have pointed out that the drift of the geomagnetic poles has accelerated and that the distance of any one station from the average location of the auroral oval has changed as a result. However, the Australian *aa* stations have drifted poleward in geomagnetic coordinates by about 2° since 1868, whereas the English *aa* stations have drifted equatorward by about 4° and thus opposite effects would have been observed in the aa_N and aa_S if this were an important factor. In addition, the sensitivity and accuracy of the magnetometers deployed have increased and the instruments have changed from analogue to digital. Many and subtle site changes are also possible, for the example the building of nearby power lines and the height of the water table.

Another check is to compare the *aa* data with long and homogeneous data series from other stations. Figure 76 shows one such comparison with data from the Sodankylä magnetometer which extends back to 1914. Svalgaard et al. [2004] suggest that the method of compilation of *aa*, via the range of variation in all 3-hour intervals, has introduced the drift erroneously (but only before 1957). These authors proposed an alternative *inter-hour variability* (IHV) index which they applied to data from the American longitude sector alone. However, Clilverd et al. [2004] have shown that application of the IHV algorithm to the data from the *aa* stations produces a variation which is very similar indeed to *aa*. In addition, these authors show that variations in both *aa*-equivalent or IHV indices using the long data series from Sodankylä (from 1914, see 76), Eskdalemuir (from 1911) and Niemegk (from 1890) also all agree very closely with *aa*.

In order to study relative drifts between two parameters A and B, it is important to look at the evolution of the residuals to the fits, (A - A_{fit}), where A_{fit} is the best linear regression fit of B to A. Figure 77 shows the plot for aa_N and aa_S (grey histogram) and for *aa* and *Am* (black line). In these plots there is no evidence for a long-term drift because the residuals oscillate around zero, and neither is there any evidence for step-like changes that could result from, for example, an uncalibrated change in either data sequence. One interesting feature is that there is a 22-year cycle in the residuals for aa_N and aa_S. This is likely to be related to known asymmetries in

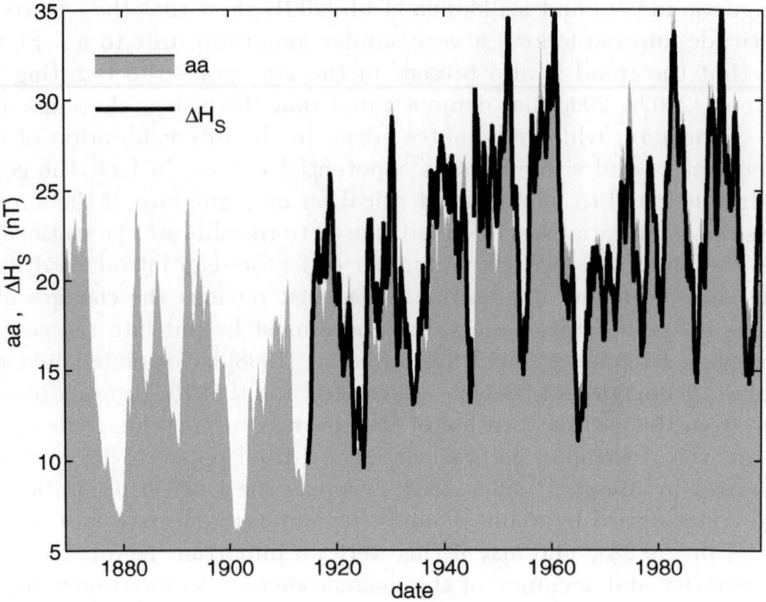

Fig. 76. Comparison of the *aa* index with the standard deviation of horizontal fluctuations ΔH_S observed at Sodankylä, Finland

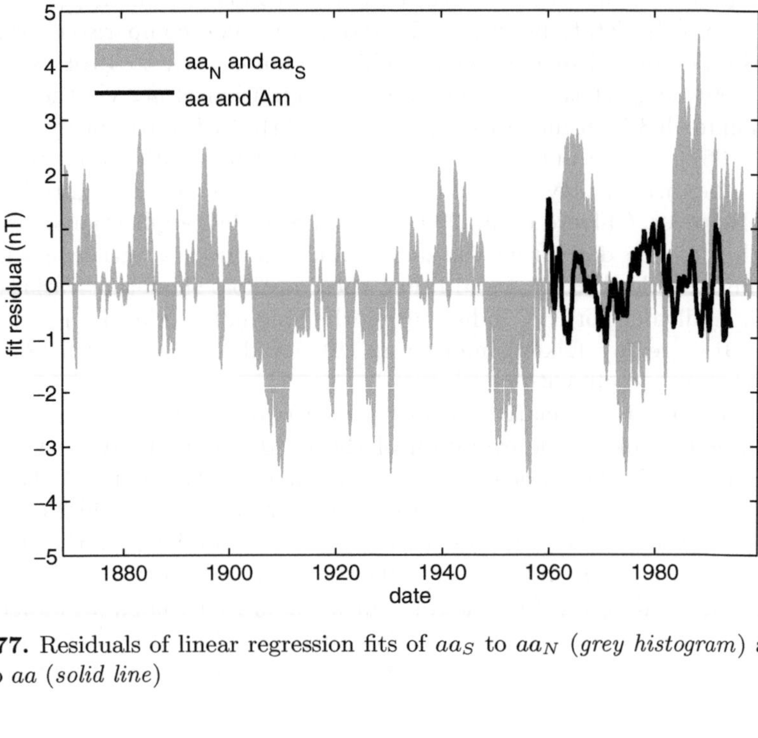

Fig. 77. Residuals of linear regression fits of aa_S to aa_N (*grey histogram*) and of A_m to aa (*solid line*)

the occurrence of IMF $[B_Y]_{GSE} > 0$ and $[B_Y]_{GSE} < 0$, an example of which can be seen in the distribution of hourly $[B_Y]_{GSE}$ values (in this case for 1964–1973) shown in the top left panel of the Fig. 78. For this interval, there are an excess of negative values. In order to look at the variation of this asymmetry with time, the top right panel of the figure shows the variation of δB_Y, the difference between the 3-monthly means of the absolute values of all the positive $[B_Y]_{GSE}$ samples and of all the negative $[B_Y]_{GSE}$ samples (so the distribution shown in the top left gives $\delta B_Y < 0$). This is significant for the *aa* index is because IMF $[B_Y]_{GSE} > 0$ (positive δB_Y) gives more-negative $[B_Z]_{GSM}$ at 10 UT, when the southern hemisphere *aa* station is close to the auroral electrojet, whereas $[B_Y]_{GSE} < 0$ (negative δB_Y) gives more-negative $[B_Z]_{GSM}$ at 22 UT, when the northern hemisphere *aa* station is in a better position to respond. Thus $\delta B_Y < 0$ favours the detection of geomagnetic activity in the northern hemisphere station (giving a positive residual in the second panel). Figure 78 shows that there is indeed an anticorrelation of the fit residual and δB_Y. The variation of δB_Y initially shows a clear Hale

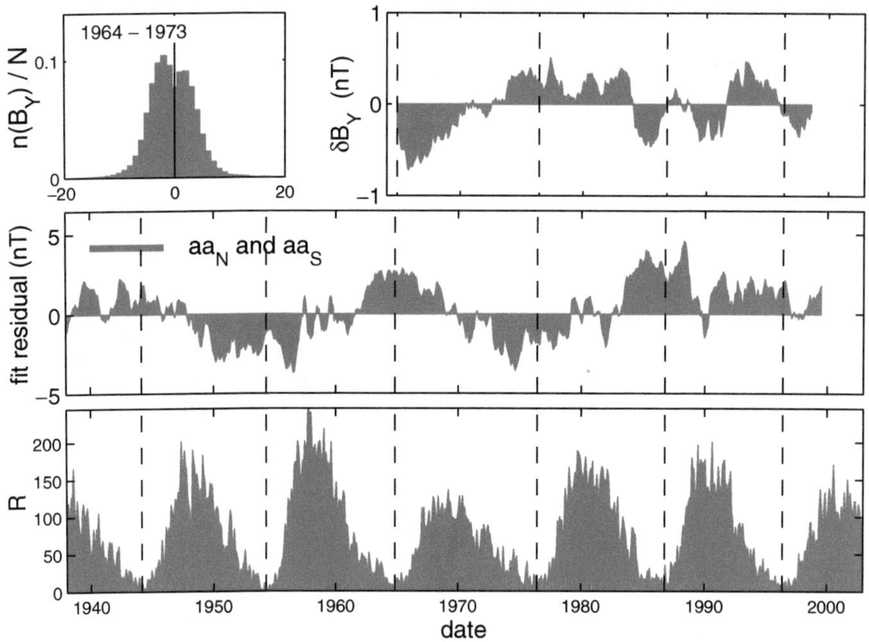

Fig. 78. Hale cycle variations in geomagnetic activity. (*Top left*) The distribution of $[B_Y]_{GSE}$ values for 1964–1973 ($n(B_Y)$ is the number of hourly averaged samples in 1 nT B_Y bins and N is the total number of such samples). (*Top right*) The variation of the asymmetry in IMF $[B_Y]_{GSE}$ values, δB_Y. (*Middle*) Residuals of linear regression fits of aa_S to aa_N, as shown in Fig. 77. (*Bottom*) The sunspot number, R. Vertical *dashed* lines mark times of sunspot minima

(22-year cycle) variation, reversing shortly after solar maximum (see bottom panel), at about the same time that the solar polar field reverses. This is less clear after 1988 but this may, at least in part, be due to many more gaps in the data in later years when interplanetary monitoring satellites were not continuously tracked.

Figure 79 shows the residuals for the fits of *Ap* to *aa* and of ΔH_S from Sodankylä, to *aa*, and thus compares the interhemispheric *aa* index with data from the northern hemisphere data only. This does provide some evidence for some spurious long-term drift in *aa* because the residual values consistently tend to be negative before 1968 and positive after it, with most rapid change between 1950 and 1970. Neither the *Ap* data nor the Sodankylä single-station data can be regarded as an absolute standard, but the combination does suggest that *aa* values may be up to about 1nT too hight since around 1960. This may be related to a station change in the northern hemisphere aa data which took place 1957. However, the comparison of aa_S and aa_N (77) indicates that the relative drift of southern and northern hemisphere stations (and that they should co-incidentally both have similar and larger drifts is highly unlikely) is less that about 0.5 nT, which would only have a 0.25nT effect on

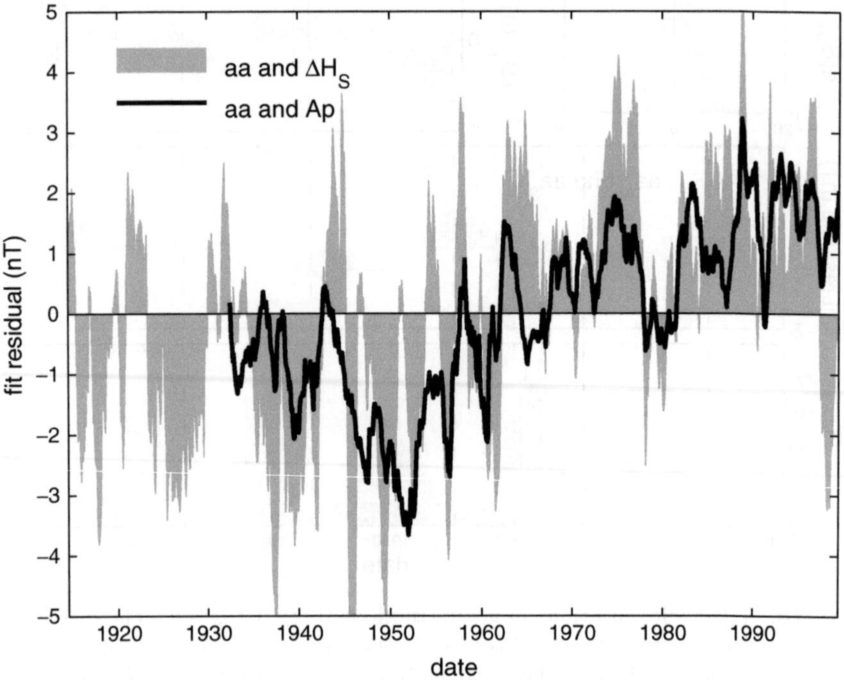

Fig. 79. Residuals of linear regression fits of A_p to *aa* (*solid line*) and of ΔH_S to *aa* (*grey histogram*) where ΔH_S is the standard deviation of horizontal fluctuations observed at Sodankylä

aa $= 0.5(aa_S + aa_N)$. The data shown in Figs. 75–79 are 12-month running means and so a calibration error causing a a discontinuity would generate a step lasting only 1 year, not the gradual change seen between 1950 and 1970. Thus either a drift in site conditions (e.g. due to water table height) or a long term change in the relative occurrence of the two polarities of IMF B_y is a more likely explanation. Values of aa since 1960 may be too high by between 0.25 and 1nT (roughly 1–4 percent). However, the main feature of the aa variation, the large rise between 1900 and 1960, is consistent with all the other data and Figs. 75 and 76 stress that these effects are small by showing similar behaviour in all indices.

5.3 Implications of the Drift in Geomagnetic Activity

The previous section discussed how geomagnetic activity is caused by energy extracted from the solar wind. The data also reveal long-term trends in geomagnetic activity over the past 150 years which mirror the trends in average sunspot numbers, cosmogenic isotopes and the occurrence of low-latitude aurora discussed in Sect. 5.1. The *aa* index has its limitations on timescales shorter than one year because of the complex interplay of station coordinates (in both geographic and geomagnetic coordinates), UT and time-of-year caused by dipole tilt effects, ionospheric conductivity variations and the MLT distribution in the deposition of solar wind energy in the ionosphere. All of these are averaged out on an annual basis by having data from two antipodal stations and annual means of *aa* reproduce trends seen in other data. Figure 80 shows the variation in annual means of *aa* and Sargent's recurrence index I_{aa}, defined for the jth 27-day Carrington rotation period as $[I_{aa}]_j = (1/13)\Sigma_{k=-6}^{+6} c_{(j+k,j+k+1)}$ where c is the correlation coefficient between two consecutive 27-day intervals of twelve-hourly *aa* values [Sargent III, 1986].

The recurrence index is seen to peak in the declining phase of each solar cycle and this peak is generally larger (in amplitude and duration) for even-numbered cycles than odd-numbered cycles. This mirrors the behaviour seen since 1964 in the mean solar wind velocity, V_{SW} [Hapgood, 1993, Cliver et al., 1996]. The recurrence index quantifies the tendency for geomagnetic activity to repeat after one solar rotation and so we can relate the declining phase peaks to the effect of corotating interaction regions (CIRs). These CIRs form on the leading edge of the fast solar wind streams that emerge from the low-latitude extensions that are features of coronal holes in the declining phase of the solar cycle (see Fig. 21). Note also that the recurrence index at times outside these peaks has decreased during the 20th century. This is a simple consequence of the rise in *aa* values shown in the top panel. (Correlation coefficients rise as the level of variation decreases, up to the limit of unity for two parameters that do not vary at all, but such a correlation has zero statistical significance).

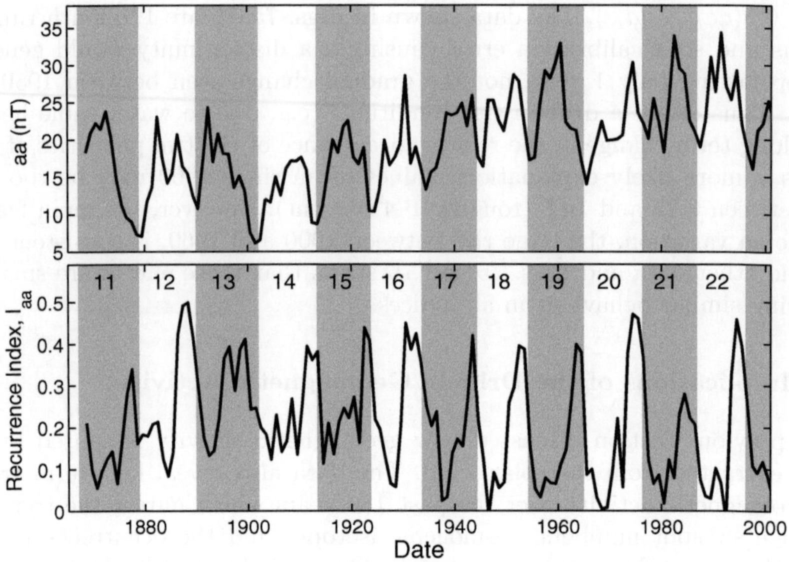

Fig. 80. (*Top*) Annual means of the *aa* index. (*Bottom*) Sargent's recurrence index I_{aa}, defined for the jth 27-day Carrington rotation period by $[I_{aa}]_j = (1/13)\Sigma_{k=-6}^{+6} c_{(j+k,j+k+1)}$ where c is the correlation coefficient between two consecutive 27-day intervals of twelve-hourly *aa* values. *Even-* and *odd*-numbered solar cycles, defined by the minimum sunspot number, are shaded *white* and *grey*, respectively

Several attempts have been made to use the *aa* data to deduce the interplanetary and solar conditions before the space age [Russell, 1975]. The success of such an extrapolation depends critically on the quality of the correlation found between the *aa* index and the combination of the interplanetary parameters (the empirical "*coupling function*") used to quantify the controlling influence of the solar wind and IMF [Baker, 1986]. An early attempt at extrapolation used data from solar cycle 20 only [Gringauz, 1981] and was based on a correlation between *aa* and V_{SW}. However, when data from solar cycle 21 were included, a much better correlation was obtained if a dependence on the southward component of the IMF was also introduced into the coupling function [Crooker and Gringauz, 1993] and this was used to look at the possible combinations of Vsw and the IMF that existed at the turn of the century [Feynman and Crooker, 1978]. More recently, Stamper et al. [1999] obtained an unprecedentedly high correlation coefficient of 0.97 (using a coupling function that is a theory-based combination of Vsw, the IMF magnitude B_{SW}, the IMF orientation, and the solar wind concentration N_{SW}), whereas the correlations for all previously proposed coupling functions were degraded by the addition of data for solar cycle 22. Importantly, the coupling function used by Stamper et al. [1999] is based on the physics of solar-wind magnetosphere coupling and not based purely on an empirical statistical relation: this enables extrapolation to be made with confidence.

Lockwood et al. [1999a,b] developed a method for estimating the IMF magnitude B_{SW} from the *aa* data using the theory of solar wind energy extraction by the magnetosphere. This exploits two strong, physics-based and extremely significant correlations between the IMF, the solar wind and the *aa* index, which Lockwood et al. derived using the data from last three solar cycles (20–22). However, there are uncertainties concerning the calibration of the early interplanetary measurements [Gazis, 1996], particularly for N_{SW} in solar cycle 20. Consequently, Lockwood and Stamper [1999] employed a different approach. They derived all correlations using data from cycles 21 and 22 only and then predictions for cycle 20 were compared with the IMF observations. Thus the cycle-20 and -23 IMF data provided an independent test of the method.

The theory by Vasyliunas et al. [1982] shows that the power delivered from the solar wind to the magnetosphere P_α is the multiplied product of three terms: (1) the energy flux density of the interplanetary medium surrounding the Earth (dominated by the kinetic energy of bulk solar wind flow); (2) the area of the target presented by the geomagnetic field (roughly circular with radius l_0); (3) the fraction t_r of the incident energy that is extracted:

$$P_\alpha = (m_{SW} N_{SW} V_{SW}^{3/2}) \times (\pi l_0^2) \times (t_r) \tag{154}$$

The dayside magnetosphere is approximately hemispherical in shape, in which case l_0 equals the stand-off distance of the nose of the magnetosphere which can, to first order, be computed from pressure balance between the Earth's dipole field and the solar wind dynamic pressure.

$$l_0 = k_1 \left(\frac{M_E^2}{P_{SW} m_0} \right)^{1/6} \tag{155}$$

where P_{SW} is the solar wind dynamic pressure ($= m_{SW} N_{SW} V_{SW}^2$), $k_1 \approx 0.89$ is a factor that allows for flow around a blunt nosed object such as the magnetosphere and M_E is Earth's magnetic moment (see Chap. 6, Kivelson and Russell, 1995 and [Merrill et al., 1996]).

The "*transfer function*" t_r must be a dimensionless quantity which includes the effect of the IMF orientation which plays a key role. Figures 69–71 show the limits of behaviour for purely southward and northward directed IMF, in the GSM frame, and the theory must allow for all IMF orientations in between. The form of the dimensionless transfer function t_r suggested by Vasyluinas et al., includes an empirical $\sin^4(\theta/2)$ dependence on the IMF clock angle θ (the angle that the IMF makes with northward in the GSM frame of reference, see Fig. 72) which allows for the role of magnetic reconnection between the IMF and the geomagnetic field [Scurry and Russell, 1991, Akasofu, 1981]. To allow for any dependence on the solar wind flow speed, the transfer function adopted also depends on the solar wind Alfvén Mach number, M_A, to the power 2α where α is called the "*coupling exponent*" and must be determined empirically.

$$t_r = k^2 M_A^{-2\alpha} \sin^4\left(\frac{\theta}{2}\right) \tag{156}$$

where k is a constant. From (154)–(156)

$$\begin{aligned} P_\alpha &= k m_{SW}^{(2/3-\alpha)} M_E^{2/3} B_{SW}^{2\alpha} \left[N_{SW}^{(2/3-\alpha)} v_{SW}^{(7/3-2\alpha)} \sin^4\left(\frac{\theta}{2}\right) \right] \\ &= k m_{SW}^{(2/3-\alpha)} M_E^{2/3} B_{SW}^{2\alpha} f = \frac{aa}{s_a} \end{aligned} \tag{157}$$

To compute α, the aa index is assumed to be proportional to the extracted power $P_\alpha = (aa/s_a)$, an assumption that is verified empirically. The optimum α, which gives the peak correlation coefficient c between P_α and aa, is then determined and the constant $s'_a = k s_a$ is then found from a linear regression fit of aa to P_α.

Stamper et al. [1999] analysed each of the terms in the best-fit coupling function given by (157) for the interval since 1963 when interplanetary monitoring began. They showed that more than half of the change in aa over the last three solar cycles was caused by an upward drift in B_{SW}. There were smaller contributions from increases in N_{SW} and V_{SW} but the average IMF clock angle θ had grown slightly less favourable for causing geomagnetic activity (because there was a slight tendency for the IMF to stay closer to the ecliptic plane). In order to use (157) to evaluate B_{SW}, the terms in the square brackets are grouped together into a single parameter f, the variation of which (on annual time scales) is dominated by that in V_{SW}. The recurrent intersections with long-lived CIRs ahead of fast, solar wind emanating from the low-latitude extension of coronal holes raise both the mean V_{SW} and the recurrence index I_{aa}. Hence both f and I_{aa} increase together in the declining phase of sunspot cycles. However, I_{aa} tends to remain high towards sunspot minimum because aa values are low and relatively constant, whereas V_{SW} and f are lower. Consequently, Lockwood and Stamper [1999] adopted a relationship for a predicted f of the form

$$f_p = s_f I_{aa}^\beta aa^\lambda + c_f \tag{158}$$

where the exponents β and λ give the optimum correlation coefficient and the constants s_f and c_f are then found from a linear regression fit of observed f against f_p. Substituting for f in (157) using f_p given by (158) allowed Lockwood et al. to compute B_{SW} from the aa data series. They employed estimates of M_E from the IGRF reference model fit to geomagnetic data and assumed the composition of the solar wind is constant with a mean ion mass of 1.15 a.m.u. The top two panels of Fig. 81 show how closely

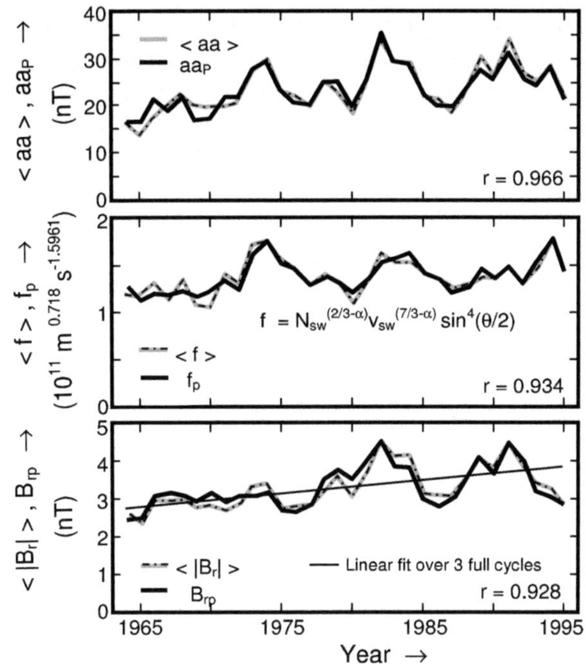

Fig. 81. The three correlations in annual means used to compute the open solar flux from the *aa* index. (*Top*) The best-fit *aa* index value predicted using (157), aa_P, and the observed annual mean of *aa*. (*Middle*) The parameter f, defined by (157), as predicted from *aa* by (158), f_p, and as measured by interplanetary satellites, f, and (*Bottom*) the radial field value observed, $|B_r|$, and that predicted from the IMF strength using a constant average garden hose angle, B_{rp}. In each case the predicted values are the *black solid* lines and the observed values are *grey*-and-*black dot–dash* lines. The correlation coefficients are 0.966, 0.934 and 0.928 which are all significant at greater than the 99.99% level

P_α from (157) matches *aa* and how the observed parameter f can be matched using (158) and the *aa* index data.

In three dimensions, Parker spiral theory [e.g. Gazis, 1996] predicts the heliospheric field in heliocentric polar coordinates (r, ϕ, ψ) will be

$$B_{SW} = \{B_r^2 + B_\phi^2 + B_\psi^2\}^{1/2} = B_r[1 + \tan^2\gamma]^{1/2}$$
$$= B_0 \left(\frac{R_0}{r}\right)^2 \left\{1 + (\omega r \cos\psi v_{SW})^2\right\}^{1/2} \quad (159)$$

where B_0 is the coronal source field at the solar source sphere, $r = R_0$ from the centre of the Sun (where the solar field becomes approximately radial), ω is the equatorial angular solar rotation velocity, and ψ is the heliographic latitude. Parker spiral theory is very successful in predicting annual means of the heliospheric field orientation around Earth [Stamper et al., 1999] because perturbing phenomena like CIRs and CMEs are averaged out. Note however, that at higher solar latitudes agreement is not so good [Smith and Bieber, 1991]. Near the ecliptic, both the observed gardenhose angle and that predicted by (159) (dashed line) remain close to 45° and, as a result, the radial heliospheric field component $|B_r|$ is proportional to B_{SW}, i.e. $|B_r| \approx |B_{rp}| = s_B B_{SW}$ to a very good degree of approximation.

The bottom panel of Fig. 81 shows the radial component of the observed IMF. It reveals an upward drift superposed on the solar cycle variation. Equation (159) tells us that these variations in $|B_r|$ reflect variations in the coronal source field, $|B_0|$. The thin straight line is a linear regression fit over three full solar cycles and reveals that the increase is by a factor of 1.3 over this interval.

Using the Ulysses result, (73), (157) and (158) yield

$$F_S = \left(\frac{1}{2}\right) 4\pi R_1^2 |B_r| = 2\pi R_1^2 s_B B_{SW}$$

$$= 2\pi R_1^2 s_B \left\{ \frac{[s'_a (s_f I_{aa}^{\beta} aa^{\lambda} + c_f) m_{SW}^{(2/3-\alpha)} M_E^{2/3}]}{aa} \right\}^{-0.5/\alpha} \qquad (160)$$

Table 6 gives all the best-fit coefficients derived from the fits shown in Fig. 81 which can be used in (160) to compute the open solar flux F_S from aa. Estimates of M_E from the IGRF model fit to geomagnetic data are used for a given date and it is assumed the composition of the solar wind gives the present-day mean ion mass of 1.15 a.m.u. at all times. Annual means of aa are required to average over the UT and time-of-year dependencies discussed in the previous section, but these can be generated on a monthly basis by moving the 12-month window forward one month at a time (thus only every 12th point is fully independent data). The results are shown in Fig. 82. Note that the extrapolation $[F_S]_{aa}$ is not a simple correlation of aa with open flux:

Table 6. Regression Fits Used to Compute F_S

Fitted Parameters	Correlation Coefficient, c	Significance Level (%)	Coefficients	Slope	Intercept
aa and aa_p	0.966	> 99.99	$\alpha = 0.3085$	$s'_a = ks_a =$ 4.7022×10^{-18}	–
f and f_p	0.934	> 99.99	$\beta = 0.2271$ $\lambda = 1.2114$	$s_f =$ 5.71×10^5	$c_f =$ 2.61×10^7
$\langle \lvert B_r \rvert \rangle$ and B_{rp}	0.928	> 99.99	–	$s_B = 0.5606$	–

units: aa_p (in nT) $= s'_a \langle M_E$ in T m$^3 \rangle^{2/3} m_{SW}^{(2/3-\alpha)} \langle N_{SW}$ in m$^{-3} \rangle^{(2/3-\alpha)} \langle v_{SW}$ in km s$^{-1} \rangle^{(7/3-2\alpha)} \langle B_{SW}$ in nT $\rangle^{2\alpha} \langle \sin^4(\theta/2) \rangle$ $f = \langle N_{SW}$ in m$^{-3} \rangle^{(2/3-\alpha)} \langle v_{SW}$ in km s$^{-1} \rangle^{(7/3-2\alpha)} \langle \sin^4(\theta/2) \rangle$ and $f_p = s_f \langle I \rangle^{\beta} \langle aa$ in nT $\rangle^{\lambda} + c_f$ B_{rp} (in nT) $= s_B \langle B_{SW}$ in nT $\rangle$

it is based on the theory of energy coupling between the solar wind and the magnetosphere and uses the recurrence index to remove the effect of fast solar wind streams. Figure 82 shows the agreement of the open flux derived from *aa*, using (160), with the estimate derived from interplanetary measurements, using (73). The agreement can be seen to be very good and both the solar cycles and the drift observed after 1964 are very well reproduced.

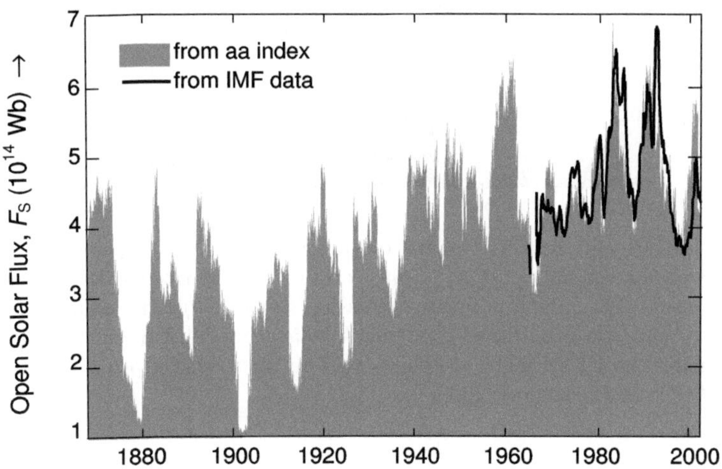

Fig. 82. The long term variation of open solar flux derived from the *aa* index $[F_S]_{aa}$ (*grey histogram*) using the procedure developed by Lockwood et al. [1999a,b] and derived from IMF measurements, $[F_S]_{IMF}$ (*solid line*)

The drift in average *aa* values is highly significant, amounting to more than a doubling in average values between 1900 and 1960. Note that the analysis of the aa index in Sect. 5.2 suggest that the open flux for up to 1960 may be a constant factor of 1-4 percent too low, compared to modern-day values. The recent trends in the data are shown in Fig. 83. The plot shows 11-year running means of various open solar flux estimates: from *aa*, $[F_S]_{aa}$; from IMF measurements, $[F_S]_{IMF}$; from a linear regression fit to the anticorrelated cosmic ray counts observed by the Moscow neutron monitor $[F_S]_M$; and from the solar magnetograms using the Potential Field Source Surface PFSS method $[F_S]_{PFSS}$ [Schatten et al., 1969, Schatten, 1999]. All methods show the same trends and all point to 1987 being a significant peak in the long-term variation of the open flux.

The two perihelion passes by the Ulysses spacecraft provide a good opportunities to test the various methods of computing the open solar flux, under solar minimum and solar maximum conditions (see Sect. 2.5). In these passes, the satellite took about 9 solar rotations to traverse from −80° heliographic latitude to +80°: with the assumption that there was little drift in the open flux during these intervals and that short term events averaged out,

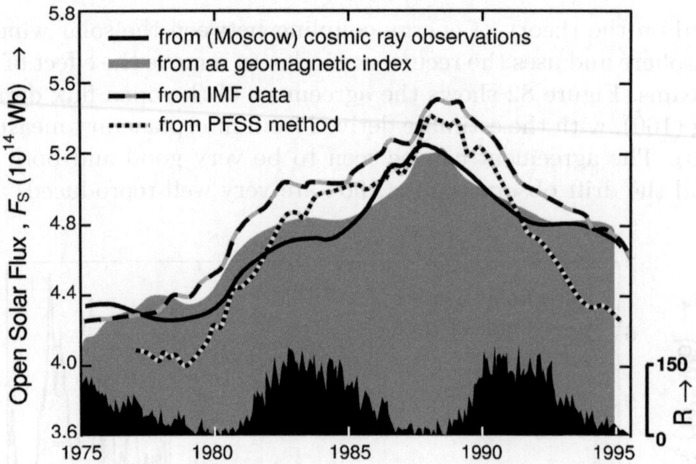

Fig. 83. Eleven-year running means of various indicators of the open solar flux: derived from the *aa* index, $[F_S]_{aa}$ (*grey area*); derived from IMF measurements, $[F_S]_{IMF}$ (*dashed line*); from a linear regression fit to cosmic ray counts observed by the Moscow neutron monitor $[F_S]_M$ (*black solid line*); and from the solar magnetograms using the PFSS method $[F_S]_{PFSS}$ The *black* histogram gives the sunspot number, R. From Lockwood [2003]

the observed radial field can be averaged to give estimates of the average open solar flux during the passes $[F_S]_U$ that include data from almost all latitudes. Lockwood et al. [2004] have tested out the near-Earth methods (from IMF data and the *aa* index) and the PFSS method against these data. They also tested the predictions of the model by Solanki et al. [2000], $[F_S]_{SM}$, bearing in mind they were made after the first perihelion pass but before the second. (This model is discussed further below). The results are shown in Table 7. If we take $[F_S]_U$ to be our best estimates, $[F_S]_{IMF}$ and $[F_S]_{aa}$ values are

Table 7. Estimates of the open solar flux During the First and Second Perihelion Passes of Ulysses

	First Perihelion Fast Latitude Scan		Second Perihelion Fast Latitude Scan	
	Open Flux, F_S (10^{14} Wb)	$\frac{\{F_S - [F_S]_U\}}{[F_S]_U}$	Open Flux, F_S (10^{14} Wb)	$\frac{\{F_S - [F_S]_U\}}{[F_S]_U}$
From Ulysses, $[F_S]_U$	4.54	0	5.05	0
From IMF, $[F_S]_{IMF}$	4.77	+5%	4.85	−4%
From *aa*, $[F_S]_{aa}$	4.31	−5%	5.01	−1%
From PFSS, $[F_S]_{PFSS}$	3.93	−13%	2.70	−47%
From model, $[F_S]_{SM}$	4.15	−9%	4.31	−15%

accurate to within 5%, but $[F_S]_{PFSS}$ to only 47% (the error being much higher for the solar maximum pass). $[F_S]_U$ for the second (solar maximum) pass was only slightly greater than for the first (at solar minimum), which would not be expected from the variation of $[F_S]_{IMF}$ seen over previous cycles. However, the Solanki et al. model does reproduce this behaviour well. It underestimates the open flux in both cases, but by only 9% and 15%. Note that this application of the model is as given by the original authors, i.e. the intitial conditions were that $[F_S]_{SM} = 0$ at the end of the Maunder minimum. To be accurate to within 15%, 300 years later gives some confidence in the predictions of the model for the intervening years.

Figure 84 provides an insight to the origins of this variation in open flux. The top panel shows the rate of change of $[F_s]_{aa}$ has been positive for most of the time since 1868, but there have been two periods in which the open flux has declined, 1890–1903 and 1957–1969. These correspond to the two longest solar cycles. The second panel shows the solar cycle length, L, determined using the autocorrelation technique of Lockwood [2001a]. (Note that there are similarities, but also considerable differences, to the well-known heavily filtered variation of L, derived from peak and minimum timing analysis by Friis-Christensen and Lassen [1991]). Other methods to derive L have been reviewed by Fligge et al. [1999]. One can interpret the top two panels of Fig. 84 as showing that a series of short cycles cause a build-up in open flux, whereas long cycles allow it to decay. Solar cycle length is also related to sunspot number [Solanki et al., 2002a] and the rise in smoothed sunspot number R_{11} is also reflected in a rise in L. Hence the relationship between L and $\mathrm{d}F_s/\mathrm{d}t$ may also depend on the rate of production of open flux E by emergence in active regions. Solanki et al. [2000] suggest that the loss of open flux is, on average, linear (i.e. at a rate F_S/τ) with a time constant τ of a few years. Such a long timeconstant has be questioned, but it should be remembered that it is an average of newly-emerged open flux in active regions and long-lived open flux in polar coronal holes. The fourth panel in Fig. 84 shows the emergence rate E computed using the simple continuity equation proposed by Solanki et al. [2000].

$$\frac{\mathrm{d}F_S}{\mathrm{d}t} = E - \left(\frac{F_S}{\tau}\right) \tag{161}$$

and time constants ranging from τ of 1.5 yr (upper line) to 3.5 yr (lower line) in steps of 0.5 yr. It can be seen the emergence rate required has a variation which is a combination of R_{11} and R (shown in the bottom panel). Solanki et al. [2000] devised a method for computing open flux emergence from sunspot number and used (161) to model the variation in F_S. They assumed that the open flux fell to zero at the end of the Maunder minimum and modelled the variation forward in time to get a good match to the results of Lockwood and Stamper [1999]. Foster and Lockwood [2001] used the spread of sunspot latitudes in the Greenwich sunspot data to model the emergence

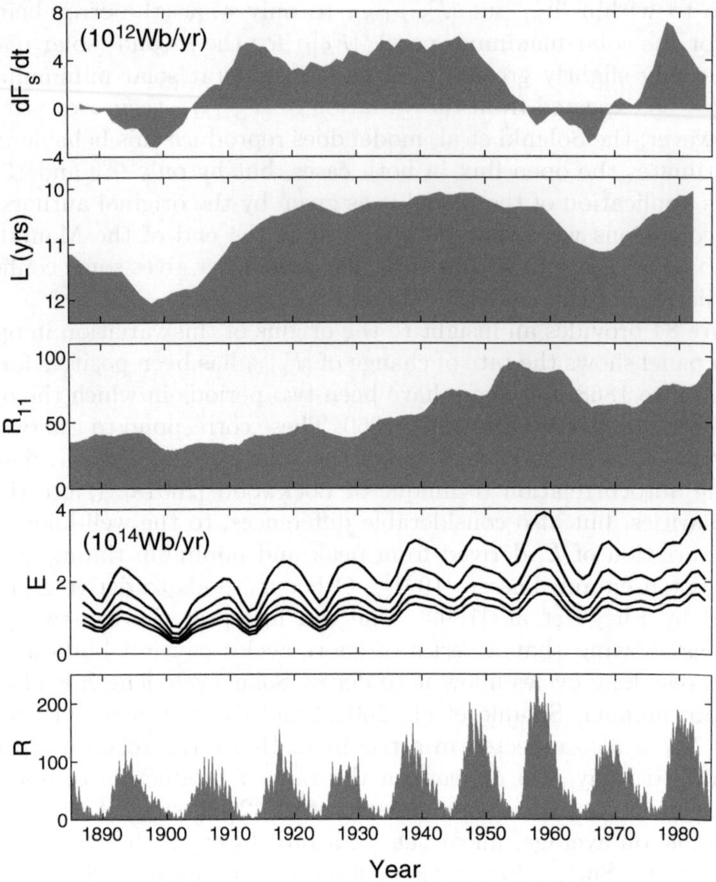

Fig. 84. (From *top* to *bottom*): The rate of change of open flux, $\mathrm{d}F_S/\mathrm{d}t$ derived from the *aa* index; the length of the solar cycle L determined from the peak of the autocorrelation function of sunspot number [see Lockwood, 2001a,b, Lockwood and Foster, 2001] – note that the L scale has been inverted; the 11-year smoothed sunspot number, R_{11}; the emergence rate E, for linear loss time constants τ of 1.5(0.5)3.5 yr; and the sunspot number, R

rate. They started the model from the observed open flux for the year 2000 and evaluated (161) backwards in time to 1874. The best-fit time constant τ is 3 years and gives the results shown in Fig. 85.

Solanki et al. [2002b] have refined this model by considering various classifications of surface solar magnetic flux, including active regions, ephemeral flux and active region remnants, and applying a continuity equation to each, forming a coupled set of equations. The best fit to $[F_S]_{aa}$ then yields a shorter time constant of nearer 1.5 years. An important point about this second model is that it predicts that although the open flux is only a small fraction of the

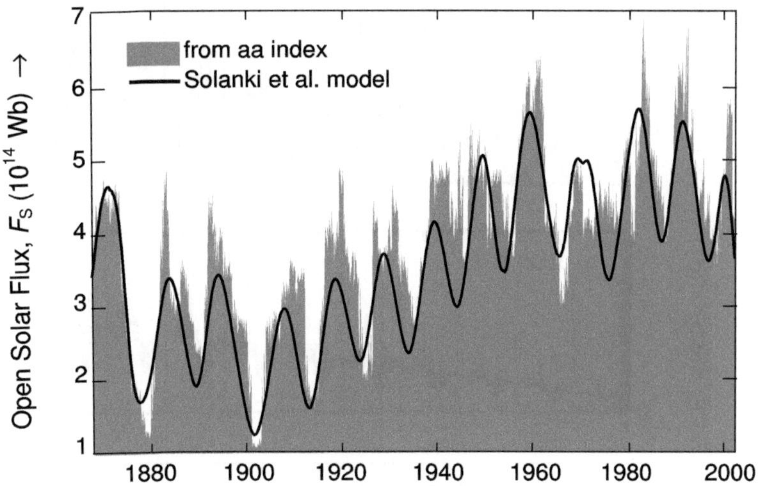

Fig. 85. Comparison of open flux derived from the *aa* index by Lockwood et al. [1999a], $[F_S]_{aa}$ *(in grey)* with model predictions $[F_S]_{SM}$ by the model of Solanki et al. [2000], as implemented by Foster and Lockwood [2001]

total surface flux, the two have similar time variations. If this prediction were to be confirmed, open flux (and hence cosmic ray fluxes and cosmogenic isotopes) would be confirmed as reliable proxies for the surface solar flux.

Numerical models of the evolution of emerged flux, under the influences of differential rotation, supergranular diffusion, and meridional circulation have also reproduced the long term drift in the open solar flux [Schrijver et al., 2002, Wang et al., 2002, Lean et al., 2002, Wang et al., 2002, Wang and Sheeley Jr., 2004]. In these models, the total newly-emerged flux and its distribution in the active region belts, is prescribed as an input to the model and is made to follow the butterfly pattern and any long-term trend indicated by the sunspot number. An example is given in Fig. 86 (from Lean et al., 2002), in which a series of solar cycles of increasing strength are simulated. The bottom panel shows the longitudinally-averaged surface magnetogram which successfully reproduces the major features of observations (compare with Fig. 14). Part (b) shows the resulting open flux variation which reveals an upward drift of a magnitude comparable to that derived by Lockwood et al. [1999a,b] from the *aa* index. Thus if we have a series of increasing solar cycles, as indeed were observed between 1900 and 1960, these simulations predict that the open solar flux will rise. Lean et al. point out that this is accompanied by only a very small rise in the solar-minimum surface flux and thus, unlike the Solanki et al. [2002b] simulation discussed above, the open flux is not a good indicator of the surface flux in these predictions.

Certainly, Fig. 86 predicts that a rise in open solar flux can occur in the almost complete absence of a corresponding rise in the surface flux. However,

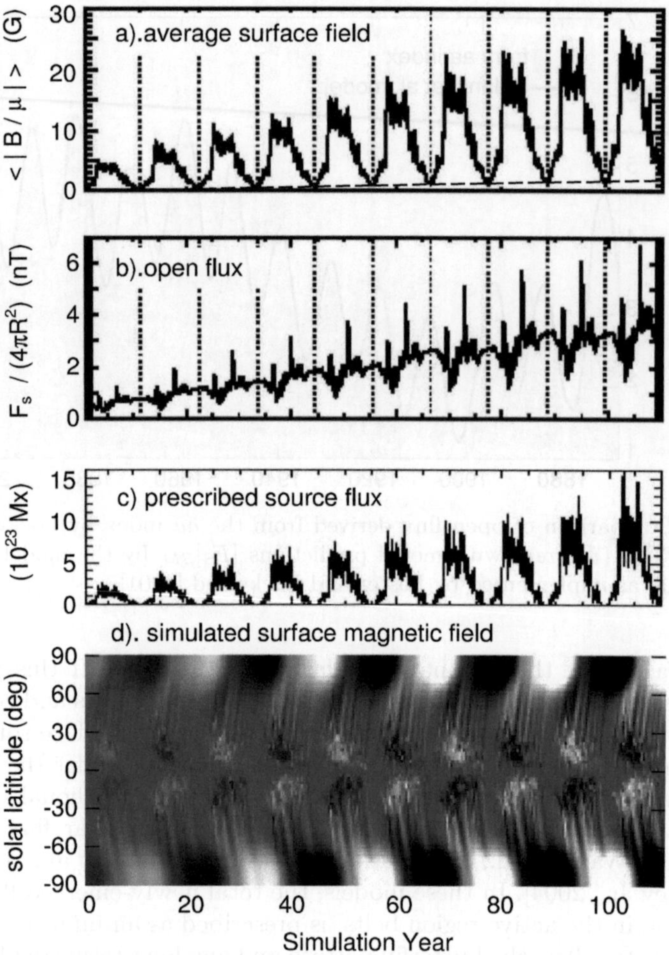

Fig. 86. Simulation of the surface magnetic flux and the open solar flux by Lean et al. [2002]. Flux emergence is prescribed in a butterfly pattern and the total emerged flux that is input into the model is shown in panel (**c**). This flux evolves to give the simulated longitudinally averaged surface magnetogram shown in (**d**). The time variations of the resulting total surface flux and the open solar flux are shown in panels (**a**) and (**b**), respectively

it should be noted that this simulation only included the output of the strong solar dynamo, with all emerged flux associated with active regions that do not overlap in successive cycles. No flux emergence from the weak dynamo and in extended solar cycles was included. Thus effects of varying degrees of cycle overlap, giving large drifts in the solar minimum surface flux, were excluded and so the inputs to the model determined that there was no drift in surface flux. The ephemeral flux produced by the weak solar dynamo may have random orientations, in which case it is unlikely to coherently add to influence the open solar flux [Wang and Sheeley Jr., 2003c], but would still contribute to the surface flux. In summary, this simulation implies that there need not be a strong relationship between the open solar flux and the surface flux, but does not exclude the possibility that there is such a relationship in practice.

5.4 Open Solar Flux, Cosmic Rays and Cosmogenic Isotopes

Section 3.2 has discussed the strong and highly significant anticorrelation between open flux derived from *aa*, $[F_S]_{aa}$, and cosmic ray observations (see Figs. 42 and 45). The longer data series of $[F_S]_{aa}$ available allows us to search for corresponding anticorrelations with cosmogenic isotopes produced by cosmic ray bombardment [O'Brien et al., 1991]. Lockwood [2001a, 2003] has found strong and significant anti-correlations between cosmogenic isotopes and the $[F_S]_{aa}$ data which begin in 1868. A summary is provided by Fig. 87 which shows the variations of the ^{10}Be isotope from the Dye-3 Greenland ice core [Beer et al., 1990, 1998, Beer, 2000] and the ^{14}C production rate derived from observed abundances in tree rings using a 2-reservoir model [Stuiver and Quay, 1980, Stuiver and Braziunas, 1989, Stuiver et al., 1988a,b]. Both have been scaled in terms of open flux by a linear regression fit to $[F_S]_{aa}$. Further evidence for the century-scale drift in open flux has been found from the ^{44}Ti cosmogenic isotope found in meteorites [Bonino et al., 1998]. In addition, Ivanov and Miletsky [2004] have shown that reconstructions of the open flux based on H-α spectroheliograms show a very similar variation. The grey area in Fig. 87 gives the predictions of the model of Solanki et al. [2000], made using sunspot number to quantify emergence rate and working backwards in time from modern-day values and fitted to the $[F_S]_{aa}$ data (black line). Note that in Fig. 87 the $[F_S]_{aa}$ sequence has been continued back in time to 1844 using the extension to the *aa* index made using the Helsinki magnetometer data by Nevanlinna and Kataja [1993]. It can be seen that the model agrees well with the cosmogenic isotope data and the linear regression fits yield average open flux values at the end of the Maunder minimum of about 1.5×10^{14} Wb, roughly one third of the present-day values. Details of the regression fit and the inferred response function of the ^{10}Be cosmogenic isotopes are given by Lockwood [2001a].

From the anticorrelation between the open flux derived from the aa index and the Dye-3 ^{10}Be cosmogenic isotope data, Lockwood [2001a] find by linear extrapolation that the average open solar flux was about a quarter of modern values. This value also agrees well with that obtained by Lockwood [2003] by running the continuity model of Solanki et al. [2000] backwards in time, starting from modern day values (see 87). Modelling by Wang and Sheeley Jr. [2003a] suggests a larger drift in open flux, by a factor of about 7. However, from cosmic ray shielding theory and cosmogenic isotopes Scherer and Fichtner [2004] have also derived the factor 4 shown in 87.

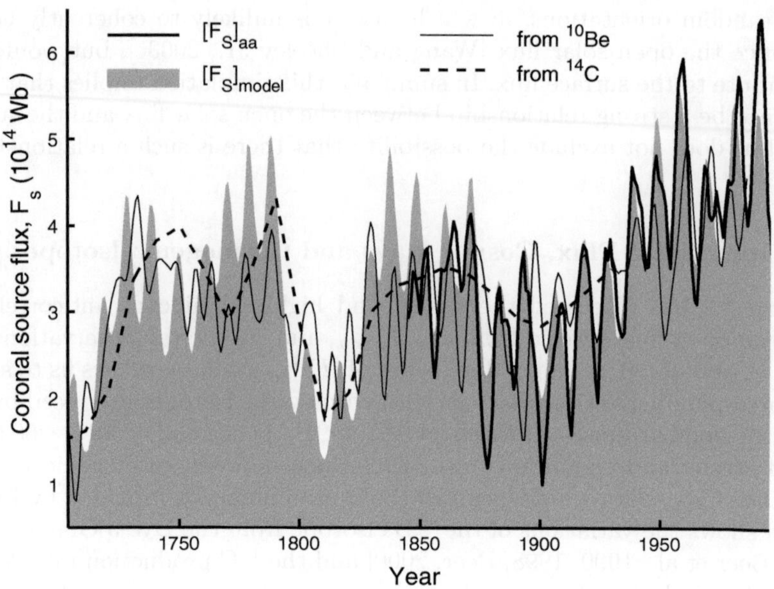

Fig. 87. The open flux derived from the *aa* index (with the Helsinki extension), $[F_S]_{aa}$ (*black solid line*) and the best fit prediction of the Solanki et al. [2000] model, as implemented by Lockwood [2003], $[F_S]_{SM}$ (*grey area*). Also shown are the best-fit linear regression fits of the Dye-3 Greenland ice core ^{10}Be abundance data (*thin line*) and the production rate of ^{14}C derived from tree ring data using a two-reservoir model to allow for atmosphere–biomass and atmosphere–oceans exchange (*dashed line*). From [Lockwood, 2003]

In addition to the Dye-3 Greenland core, a longer data sequence available from Antarctica [Raisbeck et al., 1990, Bard et al., 1997]. Both these data sequences, along with the ^{14}C data, have been used by Usoskin et al. [2003a,b] to investigate the variation of sunspot numbers. The results are considerably different from extrapolations based on statistical properties of the sunspot record since 1600 which are assumed to have been persistent. The new predictions are based on the physics of cosmic ray shielding and the production of cosmogenic isotopes, and so are much more credible. The flux of cosmic rays impinging on the Earth's atmosphere is derived from the measured ^{10}Be abundance. A quantification of the modulation of cosmic rays in heliosphere (the integrated effect of the terms in the Parker transport equation) is used to determine the Sun's open magnetic flux F_S. The model of F_S by Solanki et al. [2000], using the best fit loss rate to reproduce the results of Lockwood and Stamper [1999], is then used to compute the sunspot number (which controls the emergence rate in the model). This procedure can account for a non-linear relationship between ^{10}Be concentration and sunspot numbers and so can allow for the emergence of open flux, and associated enhanced cosmic ray shielding, that is not accompanied by sunspots (as was discussed earlier in relation to the Maunder minimum). The potential emergence of open flux without sunspots is one of the biggest uncertainties in the reconstruction. On these longer timescales it becomes increasingly important to allow for variations in the geomagnetic field as this also shields the atmosphere (to a degree that depends on latitude) from cosmic rays. The results are shown in Fig. 88.

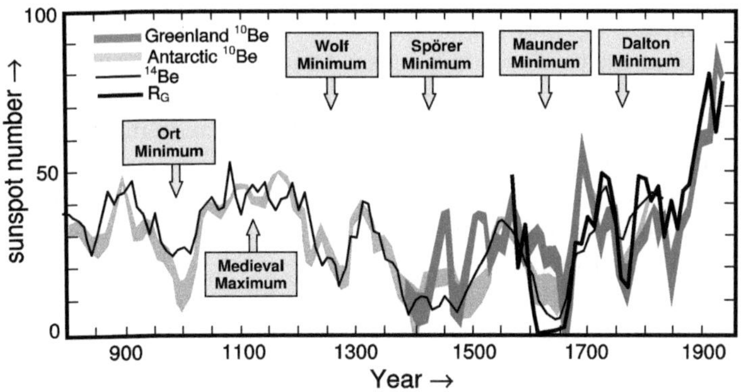

Fig. 88. Reconstruction of sunspot numbers by Usoskin et al. [2003a,b]. The *thick black* curve shows smoothed values of the observed group sunspot number, R_G, since 1610. The dark grey area is the sunspot number reconstructed from ^{10}Be concentrations in the Dye-3 Greenland ice core and the *light grey* is the corresponding reconstruction from the Antarctic data. The area is the uncertainty. The *thin black* curve gives the scaled variation of the ^{14}C concentration in tree rings, corrected for the variation in the geomagnetic field. Various maxima and minima are highlighted [Stuiver and Braziunas, 1989]. The ^{14}C record has been shifted by the optimum lag to allow for the long attenuation time for ^{14}C [Bard et al., 1997]

We now know that Earth's field has been decreasing over the past millennium [e.g. Tric, 1992, Baumgartner et al., 1998, Laj et al., 2001] and without allowance for this, the reconstruction would overestimate the heliospheric shielding and so give higher levels of solar activity in the past than is revealed by Fig. 88 [Bhattacharyya and Mitra, 1997, Damon et al., 1978].

The most striking feature of Fig. 88 is that solar activity is considerably higher now than at any time since 800 AD. There have been cycles of activity before but, for example the medieval maximum, which is clearly defined in the figure, showed considerably lower levels of solar activity (roughly half) than we are experiencing today. The rise in the open flux over the past 150 years (and the associated rise in geomagnetic and auroral activity and the fall in cosmic ray fluxes) appears to have been an relatively unusual event, at least over the past 1200 years.

6 Implications for Earth's Climate

In 1801, the astronomer William Herschel proposed a link between solar irradiance and climate [Herschel, 1801]. This was based on an apparent correlation he had found between sunspot numbers and the price of wheat. He argued that the presence of more dark spots on the Sun at sunspot maximum would give a lower solar irradiance, driving a cooling of Earth's climate and so causing wheat yields to fall. The law of supply-and-demand would then force up the price of wheat on the open market. Most parts of this ingenious combination of solar physics, climate forcing, agricultural science and economics stand up to closer inspection today. However, this is not true for the first step of the argument, for we now know that the irradiance of our Sun is not reduced at sunspot maximum. Rather, it is increased because of the dominant effect of small facular flux tubes which accompany the larger, sunspot flux tubes. This is just the first of many examples of Sun-climate studies in which insufficient attention was paid to *significance* of a derived correlation [Wilks, 1995].

In many respects, Herschel's reasoning was sound. We now know that variations in solar outputs on time scales of about 20 years and greater, (i.e. longer than the time constant for terrestrial response to changes in radiative forcing), do indeed have an influence on Earth's climate. Section 2.6 has reviewed recent advances in our knowledge of long-term changes in solar activity (in terms of sunspot number and open solar flux). The theory of cosmic ray shielding by the heliosphere, despite some uncertainties, is sufficiently mature that it gives us a good understanding of the variation in cosmic ray fluxes at Earth (and of the cosmogenic isotope record). However, the implications for other solar outputs, and in particular the total and spectral solar irradiance, are not yet understood. This section reviews the evidence for an influence on terrestrial climate associated with long-term changes in solar activity and analyses the implications.

The shortwave (wavelength λ less than about 4 μm) power input to Earth's climate system is

$$P_{in} = I_{TS} \pi R_E^2 (1 - A) \tag{162}$$

where I_{TS} is the total solar irradiance (TSI), R_E is the mean Earth radius, and A is Earth's Bond albedo, the fraction of the shortwave power incident on the Earth which is reflected back into space, integrated over all directions. The output longwave power is

$$P_{out} = 4\pi R_E^2 \sigma T_E^4 = 4\pi R_E^2 \sigma (1 - g) T_S^4 \tag{163}$$

where σ is the Stefan–Boltzmann constant, T_E is the effective temperature of the Earth and its atmosphere ($\approx 255\,\mathrm{K}$), T_S is the surface temperature of the Earth and g is the normalised greenhouse effect ($= G/(\sigma T_S^4)$), where G is the greenhouse radiative forcing (in $\mathrm{W\,m^{-2}}$). In radiative equilibrium, $P_{out} = P_{in}$ which yields

$$T_S = \left[\frac{I_{TS}(1-A)}{4\sigma(1-g)} \right]^{1/4} \tag{164}$$

Using typical values of the TSI, $I_{TS} \approx 1366.5\,\mathrm{W\,m^{-2}}$, incident on a disc of area πR_E^2, spread over a surface area of $4\pi R_E^2$, the incident SW power per unit surface area $\approx 1366.5/4 = 342\,\mathrm{W\,m^{-2}}$. The mean SW reflected power is roughly $107\,\mathrm{W\,m^{-2}}$ so the Bond Albedo, $A \approx 107/342 \approx 1/3$. The SW power reflected by clouds, aerosols etc. is near $77\,\mathrm{W\,m^{-2}}$ and so the atmospheric contribution to albedo is approximately 72%. The SW power reflected by surface is near $30\,\mathrm{W\,m^{-2}}$ and so the surface contribution to albedo is approximately 28%.

6.1 Milankovich Cycles

On timescales greater than about 10 kyr, changes in the solar climate forcing, caused by changes in the Earth's orbit, are thought to be the controlling influence which causes Earth's climate to oscillate between glacial and interglacial phases. There are three main effects that make up the "astronomical forcing" of climate:

1. Cycles in Earth's orbit *eccentricity*. These cycles cause the Earth-Sun distance, R_1, to oscillate between being almost constant during the year (near circular Earth orbit) to larger annual variations for more elliptical orbits. For a given solar luminosity, this introduces annual cycles of varying amplitude into the total solar irradiance, according to (77). This causes very small variations in the annual means of TSI. This effect introduces several variations of period around 100 kyr as well as 412 kyr and 2 Myr
2. Cycles in Earth's axial tilt (obliquity). Changes in the axial tilt of the Earth alter the pattern of insolation of the Earth, with a larger annual variation in the latitudinal distribution occurring for greater tilts. This effect causes a strong variation of period 41 kyr.
3. The *precession of the equinoxes*. The phase of the annual cycle introduced by Earth's orbital eccentricity, relative to the annual cycle introduced by Earth's axial tilt, will depend on where on the elliptical orbit the equinoxes (and hence solstices) occur. A major effect is through the lengths of the seasons. This introduces strong periodicities near 23.7, 22.4 and 19kyr.

Spectral analysis of paleoclimate indicators reveals many of the periodicities predicted for the above orbital characteristics. However, some of the predicted periodicities do not occur strongly and others are found strongly which should be very weak (for example, at 100 kyr). In addition the dominant period has changed from 41 kyr to 100 kyr without any change in the orbital characteristics. Uncertainties in the dating of the paleoclimate data used allows a certain amount of "wiggle-matching" in which the dates for one or both data sequences are adjusted to get good agreement. However, as the accuracy of the dating has improved the timing of several glacial events and the timing of the orbital changes thought to drive them have been found to be less consistent than they were previously thought to be. It is now thought that the climate system is not driven directly into glaciations in a linear, or even a weakly non-linear, manner by the Milankovich orbital cycles. Rather, there may be sudden and highly non-linear changes best understood using catastrophe theory (see review by Paillard, 2001).

Here we restrict our attention to the present interglacial warm period, the *Holocene*.

6.2 Paleoclimate Evidence During the Holocene

In recent years, a body of evidence has emerged that solar variations have had a clear and marked effect on climate throughout the Holocene, the warm period that has prevailed throughout the past ten thousand years. We here show just two examples. Figure 89 shows the average abundance of ice-rafted debris found in cores of ocean-bed sediment throughout the mid and North Atlantic [Bond et al., 2001]. These glasses, grains and crystals are gouged out in known glaciers, from which they are carved off in icebergs and deposited in the sediment when and where the icebergs melt. The sediment is dated using microfossils found at the same level in the core. The abundances of ice-rafted debris are very sensitive indicators of currents, winds and temperatures in the North Atlantic and show high, and hugely significant, correlations with cosmogenic isotopes – specifically the production rate Δ^{14}C and the abundance of ^{10}Be (correlation coefficients 0.44 and 0.59, respectively, that are both significant at greater than the 99.99% level). Note that the horizontal scale in Fig. 89 is years BP (before present) so time runs from right to left.

Figure 90 shows a second example of such paleoclimate evidence by Neff et al. [2001]. In this case, the oxygen isotope ratio δ^{18}O, as measured in stalagmites in Oman, is found to show an exceptional correspondence with the cosmogenic isotopes. U–Th (Uranium–Thorium series) dating is used on the stalagmite and the limits to allowed temporal wiggle-matching are set by experimental uncertainties and have been rigorously adhered to. We can use δ^{18}O in this case as a proxy for rainfall and the δ^{18}O depletions reveal enhanced rainfall caused by northward migrations of the inter-tropical convergence zone. Similar results are obtained using stalagmite growth and layer thickness and the ^{13}C isotope.

As discussed in Sect. 3.3, the fact that the correlations are found for both the ^{14}C and ^{10}Be isotopes is very important. Both are spallation products of galactic cosmic rays hitting atmospheric O, N and Ar atoms. However, there the similarities end because their transport and deposition into the reservoir where they are detected (ancient tree trunks for ^{14}C and ice sheets or ocean sediments for ^{10}Be) are vastly different in the two cases. We can discount the possibility that the isotope abundances in their respective reservoirs are similarly influenced by climate during their terrestrial life-history: this is because the transport and deposition of each is so vastly different. Thus we

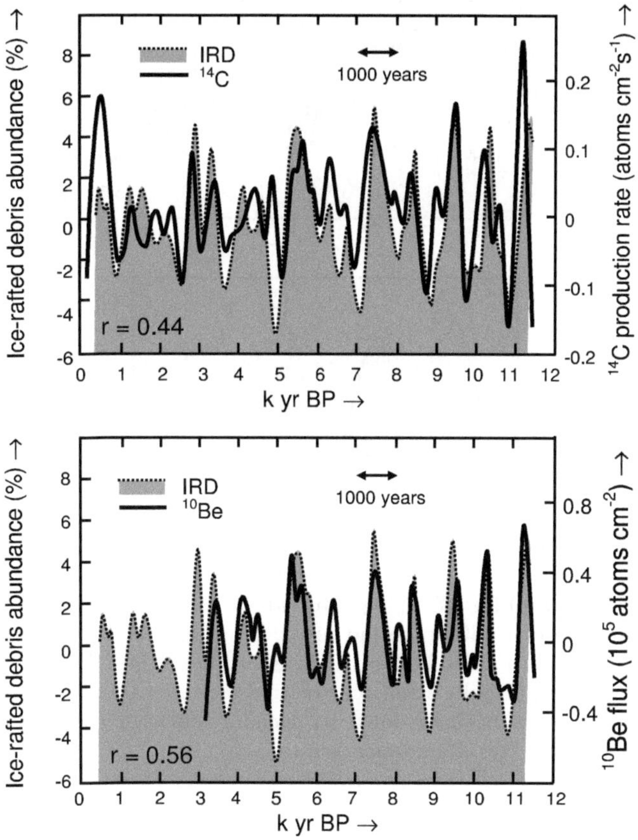

Fig. 89. The abundance of ice-rafted debris (IRD, such as quartz grains, volcanic glass and hematite-stained crystals originating from known regions and glaciers in the North Atlantic), as found in ocean-bed sediment cores. The *dotted* curve bounding the *grey* area gives the mean IRD abundance as a function of sedimentation date (in kyr before present, BP – note therefore that time runs from *right* to *left* in this plot). *Solid black* lines give the cosmogenic isotope records, the production rate of ^{14}C (*top panel*) and the abundance of ^{10}Be (*bottom panel*) [adapted from Bond et al., 2001]

can conclude that the correlation is found for both isotopes because of the one common denominator in their production, namely the incident galactic cosmic ray flux, and that this varied in close association with the climate indicators throughout the 10 kyr of the Holocene (the Atlantic–Arctic circulation pattern in the case of Fig. 89 and the latitude shifts in the tropical rainfall in Fig. 90).

The flux of the galactic cosmic rays that generate the cosmogenic isotopes is modulated by three influences: (1) the interstellar flux of GCRs incident on the heliosphere; (2) the GCR shielding by the heliosphere; and (3) the GCR

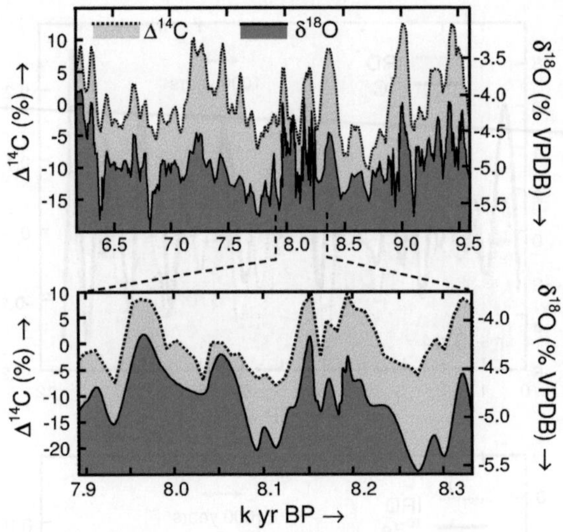

Fig. 90. The oxygen isotope ratio $\delta^{18}O$ found in stalagmites in Oman (*dark grey* bounded by *solid line*), compared to the global tree-ring $\Delta^{14}C$ record (*light grey*, bounded by *dotted line*) for 6.5–9.5 kyr BP (*upper panel*). The ratio $\delta^{18}O$ is a proxy for local rainfall. For a 430-yr period around 8.1 kyr BP, the stalagmite growth was sufficiently rapid to allow higher time resolution studies. As can be seen in the *lower* panel, the exceptionally strong correlation seen over thousand-year timescales in the top panel is maintained down to decadal scales in the *lower* panel. The correspondence with ^{10}Be abundance is similarly close [adapted from Neff et al., 2001]

shielding by the geomagnetic field. The spatial scale of interstellar GCR flux variation in our galaxy is sufficiently large compared to distances moved by our solar system through the galaxy, that we can neglect incident GCR variations on timescales of kyr and smaller (although this may become a significant factor on Myr timescales [Shaviv, 2002, 2004]). The geomagnetic field shielding has varied on timescales of 10 kyr. Mostly this variation has been gradual [Tric, 1992, Baumgartner et al., 1998] although there have been shorter-lived weakenings of the field (which may be magnetic reversal onsets that did not develop), such as the Laschamp event around 40 kyr BP [Laj et al., 2001]. These events complicate the cosmogenic isotope record [Bhattacharyya and Mitra, 1997, Damon et al., 1978] but are not consistent with the variations seen on timescales of order 1 kyr and less in Figs. 89 and 90. This being the case, most of the variations on these timescales arise from heliospheric shielding.

With these considerations, the correlations between these cosmogenic isotopes and paleoclimate indicators allow just two possible classes of explanation outlined in Table 8, which also gives some suggested mechanisms in each case.

Table 8. Implications of paleoclimate correlations with cosmogenic isotopes

Category	Explanation	Suggested Mechanisms	Terms in (164)
A	Cosmic rays induce the changes directly and thus influence climate	Air ions produced by cosmic rays seed significant numbers of cloud condensation nuclei (CCN)	A and g
		Air ions produced by cosmic rays modulate the global electric (thunderstorm) circuit	A and g
B	Cosmic ray fluxes are a proxy for another factor which influences climate	Cosmic rays are anticorrelated with total solar irradiance and cloud cover changes are associated with the changes in total radiative forcing	I_{TS} (with possible feedback to A)
		Cosmic rays are anticorrelated with UV solar irradiance which has a disproportionate effect on global climate and cloud cover via the production of stratospheric ozone	I_{TS} (with feedback to g and A)

6.3 Detection–Attribution Studies of Century-Scale Climate Change

Models of Earth's coupled ocean–atmosphere system allow simulation of the global distribution of surface temperature change. The input variations required include total solar irradiance change, volcanic aerosol loading, and anthropogenic effects (aerosol pollution, sulphates, and greenhouse gas content) [Crowley, 2000]. Over recent solar cycles, each of these inputs has varied and their effects may have had non-linear interactions with each other. This makes untangling the relative importance of the various mechanisms for influencing Earth's climate very difficult. Several important climate parameters have shown apparent solar cycle variations (in addition to the cloud cover data discussed in Sect. 6.4) – often in good agreement with the TSI record. For example, White et al. [1997] and White and Cayan [1998] find sea surface temperature variations for all the Earth's major oceans on 11-year timescales. The large thermal capacity of the oceans means that these fluctuations are not detected in global surface temperatures [Wigley and Raper, 1990, Cubasch et al., 1997]; however, caution is needed because ocean oscillations can have natural periods that can give beat periods of decades.

Figure 91 establishes some principles by looking at the possible effect of solar variability on 100-year timescales on global climate change. Fuller simulation sets from this General Circulation Model (GCM), and discussion of principles used, have been given by Tett et al. [1999] and Stott et al. [2000].

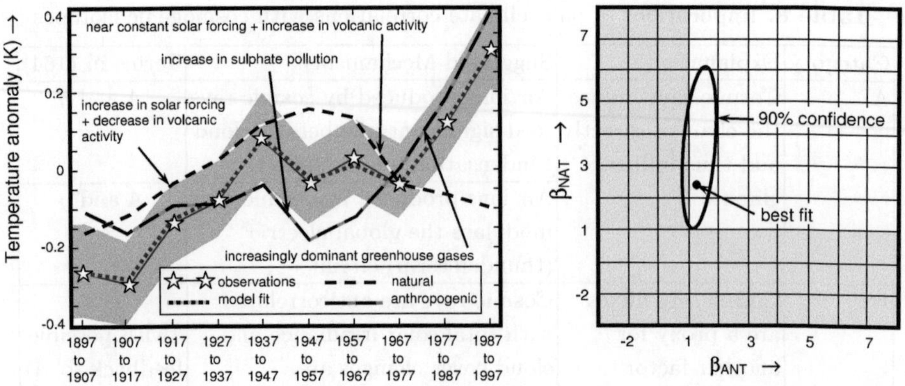

Fig. 91. An example of detection–attribution analysis of climate change over a 100-year interval. The fits to the sampled temperature record (*left*) were made on global spatial patterns of temperature anomaly, using the Hadley Centre's HAD3CM model. As well as the best fit, the contributions of natural and anthropogenic forcings are shown separately and some of the dominant effects on the variations are labelled. To obtain this best fit, the natural forcing has been amplified by a "beta factor" $\beta_{NAT} = 2.5$, relative to the predicted natural forcing. In addition, the anthropogenic forcing required was amplified by a factor $\beta_{ANT} = 1.15$. This is demonstrated by the detection–attribution diagram shown on the *right*, which also shows the ellipse formed by the coordinates of the 90% confidence limit to the best fit. Analysis shows that β_{NAT} is largely required to amplify the total solar irradiance reconstruction used as an input, which is that by Lean et al. [1995] (Courtesy P. Stott, Hadley Centre for Climate Change)

The predictions make use of the Hadley Centre's HAD3CM general circulation model, which employs 19 atmospheric levels with a grid size of 2.5° of latitude and 3.75° of longitude up to an altitude of 10 km, along with 20 ocean depths on a 1.25° × 1.25° latitude-longitude grid down to a depth of 5 km. It employs inputs representing volcanic, solar and anthropogenic forcings. For the solar forcing the TSI reconstruction by Lean et al. [1995] was used. Fits were made to the observed global spatial pattern of the temperature anomaly response to these forcings. Figure 91 is an example of the very good matches to the sampled global warming curve that can be achieved. The figure also shows separately the variations of natural and man-made effects on the temperature and the dominant causes of change in various sectors of these graphs are labelled. Of particular note is that the solar effect was mainly in the interval 1907–1947 whereas the anthropogenic contribution was dominant after 1967. Thus the solar and anthropogenic forcing increases were at different times and amplifying one will not necessarily decrease the other. In order to get the fit shown in the figure, amplification ("beta") factors were required, relative to the predictions from radiative forcing [Hansen et al., 1997]. In this case, the best fit required that the solar effect be amplified by a factor of about 2.5. This increased solar influence allowed a much better fit to the peak in the temperature curve around 1947. The work of Stott et al. [2000] finds an even larger factor of 3. At the 90% confidence level this solar amplification is between 1 and 6. Interestingly, the best fit also required an amplification of the anthropogenic greenhouse effect (by a more modest β-factor of 1.15). Thus enhanced solar contribution appears to imply an enhanced sensitivity to anthropogenic effects, but with the onset of the latter somewhat later. This means that for these 100-year simulations, the ellipses formed by the 90% confidence level tend to be oriented in "*detection-attribution*" diagrams as shown to the right of Fig. 91. Note that Fig. 91 is just one of many predictions made by the same model for almost identical input conditions because of the internal variability of the coupled system. Thus one needs to consider an ensemble of predictions. In addition, different models predict different behaviour. Nevertheless some general themes are emerging [Crowley, 2000, Bauer et al., 2003, Cubasch et al., 1997, Tett et al., 1999, Stott et al., 2000].

These global climate simulations call for a mechanism which amplifies the solar effect above what one would expect from radiative forcing by the reconstructed TSI variation. A number of possibilities have been proposed. The cosmic ray–cloud mechanism discussed in Sect. 6.4 would certainly be one. Note however that clouds have two opposing effects (depending on their height and characteristics), as they both reflect SW light, increasing A in (164), and trap in LW radiation, increasing g. Another possibility is that spectral irradiance changes in the UV have a disproportionate effect [Haigh, 1994, 1999a,b, 2001, Shindell et al., 1999, 2001, Larkin et al., 2000]. These are known to cause solar cycle variations in the stratosphere, where they modulate the quasi-biennial oscillation [Labitzke and van Loon, 1997, Gray et al., 2001] and it has been proposed that these may propagate down into the troposphere. Other possibilities may involve the modulation of the global electric (thunderstorm) circuit [Bering et al., 1998] by air ions produced by comic rays [Markson, 1981, Harrison, 2002a].

Whatever the cause, evidence is growing that the solar influence on climate over the past 150 years is somehow amplified by a factor that appears to be about 3.

6.4 Direct Cosmic-Ray Effects: Cosmic Rays and Clouds

The most controversial suggestion under category A in Table 8 is that cosmic rays directly modulate the formation of clouds [Svensmark and Friis-Christensen, 1997, Svensmark, 1998, Marsh and Svensmark, 2000a,b, 2004, Udelhofen and Cess, 2001, Kristjánsson and Kristiansen, 2000, Carslaw et al., 2002, Arnold and Neubert, 2002]. This idea is largely based on the observed correlation over recent solar cycles between galactic cosmic rays counts and the global composite of satellite cloud cover observations compiled by the International Satellite Cloud Climatology Project, ISCCP [Rossow et al., 1996].

The best correlations between cosmic rays and global cloud cover have been obtained by Marsh and Svensmark [2000a,b] from the infrared observations of clouds (10–12 μm) that make up the "D2" set compiled by ISCCP. This dataset is compiled from observations from a wide variety of spacecraft and inter-calibration of the instruments is difficult. The correlation is not found for all clouds, in fact it is only present for cloud-top pressures exceeding 680 hPa, corresponding to altitudes below about 3.2 km. Figure 92 illustrates this correlation in monthly means of the data, de-trended to remove annual variations. The light grey area shows the full cloud cover anomaly (ΔC) dataset that was available until recently (covering the interval 1983–1994) and the dotted line shows the cosmic ray flux from the Moscow neutron monitor (which detects the products of cosmic rays of rigidity 2.46 GV and above: the results for other stations are very similar). The peak correlation coefficient is $c = 0.65$, with a best-fit lag of 4 months introduced into the cloud cover data sequence. This means that $c^2 = 42\%$ of the variation in the cloud cover could be attributed to the cosmic rays. The significance of the correlation, S is high at 99% (i.e. there is only a 1% probability that this result was obtained by chance). If we introduce smoothing into the time series the correlation coefficient is greatly increased, rising to $c = 0.914$ for 12-month

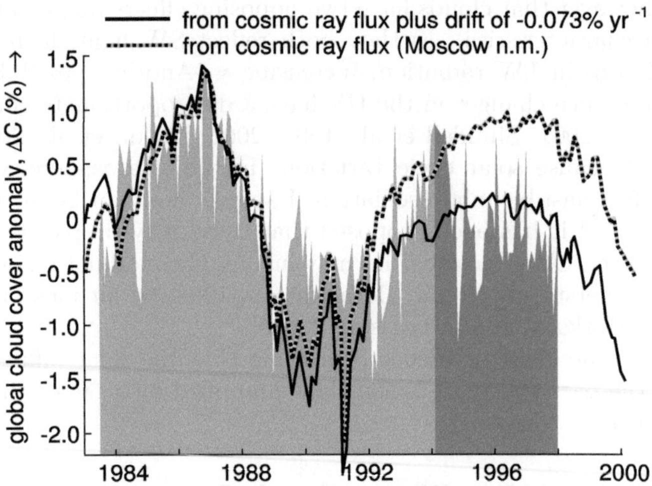

Fig. 92. The percent global cloud cover anomaly, ΔC for low altitude (< 3.2 km) cloud, seen at IR wavelengths and combined into the ISCCP D2 dataset. The *light grey shaded* area shows the original datset for which Marsh and Svensmark [2000a] and Marsh and Svensmark [2000b] discovered a decadal-scale variation, with a strong correlation with cosmic ray fluxes. The *darker grey* area is the variation of the recently-added data for after 1994. The *dotted* line shows the cosmic ray counts (scaled from the best-fit linear regression) from the Moscow neutron monitor. The *solid* line shows the best-fit combination of the cosmic ray flux variation added to a downward linear drift at a rate of $0.073\%\,\mathrm{yr}^{-1}$

running means ($c^2 = 84\%$). However, this increases the persistence in the data series and the result can no longer be considered statistically significant [Wilks, 1995, Lockwood, 2002a]. In order to achieve a significance of 99%, a correlation of this level would need to be maintained in smoothed data from a further 50 years. Some authors [e.g. Kristjánsson and Kristiansen, 2000] have questioned the validity of this correlation but Marsh and Svensmark have used other means to check its validity by producing global spatial maps of the correlation and looking at its coherence. They find that it is primarily liquid, maritime clouds, away from regions of *El-Niño events*, that correlate well. A similar conclusion has been reached by Udelhofen and Cess [2001] in ground-based data from 90 weather stations across the North American continent. Instrument relocation and changes mean that a long-term drift in these ground-based data cannot be determined, but de-trended data show a clear and persistent solar cycle variation in coastal cloud cover, of the type shown in Fig. 92, in data that extends back to 1900.

Recently, the D2 dataset has been extended to cover period the after 1994. The new data, shown in dark grey in Fig. 92, appear to show the correlation breaking down. Marsh and Svensmark [2004] argue that there may be problems in the intercalibration between the old and the new data. Their evidence for this is that there is a simultaneous sudden jump in the overlying cloud cover at greater altitudes in the combined data, and that some datasets do not show such a major and sudden decrease during 1994 as in these D2 data. In particular, these authors have made a comparison with independent observations of clouds obtained from the SSMI instrument onboard the DMSP satellites. This instrument operates at microwave wavelengths, which are able to penetrate ice and dust clouds, and thus observe liquid water clouds. This cloud data is available over the oceans for periods between July 1987–June 1990 and Jan 1992–Oct 2001. The 18 month gap is due to a problem with one of the sensor's 4 channels; however, on board calibration was maintained during this period using the 3 remaining channels. Since it is liquid clouds that give the correlation in the D2 data, this is a good data set with which to check this part of the ISCCP low cloud dataset. Differences and drifts between the two datasets do exist and Marsh and Svensmark argue that it is possible that the growing discrepancy between the new D2 data and the cosmic ray flux variation is caused by an instrumental effect.

Lockwood [2002b] has pointed out that the cloud cover variation could be made up of two components: a solar cycle variation added to a downward drift associated with anthropogenic warming. The solid line in Fig. 92 shows the best-fit multi-variable fit to the D2 low-altitude data, using the Moscow cosmic ray counts and an independent linear variation. The best fit is obtained with a decline in cloud cover at $0.073\%\,\mathrm{yr}^{-1}$ over the interval 1982–1998 – in fact, very similar to the drift predicted by a global climate simulation [Lockwood, 2002b]. With the addition of this linear decline, the correlation is improved slightly, but the significance remains roughly the same because the effects of the increased correlation and of the longer data sequence are offset by the increased number of degrees of freedom. This can be considered, at best, as only an indication of a possibility because the model does not contain any mechanisms for ion-induced CCN production and it is quite likely that if such a mechanism did exist, it would not be independent of the anthropogenic effect on cloud cover.

The studies by Marsh and Svensmark [2000a,b] and Udelhofen and Cess [2001] appear to show a solar cycle variation in cloud cover. However, we must be cautious in ascribing this variation purely to the direct effect of cosmic rays. For example, as will be discussed in the next section, cosmic ray fluxes are significantly anticorrelated with total solar irradiance [Lockwood, 2002a]. Lockwood [2001b], Lockwood and Foster [2001], Lockwood [2002b] showed that the peak correlation coefficient for the cloud cover anomaly was +0.654 with the cosmic ray data but was −0.741 with the TSI. These correlations are significant at the 99.8% and 99.6% levels. Although the correlation is marginally higher for TSI than for the cosmic rays, application of the Fisher-Z test [Lockwood, 2002a] shows that the difference between these two correlations is not significant (the significance level of the difference being only 30% so the probability that the difference arose by chance is 0.7). Therefore, although the presence of a strong and persistent solar cycle variation in cloud cover would verify a solar/heliospheric effect, from these correlations we cannot tell which of the two mechanisms implied by the paleoclimate studies is at work (or what combination of the two).

The simulation work by Yu and Turco [2000] suggest that air ions produced by cosmic rays could grow into CCNs, as postulated by this mechanism. The major debate is if such an effect would be significant compared to the many other sources of CCNs [Carslaw et al., 2002, Harrison and Carslaw, 2003].

Recent observations have added to the complexity of this debate and much controversy remains [Svensmark and Friis-Christensen, 1997, Friis-Christensen and Svensmark, 1997, Laut and Gundermann, 199, Svensmark, 1998, Marsh and Svensmark, 2000a,b, Beer, 2001, Marsh and Svensmark, 2004, Wagner et al., 2001]. Observations of cosmogenic isotopes in meteorites have been interpreted as showing that GCR fluxes impinging on the heliosphere have changed on timescales of order 150 Myr [Shaviv, 2002, 2004] and it has been argued that the times of the spiral arm crossings coincide with periods of terrestrial glaciation. (Note that such an effect could also have been caused by enhanced dust and by impacts with larger bodies within the galactic spiral arms). Because the proposed GCR modulation on 150 Myr timescales would be caused by effects outside the heliosphere and so would be independent of the Sun, this effect would require a direct effect of GCR fluxes on climate. On the other hand, other authors have reported that the periods of low geomagnetic field, such as the Laschamp event, give enhanced GCR fluxes in the Earth's atmosphere but did not influence climate in the

Greenland area [Beer, 2001], as they would have done for a direct effect of GCRs on CCNs. The CLOUD experiment at CERN has been proposed to establish the CCN growth rates and so establish if there is a direct causal effect [Kirkby et al., 2001].

6.5 Direct Cosmic-Ray Effects: The Global Electric Circuit

The production of cloud condensation nuclei by cosmic rays is not the only possibility in category A of Table 8 because cosmic rays are the source of electrical conductivity in the sub-ionospheric gap and thus are vital to the global electric thunderstorm circuit [Markson, 1981, Bering et al., 1998, Harrison and Alpin, 2001, Harrison, 2002a]. Atmospheric electric field changes could be linked to changes in global temperature, as they modulate global changes in ions and, potentially, non-thunderstorm clouds. Thunderclouds charge by collisions between ice and water moving vertically at different velocities in convective activity: in most cases, the top of the cloud becomes positively charged, the base negatively charged. Current flows from the ground to the cloud in the form of lightning and up from the cloud to the horizontally-conducting ionosphere above about 80 km. The latter is made possible by the conductivity in the sub-ionospheric gap due to air ions produced by GCRs and causes optical signatures such as Sprites, Elves, and Blue Jets. Thus thunderstorms charge the ionosphere up to a (often considerable) positive potential. Away from the thunderclouds that power the circuit, the return downward current is driven by the ionospheric potential and, again, is made possible by the GCR-induced conductivity and, at the lowest altitudes, ionisation caused by the release of radioactive gases from the ground. The fair-weather electric field corresponds to this return current. The fair-weather electric field has been shown to have fallen by about 3% per decade over the 20th century at a number of sites [Harrison, 2002b]. Given that lightning is known to be influenced by GCRs and the solar cycle [Schlegel et al., 2001, Solomon et al., 2001, Arnold and Neubert, 2002], this may be consistent with long-term modulation of sub-ionospheric conductivity caused by a predicted drop in GCRs fluxes associated with an observed rise in the heliospheric field [Carslaw et al., 2002]. Distinguishing cause from effect is very difficult in this context.

6.6 Open Solar Flux, Cosmic Rays and Solar Irradiance

For category B of Table 8, the most likely factor for which cosmic rays could be a proxy is the total solar irradiance. In fact, it is interesting to note that most paleoclimate scientists explicitly or implicitly assume this to be the case, i.e. the cosmogenic isotopes are assumed to be so highly (anti)correlated with TSI that they can be used as an index of TSI variation. As discussed in the following sections, this may turn out to be a valid assumption; however, if this is the case, it raises very interesting questions as to why the heliospheric shield should be so well correlated with TSI.

Figure 93 shows that the open solar flux $[F_S]_{IMF}$ is both strongly anti-correlated with the cosmic ray flux and strongly correlated with the TSI, as noted by Lockwood and Stamper [1999]. However, the direct anti-correlation between cosmic ray counts and TSI is weak early in the TSI data series (before 1980, see Wang et al., 2000a). This could be because the composite has underestimated the early degradation in the TSI data; however, it is more likely that this reveals the limitations of the correlation. Correlations of various cosmic ray and TSI data and their significance have been reviewed by Lockwood [2002a].

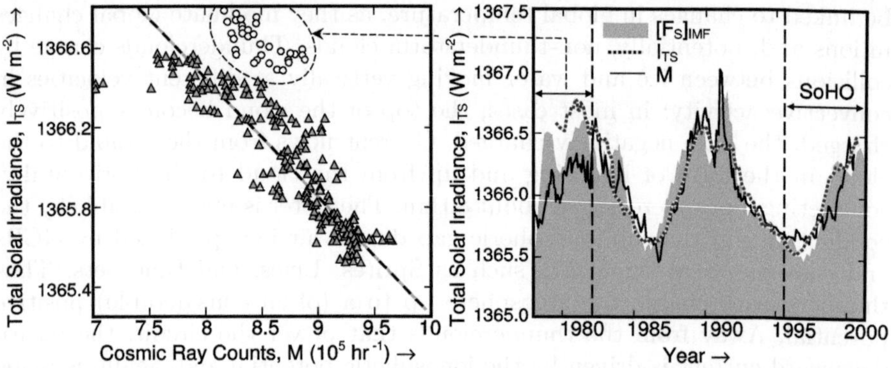

Fig. 93. The correlation between total solar irradiance, open solar flux and cosmic ray fluxes [Lockwood, 2002a,b]. The scatter plot on the *left* shows 27-day averages of TSI against the cosmic ray counts M observed at Moscow (>5 GeV). The (anti)correlation coefficient is 0.9, which is significant at the 86.5% level. The *grey-and-black dashed* line is the best-fit linear regression, and using this the M variation is scaled in terms of TSI to give the thin *black* line in the *right* hand plot, which should be compared with the observed TSI variation (*dotted line*). Agreement is good except for the earliest data (marked by the *left-hand* bar and given by open circles in scatter plot). The *grey* area in the *right-hand* plot shows the open solar flux, $[F_S]_{IMF}$ derived from observed radial IMF strength using the Ulysses result: the open flux correlates well with both TSI and cosmic ray counts. The correlation was first noted in the data from before 1996 [Lockwood and Stamper, 1999], but has continued in the TSI data from the VIRGO instrument on SoHO (marked with the *right-hand* bar)

However, that this correlation has held over the past 2 decades does not mean that it will also have held over the century and millennial timescales relevant to climate change. To make such an extrapolation we must understand any physical mechanism(s) behind the correlation. Lockwood and Stamper [1999] have extrapolated the TSI data sequence using the correlation and the result is similar, in both form and amplitude, to the TSI reconstruction by

Lean [2000] (see Fig. 94). The amplitude for the Lean reconstruction is based on comparison of the luminosity of non-cyclic Sun-like stars with the Sun in its Maunder minimum state. If this stellar analogue is valid, the extrapolation by Lockwood and Stamper shows that TSI and open flux are indeed correlated on century scales as well as decadal scales.

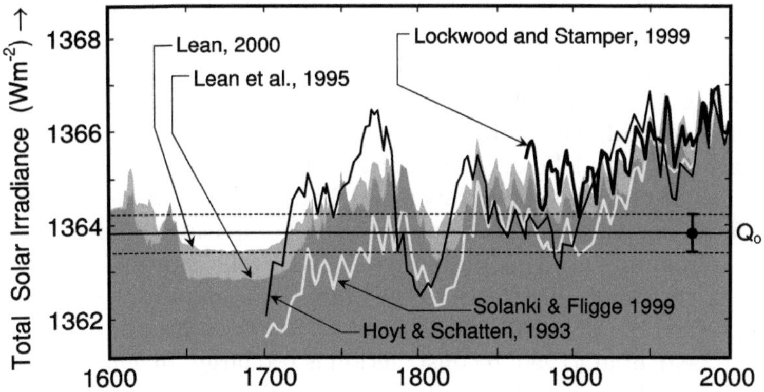

Fig. 94. Various reconstructions of past total solar irradiance. That by Hoyt and Schatten [1993] is based on the length of the solar cycle, whereas those by Lean et al. [1995] and Lean [2000] are based on a combination of sunspot number and its 11-year running mean. Solanki and Fligge [1999], Solanki and Fligge [1998] and Fligge et al. [1998] also used sunspot numbers. The amplitude of the variations is estimated from equating the Sun's Maunder minimum with the average luminosity of non-cyclic, Sun-like stars. The only reconstruction to avoid use of a stellar analogue is that by Lockwood and Stamper [1999] who used a simple correlation with open solar flux derived from geomagnetic activity $[F_S]_{aa}$. The value Q_0 is estimated by Foster [2004] for a magnetic field-free surface and so represents the minimum that could be seen in the Maunder minimum, due to the known modulation of surface emissivity by magnetic fields: any lower values require one to invoke unproven effects such as shadow effects due to magnetic fields deep in the convection zone

Any link between TSI and cosmic rays would involve emerged solar magnetic field: solar brightness variations are associated with the flux and distribution of magnetic field threading the photosphere (specifically, flux tube radii – see Sect. 2.5), whereas heliospheric shielding is linked with the magnitude and structure in the open magnetic field which leaves the top of the solar corona (see Sect. 3). The open flux is only a few percent of the surface flux and we do not understand why or if the two should have a fixed ratio on timescales of decades and greater. Furthermore TSI is concerned with the distribution of flux tube sizes in the photosphere and any link to open solar flux is certainly not obvious and not understood.

6.7 Reconstructing Past Variations in Total Solar Irradiance

A key input into the detection–attribution analysis of long-term climate change, of the type shown in Fig. 91, is a reconstruction of the total solar irradiance, TSI. Reliable, space-based measurements of TSI are only available since 1978 (see Sect. 2.5) and, in order to evaluate the relative roles of solar change, volcanoes and anthropogenic effects, it is necessary to extend the sequence to century timescales.

Various proxies have been used, giving TSI reconstructions that have similarities, but also important differences, as shown in Fig. 94. It can be argued that none of these proxies is on a firm theoretical foundation. In the Lean et al. [1995] and Lean [2000] reconstructions, the waveform used is a combination of the sunspot number and its 11-year running mean and the amplitude of this variation is determined by comparing our Sun during the Maunder minimum with the luminosity of non-cyclic Sun-like stars, assumed to also be in corresponding states to the Maunder minimum. By re-calibrating this comparison, Lean [2000] and Lean et al. [2002] argue that the long-term drift in the Lean et al. [1995] reconstruction is too large. In general, this would make the inferred β_{NAT} factor from detection–attribution climate studies (see Sect. 6.3) larger than computed using the Lean et al. [1995] reconstruction (as used in the example shown in Fig. 91). However, note that if the amplitude waveform becomes too small, these studies may fail to detect any solar signal in the natural variability of the climate system.

The reconstruction by Hoyt and Schatten [1993] uses the solar cycle length, L, a choice partly driven by the correlation with global surface temperatures by Friis-Christensen and Lassen [1991]. Leaving aside concerns about the long timescale filter used to derive the L variation and the fact that different procedures give different estimates of L (see Sect. 5.3), the problem with this is that we cannot both use this to justify the reconstruction and then use the reconstruction as an independent input into the detection–attribution process. Figure 84 shows that solar cycle length, smoothed sunspot number and open flux emergence are all related [Lockwood, 2001a] and it is this which gives the similarity to the waveforms of the various TSI reconstructions shown in Fig. 94. The combination of sunspot number R and smoothed sunspot number R_{11} used by Lean et al. [1995], Lean [2000] and Solanki and Fligge [1999] preserves the solar cycle variation seen since 1978 but also generates a long term drift. The ratio of the recent solar cycle amplitude in TSI δI_{TS} to the long term drift ($\delta I_{TS}/\Delta I_{TS}$ where ΔI_{TS} is the difference between cycle-averaged TSI now and in the Maunder minimum) is set by the choice of weighting factors used for R and R_{11}.

The above reconstructions treat TSI as a single parameter, whereas the 4-component models discussed in Sect. 2.5 showed that its variations are the sum of sunspot darkening and the brightening due to faculae that reside in the active regions, the network lanes and ephemeral flux region. Thus TSI can be expressed as

$$I_{TS} = Q_0 + f_{bn0} + \Delta f_{bn} + f_{ba} - P_{SI} \tag{165}$$

where Q_0 is irradiance the of the Sun when free of all surface magnetic features (but could vary due to effects deeper in the convection zone [Libbrecht and Kuhn, 1984, Kuhn et al., 1988, Kuhn and Libbrecht, 1991]); P_{SI} is the photometric sunspot index developed by Foukal [1981], Hudson et al. [1982] and Fröhlich et al. 1994 (see Sect. 4.12) to quantify the effect of sunspot darkening; f_{ba} is the brightening effect of faculae in active regions; and $(f_{bn0}+\Delta f_{bn})$ is the effect of faculae outside active regions (in the network and ephemeral flux regions), which has been sub-divided into a part that varies with the strong dynamo and the solar cycle, Δf_{bn}, and a part associated with the weak, turbulent dynamo that persists at solar minimum, f_{bn0}. Note that f_{bn0} need not be a constant background and may vary within the solar cycle and from solar cycle to solar cycle. A major contribution to long-term drift in f_{bn0} at sunspot minimum would be the amount of brightening associated with varying degrees of overlap of extended solar cycle phenomena.

The combined contribution of faculae to TSI ($f_b = f_{bn0} + \Delta f_{bn} + f_{ba}$) has been quantified using chromospheric line emissions as proxies. In particular, the MgII index (core-to-wing ratio of MgII line at 280 nm) has been widely used. This was originally described by Heath and Schlesinger [1986] and Donnelly [1988] and is available from UV measurements since the start of NIMBUS-7 in November 1978. Composites of the MgII index from several sources are used in proxy models of irradiance variability during the past 23 years [Lean et al., 1982, Foukal and Lean, 1986, Fröhlich and Lean, 1998a, Lean et al., 2001, Fröhlich, 2003].

Figure 95 shows the results of an analysis of TSI variability over the past 2 solar cycles by Fröhlich [2003]. In order to account for differences between the chromospheric MgII index proxy (see 17) and the facular contribution to solar irradiance variations separately on short (27-day) and long (11-year) timescales. The MgII index has been separated into these short- and long-term components and these are then linearly regressed against TSI over the period of observations, yielding TSI components P_{Fs} and P_{Fl}, respectively. To account for possible differences between the different cycles (e.g. overlap in extended cycles) this calibration procedure has been carried out for each cycle separately. This implies a long-term trend of $-0.52\,\mathrm{ppm\,yr}^{-1}$. The scaling factors for P_{Fl} are about 1.5 times larger than that for P_{Fs}. We expect a different relationship between the contrast of active region faculae and their chromospheric signature than for the network faculae because these two types of faculae have different angular dependencies (which are both quite different from that of the quiet Sun, [Unruh et al., 2000]). The MgII proxy essentially represents the projected area of the magnetic fields that produce faculae in the photosphere but it does not mimic the angular distribution of the outgoing radiance of these features. This would effect the two components P_{Fs} and P_{Fl} differently because the 27-day variability is caused by longitudinal structure

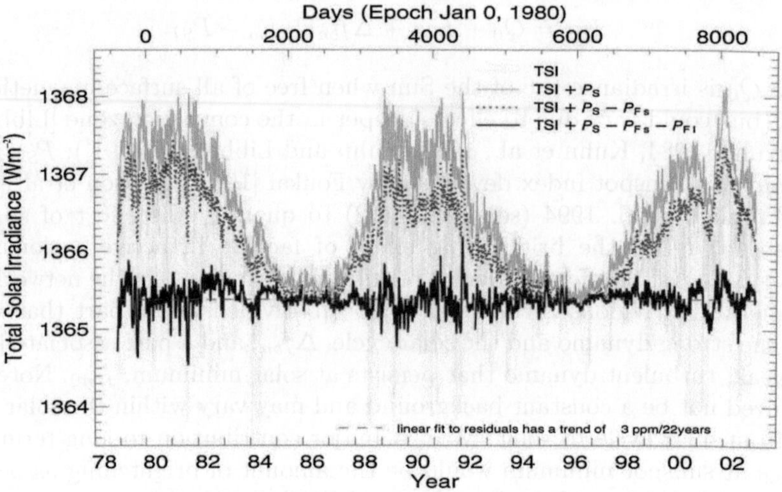

Fig. 95. Analysis of decadal TSI variability by Fröhlich [2003] using the MgII index as proxy for the facular contribution. The TSI observations, as given in Fig. 47, plus the PSI from sunspot data ($I_{TS} + P_{SI}$). The *dotted* line shows the results of taking away short term (27-day) variations in the facular brightening by showing ($I_{TS} + P_{SI} - P_{FS}$) where P_{FS} is the best-fit linear regression using the MgII index. The *solid* line is ($I_{TS} + P_{SI} - P_{FS} - P_{FI}$) which shows the results of taking away the effect of longer term (decadal) facular brightening variability again, quantified using best linear regressions with the MgII index. The linear fit to the residual (*solid line*) yields a slope of -3.8 ± 0.2 ppm/year

which is much greater for active region faculae than for network faculae which are distributed relatively evenly on the disk.

Analysis of historic observations of faculae has been interpreted as showing insufficient change in the background network faculae over the past 150 years to explain, via a change in f_{bn0}, the long-term drifts in the irradiance in the reconstructions shown in Fig. 94 [Foukal and Milano, 2001]. However, intercalibration of modern and historic data introduces very large uncertainties into thizs result. If this were to be confirmed, it would not invalidate the TSI reconstructions shown in Fig. 94, but would require them to invoke additional, as yet hypothetical, effects of sub-surface fields on the term Q_0 (such as "shadow" effects of magnetic field in the convection zone [Libbrecht and Kuhn, 1984, Kuhn et al., 1988, Kuhn and Libbrecht, 1991].

In the following sections we use (165) and look at the possible long-term variation of each term separately.

6.8 Reconstructing Past Variations in the Photometric Sunspot Index

The theory of the PSI outlined in Sect. 4.12 predicts that monthly means will depend primarily on the area of the disc covered by sunspots, with only a relatively weak dependence on the position of the spots (136). This is confirmed by Fig. 96 which shows monthly PSI as a function of the sunspot group area composite by Foster [2004]. These data are as used in Fig. 12 and are from Greenwich (1874–1976) and Mount Wilson (1982–present) observations, with the "SD" data from the former Soviet Union for (1977–1981). The SD data are also used to intercalibrate the other two datasets over the interval for which it is available (1968–1992). The Greenwich–Mt. Wilson calibration factor of 1.39 for group area used is similar to that found in previous reconstructions, but this composite by Foster [2004] also yields credible position data throughout the interval. Because the correlation with PSI is so strong, the best fit linear regression can be used to generate monthly PSI values from the composite sunspot group data from 1874 onwards. The results are shown by the black histogram in Fig. 97.

There is a systematic behaviour between monthly values of sunspot numbers and PSI, as shown by Fig. 98; however, there is much more scatter than for the sunspot group area and the variation is not linear. The best fit shown

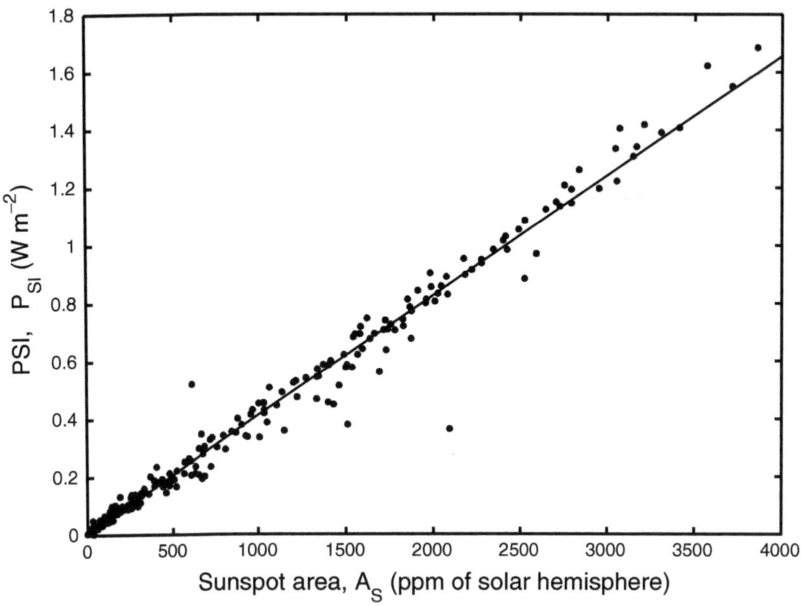

Fig. 96. Scatter plot of monthly means of the photometric sunspot index, PSI, against the surface area of sunspot groups, A_S, from the Greenwich/Mount Wilson composite of sunspot group data [Foster, 2004]

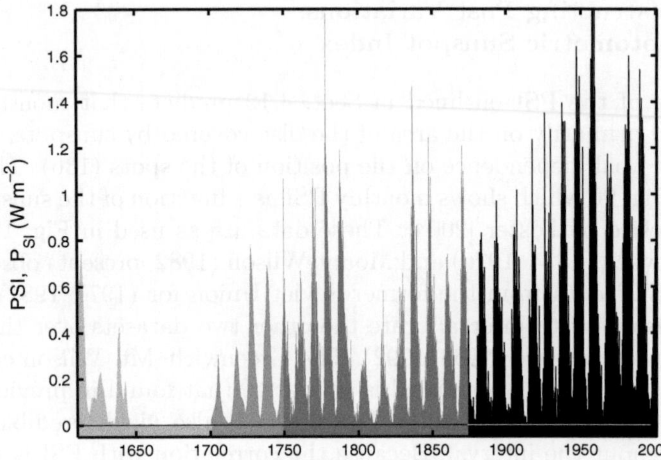

Fig. 97. Reconstruction of sunspot darkening, quantified by the photometric sunspot index. The *black* histogram gives monthly values from the Greenwich/SD/Mount Wilson composite (see scatter plot given in Fig. 96). The *darkest grey* are monthly values generated from the cubic fit to sunspot numbers (see Fig. 98), the *middle grey* are annual values derived from sunspot numbers and the *lightest grey* are annual values derived from the group sunspot area

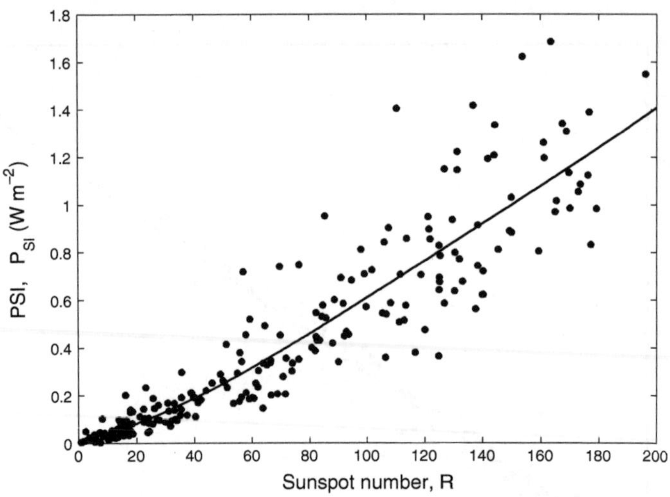

Fig. 98. Scatter plot of monthly values of photometric sunspot index as a function of the sunspot number. The line gives the best cubic regression fit

is a cubic regression, which is used to extend the PSI reconstruction back to 1740 (dark grey histogram in Fig. 97). Extension to earlier dates requires annual values and in Fig. 97 annual sunspot numbers are used back to 1704 and group sunspot numbers back to 1610.

Because of the large number of sunspot observations and the relatively simple relationships between PSI and sunspot data, the PSI reconstruction can be achieved with relatively high accuracy and confidence.

6.9 Reconstructing Past Variations in Active-Region Facular Brightening

In this section, we evaluate the contribution of faculae associated with active regions, using the composite Greenwich/SD/Mount Wilson data on sunspot groups, along with the facular contrast algorithm of Ortiz et al. [2002], discussed in Sect. (4.13). Equation (148) enables us to compute the contrast C_{MDI} for any one pixel of a full-disc MDI image in the continuum emission around 676.8 nm due to small flux tubes which contribute to facular brightening. Figure 61 shows how C_{MDI} varies with the field in the magnetogram pixel, B_{MDI} and the position on the solar disc, μ (and hence the radial field value, B_{MDI}/μ for the assumption that the field is radial). The contrasts are relative to a field-free quiet Sun intensity, corrected for limb darkening using the function derived by Neckel and Labs [1994], as shown in Fig. 56 for the MDI wavelength. The facular contrast at this wavelength is here taken to equal to the average value needed for TSI calculations: this is seen to be a good approximation from the wavelength dependence of facular contrast given by Unruh et al. [1999, 2000], from which a more accurate correction factor could be evaluated if the spectral shape of facular emission is assumed to remain constant. In general, the correction required will depend on μ.

In order to exploit the contrasts and the sunspot data on active regions, we need to know the distribution of B_{MDI}/μ values in both active regions and in the quiet Sun. The principles and difficulties are emphasised here by Fig. 99 which shows an example of an isolated active region, AR NOAA 7978, which has been studied in detail by Ortiz et al. [2000] and Ortiz et al. [2003]. This active region crossed the solar disc during the 1996 solar activity minimum and was the only one present on an otherwise almost featureless Sun. Ortiz et al. were able to study the effect of the facular brightening associated with this region on the observed total solar irradiance. Harvey and Hudson [2000] noted that a complex of active regions first appeared in Carrington Rotation (CR) 1908. During CR 1911, on 4 July, a strong new centre of magnetic activity appeared within this complex to the east of the disk centre before rotating onto the farside of the Sun. It can be seen in the variation of the daily sunspot group areas plotted in blue to the right of the figure. This AR contributed all of the observed total group area when it rotated back onto the visible disk. By 3 September, the spots had almost all disappeared, leaving faculae in the area where the active region had been. SoHO MDI magnetograms, observed when this AR was near the centre of the disk during Carrington rotations CR1911–CR1916, are also shown in the figure. It can be seen that the flux tubes of the AR become smaller and more dispersed as it evolves and spreads out.

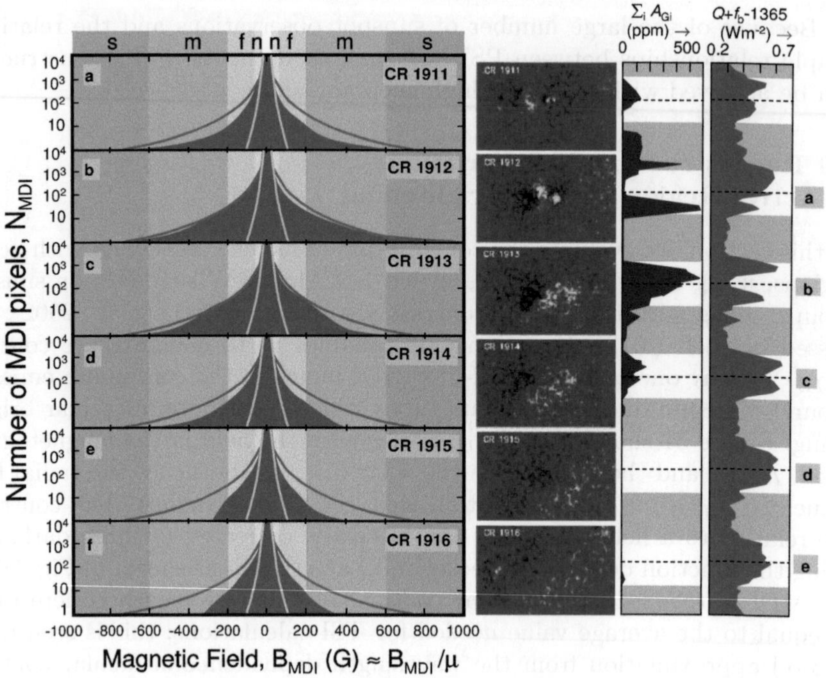

Fig. 99. Observations of an isolated active region AR NOAA 7978 for Carrington rotations CR1911–CR1916. (*Left column*) Distributions of magnetic flux observed by the MDI instrument on SoHO (*grey*) with model distributions for active regions (*red*) and the quiet Sun (*green*). Distributions show the number of MDI pixels giving the observed radial field, B_{MDI}/μ. The B_{MDI}/μ values *shaded* in various shades of *grey* denote the approximate limits of sunspots (s), micropores (m), faculae (f), and the network (n). The second column shows the MDI magnetograms of this region from which the distributions are taken. (Black is inward field, white is outward and blue is near zero). The third column gives the sunspot group area, A_G (*in blue*), and the fourth column gives the sum of the quiet Sun plus facular brightening, $Q_0 + f_b$ (minus a reference TSI of 1365 W m^{-2} – in *red*). In the two *right-hand* columns time runs down the plot and times when the AR is on the far side of the Sun are *shaded grey*

In the right hand panels of Fig. 99, the times when the centre of the AR were on the farside of the Sun are marked in grey and the times when it was close to the centre of the visible disk (about which the distributions shown to the left of the figure were observed) are given by the dashed lines. The evolution of the sunspot group described above can be seen, with almost no sunspots being observed after they fade during CR1914. The red plot to the right of Fig. 99 shows the sum of the TSI and PSI, which by (165) equals the sum of the surface-field-free and the total facular brightening ($Q_0 + f_b$). The variations with solar rotation show how this active region dominated both the sunspot group data and the irradiance variations at this time, with the irradiance elevated only when AR NOAA 7978 was on the visible disk, giving a strong 27-day signal. Initially, the facular enhancement is greatest when the active region is near the limb, giving the characteristic double-peaked temporal variation during CR1912. This is also present in CR1913, but is less marked as the region expands in area. By CR1914, the remnant of the AR has expanded to the extent that the μ-dependence of the excess emission of the individual flux tubes is smeared out by the variety of μ values present. Note that the excess facular emission persists after the sunspots have faded, but was also present before the main sunspots appeared during CR1911. During CR1912 the observed distribution was close to the model AR distribution adopted here, $n_{AR}(B_{MDI}/\mu)/N$, shown in red, and where N is the sum of n_{AR} over all B_{MDI}/μ values. As the region faded the distribution of radial field evolved back towards the model quiet sun distribution $n_{QS}(B_{MDI}/\mu)/N$, (shown in green) which is as given by Ortiz [2003]. The vertical bands filled in various shades of grey roughly demarcate (small) sunspots (s), micropores (m), faculae (f) and network (n), using the approximate criteria laid down by Ortiz, namely: pixels with radial field $(B_{MDI}/\mu) > 600\,\mathrm{G}$ are sunspots, $600\,\mathrm{G} > (B_{MDI}/\mu) > 200\,\mathrm{G}$ are micropores, $200\,\mathrm{G} > (B_{MDI}/\mu) > 60\,\mathrm{G}$ are faculae and $60\,\mathrm{G} > (B_{MDI}/\mu) > 20\,\mathrm{G}$ are network faculae.

The radial field distributions show enhanced wings when the sunspots are present. For outward field $(B_{MDI}/\mu > 0)$ this extends right out to 1000 G for CR1912, but only to near 500 G for inward field $(B_{MDI}/\mu < 0)$. The distribution is also noticeably asymmetric with more flux tubes around B_{MDI}/μ of -200 G than around $+200$ G. These asymmetries make fitting the observed active region distributions problematic; however, because intensity variations are the same for inward and outward field, we can simplify this by averaging over the positive and negative B_{MDI}/μ. Ortiz [2003] provides an approximate definition of a facula as having a radial field (in an MDI pixel) in the range 60–200 G. With this definition, the fraction of pixels within the region in question that are faculae, F_f, equals 0.2162 and 0.001 for the AR model and the quiet Sun, respectively. Thus we expect F_f to be of order 20% in active regions, but 1% in the quiet sun.

These two model distributions (AR and QS) can be combined to reproduce the annual distributions, $n_a(B_{MDI}/\mu)$, as given by Ortiz [2003], who fitted observed distributions with power law variations of the form $n_a \propto (B_{MDI}/\mu)^{-\alpha}$. Because the form of the annual distributions changes near 250 G (close to where excess pixels are seen in Fig. 99 for AR NOAA 7978), Ortiz et al. [2003] fitted different values of α above and below 250 G. The distributions are normalised to be continuous across the 250 G threshold. Because 10 G bins were used to derive the distributions, we here do not extend the annual distributions below 5 G and then normalise so that the sum of all pixels over the range 5–1000 G is N. The resulting annual distributions are shown by the thin lines in Fig. 100.

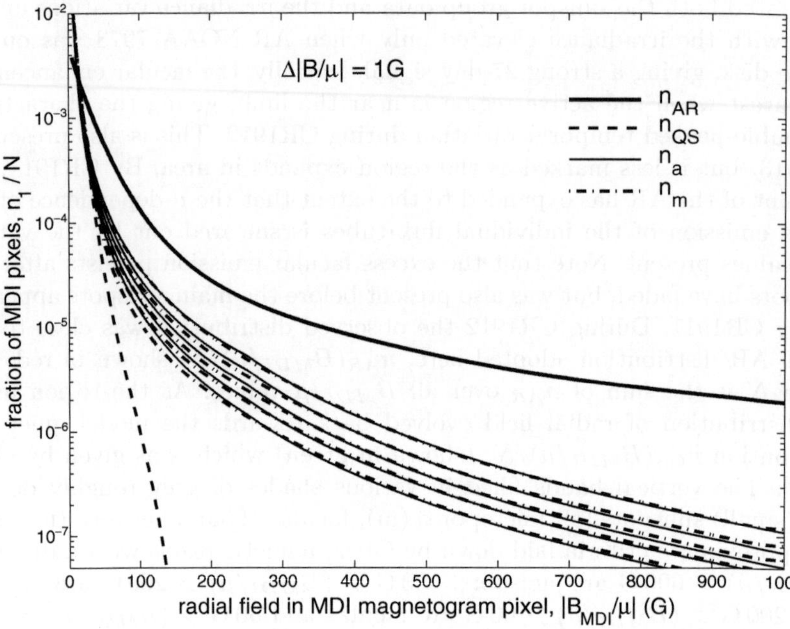

Fig. 100. The observed annual distributions $n_a(B_{MDI}/\mu)$ from Ortiz [2003] are here shown for 1996–2001 (*thin solid lines*), along with the AR and QS distributions defined in Fig. 99 (respectively $n_{AR}(B_{MDI}/\mu)$ shown by the *thick solid line* and $n_{QS}(B_{MDI}/\mu)$ shown by the *dashed line*). The *dot–dash black* lines are best fits $n_m(B_{MDI}/\mu)$, weighted combinations of n_{QS} and n_{AR}, generated using (166). For all distributions the numbers of pixels, n_1, is for 1 G bins of $|B_{MDI}/\mu|$ and N is the sum of n over all such bins

If the Sun is thought of in terms of a three component model (namely quiet sun, sunspots and faculae), to model the small flux-tube part of the distributions shown in Fig. 100 we need only the two components, the quiet sun and faculae. The overall annual distributions would then be given by the weighted sum of the AR and QS distributions where α_{AR} is the effective active region filling factor of the disc.

$$n_m = \alpha_{AR} \times n_{AR} + (1 - \alpha_{AR}) \times n_{QS} \tag{166}$$

Figure 100 shows how the model AR distribution, n_{AR}, can be combined with the quiet Sun distribution, n_{QS}, using (166) to generate the good fits to the observations, as shown by the dot-dashed black lines. The best-fit AR filling factors α_{AR} were regressed with the corresponding annual means of sunspot group surface area, the Greenwich/Mount Wilson data giving

$$\alpha_{AR} = 0.0453 + 75.5957(A_G/A_{SH}) \tag{167}$$

where A_{SH} is the area of a solar hemisphere. Using (166) and (167) we can generate the distribution of radial field values that we would expect to see around an active region of known area A_G.

Using the equations of Ortiz et al. [2002], we can compute the average facular contrast at a given μ for active regions and the quiet Sun, using the distributions of radial B values as shown in Figs. 99 and 100. If a given radial field (B_{MDI}/μ) gives a contrast C_{MDI} and has a filling factor $f_{MDI}(= n/N$ where n is the number of pixels showing radial field (B_{MDI}/μ) in an area A_D of the visible disc and N is the total number of pixels in the same area A_D). The curve fitting used by Ortiz et al. ensures that the contrasts C are zero when $(B_{MDI}/\mu) = 0$, i.e. they are relative to a solar surface that is free of magnetic field. The effect on irradiance is proportional to the product $f_{MDI}C_{MDI}$ and this is shown, as a function of B_{MDI}/μ and μ, in Fig. 101 for the model active region (AR) distribution n_{AR} and model quiet Sun (QS) distribution n_{QS}.

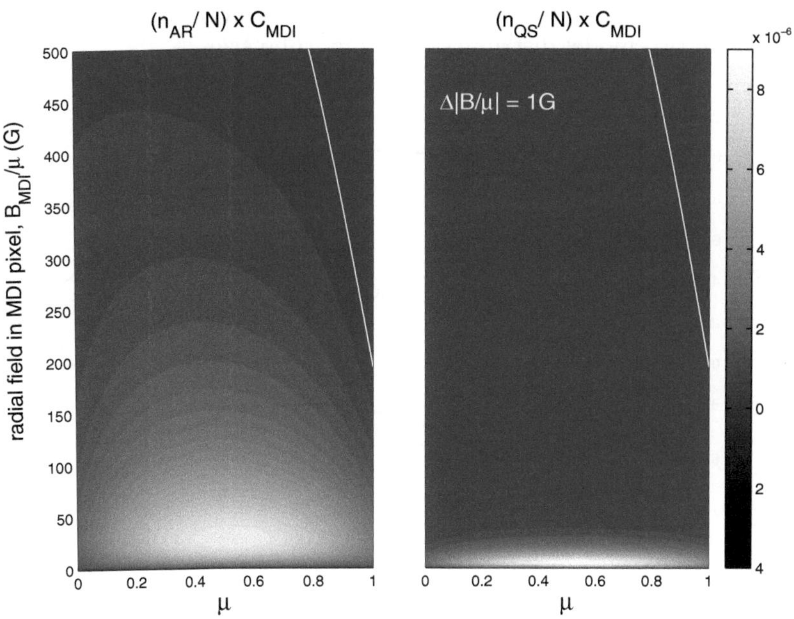

Fig. 101. The contrasts of pixels, normalised by the occurrence of MDI pixels of the radial field strength (B_{MDI}/μ) in question, $f_{MDI}C_{MDI}$ ($f_{MDI} = n/N$ where n is the number of pixels visible disc with B_{MDI}/μ in 1 G bins and N is the total number of pixels in the area A_D) – plotted here as a function of B_{MDI}/μ and μ. The *left-hand* plot is for the model active region distribution, $n = n_{AR}$, shown by the *thick* line in Fig. 100 (and the *red lines* in Fig. 99). The *right-hand* plot is for the model quiet Sun distribution, $n = n_{QS}$ shown by the *dashed* line in Fig. 100 (and the *green lines* in Fig. 99). The white contour corresponds to zero contrast

It can be seen that the brightening effect of active region flux tubes is mainly in the range $20\,\mathrm{G} < (B_{MDI}/\mu) < 200\,\mathrm{G}$. Brightening by flux tubes in the quiet sun is restricted to small flux tubes (less than about 60 G), but is also significant. The latter will be dominated by network faculae and ephemeral flux tubes. The top panel of Fig. 102, shows the averages of the distribution of contrasts at a given μ, $\langle C_{MDI} \rangle$, as a function of μ. These are shown by the solid and dashed lines for the AR and QS models, respectively. Using the definition of contrast given in (139) and of the limb darkening function $L_D(\mu) = I_{QS}(\mu)/I_O$, where the surface-field-free intensity $I_{QS}(\mu)$ equals I_O at the disc centre ($\mu = 1$), the intensity of an area of the Sun is given by

$$I(\mu) = I_O L_D(\mu)(\langle C(\mu) \rangle + 1) = I_O L_D(\mu) \langle C(\mu) \rangle + I_O L_D(\mu) \qquad (168)$$

Thus increases in intensity, over the case where there are no surface fields (everywhere $B_{MDI}/\mu = 0$), are proportional to the product of the mean contrast $\langle C \rangle$ and the limb darkening function L_D. This product is shown as a function of the disc position parameter μ in the bottom panel of Fig. 102, for the same two model distributions of flux tube sizes.

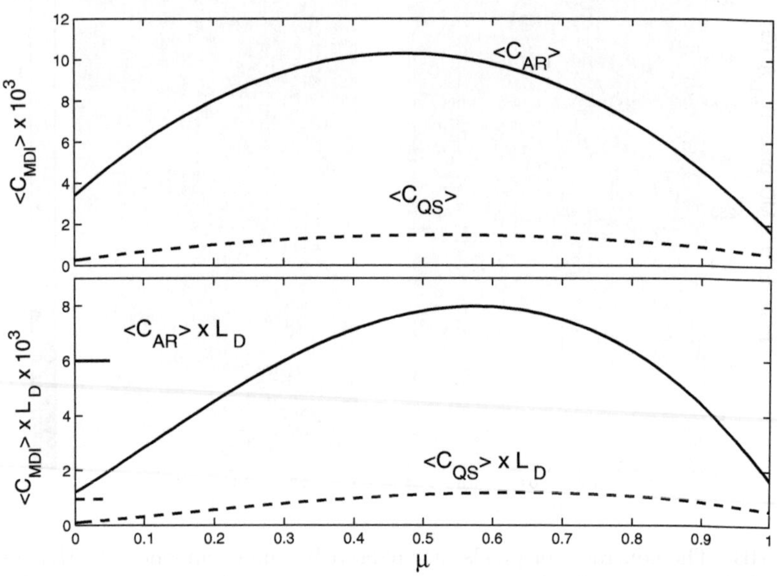

Fig. 102. (*Top*) The average contrast of MDI-sized pixels at a given disc position parameter μ, for radial field distributions: (*solid line*) for active region model, $\langle C_{AR} \rangle$, and (*dashed line*) quiet Sun distribution, $\langle C_{QS} \rangle$. (*Bottom*) The average contrasts multiplied by the limb darkening factor (the product being proportional to intensity) at a given μ. The *short horizontal* lines on the *left* give the coresponding disc-averaged values $\langle ((\langle C_{AR2} \rangle L_D)) \rangle_D = 6.086 \times 10^{-3}$ and $\langle ((\langle C_{QS} \rangle L_D)) \rangle_D = 1.057 \times 10^{-3}$

To reconstruct the brightening effect of small flux tubes in active regions, consider the effect of an active region of surface area A_{AR}. This will occupy an area μA_{AR} on the visible disc of the Sun. The region has an average contrast $\langle C_{AR}\rangle$, relative to the intensity of a magnetic-field free pixels at the same μ. The filling factor of this region on the solar disc is

$$\alpha_{AR} = \frac{\mu A_{AR}}{\pi R_s^2} = \frac{N_{AR}}{N_S} \tag{169}$$

where N_{AR} is the number of MDI pixels within the active region and N_S is the total number of MDI pixels in the solar disc. From (168), the change in total solar irradiance due to the facular brightening of one AR pixel, relative to the quiet Sun is

$$\delta f_{b1}(\mu) = I_{AR}(\mu) - I_{QS}(\mu) = I_O L_D(\mu)\left(\langle C_{AR}(\mu)\rangle - \langle C_{QS}(\mu)\rangle\right) \tag{170}$$

Summing over all pixels in the group, which is assumed small enough that $\langle C_{AR}\rangle$ and L_D are approximately constant:

$$\delta f_{bi} = I_O \sum_{i=1}^{N_{AR}} \left[\langle C_{AR}\rangle - \langle C_{QS}\rangle\right]_i L_{Di} = N_{AR} I_O \left[\langle C_{AR}\rangle - \langle C_{QS}\rangle\right]_i L_{Di} \tag{171}$$

If we sum all N_S pixels for a magnetic-free Sun, we derive a total field-free solar irradiance

$$Q_0 = I_O \sum_{j=1}^{N_S} L_{Dj} = N_S I_O \langle L_D\rangle_D \tag{172}$$

where $\langle L_D\rangle_D$ is the disc-averaged limb darkening factor, which in Fig. 56 was shown to be equal to 0.8478 for the wavelength at which MDI operates.

Substituting (172) and (169) into (171) yields

$$\delta f_{bi} = \alpha_{ARi}\left\{\frac{Q_0}{\langle L_D\rangle_D}\right\}\left[\langle C_{AR}\rangle - \langle C_{QS}\rangle\right]_i L_{Di} \tag{173}$$

If we then sum over all N active regions present on the solar disc at any one time we get the total contribution of active regions to facular brightening, relative to the quiet Sun

$$f_{ba} = \sum_{i=1}^{N} \delta f_{bi} = \left\{\frac{Q_0}{\langle L_D\rangle_D}\right\}\sum_{i=1}^{N} \alpha_{ARi}\left[\langle C_{AR}\rangle - \langle C_{QS}\rangle\right]_i L_{Di} \tag{174}$$

The upper panel of Fig. 103 shows the facular brightening in active regions, f_{ba}, estimated using (174) with the Greenwich/SD/Mt. Wilson sunspot group composite dataset. A value of $Q_0 = 1363.8 \times 0.1\,\mathrm{W\,m^{-2}}$ is used here, as estimated in the next section. The bottom panel of the figure uses 11-year running means of f_{ba} to demonstrate the effect of solar latitude of spots,

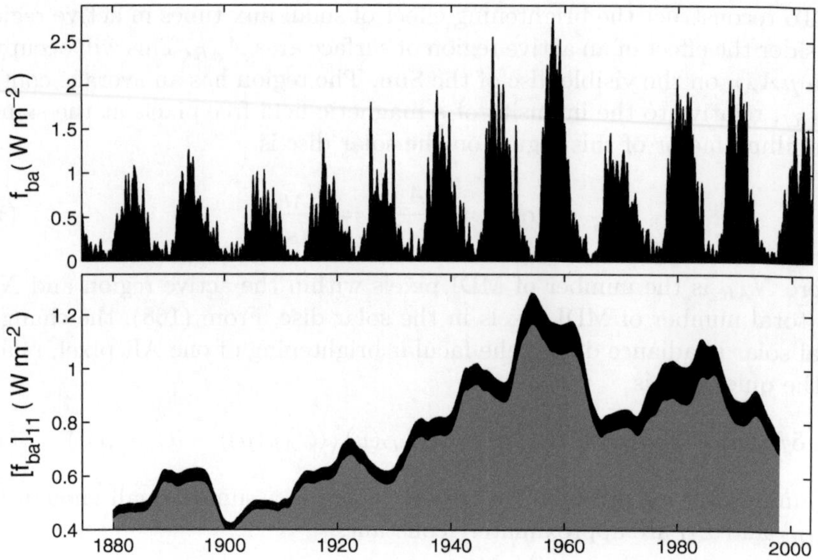

Fig. 103. (*Top*) Monthly estimates of the facular brightening in active regions, f_{ba}, computed using (174) and the Greenwich/SD/Mount Wilson sunspot group composite dataset. (*Bottom*) Eleven-year running means of f_{ba}. The *grey* area is calculated assuming that all sunspot groups are at the solar equator, whereas the upper edge of the *black* area gives the values calculated for the actual observed latitudes of the sunspot groups: thus the *black* area gives the effect of solar latitude of spots, through the effect on the μ values

through its effect on the μ values. The changing average latitude of the spots and the greater latitudinal extent of the spots [Foster and Lockwood, 2001] has added to the drift in the brightening, as seen from Earth, by about 10%.

It is interesting to note that the increased latitude of the spots means that if the Sun had been viewed from over the solar poles, the mean μ values would have increased over the 20th century (just as they have decreased when viewed from the Earth – see Knaack et al., 2001). Thus whereas, as seen from Earth, active region faculae have moved to nearer the limb on average (and so contributed to the rise in facular brightening), the same effect would have moved them away from the limb when viewed from over the poles and so contributed a decrease in facular brightening (reducing the rise caused by the increase in facular area). Thus the directional properties of the Sun will have changed over the past 150 years, changing the relationship between the TSI (and thus the disc-averaged intensity) and the luminosity.

Figure 104 shows a scatter plot of the predicted active region brightening against the sum of the TSI and the PSI (which by (165) equals the quiet Sun plus the total facular brightening). It can be seen that there is a good linear relationship, which is to be expected if the solar cycle variation in network

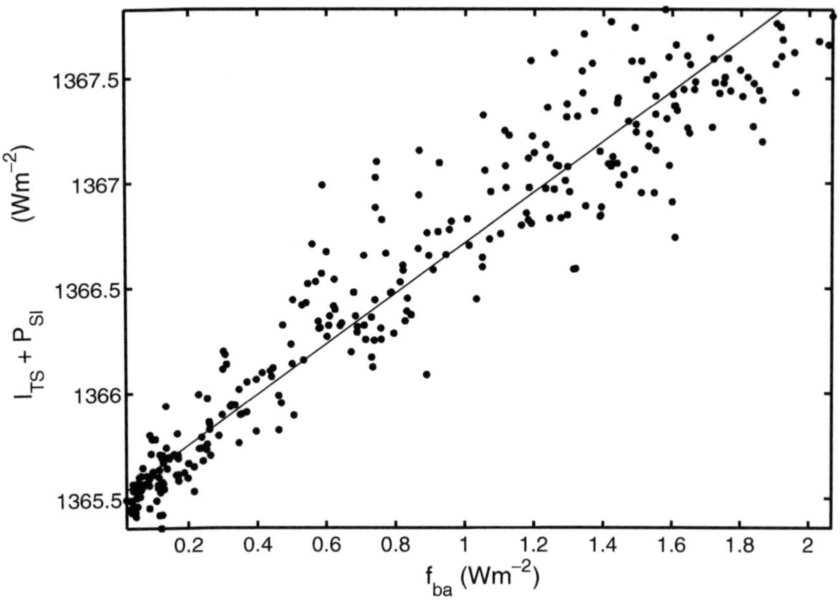

Fig. 104. Scatter plot of three-point running means of monthly values of the predicted active-region facular brightening f_{ba} (see Fig. 103) against the observed sum of the TSI and PSI, $I_{TS} + P_{SI}$. The correlation coefficient is $r = 0.953$, which is significant at the 98.0% level and explains $r^2 = 91\%$ of the variation. The *solid* line is the best linear regression fit $(I_{TS} + P_{SI}) = sf_{ba} + c$, where $s = 1.20 \pm 0.06$ and $c = 1365.5 \pm 0.1$. The best fit is obtained if the f_{ba} data (derived from the sunspot group data) is lagged by one month

facular brightening Δf_{bn} has a similar variation waveform to f_{ba}. The best fit linear regression is

$$(I_{TS} + P_{SI}) = sf_{ba} + c \tag{175}$$

where $s = 1.20 \pm 0.06$ and $c = 1365.5 \pm 0.1\,\mathrm{W\,m^{-2}}$. Using (165), this yields

$$Q_0 + f_{bn0} + \Delta f_{bn} + f_{ba} = sf_{ba} + c \tag{176}$$

The best fit is at a lag of 1 month which is consistent with the evolution of the facular brightening and sunspot group area in the example shown in Fig. 99.

6.10 Reconstructing Past Variations in the Solar Cycle Variation of Network Facular Brightening

We can apply equations equivalent to those used in the last section to the present day quiet Sun and look at the brightness increase of the quiet sun at solar minimum, relative to a magnetically-free Sun, using the average contrasts $\langle C_{QS} \rangle$. If we assume that the quiet Sun surface is homogeneous (such that $\langle C_{QS} \rangle$ is, like L_D, a function of μ only) we replace the discrete sum over all active regions with an integral over the whole disc (129). In this case, the brightening predicted is due to network faculae and any other small flux tubes that are present during recent solar minima, f_{bn0}.

$$f_{bn0} = 2\left\{\frac{Q_0}{\langle L_D \rangle_D}\right\} \int_0^1 L_D \langle C_{QS} \rangle \mu \mathrm{d}\mu = \left\{\frac{Q_0}{\langle L_D \rangle_D}\right\} \langle ((\langle C_{QS} \rangle L_D)) \rangle_D \quad (177)$$

Figure 102 shows that $\langle ((\langle C_{QS} \rangle L_D)) \rangle_D = 0.957 \times 10^{-3}$ and Fig. 56 shows that $\langle L_D \rangle_D = 0.8478$. Hence

$$\frac{f_{bn0}}{Q_0} = \frac{(1.057 \times 10^{-3})}{0.8478} = 1.247 \times 10^{-3} \quad (178)$$

Note that this value applies to recent times because it is based on the distribution of flux tube fields in the quiet Sun, as seen by SoHO. If we assume that Δf_{bn} falls to its solar minimum value at the same time as f_{ba}, (177) applied to this time yields that for 1978–2000

$$Q_0 + f_{bn0} = c = 1365.5 \pm 0.1 \mathrm{W\,m}^{-2} \quad (179)$$

From (178) and (179) $Q_0 = 1363.8 \pm 0.1\,\mathrm{W\,m}^{-2}$ and $f_{bn0} = 1.70 \pm 0.15\,\mathrm{W\,m}^{-2}$. In fact, analysis shows that a larger error in Q_0 is caused by the uncertainty in the Ortiz contrast. Using the spread of contrasts around the Ortiz polynomial fits, the uncertainty in Q_0 is found to be $\pm 0.4\,\mathrm{W\,m}^{-2}$. This Q_0 value and uncertainty is shown in Fig. 94. Note that this Q_0 value strictly applies to 1978–2000 as this is the interval of TSI data used in its derivation (Fig. 104). The Maunder minimum values of the reconstructions by Hoyt and Schatten [1993], Solanki and Fligge [1999] and Lean et al. [1995] all fall below this value and so one must invoke sub-surface magnetic effects (alpha or beta effects in the convection zone) for these reconstructions. Such effects would allow Q_0 to be lower in the Maunder minimum than at present. The Lean [2000] simulation is consistent with a constant Q_0, but is near the lower limit of the uncertainty band. If there are no subsurface effects and TSI variability is only due to surface effects, as the 3- and 4-component modelling of data since 1978 suggest (see Sect. 4.14) [Solanki and Fligge, 2002], the extrapolation based on open flux by Lockwood and Stamper [1999] is the most consistent with this Q_0 value.

In fact, $I_{TS} = Q_0$ in the Maunder minimum is unlikely to be valid, given that the ^{10}Be cosmogenic isotope record through the Maunder minimum shows a continuing solar cycle variation [Beer et al., 1990], indicating that at least some flux emergence continued.

Equation (176) yields $(\Delta f_{bn} + f_{ba}) = s f_{ba}$ where $s = 1.20 \pm 0.06$. Thus $f_{ba}/(\Delta f_{bn} + f_{ba})$ equals $(1/s) = 0.83$, in other words this analysis predicts that 83% of the solar cycle variation in facular brightening is caused by active region faculae and 17% by the network and ephemeral regions. This is very similar indeed to the ratio derived by Walton et al. [2003] from observations of faculae by San Fernando Observatory (SFO). This instrument gives 512 pixels across a solar diameter and observes the intensity of the 393.4 nm CaII K line emission (see 17). The threshold used to define a facula in the San Fernando data is a K-line contrast exceeding $4.8\%/\mu$ [Chapman et al., 2001, and references therein]. As an additional check, the total surface facular area A_f (as a fraction of a solar hemisphere) from the SFO were regressed against the total facular brightening derived here $(\Delta f_{bn} + f_{ba} + f_{bn0})$. The correlation coefficient is $r = 0.9657$, which is significant at the 75% level and explains $r^2 = 93.3\%$ of the variation. The best linear regression fit is $(\Delta f_{bn} + f_{ba} + f_{bn0}) = 0.7174$ [A_f in 10^4 ppm]+0.15.

Figure 105 shows, in black, the observed facular brightening and compares with the contributions of active regions and the network/ephemeral flux. This figure shows several peaks in the network/ephemeral facular brightening that follow immediately after peaks in the active region brightening. These

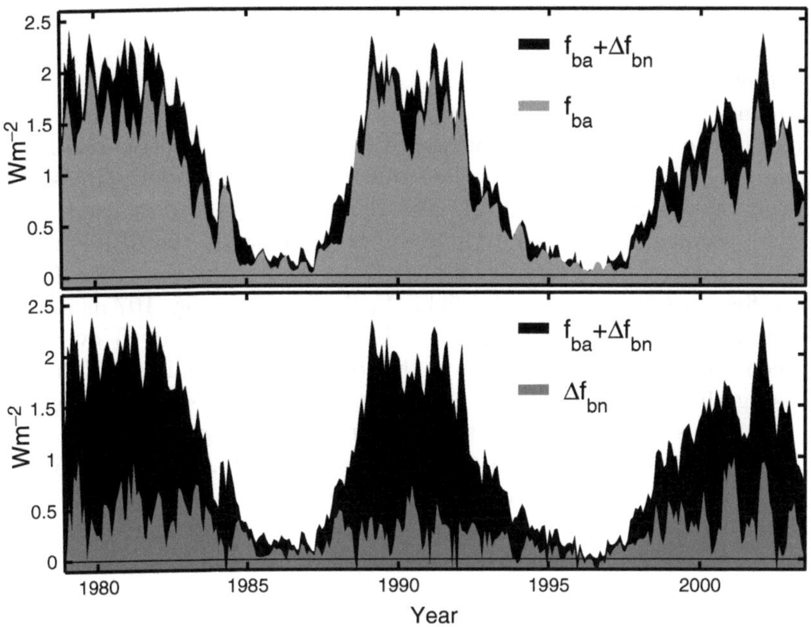

Fig. 105. The components of facular brightening over the last few solar cycles. In both panels, the *black* histogram gives the observed variation of $(\Delta f_{bn} + f_{ba})$. In addition, the *top* panel compares this with the f_{ba} variation, computed from the sunspot group data. The *lower* panel also shows the inferred solar cycle variation in the network facular brightening, Δf_{bn} (*in grey*)

appear to be faculae left over from active regions. This effect is particularly pronounced for the most recent solar cycle (23), where three large peaks in Δf_{bn} can be seen in the figure following peaks in f_{ba}. The effect is also present in cycles 21 and 22, but is much less pronounced. In several places, decreases in Δf_{bn} are simultaneous with peaks in f_{ba}, implying that f_{ba} has been overestimated.

6.11 Reconstructing Past Variations in Background Network and Ephemeral Facular Brightening

In the previous two sections we have used the sunspot group data to predict the solar cycle variation in active region and network facular brightening (respectively f_{bn} and Δf_{bn}) right back to the start of the Greenwich data in 1874. In addition, the analysis revealed the modern-day values of Q_0, the TSI of the Sun when free of surface magnetic features, and the background brightening by network and ephemeral flux that persists at sunspot minimum, f_{bn0}. Variations in f_{bn0} from minimum to minimum will be more pronounced if weak flux tubes of the extended solar cycle are a significant brightening factor and the degree of overlap varies. As yet we have little hard information on such effects.

To understand possible variations in f_{bn0}, we can model the effect of changes in the quiet Sun distribution of field, outside of active regions, n_{QS}, by assuming that it always has the same shape as in modern times (and as observed by the SoHO satellite). The distribution is then varied in width and then renormalized so that the number of pixels on the disk is constant. The resulting set of model distributions are shown in Fig. 106. The total surface flux in 1 G bins of radial field, F_1, is shown in Fig. 107 as a function of B/μ, for the distributions shown in Fig. 106. Integrating F_1 over all B/μ (between -1000 G and 1000 G) gives the total surface flux in small flux tubes, F_q. Using the Ortiz et al. [2002] contrasts with the distributions shown in Fig. 106 yields the disc-integrated facular brightening, f_{bn0}. This is found to vary linearly with the total surface magnetic flux F_q in small flux tubes (at $|B/\mu|$ below 1000 G) between $f_{bn0} = 0$ for $F_q = 0$ to $f_{bn0} = 1.71\,\mathrm{W\,m^{-2}}$ for $F_q = 3.4 \times 10^{15}$ Wb (for the present-day QS distribution shown by the thicker line in Figs. 106 and 100 and the green lines in Fig. 99). Thus [f_{bn0} in $\mathrm{W\,m^{-2}}$] $= 0.5$ [F_q in $10^{15} Wb$].

Therefore the variation in f_{bn0} depends on the variation of the surface magnetic flux in small flux tubes outside of active regions. We here assume that this varies linearly with the 11-year smoothed mean of the sunspot number, R_{11}, and define the amplitude of this variation with three assumptions given in Table 9.

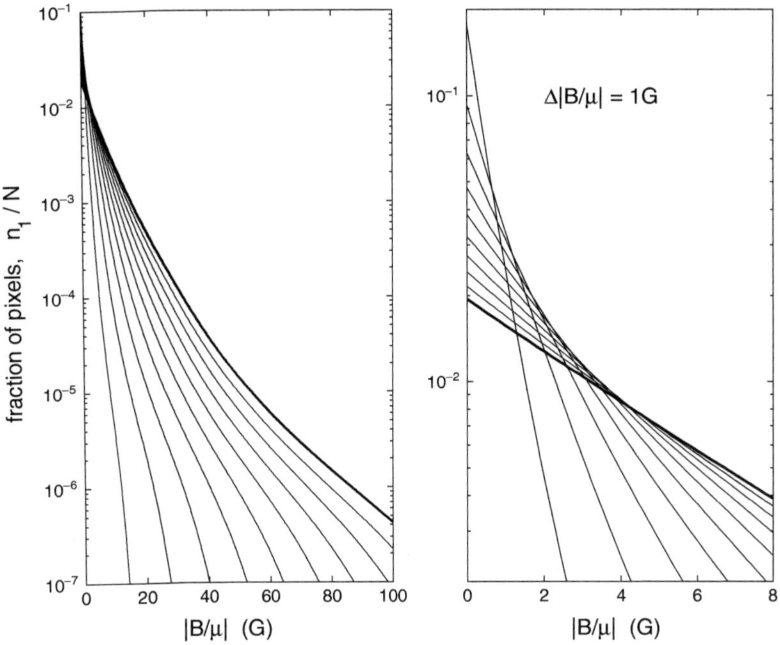

Fig. 106. Distributions of radial field values for the quiet Sun. The *heavier* line is the modern-day distribution and the others are generated by reducing the width of the distribution and renormalising to give a constant number of total pixels. The plot gives the fraction of pixels in 1 G bins, n_1/N. The *left-hand* plot covers the range of $|B/\mu|$ of 0–100 G, the *right-hand* plot covers the range 0–8 G in more detail, showing the increase in near-zero fields when the width of the distribution is reduced. The limit of zero width yields a delta function at $|B/\mu| = 0$

Table 9. Assumed Variations in the Quiet Sun between the Maunder Minimum and the Present Day

Assumption Number	Assumption	ΔF_q (10^{15} Wb)	Δf_{bn0} ($\mathrm{W\,m^{-2}}$)
1	That there has been no century-scale drift in f_{bn0} and F_q	0	0
2	That F_q fell to zero during the Maunder minimum	$F_m = 3.4$	1.7
3	That F_q fell to half present-day values, $F_m/2$, during the Maunder minimum	$F_m/2 = 1.7$	0.85

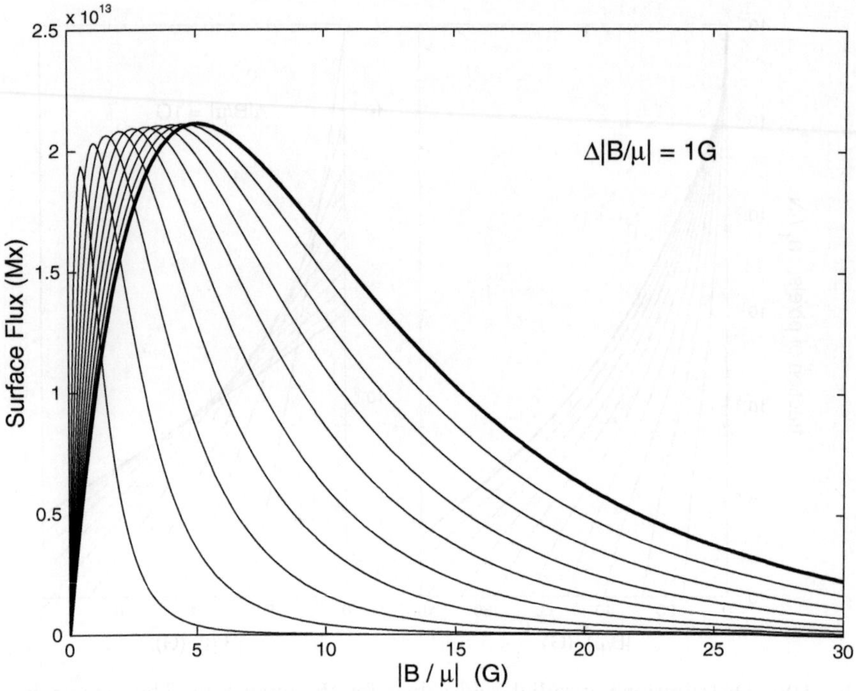

Fig. 107. The total surface flux in 1 G bins of radial field, F_1, as a function of $|B/\mu|$ for the family of quiet Sun distributions shown in Fig. 106

6.12 Reconstructing Past Variations in Total Solar Irradiance from the Sum of All Effects

Sections 6.8–6.11 give ways of computing all of the terms in (165). By adding these terms together, we can generate a reconstruction of TSI that is independent of stellar analogues. Using the Greenwich sunspot data and facular contrasts developed from observations by the SoHO spacecraft, reconstructions of each of the separate terms can be made without any major assumptions. The two major exceptions to this are the background quiet-Sun brightening f_{bn0} and the surface field-free Q_0. We here assume that irradiance variability is all due to surface effects (i.e. there are no effects deep in the CZ) so that Q_0 is constant at the value deduced in Sect. 6.10 for 1978–2003 applies at all times. To include the variation in the f_{bn0} term, we make the three assumptions listed in Table 9. Assumptions 1 and 2 give limits of behaviour and assumption 3 gives behaviour that is halfway between the two.

It is useful to visualise the magnetic field distributions that these assumptions imply. These are computed using (166) and (167) from the observed sunspot group area A_G. The distribution of active region radial field values (n_{AR}) is kept constant and so the change in active region fields is all due to the change in the area A_G. To allow for any long-term changes in the quiet Sun, the width distribution n_{QS} is varied as in Fig. 104, using the total flux, F_q, the long term variation of which is assumed to follow the same form as smoothed sunspot numbers R_{11}, with amplitude set by the assumptions given in Table 9. The results are given in Fig. 108.

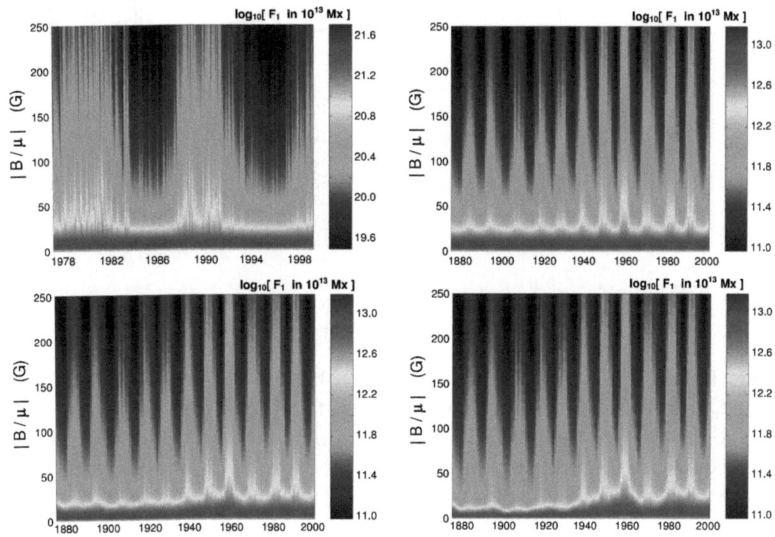

Fig. 108. Distributions of radial surface flux. The logarithm to base 10 of magnetic flux in 1 G bins, $\log_{10}(F_1$ in 10^{13} Mx), is colour-coded as a function of year and radial field $|B/\mu|$. The *top left* panel gives monthly mean distributions derived from monthly means of the sunspot group area from 1978 to 2000. The remaining three panels show annual mean distributions derived from annual means of the sunspot group area data from 1874 to 2001 and using assumptions 1, 2 and 3 of Table 9 (for *top right*, *bottom right* and *bottom left*, respectively)

The top left panel of Fig. 108 shows the monthly distributions of magnetic flux in 1 G bins, F_1, for 1978–2000 (for which TSI data are available). In fact, this plot uses assumption 3, but because the smoothed sunspot number is relatively constant during this interval, the other two assumptions give almost identical results. The solar cycle is seen as the rise and fall in the number of large $|B/\mu|$ pixels, with a corresponding slight drop in QS pixels with $|B/\mu|$ near 10 G. The other three panels show the reconstructed field distributions from annual means of the sunspot group area for 1874–2001. The effect of the three assumptions can clearly be seen at $|B/\mu|$ below about 25 G. Figures 109–111 show the corresponding TSI reconstructions.

Figure 109 shows the reconstruction for assumption 1. The thin black line shows the observed values after 1978 and the thick black line shows 11-year running means. There is an upward trend in the reconstructed TSI over the

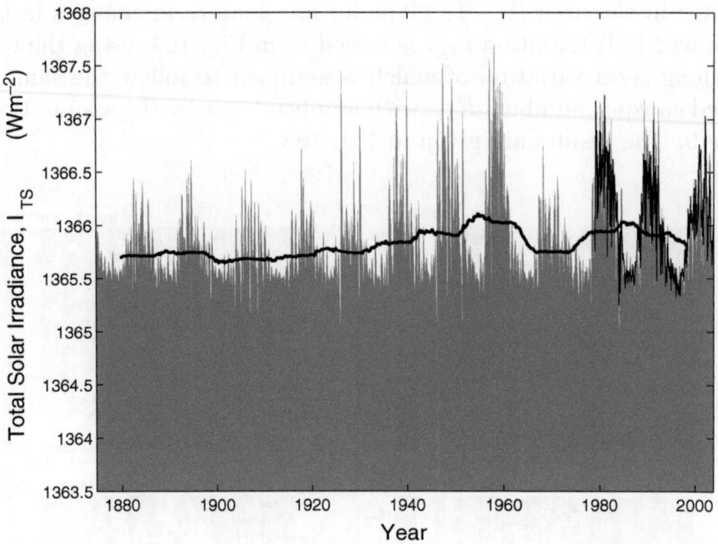

Fig. 109. Reconstruction 1 of the TSI, I_{TS}, based on the assumption that the quiet Sun Q_0 and the solar-minimum network facular brightening f_{bn0} have both remained constant over the past 150 years (assumption 1 of Table 9). The *grey* area gives the reconstructed monthly values and the *thin black* line gives the observed values from the PMOD TSI composite. The *thick black* line is the 11-year running mean

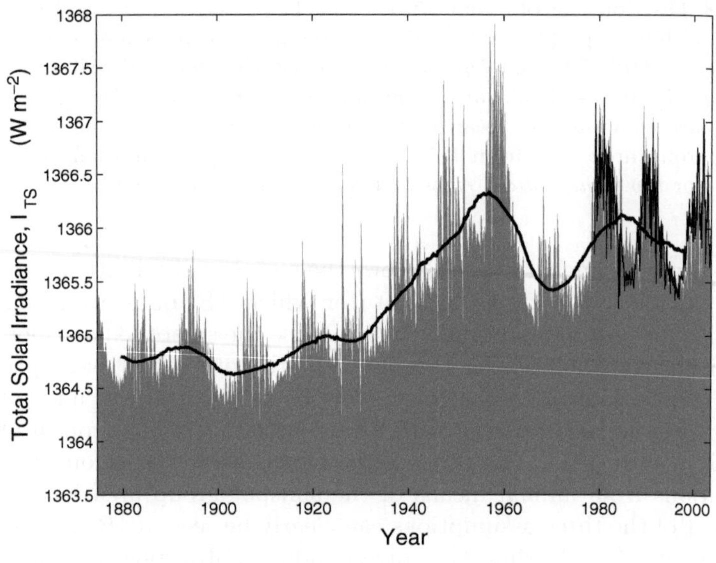

Fig. 110. The same as Fig. 109 but assuming that the magnetic flux threading the photosphere fell to zero by the end of the Maunder minimum (assumption 2 in Table 9)

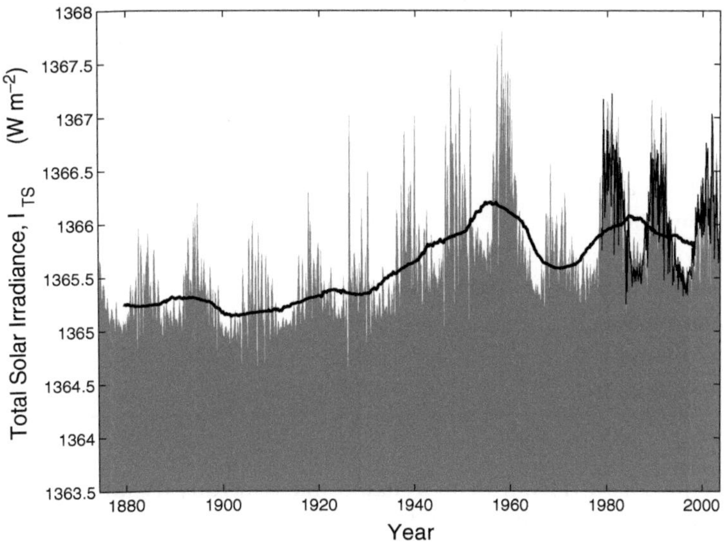

Fig. 111. The same as Fig. 109 but assuming that the total flux in small flux tubes at sunspot minimum fell to half present-day values during the Maunder minimum (assumption 3 in Table 9)

past 150 years, in this case only because of the increasing amplitude of the solar cycles. In this case, the ratio of the long-term drift (quantified by ΔI_{TS}, the difference between average TSI values observed since 1978 and inferred for the Maunder minimum) and the average amplitude of solar cycles since 1978 (δI_{TS}) is $(\Delta I_{TS}/\delta I_{TS}) = 0.5$. Figures 109 and 110 show the corresponding plots for, respectively, assumptions 2 and 3 of Table 9. If the surface field fell to zero in the Maunder minimum, the ratio $(\Delta I_{TS}/\delta I_{TS})$ would be 2.21, whereas if it fell to half present day values $(\Delta I_{TS}/\delta I_{TS}) = 1.45$. Table 10 compares the $(\Delta I_{TS}/\delta I_{TS})$ values for the reconstructions presented in this section with those previously published. In addition, the ratio to the value for the Lean et al. [1995] reconstruction is presented (because that was used in the detection–attribution studies presented in Sect. 6.3).

The likelihood of the three scenarios given by Figs. 109–111 depends on the coupling between the strong and the weak dynamos and the extent to which the small flux tubes outside active regions are the remnants of active regions, as opposed to emerged through regions outside active regions. There are a number of unknowns about the Maunder minimum behaviour. Firstly does the lack of sunspots show a complete shutdown of the strong dynamo or did it continue, but only produce smaller flux tubes? If it did completely or almost completely shut down, the quiet Sun facular brightening due to the network and remnants of active regions would be lost, in addition to the active region brightening. The extended solar cycle features evolve as they

Table 10. Comparison of the long-term drift in various TSI reconstructions

Reconstruction	Maunder Minimum I_{TS} (W m^2)	$\Delta I_{TS}/\delta I_{TS}$	$\frac{\Delta I_{TS}/\delta I_{TS}}{L_{EA}}$
Lean et al. [1995]	1362.8	$L_{EA} = 3.2$	1
Lean [2000]	1363.4	2.6	0.81
Hoyt and Schatten [1993]	1362.0	4.0	1.25
Solanki and Fligge [1999]	1361.5	4.5	1.41
Assumption 1	1365.5	0.5	0.16
Assumption 2	1363.8	2.2	0.69
Assumption 3	1364.5	1.5	0.47

migrate equatorward from weak dynamo-like to strong dynamo-like, revealing that the two are strongly coupled. Thus if the strong dynamo were to cease to operate, it is quite likely that the weak dynamo would cease also. On the other hand, the ^{10}Be record indicates that the flux continued to emerge during the Maunder minimum, which could argue that at least the weak dynamo continued. However, only if emerged flux in ephemeral regions is consistently aligned will it contribute to the open flux and so contribute to the modulation of ^{10}Be [Wang and Sheeley Jr., 2003c]. Without such an effect, the ^{10}Be data from the Maunder minimum require that the strong dynamo continued to be active but at reduced strength, such that the BMRs produced did not cause sunspots.

Resolution of these issues, in relation to TSI reconstruction, will require longer data sequences on TSI than we have available at present. One possible approach may be to try to exploit the longer data series of ground-based solar irradiance measurements, if sufficiently accurate atmospheric corrections can be developed. If such approach is not possible, there may be no alternative other than to wait until the space-based TSI record is long enough. We do now have almost 2.5 solar cycles of reliable TSI data and the remainder of this section compares the trend in these data with those predicted by the reconstructions presented here.

Figure 112 compares the smoothed mean variations shown in Figs. 109–111 with several other reconstructions. The mauve, green and cyan lines are for assumptions 1,2 and 3 in Table 10 (changes in quiet Sun flux since the Maunder minimum of $\Delta F_q = 0$, F_m and $F_m/2$, where F_m is the value for solar cycle 22). The orange line shows the corresponding 11-year means from the reconstruction by Lean [2000] and the grey dashed line is the extrapolation based on the open flux estimate $[F_S]_{aa}$ by Lockwood and Stamper [1999]. Lockwood and Stamper assumed that the regression they derived on decadal timescales also applies on century timescales.

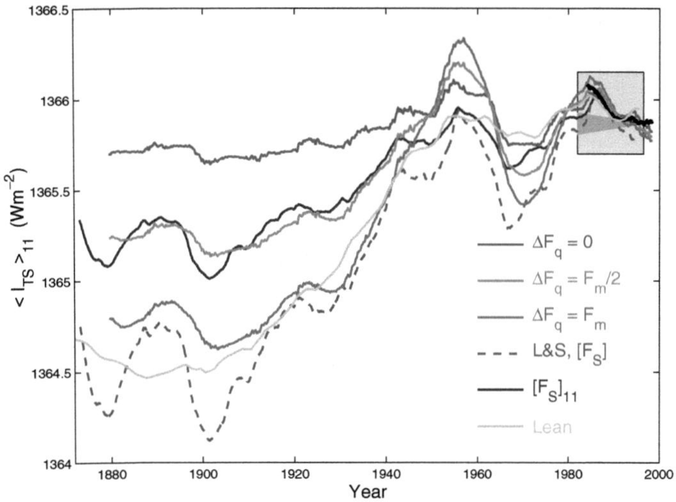

Fig. 112. Comparison of 11-year running means in various reconstructions of the TSI. The mauve, cyan and *green* lines are for assumptions 1,2 and 3 in Table 10 (changes in quiet Sun flux since the Maunder minimum of $\Delta F_q = 0$, $F_m/2$ and F_m, where F_m is the value for solar cycle 22). The *orange* line is the variation from the reconstruction by Lean [2000] and the *grey dashed* line is the extrapolation based on the open flux estimate $[F_S]_{aa}$ by Lockwood and Stamper [1999]. The *blue line* shows the result of a linear regression of 11-year running means of $[F_S]_{aa}$ with the corresponding means of TSI (*black line*). The region in the box is reproduced in detail in Fig. 113

The blue line shows the result of a linear regression of 11-year running means of $[F_S]_{aa}$ with the corresponding 11-year running means of TSI (the latter shown in the figure by the black line). The correlation coefficient is 0.91 but this has very low statistical significance because of the heavy smoothing employed. The resulting reconstruction (in blue) is very similar to that for $F_q = F_m/2$ (in cyan).

The region in the box in Fig. 112 is reproduced in detail in Fig. 113. It can be seen that the observed drift in average TSI has been downward since the start of observations in 1978. This drift is greater than the $\pm 3\,\mathrm{ppm\,yr}^{-1}$ deliniated by the grey wedge: in fact this uncertainty estimate is pessimistic and a more likely value is $\pm 1\,\mathrm{ppm\,yr}^{-1}$ [Fröhlich, 2003]. All the reconstructions also show this downward drift. That by Lockwood and Stamper [1999] and for $\Delta F_q = F_m$ are larger than from the observations, whereas those from Lean [2000] and $\Delta F_q = 0$ are smaller. The closest agreement is for the $\Delta F_q = F_m/2$ case.

With the caveat that there is not yet enough data to give statistical significance to the correlation between smoothed open flux estimates of open flux and TSI, Figs. 112 and 113 have some interesting implications. All three

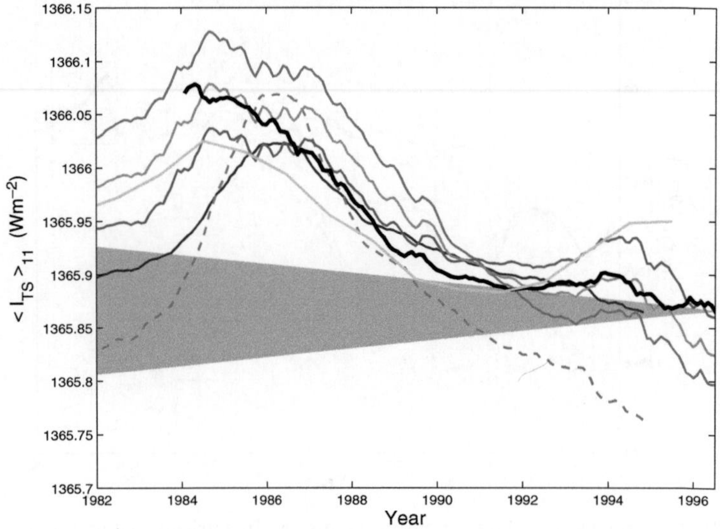

Fig. 113. Detail of the *shaded* box in Fig. 112. The *grey* wedge shows a maximum uncertainty in the long-term drift of the observed TSI composite of $\pm 3\,\text{ppm}\,\text{yr}^{-1}$ (The actual uncertainty on these timescales is more likely to be $\pm 1\,\text{ppm}\,\text{yr}^{-1}$)

of the new reconstructions presented here (for $\Delta F_q = 0$, $\Delta F_q = F_m$ and $\Delta F_q = F_m/2$) have a similar waveform to both the two reconstructions based on open flux (using the regression of monthly data and the 11-year smoothed means). This provides evidence that there is indeed a correlation between TSI and open flux on century timescales and so justifies the use of cosmogenic isotopes as a proxy for TSI. However, notice that the regression is different for the monthly or annual means (dominated by the solar cycle variation) than for the 11-year running means (dominated by the century-scale drift) and that the regression based on the former gives a larger long-term drift than the latter.

From Fig. 113, the observed long-term drift is most closely matched by the reconstruction for $\Delta F_q = F_m/2$ (assumption 3). Table 10 shows that the long-term drift for this reconstruction is half that in the Lean et al. [1995] reconstruction. If we adopt this assumption, we can use regressions with cosmogenic isotopes to reconstruct the irradiance variation over millenial timescales. Crowley [2000] used cosmogenic isotope variations with the Lean et al. reconstruction to compute the TSI variation over the last millenium. The cosmogenic isotopes variations used were the ^{10}Be abundance record from ice cores by Bard et al. [1997] and Bard et al. [2000], the ^{14}C production rate from tree rings [Stuiver et al., 1988a,b, Stuiver and Braziunas, 1989] and the ^{14}C inferred from ^{10}Be data [Bard et al., 1997, Beer, 2000]. In Fig. 114 these variations have been re-scaled using the factor 0.5 found in Table 10 for the reconstruction made using assumption 3.

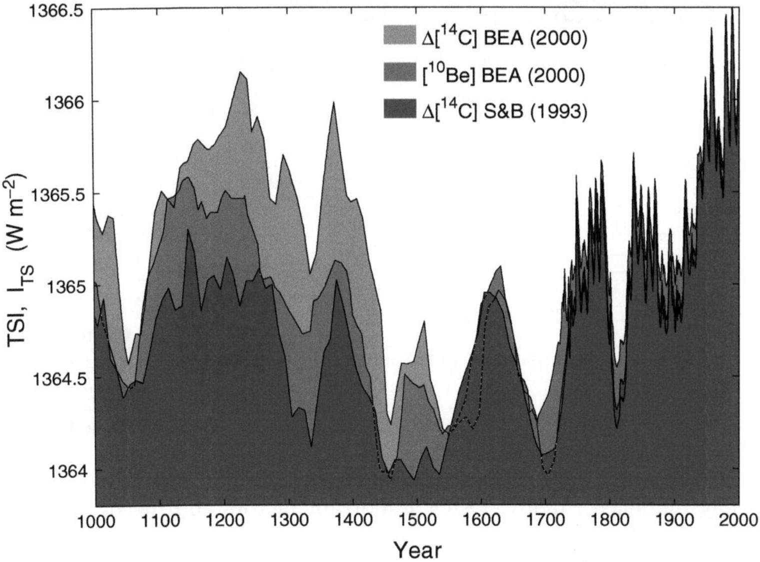

Fig. 114. The irradiance reconstruction over the last millenium based on cosmogenic isotopes and assumption 3 of Table 9 ($\Delta F_q = F_m/2$). The cosmogenic isotope records used are: the ^{10}Be abundance record from ice cores, as published by Bard et al. [1997, 2000] – labelled [^{10}Be]BEA(2000), the ^{14}C production rate from tree rings [Stuiver and Braziunas, 1989] – labelled Δ[^{14}C]S&B(1993), and the ^{14}C inferred from ^{10}Be data [Bard et al., 1997, 2000] – labelled Δ[^{14}C]BEA(2000)

7 Conclusions and Implications

The preceding sections have reviewed the physics of solar outputs and our best current understanding of how they have varied on the century timescales. This is important for global climate change, over the past century and on the millenium timescales relevant to the climate changes during the Holocene as revealed by paleoclimate studies. Figure 115 summarises schematically potential solar–climate interaction chains. The figure makes no attempt to evaluate the contribution of any one mechanism: we know already that several of them are extremely important, others will prove to be less important, some may be negligible and some may even prove to be invalid. The figure also stresses that these solar variability effects must be considered alongside other known effects, such as those due to volcanoes, human activities (including greenhouse gas emissions, sulphate pollution, and deforestation) and changes in Earth's orbital characteristics.

Detection–attribution techniques allow us to evaluate the effects of these inputs and hence allow some prediction of the future evolution of our climate. However, such studies depend on the variations used as inputs to the general circulation models of the coupled atmosphere–ocean system. Many of these

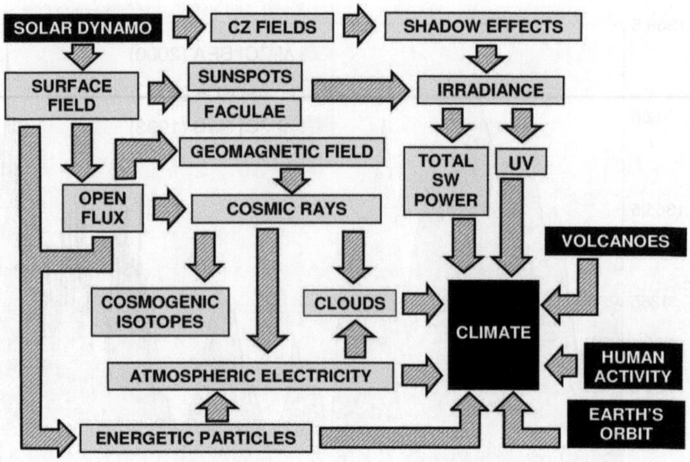

Fig. 115. Schematic view of potential links and mechanisms by which solar variability may influence climate. Other important factors and mechanism chains are summarised by *black* boxes

studies have assumed that the only important solar output is the total solar irradiance, the variation of which has often been quantified by using the Lean et al. [1995] reconstruction. They have revealed that the solar effect is larger than expected from radiative forcing (i.e the solar β-factor exceeds unity), pointing to either a correlated parallel mechanism (for example, direct cosmic ray effects) or to some feedback mechanism that amplifies the effect of TSI variations (for example, TSI-induced albedo changes).

Paleoclimate studies reveal a link between cosmogenic isotopes and climate indicators throughout the Holocene, again showing that either there is a direct effect of cosmic rays on climate, or that cosmic rays vary in synchronisation with another factor, such that they are not the cause of the effect but their fluxes are nevertheless a good index (or "proxy") for it. The most likely correlated factor is the TSI (or UV irradiance). However, this raises difficulties in understanding why the open solar magnetic flux, which modulates cosmic ray fluxes, is so closely related to the surface flux features which modulate TSI. Section 6 presented some new evidence that open solar flux is indeed related to TSI on century, as well as decadal, timescales and so provides the first firm justification for the use of cosmogenic isotopes as a proxy for TSI.

Previous reconstructions of TSI have employed stellar analogues to quantify the long-term drift (usually quoted as the change in average TSI between the Maunder minimum and the present day). The problem with this is that the drift derived depends on the assumptions made about our Sun's place in the observed distributions of Sun-like stars. In Sect. 6, the various brightening and darkening terms in the full expression for TSI were calculated separately without using stellar analogues. This is a first attempt to do this and is based on reconstructing the distribution of surface flux tube sizes using data from the SoHO spacecraft, covering little more than half of a solar cycle and for one (visible, red) wavelength only. Therefore it is to be expected that these estimates will need revision as more data become available.

A clear downward drift in 11-year running means of TSI has been observed since space-based TSI observations began in 1978. This drift exceeds the instrumental uncertainties and is matched by a reconstruction for which the quiet-Sun magnetic flux falls to about half of present-day values in the Maunder minimum. This implies that the long-term drift may be roughly half that predicted in the Lean et al. [1995] reconstruction used to generate the β factors discussed above. In turn, this suggests that the solar amplification β factor may be roughly twice the magnitude that we thought previously (i.e. of order 5 or 6). The implication is that although the century-scale drift in solar irradiance may be smaller in amplitude than previously envisaged, the sensitivity of the Earth's climate to these variations would be correspondingly greater.

Acknowledgements. The author is grateful to a great many scientists that have contributed to this text through discussions, reprints and preprints. He particularly thanks four PhD students at the University of Southampton: Alexis Rouillard, for insights and contributions to this review, relating to the shielding of cosmic rays by the heliosphere; Simon Foster, for his work on the sunspot group composite and total solar irradiance reconstruction; Ivan Finch for his analysis of seasonal and daily variations in geomagnetic activity and Steve Morley for proof-reading and formatting this text.

References

H. S. Ahluwalia. Galactic cosmic ray intensity variations at a high-latitude sea-level site 1937-1994. *J. Geophys. Res.*, 102:24229–24236, 1997.

S. I. Akasofu. Energy coupling between the solar wind and the magnetosphere. *J. Geophys. Res.*, 28:121, 1981.

A. Allen. *Astrophysical Quantities.* Athlone Press, London, 1973.

R. C. Altrock. An 'extended solar cycle' as observed in Fe XIV. *Sol. Phys.*, 170: 411–423, 1997.

S. K. Antiochos, C. R. DeVore, and J. A. Klimchuk. A model for solar coronal mass ejections. *Astrophys. J.*, 510:485–493, 1999.

E. Antonucci, J. T. Hoeksema, and P. H. Scherrer. Rotation of the photospheric magnetic fields: a north–south asymmetry. *Astrophys. J.*, 360:296–304, 1990.

N. Arnold and T. Neubert. Cosmic influences on the atmosphere. *Astron. Geophys.*, 43:5189–5201, 2002.

R. L. Arnoldy. Signature in the interplanetary medium for substorms. *J. Geophys. Res.*, 43:5189–5201, 1971.

M. J. Aschwanden and A. M. Title. Solar magnetic loops observed with TRACE and EIT. In *Stars as Suns: Activity, Evolution, and Planets*, volume 219, pages 503–515, 2004.

H. W. Babcock. The topology of the sun's magnetic field and the 22-year cycle. *Astrophys. J.*, 133:572–587, 1961.

J. N. Bahcall. *Encyclopedia of Astronomy And Astrophysics*, chapter Solar Interior: Neutrinos, pages 1–9. Nature Publishing Group, 2001.

D. Baker. Statistical analyses in the study of Solar Wind–Magnetosphere coupling. In Y. Kamide and J. A. Slavin, editors, *Solar Wind–Magnetosphere Coupling*, pages 17–38. Terra Scientific, Tokyo, 1986.

A. Balogh, E. J. Smith, B. T. Tsurutani, D. J. Southwood, R. J. Forsyth, and T. S. Horbury. The heliospheric field over the south polar region of the sun. *Science*, 268:1007–1010, 1995.

E. Bard, G. M. Raisbeck, F. Yiou, and J. Jouzel. Solar modulation of cosmogenic nuclide production over the last millenium: comparison between 14C and 10Be records. *Earth and Planet. Sci. Lett.*, 150:453–462, 1997.

E. Bard, G. M. Raisbeck, F. Yiou, and J. Jouzel. Solar irradiance during the last 1200 years based on cosmogenic nuclides. *Tellus*, 52B:985–992, 2000.

L. F. Bargatze, R. L. McPherron, and D. N. Baker. Solar wind–magnetosphere energy input functions. In Y. Kamide and J. A. Slavin, editors, *Solar Wind–Magnetosphere Coupling*, pages 101–109. Terra Scientific, Tokyo, 1986.

E. Bauer, M. Claussen, V. Brovkin, and A. Huenerbein. Assessing climate forcings of the earth system for the past millennium. *Geophys. Res. Lett.*, 30(6):1276–1279, doi.10.1029/2002GL016639, 2003.

S. Baumgartner, J. Beer, and H. A. Synal. *Science*, 279:1330, 1998.

W. Baumjohann. Merits and limitations of the use of geomagnetic indices in solar wind–magnetosphere coupling studies. In Y. Kamide and J. A. Slavin, editors, *Solar Wind–Magnetosphere Coupling*, pages 101–109. Terra Scientific, Tokyo, 1986.

J. G. Beck, T. L. Duvall, and P. H. Scherrer. Long-lived giant cells detected at the solar surface. *Nature*, 394:653–655, 1998.

J. Beer. Long-term indirect indices of solar variability. *Space Sci. Rev.*, 94:53–66, 2000.

J. Beer. Ice core data on climate and cosmic ray changes. In J. Kirkby and S. Mele, editors, *Proc. Workshop on Ion–Aerosol–Cloud interactions, CERN, 18–20 April 2001, CERN, , CERN Yellow Report, CERN-2001-007 (ISSN 0007-8328, ISBN 92-9083-191-0)*, pages 3–11, 2001.

J. Beer, A. Blinov, G. Bonani, R. C. Finkel, H. J. Hofmann, B. Lehmann, H. Oeschger, A. Sigg, J. Schwander, T. Staffelbach, B. Staufer, M. Suter, and W. Wolfi. Use of ^{10}Be in polar ice to trace the 11-year cycle of solar activity. *Nature*, 347:164–166, 1990.

J. Beer, S. Tobias, and N. Weiss. An active sun throughout the maunder minimum. *Sol. Phys.*, 181:237–249, 1998.

A. Belov. Large-scale modulation: view from the earth. *Space Sci. Rev.*, 93:79–105, 2000.

T. Berger, M. G. Löfdahl, G. Scharmer, and A. M. Title. Observations of magnetoconvection in sunspots with 100 km resolution. In *34th Meeting of the Solar*

Physics Division of the American Astronomical Society, June 17, Laurel, MD, 2003.

E. A. Bering, A. A. Few, and J. R. Benbrook. The global electric circuit. *Phys. Today*, 51(10):24–30, 1998.

A. Bhattacharyya and B. Mitra. Changes in cosmic ray cut-off rigidities due to secular variations of the geomagnetic field. *Ann. Geophys.*, 15:734–739, 1997.

L. Bierman. *Vierteljahrsschr. Ast. Ges.*, 76:194, 1941.

G. Bond, B. Kromer, J. Beer, R. Muscheler, M. N. Evans, W. Showers, S. Hoffman, R. Lotti-Bond, I. Hajdasand, and G. Bonani. Persistent solar influence on north atlantic climate during the holocene. *Science*, 294:2130–2136, 2001.

G. Bonino, G. Cini Castognoli, N. Bhabdari, and C. Taricco. Behavior of the heliosphere over prolonged solar quiet periods by ^{44}Ti measurements in meteorites. *Science*, 270:1648–1650, 1998.

G. Bonino, G. Cini Castognoli, D. Cane, C. Taricco, and N. Bandahri. Solar modulation of the galactic cosmic ray spectra since the Maunder minimum. In *Proc. ICRC*, pages 3769–3772. Copernicus Gesellschaft, 2001.

L. F. Burlaga. Large-scale fluctuations in B between 13 and 22AU and their effects on cosmic rays. *J. Geophys. Res.*, 92:13647–13652, 1987.

H. V. Cane. Coronal mass ejections and forbush decreases. *Space Sci. Rev.*, 93: 55–77, 2000.

H. V. Cane, G. Wibberenz, I. G. Richardson, and T. T. von Rosenvinge. Cosmic ray modulation and the solar magnetic field. *Geophys. Res. Lett.*, 26:565–568, 1999.

K. S. Carslaw, R. G. Harrison, and J. Kirkby. Cosmic rays, clouds and climate. *Science*, 298:1732–1737, 2002.

F. Cattaneo and D. W. Hughes. Solar dynamo theory: a new look at the origin of small-scale magnetic fields. *Astron. & Geophys.*, 42:3.18, 2001.

G. A. Chapman, A. M. Cookson, J. J. Dobias, and S. R. Walton. An improved determination of the area ratio of faculae to sunspots. *Astrophys. J.*, 555:462–465, 2001.

G.A. Chapman, A.M. Cookson, and J.J. Dobias. Solar variability and the relation of facular to sunspot areas during cycle 22. *Astrophys. J.*, 842:541–545, 1997.

J. Christensen-Dalsgaard, W. Däppen, S. V. Ajukov, E. R. Anderson, H. M. Antia, S. Basu, V. A. Baturin, G. Berthomieu, B. Chaboyer, S. M. Chitre, A. N. Cox, P. Demarque, J. Donatowicz andW. A. Dziembowski, M. Gabriel, D. O. Gough, D. B. Guenther, J. A. Guzik, J. W. Harvey, F. Hill, G. Houdek, C. A. Iglesias, A. G. Kosovichev, J. W. Leibacher, P. Morel, C. R. Proffitt, J. Provost, J. Reiter, E. J. Rhodes Jr., F. J. Rogers, I. W. Roxburgh, M. J. Thompson, and R. K. Ulrich. The current state of solar modelling. *Science*, 272:1286–1292, 1996.

M. A. Clilverd, T. D. G. Clark, E. Clarke, and H. Rishbeth. Increased magnetic storm activity from 1868 to 1995. *J. Atmos. Sol. -Terr. Phys.*, 60:1047–1056, 1998.

M. A. Clilverd, E. Clarke, T. Ulrich, J. Linthe, and H. Rishbeth. Reconstructing the long-term aa index. *J. Geophys. Res.*, in press, 2004.

E. W. Cliver. Solar activity and geomagnetic storms. *EOS*, 75:569, 1994.

E. W. Cliver, V. Boriakoff, and K. H. Bounar. The 22-year cycle of geomagnetic activity. *J. Geophys. Res.*, 101:27091–27109, 1996.

E. W. Cliver, V. Boriakoff, and K. H. Bounar. Geomagnetic activity and the solar wind during the Maunder minimum. *Geophys. Res. Lett.*, 25:897–900, 1998.

E. W. Cliver, Y. Kamide, and A. G. Ling. Mountains versus valleys: Semiannual variation of geomagnetic activity. *J. Geophys. Res.*, 105:2143–2424, 2000.

E. W. Cliver and A. G. Ling. Secular change in geomagnetic indices and the solar open magnetic flux during the first half of the twentieth century. *J. Geophys. Res.*, 107:doi.10.1029/2001JA000505, 2002.

R. M. Close, C. E. Parnell, D. H. Mackay, and E. R. Priest. Statistical flux tube properties of 3d magnetic carpet fields. *Sol. Phys.*, 212:251–275, 2003.

D. A. Couzens and J. H. King. Interplanetary medium data book – supplement 3. Technical report, National Space Science Data Center, 1986.

S. W. H. Cowley. Acceleration and heating of space plasmas: basic concepts. *Ann. Geophys.*, 9:176–187, 1991.

S. R. Cranmer. Coronal holes and the high-speed solar wind. *Space Sci. Rev.*, 101: 229–294, 2002.

N. U. Crooker and K. I. Gringauz. On the low correlation between long-term averages of the solar wind speed and geomagnetic activity after 1976. *J. Geophys. Res.*, 98:59–62, 1993.

T. J. Crowley. Causes of climate change over the past 1000 years. *Science*, 289: 270–277, 2000.

U. Cubasch, R. Voss, G. C. Hegerl, J. Waszkewitz, and T. J. Crowley. Simulation of the influence of solar radiation variations on the global climate with an ocean–atmosphere general circulation model. *Clim. Dyn.*, 13:757–767, 1997.

P. E. Damon, J. C. Lerman, and A. Long. Temporal fluctuations of atmospheric ^{14}C: causal factors and implications. *Ann. Rev. Earth Planet. Sci.*, 6:457, 1978.

W. Deinzer, G. Hensler, M. Schüssler, and E. Weisshaar. Model calculations of magnetic flux tubes, I. equations and method. *Astron. Astrophys.*, 139:426–434, 1984a.

W. Deinzer, G. Hensler, M. Schüssler, and E. Weisshaar. Model calculations of magnetic flux tubes, II. stationary results for solar magnetic elements. *Astron. Astrophys.*, page 435, 1984b.

R. F. Donnelly. *Adv. Space Res.*, 8:77, 1988.

L. I. Dorman, G. Villoresi, I. V. Dorman, N. Iucci, and M. Parisi. High rigidity CR–SA hysterisis phenomenon and dimension of modulation region in the heliosphere in dependence of particle rigidity. volume 2, pages 69–72, 1997.

W. Droge. Solar particle transport in a dynamical quasi-linear theory. *Astrophys. J.*, 589:1027–1039, 2003.

C. S. Dyer and P. R. Truscott. Cosmic radiation effects on avionics. *Radiat. Prot. Dosim.*, pages 337–342, 1999.

J. A. Eddy. The Maunder minimum. *Science*, 192:1189, 1976.

J. A. Eddy. *The Ancient Sun*, chapter The historical record of solar activity, page 119. Pergamon Press, 1980.

S. E. S. Ferreira, M. S. Potgieter, B. Heber, and H. Fichtner. Charge-sign dependent modulation in the heliopshere over a 22-year cycle. *Ann. Geophys.*, 21:1359–1366, 2003.

J. Feynman and N. U. Crooker. The solar wind at the turn of the century. *Nature*, 275:626–627, 1978.

J. Feynman and S. B. Gabriel. Period and phase of the 88-year solar cycle and the Maunder minimum: evidence for the chaotic sun. *Sol. Phys.*, 127:393–403, 1990.

L. A. Fisk. An overview of the transport of galactic and anomalous cosmic rays in the heliosphere: theory. *Adv. Space Res.*, 23:415–423, 1999.

L. A. Fisk and N. A. Schwadron. The behaviour of the open magnetic flux of the Sun. *Astrophys. J.*, 560:425–438, 2001.

G. F. FitzGerald. Sunspots and magnetic storms. *The Electrician*, 30:48, 1892.

M. Fligge, S. K. Solanki, and J. Beer. Determination of solar cycle length using the continuous wavelet transform. *Astron. Astrophys.*, 346:313, 1999.

M. Fligge, S. K. Solanki, Y. C. Unruh, C. Fröhlich, and C. Wehrli. A model of solar total and spectral irradiance variations. *Astron. Astrophys.*, 335:709–718, 1998.

S. E. Forbush. Cosmic ray intensity variations during two solar cycles. *Geophys. Res.*, 63:651–669, 1958.

R. J. Forsyth, A. Balogh, E. J. Smith, G. Erdös, and D. J. McComas. The underlying Parker spiral structure in the Ulysses magnetic field observations. *J. Geophys. Res.*, 1995.

S. S. Foster. *Reconstruction of solar irradiance variations, for use in studies of global climate change: application of recent SoHO observations with historic data from the Greenwich Observatory.* PhD thesis, University of Southampton (School of Physics and Astronomy), 2004.

S. S. Foster and M. Lockwood. Long-term changes in the solar photosphere associated with changes in the coronal source flux. *Geophys. Res. Lett.*, 28:1443–1446, 2001.

P. V. Foukal. In L. E. Cram and J. H. Thomas, editors, *The Physics of Sunspots*, pages 391–398. Sacramento Peak Observatory, New Mexico, 1981.

P. V. Foukal and J. Lean. *Astrophys. J.*, 302:826, 1986.

P. V. Foukal and L. Milano. A measurement of the quiet network contribution to solar irradiance variation. *Geophys. Res. Lett*, 28:883–886, 2001.

E. Friis-Christensen and K. Lassen. Length of the solar cycle: an indicator of solar activity closely associated with climate. *Science*, 245:698–700, 1991.

E. Friis-Christensen and H. Svensmark. What do we really know about the sun-climate connection? *Adv in Space Res.*, 20 (4/5):913–920, 1997.

C. Fröhlich. Observations of irradiance variations. *Space Sci. Rev.*, 94:15–24, 2000.

C. Fröhlich. Solar irradiance variations. In *Proc. ISCS-2003 Symposium, Tatransk Lomnica, Slovakia, ESA-SP 535*, pages 183–193, 2003.

C. Fröhlich and J. Lean. Total solar irradiance variations. In *New Eyes to see inside the Sun and Stars*, pages 89–102, 1998a.

C. Fröhlich and J. Lean. The sun's total irradiance: Cycles, trends and related climate change uncertainties since 1976. *Geophys. Res. Lett.*, 25:4377–4380, 1998b.

C. Fröhlich, J. M. Pap, and H. S. Hudson. Improvement of the photometric sunspot index and changes of the disk-integrated sunspot contrast with time. *Sol. Phys.*, 152:111–118, 1994.

P. R. Gazis. Solar cycle variation of the heliosphere. *Rev. Geophys.*, 34:379–402, 1996.

P. M. Giles, T. L. Duvall, P. H. Scherrer, and R. S. Bogart. A subsurface flow of material from the sun's equator to its poles. *Nature*, 390:52, 1997.

P. A. Gilman and J. Miller. Nonlinear convection of a compressible fluid in a rotating spherical shell. *Astrophys. J. Supplement*, 61:585, 1986.

V. L. Ginzburg. Cosmic ray astrophysics (history and general review). *Physics-Uspekhi*, 39:155–168, 1996.

W. Gleissberg. A table of secular variations of the solar cycle. *J. Geophys. Res.*, 49:243–244, 1944.

M. N. Gnevyshev. On the 11-years cycle of solar activity. *Sol. Phys.*, 1:107, 1967.

M. N. Gnevyshev. Essential features of the 11-year solar cycle. *Sol. Phys.*, 51:175, 1977.

B. E. Goldstein. Ulysses observations of solar wind plasma parameters in the ecliptic from 1.4 to 4.5 AU and out of the ecliptic. *Space Sci. Rev.*, 72:113, 1994.

D. O. Gomez, P. A. Dmitruk, and L. J. Milano. Recent theoretical results on coronal heating. *Sol. Phys.*, 195:299–318, 2000.

J. T. Gosling, J. Birn, and M. Hesse. Three dimensional magnetic reconnection and the magnetic topology of coronal mass ejections. *Geophys. Res. Lett.*, 22: 869–872, 1995.

L. J. Gray, S. J. Phipps, T. J. Dunkerton, M. P. Balwin, E. F. Drysdale, and M. R. Allen. A data study of the influence of the equatorial upper stratosphere on northern hemisphere stratospheric sudden warmings. *Quart. J. Roy. Meteorol. Soc.*, 127:1985–2003, 2001.

K. I. Gringauz. Average characteristics of the solar wind and its variation during the solar cycle. In H. Rosenbauer, editor, *Solar Wind 4, Report MPAE-W-100-81-31.* MPI für Aeronomie, Lindau, Germany, 1981.

J. D. Haigh. The role of stratospheric ozone in modulating the solar radiative forcing of climate. *Nature*, 370:544–546, 1994.

J. D. Haigh. A GCM study of climate change in response to the 11-year solar cycle. *Quart. J. Roy. Meteorol. Soc.*, 125:871–892, 1999a.

J. D. Haigh. Modelling the impact of solar variability on climate. *J. Atmos. Sol. -Terr. Phys.*, 61:63–72, 1999b.

J. D. Haigh. Climate variability and the influence of the Sun. *Science*, 294:2109–2111, 2001.

J. Hansen, M. Sato, and R. Ruedy. Radiative forcing and climate response. *J. Geophys. Res*, 102:6831–6864, 1997.

W. B. Hanson, W. R. Coley, R. A. Heelis, N. C. Maynard, and T. L. Aggson. A comparison of in situ measurements of E and -V x B from Dynamics Explorer 2. *J. Geophys. Res.*, 98:21501–21516, 1994.

M. A. Hapgood. A double solar-cycle variation in the 27-day recurrence of geomagnetic activity. *Ann. Geophys.*, 11:248, 1993.

M. A. Hapgood, G. Bowe, M. Lockwood, D. M. Willis, and Y. Tulunay. Variability of the interplanetary magnetic field at 1 A.U. over 24 years: 1963–1986. *Planet. Space Sci.*, 39:411–423, 1991.

R. G. Harrison. Radiolytic particle production in the atmosphere. *Atmos. Environ.*, 36:160–169, 2002a.

R. G. Harrison. Twentieth century secular decrease in the atmospheric electric circuit. *Geophys. Res. Lett.*, 29(14):doi:10.1029/2002GL014878, 2002b.

R. G. Harrison. Long-term measurements of the global atmospheric electric circuit at Eskdalemuir, Scotland, 1911–1981. *Atmos Res*, 70 (1):1–19, 2003.

R. G. Harrison and K. L. Alpin. Atmospheric condensation nuclei formation and high-energy radiation. *J. Atmos. Sol. -Terr. Phys.*, 63:1811–1819, 2001.

R. G. Harrison and K. S. Carslaw. Ion-aerosol-cloud processes in the lower atmosphere. *Rev. Geophys.*, 41:(2)1–(2)26, 2003.

K. L. Harvey. In *The Solar Cycle*, volume ASP Conf. Series Vol. 27, pages 335–367, 1992.

K. L. Harvey. In J. Pap, C. Fröhlich, H. S. Hudson, and S. K. Solanki, editors, *The Sun as a Variable Star: Solar and Stellar Irradiance Variations*, volume IAU Col. 143, page 217. Cambridge University Press, 1994.

K. L. Harvey. The solar activity cycle and sun-as-a-star variability in the visible and infrared. In *Solar Analogs: Characteristics and Optimum Candidates, Proc. 2nd. Annual Lowell Observatory Fall Workshop*, 1997.

K. L. Harvey and H. S. Hudson. Solar activity and the formation of coronal holes. *Adv. Space Res.*, 25(9):1735–1738, 2000.

K. L. Harvey, H. P. Jones, C. J. Schrijver, and M. J. Penn. Does magnetic flux submerge at flux cancelation sites? *Sol. Phys.*, 190:35–44, 1999.

K. L. Harvey and C. Zwaan. Properties and emergence of bipolar active regions. *Sol. Phys.*, 148:85–118, 1993.

D. F. Heath and B. M. Schlesinger. *J. Geophys. Res.*, 91:8672, 1986.

B Heber, P. Ferrando, A. Raviart, G. Wibberenz, R. Mueller-Mellin, H. Kunow, H. Sierks, V. Bothmer, A. Posner, C. Paizis, and M. S. Potgieter. Differences in the temporal variations of galactic cosmic ray electrons and protons: implications from ulysses at sunspot minimum. *Geophys. Res. Lett.*, 26(14):2133–2136, 1999.

P. C. Hedgecock. Measurements of the interplanetary magnetic field in relation to the modulation of cosmic rays. *Sol. Phys.*, 42:497–527, 1975.

W. Herschel. Observations tending to investigate the nature of the sun, in order to find the causes or symptoms of its variable emission of light and heat; with remarks on the use that may possibly be drawn from solar observations. *Phil. Trans. R. Soc. London*, 91:265–318, 1801.

J. Hirzberger, M. Koschinsky, F. Kneer, and C. Ritter. High resolution 2d-spectroscopy of granular dynamics. *Astronomy and Astrophysics*, 367:1011–1021, 2001.

R. Howe, J. Christensen-Dalsgaard, F. Hill, R. W. Komm, R. M. Larsen, J. Schou, M. J. Thompson, and J. Toomre. Dynamic variations at the base of the convection zone. *Science*, 287:2456, 2000a.

R. Howe, J. Christensen-Dalsgaard, F. Hill, R. W. Komm, R. M. Larsen, J. Schou, M. J. Thompson, and J. Toomre. Deeply penetrating banded zonal flows in the solar convection zone. *Astrophys. J.*, 533, 2000b.

D. Hoyt and K. Schatten. A discussion of plausible solar irradiance variations 1700–1992. *J. Geophys. Res.*, 98:18895–18906, 1993.

D. Hoyt and K. A. Schatten. Group sunspot numbers: a new solar activity reconstruction. *Sol. Phys.*, 181:491–512, 1998.

H. S. Hudson, S. Silva, M. Woodward, and R. C. Willson. The effects of sunspots on solar irradiance. *Sol. Phys.*, 76:211–219, 1982.

A. J. Hundhausen. *Introduction to Space Physics*, chapter The Solar Wind, pages 91–128. Cambridge University Press, 1995.

V.G. Ivanov and E. V. Miletsky. Reconstruction of the open solar magnetic flux and interplanetary magnetic field in the 19th and 20th centuaries. *Astron. and Astrophys.*, in press, 2004.

J. R. Jokipi. *The Sun in Time*, chapter Variations of the cosmic ray flux with time, pages 205–221. Univ. of Arizona Press, 1991.

J. R. Jokipii, E. H. Levy, and W. B. Hubbard. Effects of particle drift on cosmic ray transport. I. general properties, application to solar modulation. *Astrophys. J.*, 213:861–868, 1977.

P. D. Jones, T. J. Osborn, and K. R. Briffa. The evolution of climate over the last millenium. *Science*, 292:662–667, 2001.

J. Kirkby et al. Cloud: A particle beam facility to invetsigate the influence of cosmic rays on clouds. In J. Kirkby and S. Mele, editors, *Proc. Workshop on Ion-Aerosol-Cloud Interactions, CERN, 18-20th April 2001, CERN Yellow Report, CERN-2001-007 (ISSN 0007-8328, ISBN 92-9083-191-0)*. Pergamon, New York, 2001.

M. G. Kivelson and C. T. Russell, editors. *Introduction to Space Physics*. Cambridge University Press, 1995.

J. A. Klimchuk. Theory of coronal mass ejections. *J. Geophys. Res.*, page in press, 2003.

R. Knaack, M. Fligge, S. K. Solanki, and Y. C. Unruh. The influence of an inclined rotation axis on solar irradiance variation. *Astron. and Astrophys.*, 2001.

M. Knölker, M. Schüssler, and E. Weisshaar. Model calculations of magnetic flux tubes. III. properties of solar magnetic elements. *Astron. Astrophys.*, 194:257–267, 1988.

G. E. Kocharov, V. M. Ostryakov, A. N. Peristykh, and V. A. Vasil'ev. *Sol. Phys.*, 159:381, 1995.

A. S. Krieger, A. F. Timothy, and E. C. Roelof. A coronal hole and its identification as the source of a high velocity solar wind stream. *Sol. Phys.*, 29:505, 1973.

S. M. Krimigis, R. B. Decker, M. E. Hill, T. P. Armstrong, G. Gloeckler, D. C. Hamilton, L. J. Lanzerotti, and E. C. Roelof. Voyager 1 exited the solar wind at a distance of 85 au from the sun. *Nature*, 426:45–48, 2003.

J. E. Kristjánsson and J. Kristiansen. Is there a cosmic ray signal in recent variations in global cloudiness and cloud radiative forcing? *J. Geophys. Res.*, 105:11851–11863, 2000.

N. A. Krivova and S. K. Solanki. Effect of spatial resolution on estimating the Sun's magnetic flux. *Astron. Astrophys.*, 417:1125–1132, doi.10.1051/0004-6361:20021340, 2004.

N. A. Krivova, S. K. Solanki, and J. Beer. Was one sunspot cycle in the 18th century really lost? *Astron. Astrophys.*, 396:235–242, doi.10.1051/0004–6361:20021340, 2002a.

N. A. Krivova, S. K. Solanki, and M. Fligge. Total solar magnetic flux: dependence on spatial resolution of magnetometers. In *From Solar Min to Max: Half a solar cycle with SoHO*, volume Proc. SoHO 11 Symposium, pages 155–158, 2002b.

N. A. Krivova, S. K. Solanki, M. Fligge, and Y. C. Unruh. Reconstruction of solar irradiance variations in cycle 23: Is solar surface magnetism the cause? *Astron. Astrophys.*, 399:L1–L4, doi.10.1051/0004–6361:20030029, 2003.

J. R. Kuhn and K. G. Libbrecht. Non-facular solar luminosity variations. *Astrophys. J.*, 381:L35–L37, 1991.

J. R. Kuhn, K. G. Libbrecht, and R. H. Dicke. The surface temperature of the sun and changes in the solar constant. *Science*, 242:908–911, 1988.

K. Labitzke and H. van Loon. The signal of the 11-year sunspot cycle in the upper troposphere–lower stratosphere. *Space Sci. Rev.*, 80:393–410, 1997.

C. Laj et al. North Atlantic paleointensity stack since 75 ka (NAPIS-75) and the duration of the Laschamp event. *Phil. Trans. R. Soc. Lond.*, 358:1009–1025, 2001.

A. Larkin, J. D. Haigh, and S. Djavidnia. The effect of solar UV irraadiance variations on the Earth's atmosphere. *Space Sci. Rev.*, 94(1/2):199–214, 2000.

D. E. Larson, R. P. Lin, J. M. McTiernan, J. P. McFadden, R. E. Ergun, M. McCarthy, H. Réme, T. R. Sanderson, M. Kaiser, R. P Lepping, and J. Mazur. Tracing the topology of the october 18–20, 1995, magnetic cloud with -0.1–10^2 kev electrons. *Geophys. Res. Lett.*, 24:1911–1914, 1997.

P. Laut and J. Gundermann. Does the correlation between cycle lengths and northern hemisphere land temperatures rule ot any significant global warming from greenhouse gasses? *J. atmos. sol.-terr. Phys.*, 60:1–3, 199.

J. Lean. Variations in the sun's radiative output. *Rev. Geophys.*, 29:505–535, 1991.

J. Lean, J. Beer, and R. Bradley. Reconstruction of solar irradiance since 1610: implications for climate change. *Geophys. Res. Lett.*, 22:3195–3198, 1995.

J. L. Lean. Evolution of the Sun's spectral irradiance since the Maunder minimum. *Geophys. Res. Lett.*, 27:2425–2428, 2000.

J. L. Lean, W. C. Livingston, and D. F. Heath. *J. Geophys. Res.*, 87:10307, 1982.

J. L. Lean, Y. M. Wang, and N. R. Sheeley Jr. The effect of increasing solar activity on the sun's total and open magnetic flux during multiple cycles: Implications for solar forcing of climate. *Geophys. Res. Lett.*, 29:2224–2227, doi.10.1029/2002GL015880, 2002.

J. L. Lean, O. R. White, W. C. Livingston, and J. M. Picone. Variability of a composite chromospheric irradiance index during the 11-year activity cycle and over longer time periods. *J. Geophys. Res.*, 106:10645–10,658, 2001.

J. P. Legrand and P. A. Simon. Some solar cycle phenomena related to the geomagnetic activity from 1868 to 1980, i. the shock events of the interplanetary expansion of the toroidal field. *Astron. and Astrophys.*, 152:199–204, 1985.

J. P. Legrand and P. A. Simon. A 3-component solar-cycle. *Solar Phys.*, 131: 187–209, 1991.

R.B. Leighton. A magneto-kinematic model of solar cycle. *Astrophys. J.*, 156:1–26, 1969.

V. Letfus. Sunspot and auroral activity during the maunder minimum. *Solar. Phys.*, 197:203–213, 2000.

K. G. Libbrecht and J. R. Kuhn. A new measurement of the facular contrast near the solar limb. *Astrophys. J.*, 277:889, 1984.

M. Lockwood. Long-term variations in the magnetic fields of the sun and the heliosphere: their origin, effects and implications. *J.Geophys. Res.*, 106:16021–16038, 2001a.

M. Lockwood. Long-term variations in cosmic ray fluxes and total solar irradiance and global climate change. In J. Kirkby and S. Mele, editors, *Proc. Workshop on Ion–Aerosol–Cloud interactions, CERN, 18–20 April 2001, CERN, CERN Yellow Report, CERN-2001-007 (ISSN 0007-8328, ISBN 92-9083-191-0)* , pages 3–11, 2001b.

M. Lockwood. An evaluation of the correlation between open solar flux and total solar irradiance. *Astron. Astrophys.*, 382:678–687, 2002a.

M. Lockwood. Long-term variations in the open solar flux and links to variations in earth's climate. In *From Solar Min to Max: Half a solar cycle with SoHO*, volume Proc. SoHO 11 Symposium, Davos, Switzerland, ESA-SP-508, pages 507–522. ESA Publications, Noordwijk, The Netherlands, 2002b.

M. Lockwood. Relationship between the near-earth interplanetary field and the coronal source flux: Dependence on timescale. *J. Geophys. Res.*, 107:doi. 10.1029/2001JA009062, 2002c.

M. Lockwood. Twenty-three cycles of changing open solar flux. *J. Geophys. Res.*, 108:doi 10.1029/2002/JA009431, 2003.

M. Lockwood, R. B. Forsyth, A. Balogh, and D. J. McComas. The accuracy of open solar flux estimates from near-earth measurements of the interplanetary magnetic field: analysis of the first two perihelion passes of the ulysses spacecraft. *Ann. Geophys.*, 22:1395–1405, 2004.

M. Lockwood and S. S. Foster. Long-term variations in the magnetic field of the sun and possible implications for terrestrial climate. volume Proc. 1st. Solar and Space Weather Euroconference, ESP SP-463, pages 85–94, 2001.

M. Lockwood and R. Stamper. Long-term drift of the coronal source magnetic flux and the total solar irradiance. *Geophys. Res. Lett.*, 26:2461–2464, 1999.

M. Lockwood, R. Stamper, and M. N. Wild. A doubling of the sun's coronal magnetic field during the last 100 years. *Nature*, 399:437–439, 1999a.

M. Lockwood, R. Stamper, M. N. Wild, A. Balogh, and G. Jones. Our changing sun. *Astron. Geophys.*, 40:4.10–4.16, 1999b.

D. H. Mackay and M. Lockwood. The evolution of the Sun's open magnetic flux: II. full solar cycle simulations. *Sol. Phys.*, 209:287–309, 2002.

D. H. Mackay, E. R. Priest, and M. Lockwood. The evolution of the Sun's open magnetic flux: I. a single bipole. *Sol. Phys.*, 207:291–308, 2002.

M. E. Mann, R. S. Bradley, and M. K. Hughes. Northern hemisphere temperatures during the past millenium: inferences, uncertainties, and limitations. *Geophys. Res. Lett.*, 26:759–762, 1999.

D. Maravilla, A. Lara, J. F. Valdés-Galicia, and B. Mendoza. An analysis of polar coronal hole evolution: Relations to other solar phenomena and heliospheric consequences. *Sol. Phys.*, 203:27–38, 2001.

R. Markson. Modulation of the Earth's electric field by cosmic radiation. *Nature*, 291:304–308, 1981.

N. Marsh and H. Svensmark. Low cloud properties influenced by cosmic rays. *Phys. Rev. Lett.*, 85:5004–5007, 2000a.

N. Marsh and H. Svensmark. Cosmic rays, clouds and climate. *Space Science Rev.*, 94(1/2):215–230, 2000b.

N. Marsh and H. Svensmark. GCR and ENSO trends in ISCCP-D2 low cloud properties. *J. Geophys. Res.*, 2004.

P. N. Mayaud. Une mésure planetaire d'activity magnetique, basée sur deux observatories antipodaux. *Annales de Geophysique*, 27:67–70, 1971.

P. N. Mayaud. The aa indices: A 100-year series characterising the magnetic activity. *J. Geophys. Res.*, 77:6870–6874, 1972.

P. N. Mayaud. The derivation of geomagnetic indices. 1976.

D. J. McComas, S. J. Bame, B. L. Barraclough, W. C. Feldman, H. O. Funsten, J. T. Goslingand P. Riley, R. Skoug, A. Balogh, R. Forsyth, B. E. Goldstein, and

M. Neugebauer. Ulysses' return to the slow solar wind. *Geophys. Res. Lett.*, 25: 1–4, 1998.

D. J. McComas, H. A. Elliott, J. T. Gosling, D. B. Reisenfeld, R. M. Skoug, B. E. Goldstein, M. Neugebauer, and A. Balogh. Ulysses' second fast-latitude scan: Complexity near solar maximum and the reformation of polar coronal holes. *Geophys. Res. Lett.*, 29:10.1029/2001GL014164, 2002b.

D. J. McComas, H. A. Elliott, N. A. Schwadron, J. T. Gosling, R. M. Skoug, and B. E. Goldstein. The three-dimensional solar wind around solar maximum. *Geophys. Res. Lett.*, 30:10.1029/2003GL017136, 2003.

D. J. McComas, H. A. Elliott, and R. von Steiger. Solar wind from high-latitude coronal holes at solar maximum. *Geophys. Res. Lett.*, 29:1029/2001GL013940, 2002a.

D. J. McComas, J. L. Phillips, A. J. Hundhausen, and J. T. Burkepile. Observations of disconnection of open coronal magnetic structures. *Geophys. Res. Lett.*, 18: 73–76, 1991.

K. G.. McCracken. Geomagnetic and atmospheric effects upon ice. *J. Geophys. Res.*, 109:doi.10.1029/2003,JA010060, 2004.

K. G. McCracken and F. B. McDonald. The long-term modulation of the galactic cosmic radiation, 1500–2000. In *Proc. 27th. Int. Cosmic Ray Conference, Hamburg, 2001*, 2001.

F. B. McDonald, N. Lal, and R. E. McGuire. The role of drifts and global merged interaction regions in the long-term modulation of cosmic rays. *J. Geophys. Res.*, 98:1243–1256, 1993.

F. B. McDonald, E. C. Stone, A. C. Cummings, B. Heikkila, N. Lal, and W. R. Webber. Enhancements of energetic particles near the heliospheric termination shock. *Nature*, 426:48–51, 2003.

R. Merrill, M. McElhinny, and J. McFadden. *The magnetic Field of the Earth.* Academic press, NBerw York, 1996.

H. Moraal, C. D. Steenberg, and G. P. Zank. Simulations of galactic and anomalous cosmic ray transport in the heliosphere. *Adv. Space Res.*, 23:425–436, 1999.

D. Nandy. Reviewing solar magnetic field generation in the light of helioseismology. In *Local and global helioseismology: The present and the future*, volume Proc SoHO12/GONG meeting, ESA SP-5176, page 213, ESTEC, Noordwijk, The Netherlands, 2003. ESA Publications Division.

D. Nandy and A. R. Choudhuri. Explaining the latitudinal distribution of sunspots with deep meridional flow. *Science*, 296:1671, 2002.

V. Narain and U. Ulmschneider. Chromospheric and coronal heating mechanisms. *Space Sci. Rev.*, 54:337, 1990.

V. Narain and U. Ulmschneider. Chromospheric and coronal heating mechanisms II. *Space Sci. Rev.*, 75:453–509, 1996.

H. Neckel and D. Labs. Solar limb darkening 1986–1990 (303 to 1099nm). *Sol. Phys.*, 153:91–114, 1994.

U. Neff, S. J. Burns, A. Mangini, M. Mudelsee, D. Fleitmann, and A. Matter. Strong coherence between solar variability and the monsoon in oman between 9 and 6 kyrs ago. *Nature*, 411:290–293, 2001.

H. V. Neher, V. Z. Peterson, and E. A. Stern. Fluctuations and latitude effect of cosmic rays at high altitudes and latitudes. *Phys. Rev.*, 90:655–674, 1953.

H. Nevanlinna. Results of the Helsinki magnetic observatory. *Ann. Geophys.*, in press, 2004.

H. Nevanlinna and E. Kataja. An extension to the geomagnetic activity index series aa for two solar cycles. *Geophys. Res. Lett.*, 20:2703–2706, 1993.

F. Noël. Solar cycle dependence of the apparent radius of the sun. *Astron. Astrophys.*, 413:725–732, doi.10.1051/0004–6361:20031573, 2004.

R. W. Noyes. *The Sun our Star.* Harvard University Press, Cambridge, Mass., 1982.

K. O'Brien. Secular variations in the production of cosmogenic isotopes in the earth's atmosphere. *J. Geophys. Res.*, 84:423–431, 1979.

K. O'Brien, A. de la Zerda, M. A. Shea, and D. F. Smart. The production of cosmogenic isotopes in the Earth's atmosphere and their inventories. In C. P. Sonnet, M. S. Giapapa, and M. S. Matthews, editors, *The Sun in Time*, pages 317–342. University of Arizona Press, 1991.

A. Ortiz. *Solar irradiance variations induced by faculae and small magnetic elements in the photosphere.* PhD thesis, Universitat de Barcelona (Department d'Astronomia i Meteorologia), 2003.

A. Ortiz, V. Domingo, B. Sanahuja, , and C. Fröhlich. Excess facular emission from an isolated active region during solar minimum: the example of noaa ar 7978. *J. atmos. sol.-terr. Phys.*, page in press, 2003.

A. Ortiz, V. Domingo, B. Sanahuja, and L. Sánchez. An example of isolated active region energy evolution: NOAA AR 7978. In *Proc. 1st Solar and Space Weather Euroconference: "The Solar Cycle & Terrestrial Climate"*, volume ESA SP-463, pages 340–395. ESA Publications, ESTEC, Noordwijk, The Netherlands, 2000.

A. Ortiz, S. K. Solanki, V. Domingo, M. Fligge, and B. Sanahuja. On the intensity contrast of solar photospheric faculae and network elements. *Astron. Astrophys.*, 388:1036–1047, 2002.

D. Paillard. Glacial cycles, towards a new paradigm. *Rev. Geophys.*, 39:325–346, 2001.

S. Parhi, R. A. Burger, J. W. Bieber, and W. H. Matthaeus. Challenges for an 'ab-initio' theory of cosmic ray modulation. In *Proceedings of ICRC*, page 3670, 2001.

E. N. Parker. Hydromagnetic dynamo models. *Astrophys. J.*, 122:293–314, 1955.

E. N. Parker. Dynamics of the interplanetary gas and magnetic fields. *Astrophys. J.*, 128:664–676, 1958.

E. N. Parker. *Interplanetary Dynamical processes.* Interscience/Wiley, New York, 1963.

E. N. Parker. The passage of energetic charged particles through interplanetary space. *Planet. Space Sci.*, 13:9, 1965.

J. S. Perko and L. A. Fisk. Solar modulation of galactic cosmic rays. v – time-dependent modulation. *J. Geophys. Res.*, 88:9033–9036, 1983.

S. R. O. Ploner, S. K.Solanki, and A. S.Gadun. Is solar mesogranulation a surface phenomenon? *Astron. Astrophys.*, 356:1050–1054, 2000.

G. Poletto. Origin and acceleration of fast and slow solar wind. In *Stars as Suns: Activity, Evolution, and Planets, IAU Symposium, Vol. 219*, pages 563–574, 2004.

M.S. Potgieter. The modulation of galactic cosmic rays in the heliosphere: theory and models. *Space Sci. Rev.*, 83:147–158, 1998.

T. I. Pulkkinen, H. Nevanlinna, P. J. Pulkkinen, and M. Lockwood. The Earth–Sun connection in time scales from years to decades to centuries. *Space Sci. Rev.*, 95: 625–637, 2001.

G. M. Raisbeck, F. Y. Yiou, J. Jouzel, and J. R. Petit. *Phil. Trans. R. Soc. Lond. A*, 330:436, 1990.

M. P. Rast, P. A. Fox, H. Lin, B. W. Lites, R. W. Meisner, and O. R. White. Bright rings around sunspots. *Nature*, 401:678–679, 1999.

M. P. Rast, R. W. Meisner, B. W. Lites, P. A. Fox, and O. R. White. Sunspot bright rings: evidence from case studies. *Astrophys. J.*, 557:864–879, 2001.

M. J. Reiner, J. Fainberg, M. L. Kaiser, and R. G. Stone. Type III radio source located by Ulysses/Wind triangulation. *J. Geophys. Res.*, 103(A2):1923–1932, 1998.

D. Rind and J. Overpeck. Hypothesized causes of decade-to-decade climate variability: climate model results. *Quaternary Sci. Rev.*, 12:357–374, 1993.

W. B. Rossow, A. W. Walker, D. E. Beuschel, and M. D. Roiter. International satellite cloud climatology project (ISCCP): Documentation of new datasets. Technical Report WMO/TD 737, World Meteorol. Organ., Geneva, 1996.

A. P. Rouillard and M. Lockwood. Oscillations in the open solar magnetic flux with period 1.68 years: imprint on galactic cosmic rays and implications for heliospheric shielding. *Ann. Geophys.*, in press, 2004.

C. T. Russell. On the possibility of deducing interplanetary and solar parameters from geomagnetic records. *Sol. Phys.*, 42:259–269, 1975.

C. T. Russell and R. L. McPherron. Seasonal variation of geomagnetic activity. *J. Geophys. Res.*, 78:92–108, 1973.

M. Sánchez Cuberes, M. Vázquez, J. A. Bonet, and M. Sobotka. Infrared photometry of photospheric solar structures II. centre-to-limb variation of active regions. *Astrophys. J.*, 570:886–899, 2002.

H. H. Sargent III. The 27-day recurrence index. In Y. Kamide and J. A. Slavin, editors, *Solar Wind–Magnetosphere Coupling*, pages 143–148. Terra Scientific, Tokyo, 1986.

K. H. Schatten. *Sun-Earth plasma connections*, volume AGU Geophysical Monograph 109, chapter Models for Coronal and Interplanetary magnetic fields: a critical commentary. American Geophysical Union, Washington, 1999.

K. H. Schatten, H. G. Mayr, K. Omidvar, and E. Maier. A hillock and cloud model for faculae. *Astrophys. J.*, 311:460–473, 1986.

K. H. Schatten, J. M. Wilcox, and N. F. Ness. A model of interplanetary and coronal magnetic fields. *Sol. Phys.*, 6:442–455, 1969.

K. Scherer and H. Fichtner. Constraints on the heliospheric magnetic field variation during the maunder minimum from cowsmic ray modulation modelling. *Astron. and Astrophys.*, 413:L11–L14, 2004.

K. Schlegel, G. Diendorfer, S. Thern, and M. Schmidt. Thunderstorms, lightning and solar activity – Middle Europe. *J. Atmos. Sol. -Terr. Phys.*, 63:1705–1713, 2001.

D. Schmitt. The solar dynamo. In *The Cosmic Dynamo*, volume Proc. IAU-Symp. 157, page 1. Kluwer, Dordrecht, 1993.

C. J. Schrijver, M. L. DeRosa, and A. M. Title. What is missing from our understanding of long-term solar and heliospheric activity? *Astrophys. J.*, 5771: 1006–1012, 2002.

C. J. Schrijver and A. M. Title. The dynamic nature of the solar magnetic field. In *Solar and Stellar Activity: Similarities and Differences*, volume ASP Conference Series, Vol. 1000, pages 15–26, 1999.

C. J. Schrijver, A. M. Title, A. A. van Ballegooijen, H. J. Hagenaar, and R. A. Shine. Sustaining the quiet photospheric network: The balance of flux emergence, fragmentation, merging, and cancellation. *Astrophys. J.*, 487:424, 1997.

C. J. Schrijver and C. Zwaan. *Solar and stellar magnetic activity.* Cambridge University Press, 2000.

P. Schröder, R. Smith, and K. Apps. Solar evolution and the distant future of earth. *Astron. Geophys.*, 42:6.26–6.29, 2001.

M. Schüssler, D. Schmidt, and A. Ferriz-Mas. Long-term variation of solar activity by a dynamo based on magnetic field lines. In *Advances in the physics of sunspots*, volume 1st Euroconference on Advances in Solar Physics, pages 39–44, 1997.

K. Schwarzschild. Über das gleichgewicht der sonnenatmosphäre. *Nach. Kön. Gesellsch. d Wiss., Göttingen*, 195:41, 1906.

L. Scurry and C. T. Russell. Proxy studies of energy transfer to the magnetosphere. *J. Geophys. Res.*, 96:9541–9548, 1991.

N. J. Shaviv. Cosmic ray diffusion from the galactic spiral arms, iron meteorites and possible climatic connection? *Phys. Rev. Lett.*, 89:51102, 2002.

N. J. Shaviv. The spiral structure of the milky way, cosmic ray and ice age epochs on earth. *New Astronomy*, in press, 2004.

M. A. Shea and D. F. Smart. Cosmic ray implications for human health. *Space Sci. Rev.*, 93:187–205, 2000.

D. T. Shindell, D. Rindt, N. Balachandran, J. Lean, and P. Lonergan. Solar cycle variability, ozone and climate. *Science*, 284:305, 1999.

D. T. Shindell, G. A. Schmidt, M. E. Mann, D. Rindt, and A. Waple. Solar forcing of regional climate change during the Maunder minimum. *Science*, 294:2149–2152, 2001.

S. M. Silverman. Secular variation of the aurora for the past 500 years. *Rev. Geophys.*, 30:333–351, 1992.

S. M. Silverman and R. Shapiro. Power spectral analysis of auroral occurrence frequency. *J. Geophys. Res.*, 88:6310, 1983.

G. Simmnet. LASCO observations of disconnected magnetic structures out to 28 solar radii during coronal mass ejections. *Sol. Phys.*, 175:685–698, 1997.

P. A. Simon and J. P. Legrand. Some solar cycle phenomena related to geomagnetic activity from 1868 to 1980. *Astron and Astrophys.*, 182:329–336, 1987.

E. J. Smith and A. Balogh. Ulysses observations of the radial magnetic field. *Geophys. Res. Lett.*, 22:3317–3320, 1995.

E. J. Smith and A. Balogh. Open magnetic flux: variation with latitude and solar cycle. In M. Velli, R. Bruno, and F. Malara, editors, *Solar Wind Ten: Proceedings of the tenth international solar wind conference*, pages 67–70, 2003.

E. J. Smith, A. Balogh, R. J. Forsyth, and D. J. McComas. Ulysses in the south polar cap at solar maximum: heliospheric magnetic. *Geophys. Res. Lett.*, 28: 4195–4162, 2001.

E. J. Smith and J. W. Bieber. Solar cycle variation of the interplanetary magnetic field spiral. *Astrophys. J.*, 370:453–441, 1991.

S. K. Solanki and M. Fligge. Solar irradiance since 1874 revisited. *Geophys. Res. Lett.*, 25:341–344, 1998.

S. K. Solanki and M. Fligge. A reconstruction of total solar irradiance since 1700. *Geophys. Res. Lett.*, 26:2465–2468, 1999.

S. K. Solanki and M. Fligge. How much of the solar irradiance variations is caused by the magnetic field at the solar surface? *Adv. Space Res.*, 29:1933–1940, 2002.

S. K. Solanki and N. A. Krivova. Can solar variability explain global warming since 1970? *J. Geophys. Res.*, 108(A5):1200, doi:10.1029/2002JA009753, 2003.

S. K. Solanki, N. A. Krivova, M. Schüssler, and M. Fligge. Search for a relationship between solar cycle amplitude and length. *Astron. and Astrophys.*, 396:1029–1035, doi. 10.1051/0004–6361:20021436, 2002a.

S. K. Solanki, M. Schüssler, and M. Fligge. Secular evolution of the Sun's magnetic field since the Maunder minimum. *Nature*, 480:445–446, 2000.

S. K. Solanki, M. Schüssler, and M. Fligge. Secular variation of the sun's magnetic flux. *Astron. Astrophys.*, 383:706–712, 2002b.

R. Solomon, V. Schroeder, and M. B. Baker. Lightning initiation – conventional and runaway – breakdown hypotheses. *Quart. J. Roy. Meteorol. Soc.*, 127:2683–2704, 2001.

E. A. Spiegel and J. P Zahn. The solar tachocline. *Astron. Astrophys.*, 265:106–114, 1992.

H. C. Spruit. Pressure equilibrium and energy balance of small photospheric flux-tubes. *Sol. Phys.*, 50:269, 1976.

H. C. Spruit. *The Sun as a star*, volume NASA publication SP-450, chapter Magnetic flux tubes. NASA, 1981.

H. C. Spruit. Theory of solar irradiance variations. *Space Sci. Rev.*, 94:113–126, 2000.

H.C. Spruit. *The Sun in Time*, chapter Theory of luminosity and radius variations, pages 118–159. Univ. of Arizona Press, 1991.

R. Stamper, M. Lockwood, M. N. Wild, and T. D. G. Clark. Solar causes of the long term increase in geomagnetic activity. *J. Geophys. Res.*, 104:28325–28342, 1999.

O. Steiner, U. Grossmann-Doerth, M. Schüssler, and M. Knölker. Polarized radiation diagnostics of magnetohydrodynamic models of the solar atmosphere. *Sol. Phys.*, 164:223–242, 1996.

O. Steiner, U. Grossmann-Doerth, M. Schüssler, and M. Knölker. Dynamical interaction of solar magnetic elements and granular convection: results of numerical simulation. *Astrophys. J.*, 495:468, 1998.

P. A. Stott et al. External control of 20th century temperature by natural and antropogenic forcings. *Science*, 290:2133–2137, 2000.

M. Stuiver and T. F. Braziunas. Atmospheric C-14 and century-scale oscillations. *Nature*, 338:405–407, 1989.

M. Stuiver and P. D. Quay. Changes in atmospheric carbon-14 attributed to a variable Sun. *Science*, 207:11, 1980.

M. Stuiver, P. J. Reimer, E. Bard, J. W. Beck, J. S. Burr, K. A. Hughen, B. Kromer, G. McCormac, J. van der Plicht, and M. Spurk. Intcal98 radiocarbon age calibration 24,000 cal bp. *Radiocarbon*, 40:1041–1083, 1988a.

M. Stuiver, P. J. Reimer, and T. F. Braziunas. High precision radiocarbon age calibration for terrestrial and marine samples. *Radiocarbon*, 40:1127–1151, 1988b.

S. T. Suess. The solar wind – inner heliosphere. *Space Sci. Rev.*, 83:75–86, 1998.

S. T. Suess and E. J. Smith. Latitudinal dependence of the radial IMF component – coronal imprint. *Geophys. Res. Lett.*, 23:3267–3270, 1996a.

S. T. Suess, E. J. Smith, J. Phillips, B. E. Goldstein, and S. Nerney. Latitudinal dependence of the radial IMF component – interplanetary imprint. *Astron. Astrophys.*, 316:304–312, 1996b.

L. Svalgaard, E. W. Cliver, and P. Le Sager. IHV index: Reconstruction of aa index back to 1901. *Adv. Space Res.*, in press, 2004.

H. Svensmark. Influence of cosmic rays on Earth's climate. *Phys. Rev. Lett.*, 81: 5027–5030, 1998.

H. Svensmark and E. Friis-Christensen. Variation of cosmic ray flux and global cloud coverage: A missing link in solar–climate relationships. *J. Atmos. Sol. -Terr. Phys.*, 59:1225–1232, 1997.

S. F. B. Tett, P. A. Stott, M. R. Allen, W. J. Ingram, and J. F. B. Mitchell. Causes of twentieth century temperature change near the Earth's surface. *Nature*, 399: 569–572, 1999.

W. (Lord Kelvin) Thompson. Presidential address (november 1892). *Proc. Roy. Soc. (London)*, 52:299, 1893.

A. M. Title and C. J. Schrijver. The Sun's magnetic carpet. In *Cool Stars, Stellar Systems and the Sun*, volume Vol 154, pages 345–358, 1998.

K. P. Topka, T. D. Tarbell, and A. M. Title. Smallest solar magnetic elements. II. observations versus hot wall models of faculae. *Astrophys. J.*, 484:479–486, 1997.

E. Tric. Paleointensity of the geomagnetic field during the last 80,000 years. *J. Geophys. Res.*, 97:9337–9351, 1992.

P. M. Udelhofen and R. D. Cess. Cloud cover variations over the United States: an influence of cosmic rays or solar variability. *Geophys. Res. Lett.*, 28:2617–2620, 2001.

Y. C. Unruh, S. K. Solanki, and M. Fligge. The spectral dependence of facular contrast and solar irradiance variations. *Astron. Astrophys.*, 345:635–642, 1999.

Y. C. Unruh, S. K. Solanki, and M. Fligge. *Space Sci. Rev.*, 94:145, 2000.

I. G. Usoskin, K. Mursula, and G. A. Kovaltsov. Reconstruction of monthly and yearly group sunspot numbers from sparse daily observations. *Sol. Phys.*, 218: 295–305, 2003a.

I. G. Usoskin, K. Mursula, and G. A. Kovaltsov. The lost sunspot cycle: Reanalysis of sunspot statistics. *Astron. Astrophys.*, 403:743–748, doi.10.1051/0004-6361:20030398, 2003b.

I. G. Usoskin, K. Mursula, S. Solanki, M. Schüssler, and K. Alanko. Reconstruction of solar activity for the last millennium using 10be data. *Astron. Astrophys.*, 413: 745–751, 2004.

I. G. Usoskin, S. K. Solanki, M. Schüssler, K. Mursula, and K. Alanko. Millennium-scale sunspot number reconstruction: Evidence for an unusually active sun since the 1940s. *Phys. Rev. Lett.*, 91(21):211101, 1–4, 2003c.

J. F. Valdés-Galicia and B. Mendoza. On the role of large scale solar photospheric motions in the cosmic-ray 1.68-yr intensity variation. *Sol. Phys.*, 178:183–191, 1998.

J. F. Valdés-Galicia, R. Pérez-Enríquez, and J. A. Otaola. The cosmic ray 1–68-year variation: a clue to understand the nature of the solar cycle? *Sol. Phys.*, 167:409–417, 1996.

V. M. Vasyliunas, J. R. Kan, G. L. Siscoe, and S. I. Akasofu. Scaling relations governing magnetospheric energy transfer. *Planet. Space Sci.*, 30:359–365, 1982.

G. Wagner et al. Some results relevant to the discussion of a possible link between cosmic rays and earth's climate. *J. Geophys. Res.*, 106:3381–3388, 2001.

M. Waldmeier. *Die Sonnenkorona 2.* Verlag Birkhäuser, Basel, 1957.

M. Waldmeier. The coronal hole at the 7 march 1970 solar eclipse. *Sol. Phys.*, 40: 351, 1975.

S. R. Walton, D. G. Preminger, and G. A. Chapman. The contribution of faculae and network to long-term changes in the total solar irradiance. *Astrophys. J.*, 590:1088–1094, 2003.

Y. M. Wang. Empirical relationship between the magnetic field and the mass and energy flux in the source regions of the solar wind. *Astrophys. J.*, 499:L157–L160, 1995.

Y. M. Wang, S. H. Hawley, and N. R. Sheeley Jr. The magnetic nature of coronal holes. *Science*, 271:464–469, 1996.

Y. M. Wang, J. Lean, and N. R. Sheeley Jr. The long-term evolution of the Sun's open magnetic flux. *Geophys. Res. Lett.*, 27:505–508, 2000b.

Y. M. Wang, J. Lean, and N. R. Sheeley Jr. Role of a variable meridional flow in the secular evolution of the sun's polar fields and open flux. *Astrophys. J.*, 577: L53–L57, 2002.

Y. M. Wang and N. R. Sheeley Jr. Solar wind speed and coronal flux-tube expansion. *Astrophys. J.*, 355:726, 1990.

Y. M. Wang and N. R. Sheeley Jr. Solar implications of ulysses interplanetary field measurements. *Astrophys. J.*, 447:L143–L146, 1995.

Y. M. Wang and N. R. Sheeley Jr. Modelling the sun's large-scale magnetic field during the maunder minimum. 591:1248–1256, 2003a.

Y. M. Wang and N. R. Sheeley Jr. On the fluctuating component of the sun's large-scale magnetic field. *Astrophys. J.*, 590:1111–1120, 2003b.

Y. M. Wang and N. R. Sheeley Jr. On the topological evolution of the coronal magnetic field during the solar cycle. *Astrophys. J.*, 599:1404–1417, 2003c.

Y. M. Wang and N. R. Sheeley Jr. Sunspot activity and the long-term variation of the Sun's open magnetic flux. *J. Geophys. Res.*, in press, 2004.

Y. M. Wang, N. R. Sheeley Jr., R. A. Howard, and O. C. St. Cyr. Coronograph observations of inflows during high solar activity. *Geophys. Res. Lett.*, 26:1203–1206, 1999a.

Y. M. Wang, N. R. Sheeley Jr., R. A. Howard, and N. B. Rich. Streamer disconnection events observed with the lasco coronograph. *Geophys. Res. Lett.*, 26: 1349–1352, 1999b.

Y. M. Wang, N. R. Sheeley Jr., and J. Lean. Understanding the evolution of Sun's magnetic flux. *Geophys. Res. Lett.*, 27:621–624, 2000a.

Y. M. Wang, N. R. Sheeley Jr., and N. B. Rich. Evolution of coronal streamer structure during the rising phase of solar cycle 23. *Geophys. Res. Lett.*, 27: 149–152, 2000c.

D. F. Webb and E. W. Cliver. Evidence for magnetic disconnection of mass ejections in the corona. *J. Geophys. Res.*, 100:5853–5870, 1995.

N. O. Weiss. Solar and stellar dynamos. In *Lectures on solar and planetary dynamos*, page 59. Cambridge University Press, 1994.

T. Wenzler, S.K. Solanki, N. A. Krivova, and D. M. Fluri. Comparison between kpvt/spm and soho/mdi magnetograms with an application to solar irradiance reconstructions. *Astron. and Astrophys.*, in press, 2004.

W. B. White and D. R. Cayan. Quasi-periodicity and global symmetries in interdecadal upper ocean temperature variability. *J. Geophys. Res.*, 103:21355–21354, 1998.

W. B. White, J. Lean, D. R. Cayan, and M. D. Dettinger. Response of global upper ocean temperature to changing solar irradiance. *J. Geophys. Res.*, 102: 3255–3266, 1997.

G. Wibberenz and H. V. Cane. Simple analytical solutions for propagating diffusive barriers and application to the 1974 minicycle. *J. Geophys. Res.*, 105:18315–18325, 2000.

G. Wibberenz, I. G. Richardson, and H. V. Cane. A simple concept for modelling cosmic ray modulation in the inner heliosphere during solar cycles 20–23. *J. Geophys. Res.*, 107:1353–1368, 2002.

T. M. L. Wigley and S. C. B. Raper. Climatic change due to solar irradiance changes. *Geophys. Res. Lett.*, 17:2169–2172, 1990.

D. S. Wilks. *Statistical methods in the atmospheric sciences.* Academic Press, San Diego, California, USA, 1995.

R. C. Willson. Total solar irradiance trend during cycles 21 and 22. *Science*, 277: 1963–1965, 1997.

R. C. Willson, H. S. Hudson, and G. A. Chapman. Observations of solar irradiance variability. *Science*, 211:700–702, 1981.

R. C. Willson and A. V. Mordvinov. Secular total solar irradiance trend during solar cycles 21–23. *Geophys. Res. Lett.*, 30:1199–1202, doi.10.1029/2002GL016038, 2003.

P. R. Wilson, R. C. Altrock, K. L. Harvey, S. F. Martin, and H. B. Snodgrass. The extended solar activity cycle. *Nature*, 333:748–750, 1988.

F. Yu and R. P. Turco. Ultrafine aerosol formation via ion-mediated nucleation. *Geophys. Res. Lett.*, 27:883–886, 2000.

X. P. Zhao, J. T. Hoeksema, and P. H. Scherrer. Modeling boot-shaped coronal holes using SoHO–MDI magnetic measurements. In *Proceedings of The Fifth SOHO Workshop in Oslo, ESA SP-404*, pages 751–756. ESA Publications, ESTEC, Noordwijk, The Netherlands, 1997.

C. Zwaan. The emergence of magnetic flux. *Solar Physics*, 100:397–414, 1985.

C. Zwaan. *Ann. Rev. Astron. Astrophys.*, 25:83, 1987.

Stellar Analogs of Solar Activity: The Sun in A Stellar Context

M. Giampapa

National Solar Observatory/NOAO, Tucson, Arizona, USA
giampapa@noao.edu

1 Introduction

I will offer perspectives on solar variability based on observations of magnetic field-related activity and variability in stars. The stars themselves may be analogs of the Sun or they can be quite different from the Sun in their photospheric properties or in their evolutionary state. In either case, the motivation is to utilize stellar observations to extend the range in parameter space that is unavailable or not practically feasible through observations of the Sun alone. In this way, we can gain insight on the factors that control solar and solar-like variability on short time scales ranging from days to years; decadal time scales that correspond to cycle periods, similar to that of the solar activity cycle; century-long time scales during which Maunder-minima-like episodes can occur; and, evolutionary time scales that can influence the evolution of the solar system.

I will begin with a comparison of the properties of the solar cycle, as seen in various diagnostics, with what has been observed in stars, including an overview of the basic properties of kinematic dynamos and the observed correlations among dynamo parameters. I will then review the results of observations of stellar brightness changes and their joint variation with activity. I will examine the distribution of activity in samples of stars that are claimed to be analogs of the Sun based on key parameters or properties. The results of these observations of solar-type stars have been used extensively for the forecasting and "hindcasting" of solar variability.

Next I will focus on the evidence for, and the nature of, stellar surface inhomogeneities and magnetic structures on late-type, dwarf stars. Inhomogeneities, as delineated by magnetic field structures, are fundamental properties of the solar atmosphere and stellar atmospheres. It is these magnetic regions that are the origin of variability in the Sun and sun-like stars. I will also investigate the nature of coronal magnetic structures and coronal heating in the cool M dwarf stars. While not analogs of the Sun, these low mass dwarfs represent a regime in parameter space in which dynamo operation and magnetic field-related activity can be investigated in the limits of thick convection zones and ultracool photospheric temperatures, respectively.

In Sect. 7, I review how solar-like activity changes on evolutionary time scales. Through the study of young precursors to solar-type stars, we can gain insight on the early evolution of the ambient radiative and energetic particle environments of the solar system in the eras during which planetary atmospheres evolved. I will examine the relationship between stellar age, activity and rotation, including the evolution of activity diagnostics. I will also summarize the evidence for an active, early Sun.

Finally I will discuss the influence of solar-like, stellar activity on the indirect methods for the detection of extrasolar planetary systems. The detection of extrasolar planetary systems is relevant to the emerging field of astrobiology. The search for extrasolar planets will reveal whether our solar system is a common occurrence or a rare event. Ultimately, the discovery of similar solar systems will enable comparative studies of solar systems. In this search, it is important to understand how stellar variability can contribute errors, or "stellar noise", and impose limitations on the principal indirect methods currently in use to discover new worlds around other suns.

2 Stellar Cycles

Our longest record of measurement of solar activity from direct observations of the Sun is that of the solar cycle as seen in sunspot number (Fig. 1).

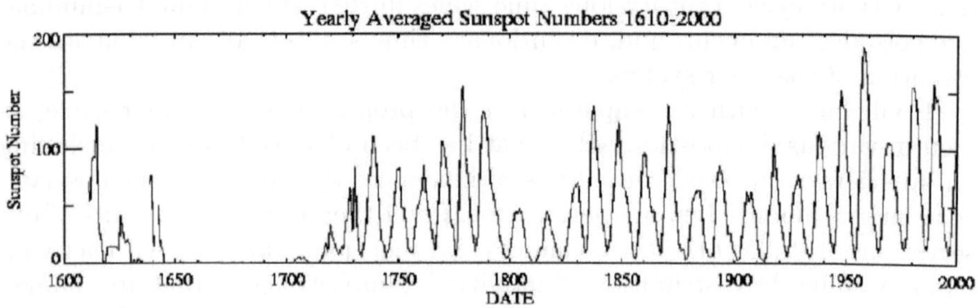

Fig. 1. The historical Sunspot Number record smoothed by yearly averages. Sunspots were observed avidly with telescopes soon after Galileo called attention to their existence. From an historical perspective, I note that if the telescope had been invented about 100 years later, the Maunder Minimum episode would have been missed entirely

The sunspot number record is the longest record of activity we have for any late-type star. The salient features of the sunspot cycle include an approximately 11-year (peak-to-peak) period interspersed with prolonged episodes of noticeably reduced sunspot number, such as the Maunder and Dalton minima, respectively. Other features that are apparent from inspection of Fig. 1 are an approximately 80-year modulation of the $\sim$11 year sunspot cycle and a secular trend of increasing peak values to the present era.

According to Hoyt et al. (1994), the latter is a real trend and not an artifact of observational methods that have evolved during the past ~400 years. I note, however, that the so-called "K-corrections" for the normalization of sunspot counts by different observers are now being re-examined. In brief review, the sunspot number, R, was defined by Wolf (1856) as

$$R = F + 10 \times G \; , \tag{1}$$

where F = the number of individual spots and G = the number of groups. For each Secondary Observer (SEC), their individual counts were scaled to the Standard Observer (SO = Wolf himself) with a scaled sunspot count of

$$R_{SEC} = K(F + 10 \times G) \; , \tag{2}$$

where K = $\langle C_{SO}/C_{SEC} \rangle$. Various Secondary Observers can exhibit changes in their K values by up to factors of 2 on all time scales. Therefore, the sunspot number record may be in need of correction for variations of the K-coefficients for the Standard Observers.

Modern observations of the Sun have provided additional signatures of the solar cycle, as illustrated in Fig. 2. The diagnostics in Fig. 2 include the total solar irradiance (TSI), the He 1083 nm line (He), the unsigned total surface magnetic flux (Mag) and the strength of the K line of singly ionized calcium (CaK) with its central line core formation in the solar chromosphere. In addition to these diagnostics, the solar cycle is also seen prominently in the X-ray and radio regimes. In each of these features, a high level of short-term variability is present at solar maximum and only a low level of variability is seen at solar minimum.

In addition to the aforementioned diagnostics, Hall & Lockwood (2000) measured the time variation of the equivalent widths of the numerous photospheric line features in the solar spectrum as recorded in the spatially integrated light of the Sun (i.e., the sun viewed as a star) using a fiber-fed, stellar spectrograph. In their power spectrum of the results (Fig. 3), the rotation of the Sun as a star is detected at a period of 27.4 days (along with a harmonic at 13.7 days).

This result demonstrates the potential utility of photospheric lines in the study of solar variability. I would also note that measurable variations in the photospheric line strengths should lead to apparent changes in the color of the Sun as measured in standard photometric bands (e.g., Johnson *UBV* or Strömgren *uvby*) that are commonly utilized for stellar observations. The quantitative magnitude of this potential effect on photometric bands has not been investigated.

2.1 Ca II H and K Line Observations

The signature of magnetic activity on solar-type stars is most prominently manifested in the visible portion of the electromagnetic spectrum in the enhanced radiation it produces in the emission cores of the H and K resonance

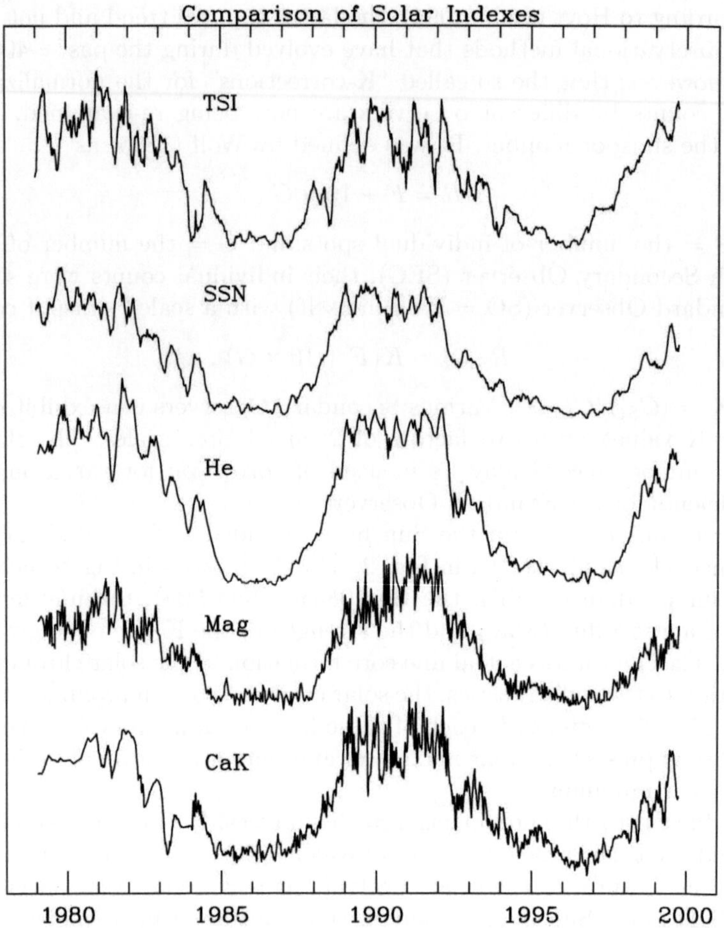

Fig. 2. Various diagnostics of the solar cycle in the Sun seen as a star. Included here are the smoothed variations of the total solar irradiance (TSI), the sunspot number (SSN), the He 1083 nm absorption equivalent width variation (He), the unsigned total surface magnetic flux (Mag) and the strength of the chromospheric K line of singly ionized calcium (CaK) within a 0.1 nm bandpass centered on the line core (*courtesy of W. C. Livingston, National Solar Observatory, Tucson, Arizona, USA*)

lines of singly ionized calcium near 400 nm. I display in Fig. 4 a spectrum of this feature as seen in the integrated spectrum of the Sun, recorded with the new SOLIS Integrated Sunlight Spectrometer (ISS).

Weak emission peaks are seen with an apparent central absorption. The latter feature is not a real absorption feature. Rather, it is a consequence of the non-LTE line formation conditions that lead to a relative decline in the source function, as first pointed out by Thomas (1957). The central feature is referred to as K_3 while the dual emission peaks are designated K_2. The

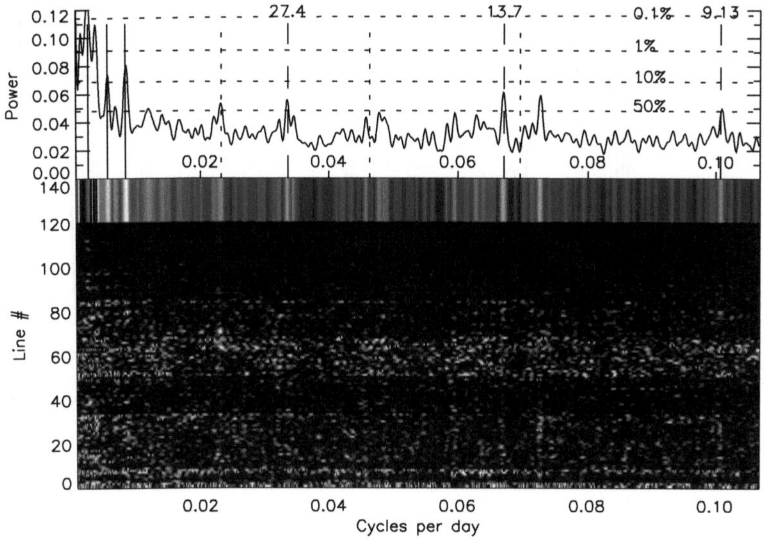

Fig. 3. The power spectrum of the temporal series of the equivalent widths of solar photospheric lines. Significant power can be seen at the canonical solar rotation period. This period approximately corresponds to mid-latitude regions where active region complexes appear, especially near the peak of the solar activity cycle (*from Hall & Lockwood 1998*)

local minima immediately before the K_2 peaks are the K_1 minima. In total, the source function of the Ca II K core essentially represents a mapping of the temperature profile of the solar (or stellar) chromosphere.

In addition to its utility as a diagnostic of the thermal structure of the chromosphere, the Ca II resonance features are spatially correlated with magnetic active regions on the Sun and, by extension, the surfaces of solar-type stars. This correlation is illustrated in Fig. 5 where bright Ca II chromospheric emission is spatially coincident with solar active regions. Indeed, Skumanich et al. (1975) found a direct correlation between magnetic field flux (as measured by a magnetograph) and relative Ca II core strength in solar plages, thus demonstrating the presumably causal relationship between magnetic fields and chromospheric heating (Fig. 6).

The appearance of emission cores in photographic stellar spectra naturally suggested that chromospheres, analogous to the solar chromosphere, must also occur in late-type stars. As illustrated in Fig. 7, the stellar H & K emission core strengths spans a range from barely perceptible, as seen in the integrated spectrum of the Sun at echelle resolutions, to significantly enhanced relative to the Sun.

Olin Wilson thus asked the obvious question whether Ca II emission in late-type stars exhibited long-term variations similar to that of the solar cycle.

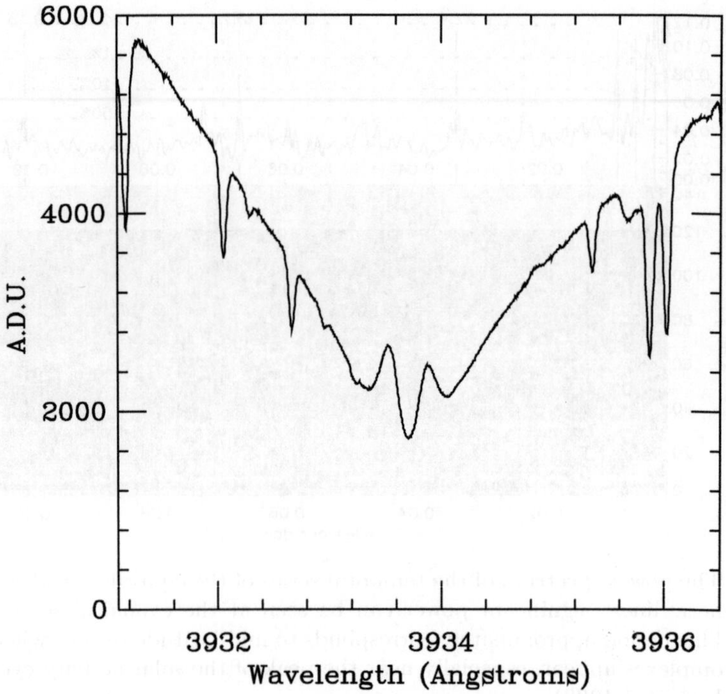

Fig. 4. The spectrum of the Ca^{+} K line in the Sun-as-a-star, as recorded by the Integrated Sunlight Spectrometer (ISS) of the new Synoptic Optical Long-term Investigations of the Sun (SOLIS) instrument. The resolving power is approximately $\lambda/\Delta\lambda = 300{,}000$. The approximate central wavelength of the feature is 3933.66 Å. The ISS utilizes an 8-mm diameter fiber to scramble the sunlight so that the Sun appears as a star at the entrance slit to the bench spectrograph. The spectrum shown here is one of the laboratory test spectra obtained with the SOLIS/ISS of the National Solar Observatory

2.2 The Mt. Wilson Program

In order to answer this question, Olin Wilson implemented a long-term program of monthly observations of a sample of stars in the solar neighborhood. Wilson utilized the telescope facilities at Mt. Wilson Observatory in conjunction with a spectrophotometric instrument to record the relative strengths of the stellar Ca II features. Wilson continued his program until publication of the results since initiation (Wilson 1978). Thereafter, the "Mt. Wilson program" of synoptic stellar H & K observations was continued by Sallie L. Baliunas and her collaborators.

An example of representative Ca II H&K time series for Sun-like stars from the Mt. Wilson program is given in Figs. 8–9. The Mt. Wilson Ca II measurements are expressed in terms of the "S-index". This index is based on the instrumental characteristics of the HK photometer that has been used

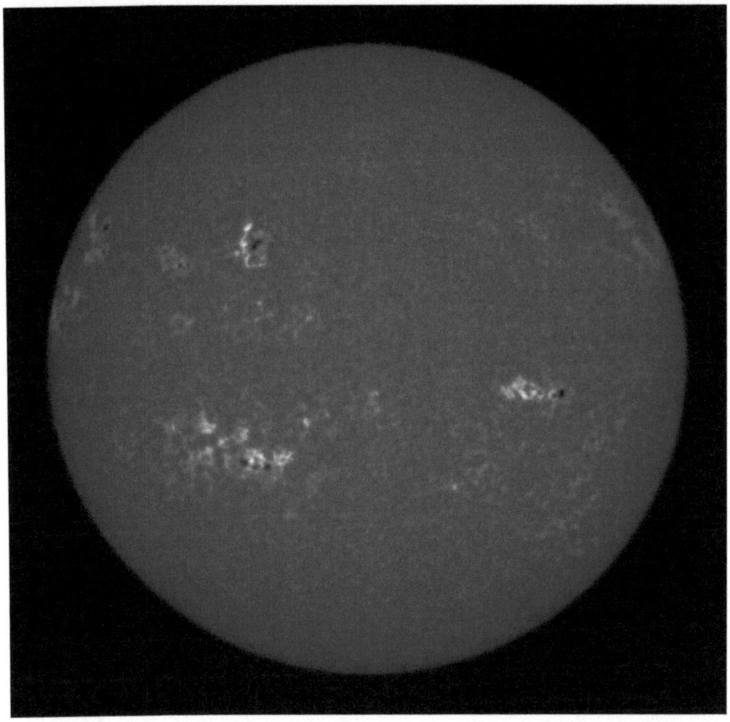

Fig. 5. A spectroheliogram in the Ca II K line, which is formed in the solar chromosphere and the chromospheres of solar-type and other late-type stars. The *bright* regions correspond to enhance K line emission associated with concentrations of magnetic flux. The *dark* regions are sunspots *National Solar Observatory, Tucson, Arizona, USA*

to acquire the data. In particular, the S-index is the ratio of the H&K counts in tunable 0.1 nm bandpasses centered at the Ca II H & K resonance lines to the total counts in nearby continuum bandpasses. Thus, it is a measure of Ca II strengths in the cores of the H & K lines normalized by the local continuum.

Other measures of stellar chromospheric Ca II emission strengths in late-type stars include the sum of the total flux in the lines between the K_1 and H_1 minima, respectively, normalized by the total stellar bolometric flux, or

$$R_{HK} = \frac{F(H+K)}{\sigma {T_{eff}}^{4}} , \tag{3}$$

where T_{eff} is the stellar effective temperature and $F(H+K)$ is the total Ca II H and K flux between the K_1 (H_1) minima above the zero flux level. The index denoted by R'_{HK} is R_{HK} corrected for the radiative equilibrium (i.e., non-chromospheric) contribution by the photosphere to the central core flux above the zero flux level. Thus, it is a model-dependent quantity. In addition

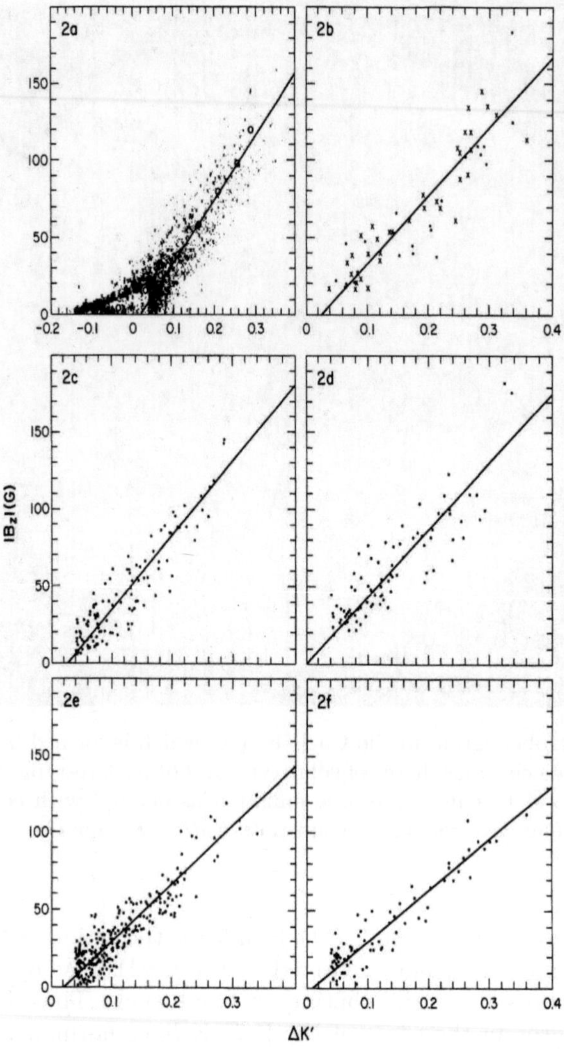

Fig. 6. The correlation between magnetic field strength and residual Ca II K line core emission in plage regions and bright elements on the Sun. The field strength is actually deduced from a magnetograph signal. The magnetograph measures the amount of magnetic *flux*, i.e., field strength × area, within an observing aperture. Essentially, the field strength is deduced by dividing the magnetograph signal by the area of the viewing aperture. Because of seeing effects, the aperture is not entirely filled with magnetic field. Therefore, the inferred magnetic field strength is really a lower limit to the true field strength (*from Skumanich et al. 1975*)

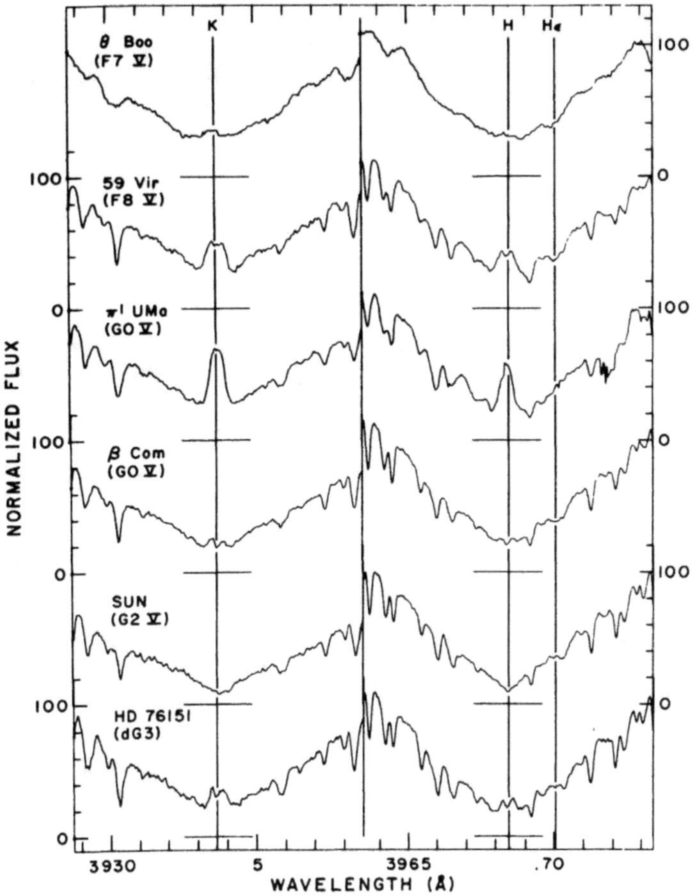

Fig. 7. The Ca II K and H line profiles in selected solar-type stars. The solar spectrum is obtained from the observation of reflected moon light in this example. A range of chromospheric Ca II resonance emission line strengths is seen here, from relatively weak (e.g., the Sun) to strong (*from Linsky, McClintock, Robertson & Worden 1979*)

to R_{HK}, the HK index is a calibrated quantity that can be related to the S-index. The HK index is defined as the calibrated relative intensity in 0.1 nm bandpasses centered on the H & K cores, respectively. Current efforts are concentrating on the use of flux-calibrated scales (see Hall & Lockwood 1995 for a detailed discussion of the calibration of the various indices in use).

2.3 Stellar Cycle Properties

The Mt. Wilson program includes stars that span the broad spectral-type range from F to K. Within the Mt. Wilson sample and project time span,

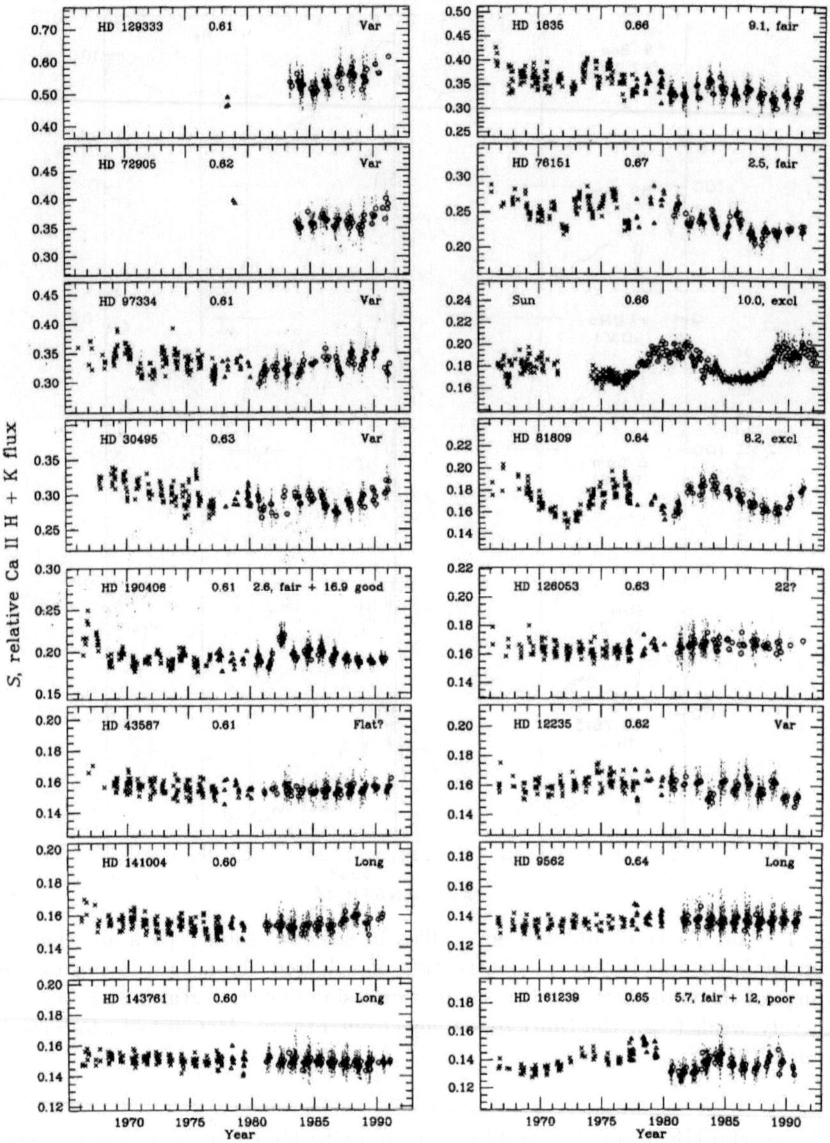

Fig. 8. Examples of stellar cycles as seen in the variation of the Ca II H+K relative strengths. The $B-V$ color is shown for each star. The mean color for the Sun is $(B-V) = 0.65$. The inferred cycle period in years is also given along with an assessment of the quality of the period determination. In some cases, regular periods analogous to the solar cycle are obvious such as the example of HD 81809. In other cases, the variation is irregular with no clear discernible period while others are simply flat, i.e., essentially constant in H+K relative strength. In some cases, a long-term secular trend may be present along with a shorter cycle period, such as HD 1835 (*from Baliunas et al. 1995*)

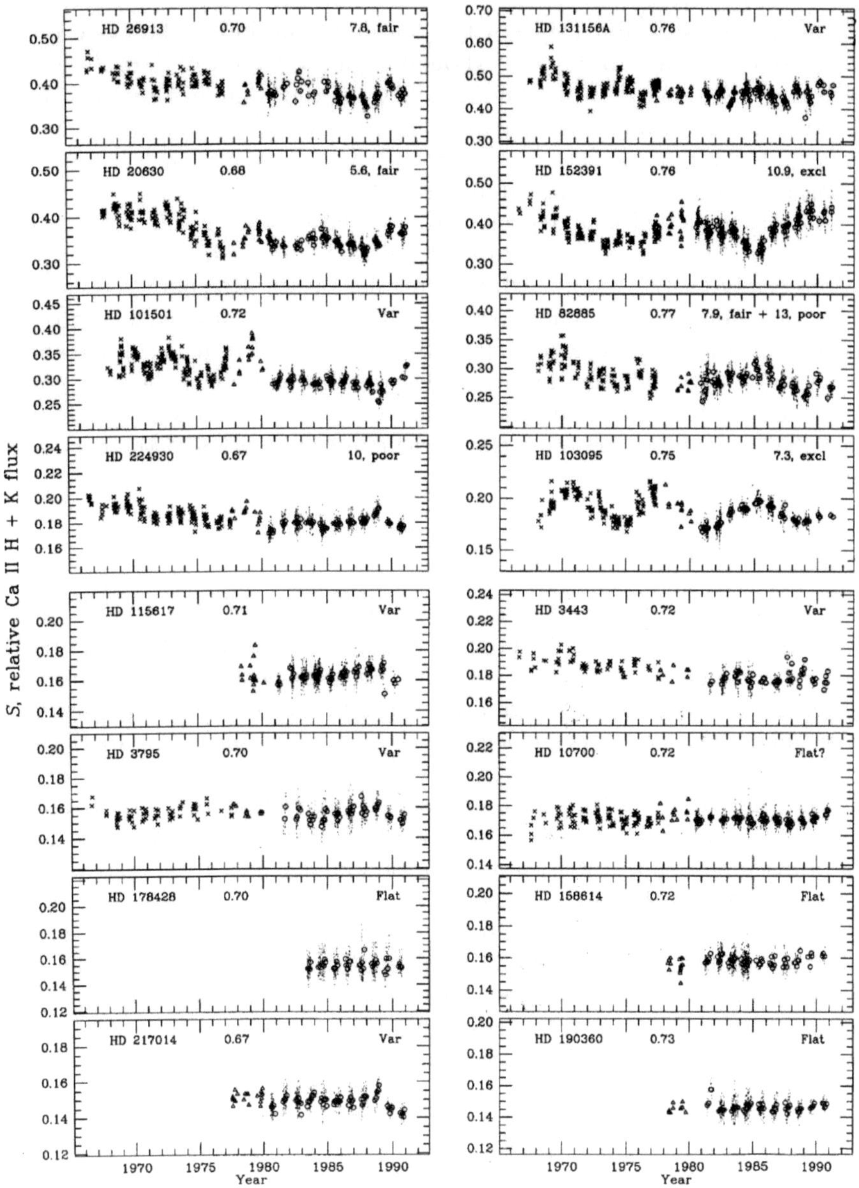

Fig. 9. Same as Fig. 8. High amplitude, regular cycles are clearly present in this set of solar-type stars that are somewhat cooler than the Sun. Stars that exhibit essentially constant values of their Ca II H & K strengths are designated as "Flat". These could be stars in a "Maunder minimum" or they could be somewhat evolved and older than the Sun, and hence magnetically quiet. The evolutionary status of these chromospherically flat stars is a critical question, as we will see in §4 (*from Baliunas et al. 1995*)

60% of the stars exhibit cycles with identifiable periods. The cycle periods range from 2.5 years to 25 years though cycle periods less than 7 years are much less well-defined. Approximately 25% of the sample stars are variable in their S-indices while 15% display a flat (non-variable) time series in H&K strength. Cycles are present in all spectral types but are much less common among F stars while widespread among K stars.

The properties of stellar cycles are correlated with mean activity level. More specifically, those stars characterized by high S-index values, i.e., chromospherically "active" stars, exhibit irregular cycles. Regular cycles are seen among low S-index, i.e., chromospherically "quiet" stars, such as the Sun. A key question within the context of dynamo theory that has yet to be addressed observationally is whether cycles are also present in the fully convective dwarf M stars, i.e., stars that no longer have a transition layer between the radiative zone surrounding the nuclear burning core and the base of the outer convection zone.

2.4 The Vaughan-Preston Gap

A feature of the distribution of the mean values of the S-index in the Mt. Wilson sample was pointed out by Vaughan & Preston (1980). In particular, a diagram of S index vs. $B-V$ color for solar neighborhood stars shows an apparent discontinuity, or gap, resulting in an apparent segregation between "high-activity" stars and low-activity, or quiet (sun-like) stars (Fig. 10).

I am not aware of a formal statistical verification of the gap as, say, reflected in a bimodal distribution in log S. I note for the interested reader that a discussion of a statistical test for bimodality in a given distribution is developed by Wolf & Sumner (2001).

The physical origin of the gap is unclear though it has been suggested that it is a consequence of two different modes of dynamo action: one mode operating in active stars and the other mode in quiet stars. Conversely, the gap has been attributed to a discontinuity in the local star formation rate combined with a continuous relation between Ca II strength and stellar age. Observations of clusters of intermediate age between that of young clusters and clusters with ages similar to the age of the Sun will shed light on the true origin of the Vaughan-Preston Gap.

2.5 Cycles, Rotation and Activity

Wilson (1978) indicated that short-term variations related to the rotational modulation of active complexes may be present in his long-term record of cycle observations. I show in Fig. 11 a schematic illustration of how concentrated complexes of magnetic field regions that are bright in Ca II H & K emission and asymmetrically distributed in longitude on the stellar surface

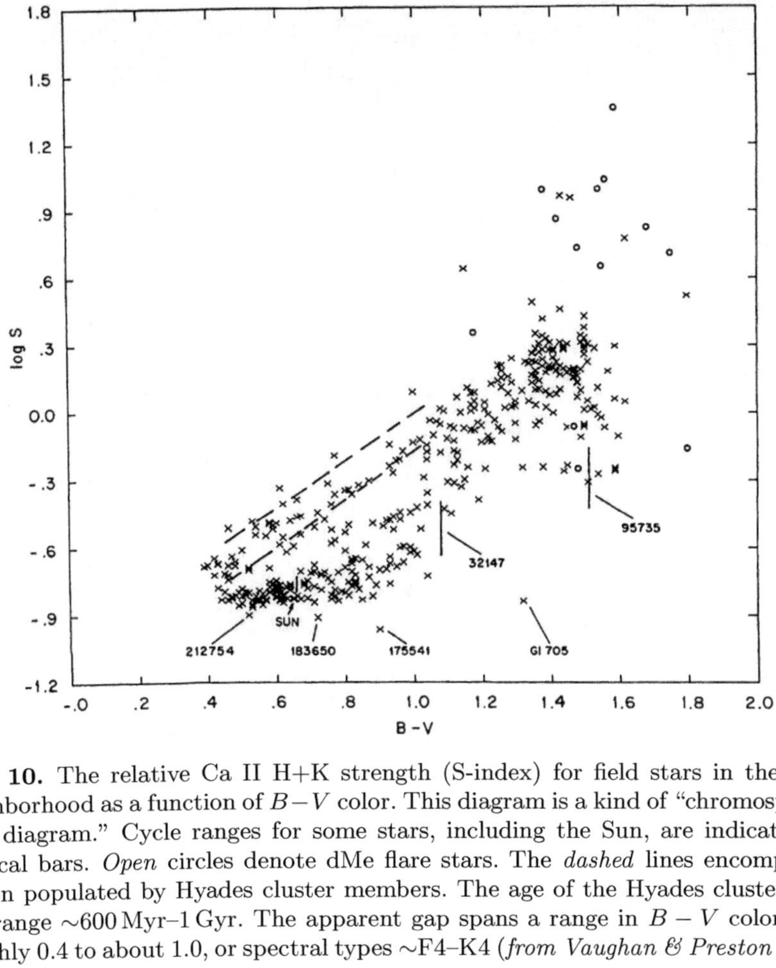

Fig. 10. The relative Ca II H+K strength (S-index) for field stars in the solar neighborhood as a function of $B-V$ color. This diagram is a kind of "chromospheric H-R diagram." Cycle ranges for some stars, including the Sun, are indicated by vertical bars. *Open* circles denote dMe flare stars. The *dashed* lines encompass a region populated by Hyades cluster members. The age of the Hyades cluster is in the range ~600 Myr–1 Gyr. The apparent gap spans a range in $B-V$ color from roughly 0.4 to about 1.0, or spectral types ~F4–K4 (*from Vaughan & Preston 1980*)

could give rise to the modulation of the strength of the Ca II emission at the near-equatorial rotation period of the star. Stimets & Giles (1980) first applied an autocorrelation analysis to Wilson's extensive, though sparsely sampled, data set to estimate rotation periods for 10 stars in the sample.

In the continuation of the Mt. Wilson program by Sallie Baliunas and her collaborators, intensive observing campaigns were conducted to achieve higher temporal resolutions and thereby obtain more accurate measurements of stellar rotation periods. An example of the results is shown in Fig. 12 for selected stars spanning a range of cycle amplitudes and periods. The corresponding, short-term rotational modulation of chromospheric emission is evident in some of the stars, as seen in the right panel of Fig. 12. This powerful, though observationally intensive, approach yields actual stellar rotation periods independent of ambiguities due to inclination effects that are inherent

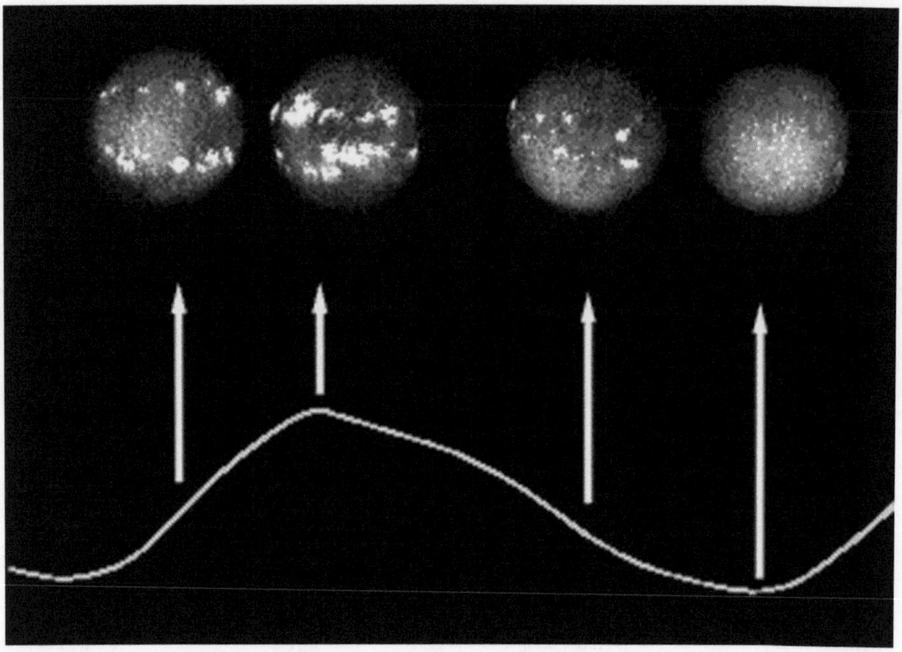

Fig. 11. A schematic illustration of how complexes of activity that are concentrated in longitude can lead to the rotational modulation of the observed Ca II H & K emission. This observationally intensive but powerful approach yields actual rotation periods on late-type stars

to spectroscopic techniques that yield projected rotational velocities. Thus, the empirical relationships between stellar rotation and magnetic field-related activity could now be determined more accurately.

I give one example of the relationship between activity and rotation in Fig. 13. Inspection of Fig. 13 reveals that activity, as measured by normalized chromospheric Ca II H & K emission, is high at short rotation periods and declines with increasing rotation period. Other diagnostics of magnetic activity, such as coronal X-ray emission, also exhibit a correlation with rotation in the sense that more rapidly rotating stars have higher levels of emission in chromospheric and coronal diagnostics than do more slowly rotating stars.

The stellar surface magnetic flux that ultimately leads to chromospheric and coronal emission arises from dynamo action. Dynamo efficiency depends on rotation rate and convective overturn time. The kinematic "$\alpha - \omega$ dynamos" define a class of dynamo models that are thought to reproduce some features of the solar magnetic field. In these kinematic dynamo models, the ω-effect is responsible for the production of toroidal flux arising from the interaction of poloidal field with azimuthal flows. The α-effect is the interaction of buoyant toroidal flux with cyclonic fluid motions in the convection zone that then leads to more poloidal flux. In this way, regenerative dynamo action

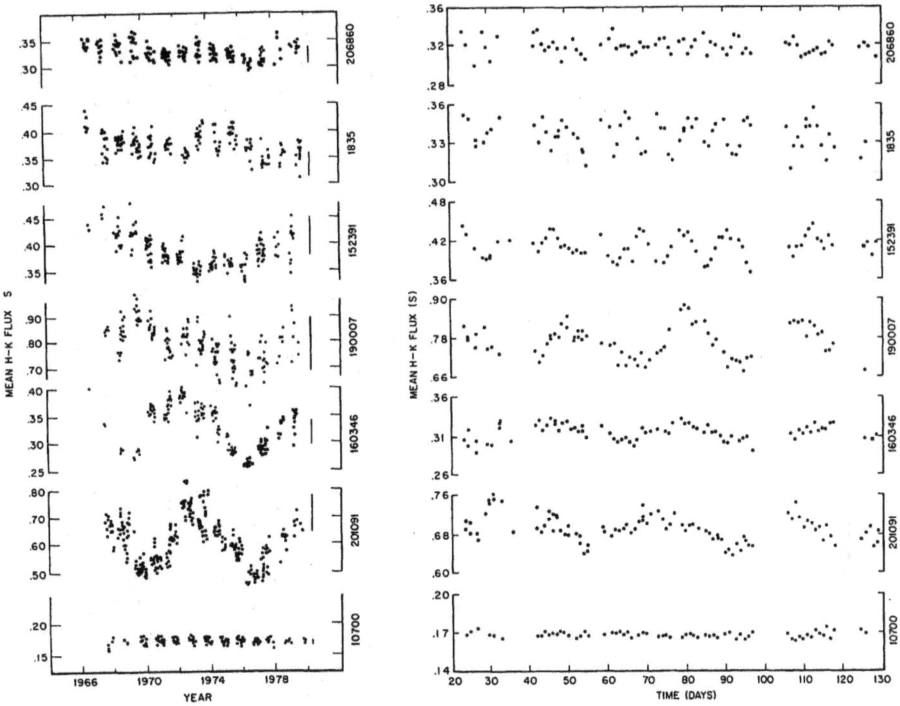

Fig. 12. Late-type stars with cycle periods (in years) that also exhibit rotational modulation of Ca II emission with periods measured in days (*from Vaughan et al. 1981*)

occurs. In stellar observations, the correlation between rotation and activity is highlighted because these are easily measurable quantities. However, it is differential rotation and, more specifically, the differential rotation profile of the star and the resulting production of buoyant magnetic flux that eventually emerges to the surface that are the key parameters in dynamo theory. However, neither differential rotation nor total magnetic flux are easily measured in the spatially unresolved spectrum of stars. The principal geometries for the kinematic $\alpha - \omega$ dynamos are illustrated schematically in Fig. 14 from Rosner (1980). The "shell" dynamo is regarded as the principal geometry that characterizes the solar dynamo. The "distributed", or turbulent, dynamo may also operate in the surface layers of stars with radiative cores. Of course, the distributive dynamo may be the principal mechanism operative in the interiors of wholly convective stars such as the late M dwarfs and the pre-main sequence T Tauri stars. However, this mean-field, kinematic approach is not yet fully developed for the circumstances of the solar magnetic field. In view of this, and the experimental difficulties in obtaining direct measurements in stars of the physical quantities that are most relevant to stellar and solar dynamo theory, current observational work emphasizes parameterizations

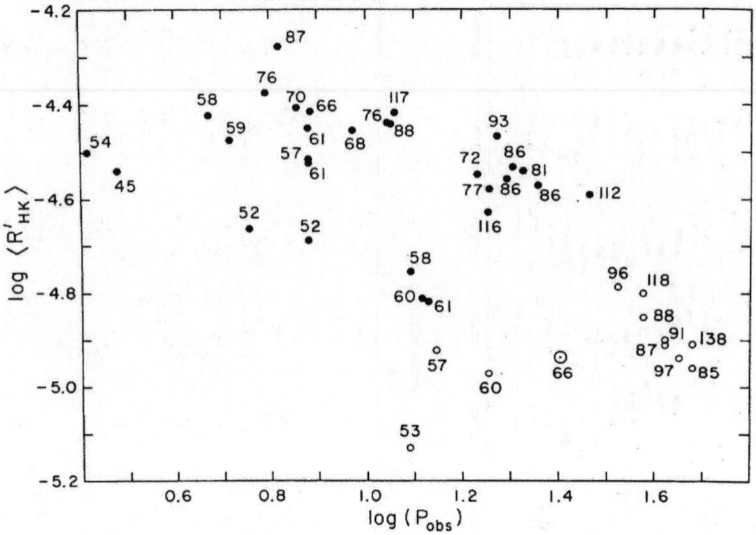

Fig. 13. The relationship between chromospheric Ca II emission and rotation in solar-type stars. The direction toward lower Ca II activity and longer rotation periods is also equivalent to increasing stellar age, in general (*from Noyes et al. 1984*)

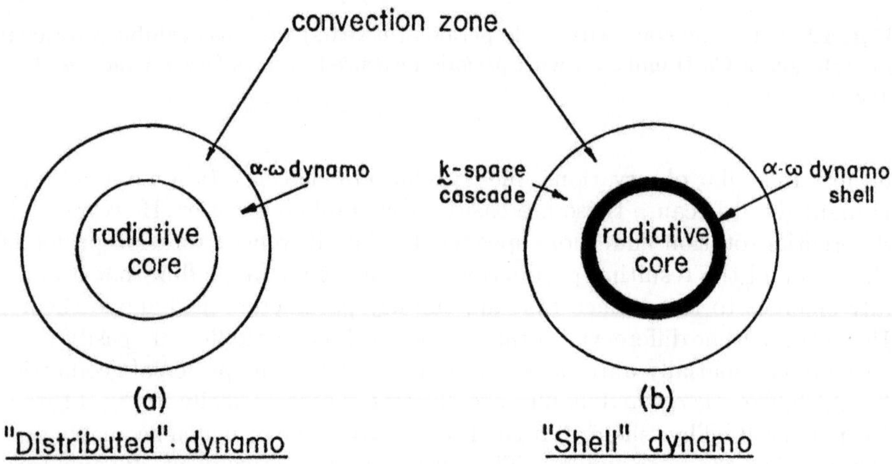

Fig. 14. A schematic representation of the two principal kinds of $\alpha - \omega$ dynamos. In the (**a**) distributed dynamo, the α and ω effects are co-spatial, occurring throughout the bulk of the convection zone. By contrast, the ω effect in the (**b**) shell dynamo is localized in the interface region between the base of the convection zone and the radiative interior (*from Rosner 1980*)

that seem to best characterize the data, i.e., minimizes the dispersion in the correlation of relevant quantities.

An example of this approach is given in Fig. 15 in which the correlation between normalized H & K chromospheric emission and the ratio of rotation period to convective turnover time, or Rossby number, is depicted for stars in the Mt. Wilson sample (Noyes et al. 1984). The Rossby number is a hydrodynamic parameter that incorporates two quantities relevant to dynamo theory, namely, rotation and convection zone depth. The convective turnover time is a model-dependent parameter. Two realizations of the correlation between activity and Rossby number are shown in Fig. 15 for two different estimates of the turnover time. The scatter in the empirical correlation is significantly reduced in Fig. 15(b) but only if an unusually large ratio of mixing length to scale height, which may not be physical, is assumed.

Another approach to relating cycle and rotation periods is given by Saar (2002) who compares cycle period with stellar rotation period (Fig. 16–upper panel). Saar separates the sample according to activity with inactive stars forming one branch and active stars forming another. In both cases, cycle periods decrease with increasing rotation rates. The relationship between the ratio of cycle frequency to rotation frequency versus inverse Rossby number is provided by Saar (2002) in the lower panel of Fig. 16 with the sample again segregated according to quiet, or solar-like, and active stars. A correlation appears to be present though not as strong as in the upper panel of Fig. 16. While the correlations in Fig. 16 would persist if the sample was considered in the aggregate, the scatter in the relations would, of course, substantially increase.

I give a depiction of the Vaughan-Preston Gap in Fig. 17, now in terms of normalized H & K emission flux versus $B - V$ color, taken from Noyes et al. (1984). The gap is even more vivid than in the original publication by Vaughan & Preston (1980) based on S-index. The results illustrated in Fig. 17 appear to provide corroborative evidence for a dynamo origin for the Vaughan-Preston Gap. However, it is important to note that the same observational selection effect arising from a possible discontinuity in the local star formation rate may nevertheless be present.

2.6 Future Directions

The course of future research in stellar dynamos will require the further development of solar and stellar dynamo theory combined with the development of sophisticated observational methods to directly measure the most physically relevant parameters. In the latter category, long-term measurements of the cycle variation of magnetic flux in late-type stars that exhibit Ca II cycles, combined with differential rotation measurements (e.g., Donahue et al. 1996), will lead to important advances in our understanding of stellar dynamos. The addition of long-term observations in other spectral diagnostics, particularly

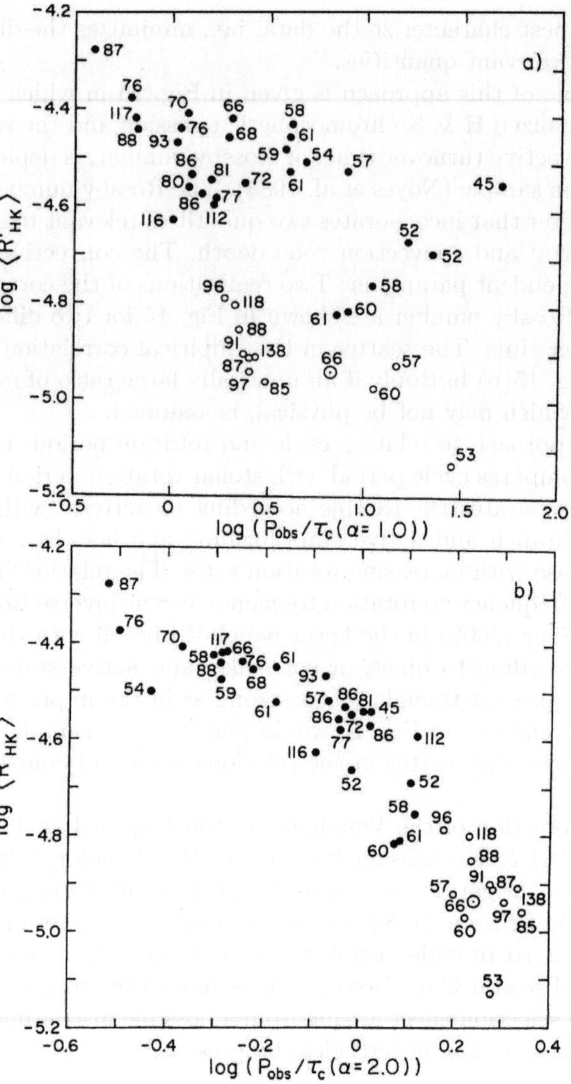

Fig. 15. Normalized chromospheric Ca II emission vs. Rossby number for convective turnover times corresponding to an interior model with a conventional ratio of mixing length to pressure scale height (*upper*) and an usually large value of this ratio (*lower*). The stars are labeled with their $B - V$ color. The Sun is indicated (*from Noyes et al. 1984*)

the X-ray, will enable the detection of cycle variations (if present) in especially quiet stars because of the high contrast of X-ray emission against the cool stellar photosphere.

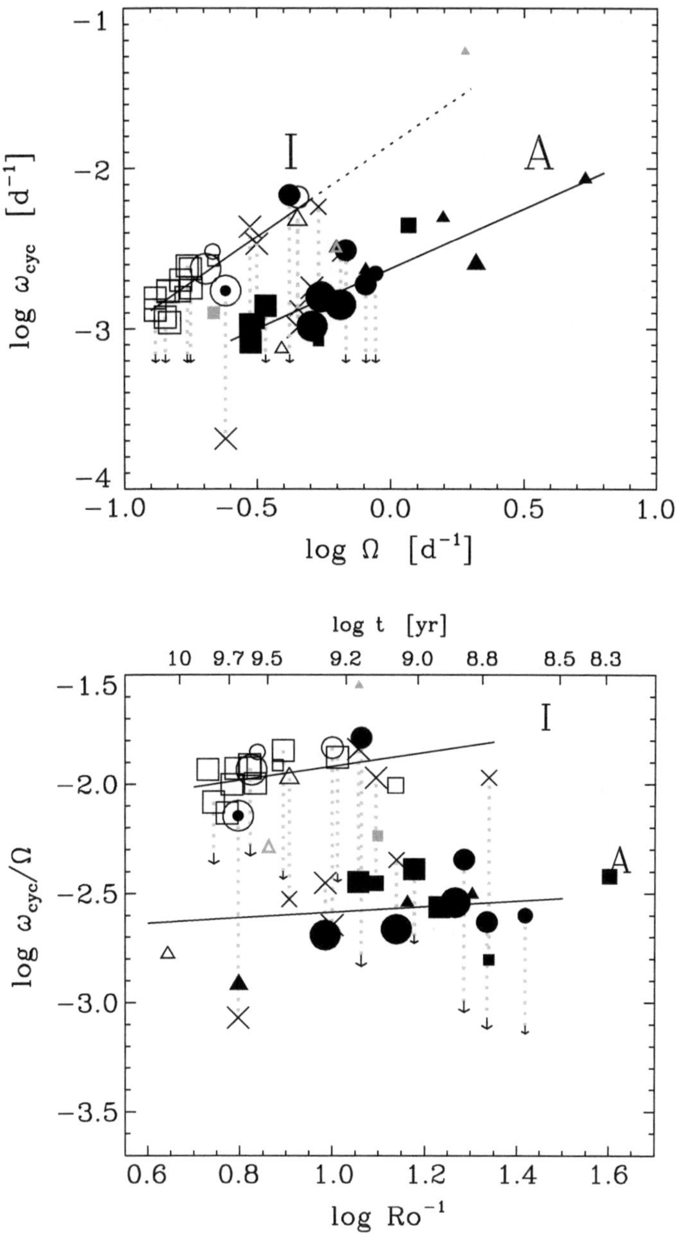

Fig. 16. Stellar cycle frequency vs rotation frequency (*upper*) and the ratio of cycle frequency to rotation frequency vs. inverse Rossby number (*lower*). The Sun is indicated. Two branches divided according to chromospherically inactive (I) and active (A) stars appear to yield the best fits (*from Saar 2002*)

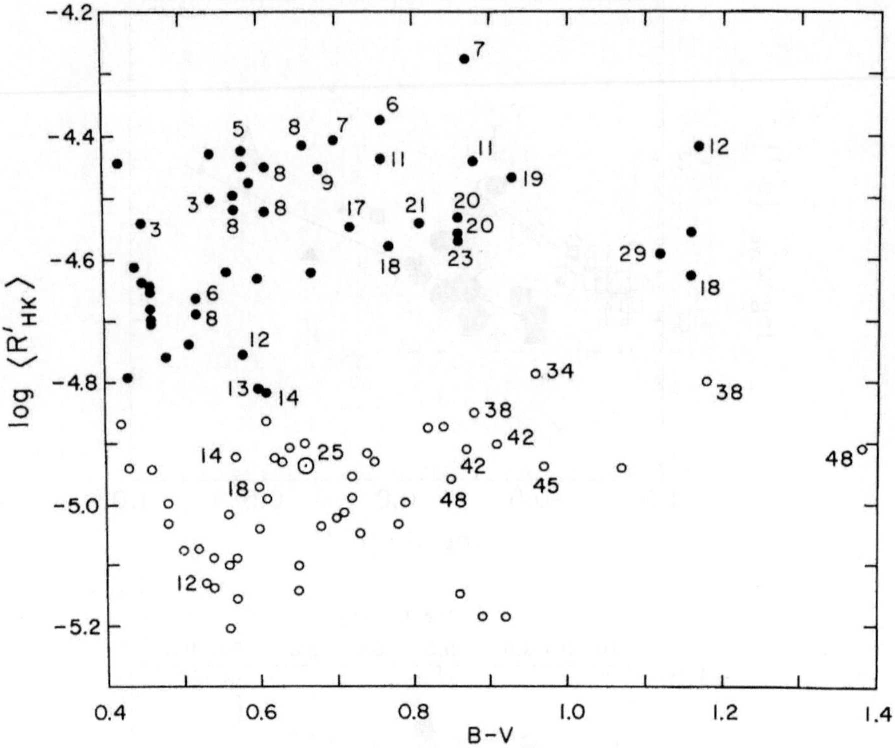

Fig. 17. The Vaughan-Preston gap but in terms of normalized chromospheric flux (instead of S) vs. color (*from Noyes et al. 1984*)

The identification and synoptic study of solar twins will reveal whether the Sun is unique among stars with otherwise extremely similar characteristics. The extension of Mt. Wilson-like programs to homogeneous stellar samples drawn from clusters will remove some of the ambiguities that are associated with heterogeneous samples of field stars. As concluded by Hall & Lockwood (2004) in their report on the results of a Workshop on the Future of Stellar Cycles Research, held at Lowell Observatory in Flagstaff, Arizona, during October 2003, each of these approaches is technically feasible today though dedicated facilities likely will be required.

3 Brightness Changes in Solar-Type Stars

3.1 Solar Irradiance Variability: A Recapitulation

A fundamental discovery of solar physics in the 20th century is that the Sun exhibits subtle but measurable changes in its bolometric irradiance on both short (~days–months) and long (~years–decades) time scales. The chapter by M. Lockwood includes an extensive review of solar irradiance variability and we only briefly recapitulate that discussion by way of introduction to the investigation of the variability of sun-like stars. In this regard, we note that solar *irradiance* variations are direction-dependent and measured in the plane of the ecliptic–not for the entire Sun. By contrast, the results of observations of stellar variability refer to the spatially integrated *brightness* changes of the star as seen in the direction of the line-of-sight to the earth, and at a full range of possible inclination angles for a given sample of stars.

The irradiance variability of the Sun is associated with surface magnetic structures, such as sunspot groups. An example of a particularly large complex of sunspots is given in Fig. 18.

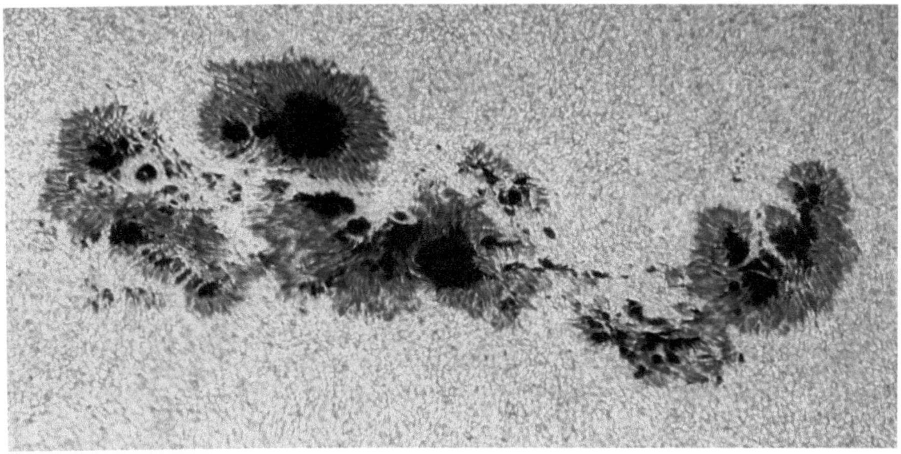

Fig. 18. A large spot complex observed with the McMath-Pierce Solar Telescope of the National Solar Observatory on Kitt Peak, Arizona, during April 2002

An example of the slight but real diminution in the solar irradiance that occurs during disk passage as the Sun rotates through the line-of-sight. Over the relatively longer time scale of a solar cycle, the Sun exhibits peak-to-peak brightness changes of ~0.1% in phase with the solar cycle. More specifically, the Sun is slightly brighter at solar maximum and slightly fainter at solar minimum. This behavior and its correlation with the solar cycle of magnetic field-related activity, as observed in the Ca II resonance features recorded for the Sun seen as a star, is depicted in Fig. 19 Note that the scatter in both panels is real and at a maximum at the peak of the 11-year solar activity cycle. The intrinsic short-term variations are minimized at solar minimum.

As in the case of cycles of magnetic activity, the detection of solar irradiance changes immediately raised the question of whether solar-type stars also exhibited low-amplitude variability analogous to that seen in the Sun. A series of investigations by G. Lockwood and R. Radick addressed this issue

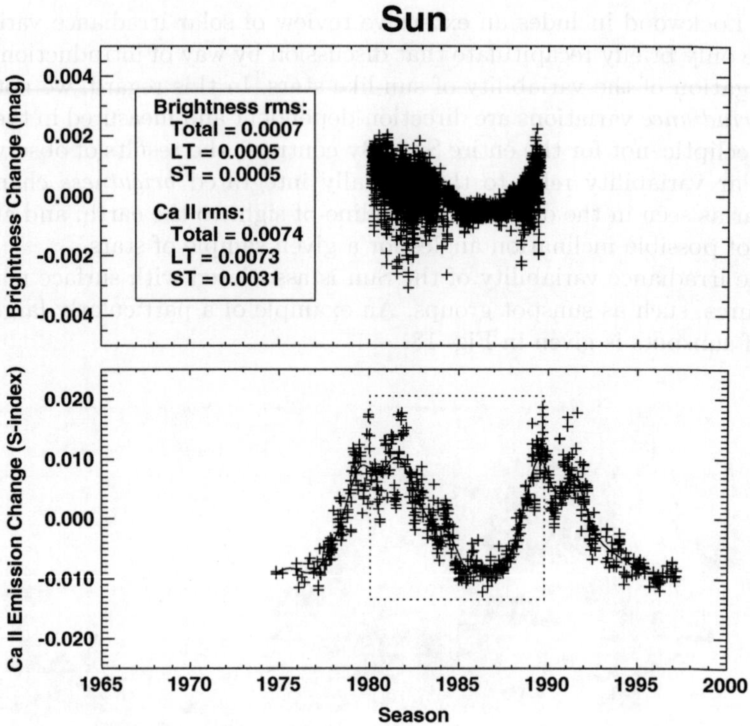

Fig. 19. The joint variation of total solar irradiance (*upper*) and Ca II emission (*lower*) during the solar cycle (*from Radick et al. 1998*)

through the application of extremely careful, high-precision photometric observations of selected samples of solar-type stars, ranging from field stars to members of young clusters.

3.2 Stellar Photometric Variability

Unlike the bolometric measurements of solar irradiance variability, the ground-based stellar observations are confined to visible wavelengths. In particular, Radick, Lockwood, and their collaborators utilized the Strömgren *ubvy* intermediate-width bands to record photometric variability as seen in selected solar-type stars. Whenever possible, these investigators sought to correlate and compare their photometric results with the results of Ca II monitoring carried out by the Mt. Wilson group for stars in common to both samples. Representative examples of the results of these dual, independent investigations are given in Figs. 20–21. The upper panels are the photometric brightness change in magnitudes and the lower panels are the change in the Ca II emission as expressed in the Mt. Wilson S-index. The statistics of the

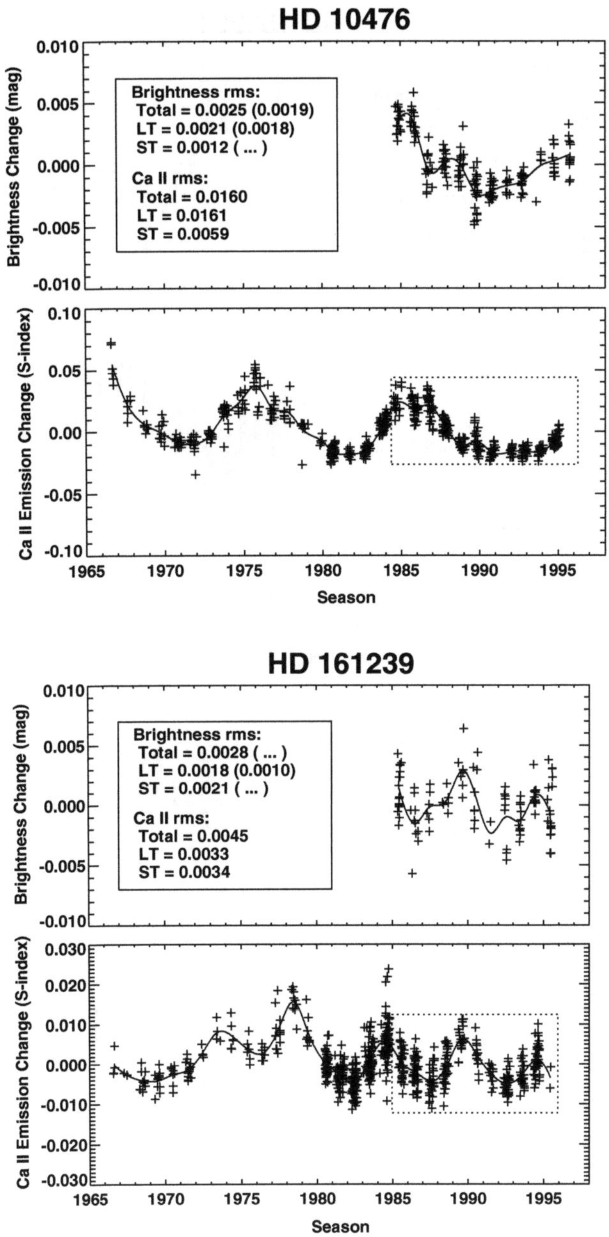

Fig. 20. The joint variation of stellar brightness and Ca II emission in solar-type stars that exhibit correlated, sun-like behavior between activity and brightness changes (*from Radick et al.* 1998)

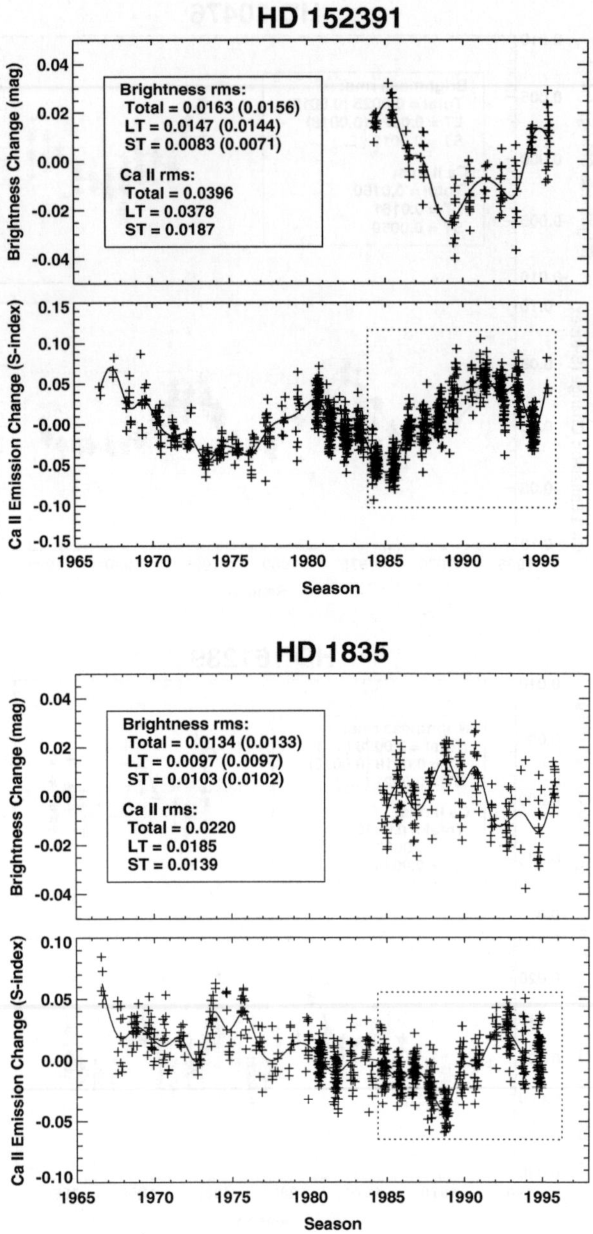

Fig. 21. The same as Fig. 20 but these active solar-type stars exhibit an inverse correlation between brightness changes and variability in chromospheric Ca II activity (*from Radick et al. 1998*)

variations in both brightness and Ca II are summarized in the inset in the upper panels.

Before we discuss the salient features of Figs. 20–21, we will summarize the relevant time scales for stellar variability. These include:

- The rotational modulation of active regions through the line-of-sight (~days)
- Active region evolution (~days to weeks)
- Active region lifetimes (~weeks to months)
- Activity cycles (~years)

These aspects of both short- and long-term variability are reflected in Fig. 21 in the form of intrinsic stellar scatter (short-term) superimposed on seasonal variations (long-term). In Fig. 20 (upper), the ~10-year, solar-like cycle of magnetic activity is clearly evident in the variation of S-index. In parallel with the Ca II variability is a photometric variation with a smoothed seasonal variation that is correlated with the Ca II index. In particular, the mean brightness of the star changes in phase with the long-term, chromospheric variability. This basic pattern of variation is repeated in the sun-like star in Fig. 20 (lower). In particular, as activity increases and declines, the brightness of the star increases and decreases. The amplitude of the long-term stellar brightness changes is roughly in the range of 0.15%–0.2%, or higher than that seen in the contemporary Sun. In these examples, the mean level of Ca II activity is similar to that of the mean Sun.

In contrast to the examples in Fig. 20, the more active stars in Fig. 21 exhibit higher rms scatter in both their Ca II and photometric variability. Significantly, the observed brightness changes are anticorrelated with Ca II activity. That is, as activity increases, the star dims slightly. As activity declines, the star becomes slightly brighter in the photometric bands. This result is encapsulated in Fig. 22, which shows mean relative chromospheric emission ratio R'_{HK} versus $B-V$ color (equivalent to stellar effective temperature). The filled circles are solar-type stars that exhibit an anticorrelation between long-term brightness changes and activity while the open circles represent stars that are solar-like, i.e., they exhibit correlated changes between long-term brightness and activity. The significance of the correlation (expressed as a false-alarm probability) is indicated.

Inspection of Fig. 22 reveals that stars with relatively higher levels of mean chromospheric emission are characterized by anticorrelated changes between photometric brightness and activity. Conversely, stars with more solar-like levels of mean chromospheric activity exhibit changes in mean seasonal brightness that are positively correlated with long-term variations in activity. The placement of the Hyades isochrone, corresponding to an age of ~600 Myr–1 Gyr, reflects the general correlation of chromospheric activity with age: young stars are generally characterized by higher mean levels of chromospheric and coronal emission than are solar-age (and older) stars.

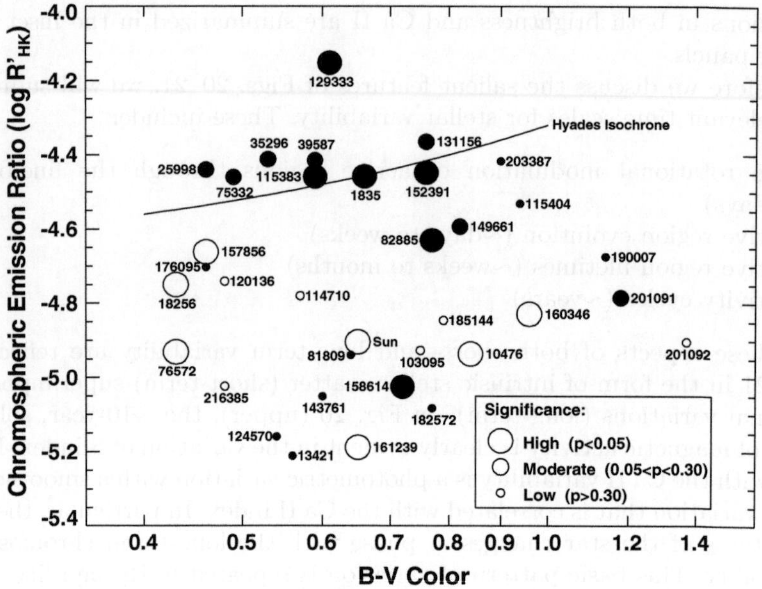

Fig. 22. The correlation between long-term changes in brightness and chromospheric Ca II activity in solar-type stars. The *filled* symbols, associated with active solar-type stars, represent anticorrelated behavior while the *open* symbols denote correlated behavior, similar to that of the contemporary Sun (*from Radick et al. 1998*)

The various aspects of the correlation between activity and stellar age will be discussed in Sect. 4 on Solar Analogs. Recalling the discussion of the previous section on Stellar Cycles (§2), we can infer that this diagram also represents a correlation with total surface magnetic flux in the sense that as the total amount of magnetic flux on the stellar surface increases, chromospheric emission, or R'_{HK} increases. Thus, the qualitative change in behavior between correlated and anticorrelated variations of brightness and activity, respectively, is quantitatively related to the total stellar magnetic surface flux and its evolution from precursors of solar analogs to stars that are more nearly like the Sun in their mean activity level.

In summary to this point, luminosity variations of active stars are anticorrelated with chromospheric variations on activity cycle time scales. By contrast, less active, sun-like stars are characterized by a positive correlation between brightness changes and chromospheric variations during a magnetic cycle. Close inspection of the temporal series of brightness and activity reveal that, on short time-scales (e.g., rotational time scales), luminosity variations and chromospheric variability are nearly always anti-correlated, as is the case for the Sun.

Amplitude of Stellar Brightness Variability

I have discussed the nature of the correlation between stellar photometric variability and chromospheric variability. I will now briefly examine the amplitude of stellar brightness changes as a function of activity. The mean amplitude of the long-term brightness variation as a function of normalized chromospheric emission strength is given in Fig. 23.

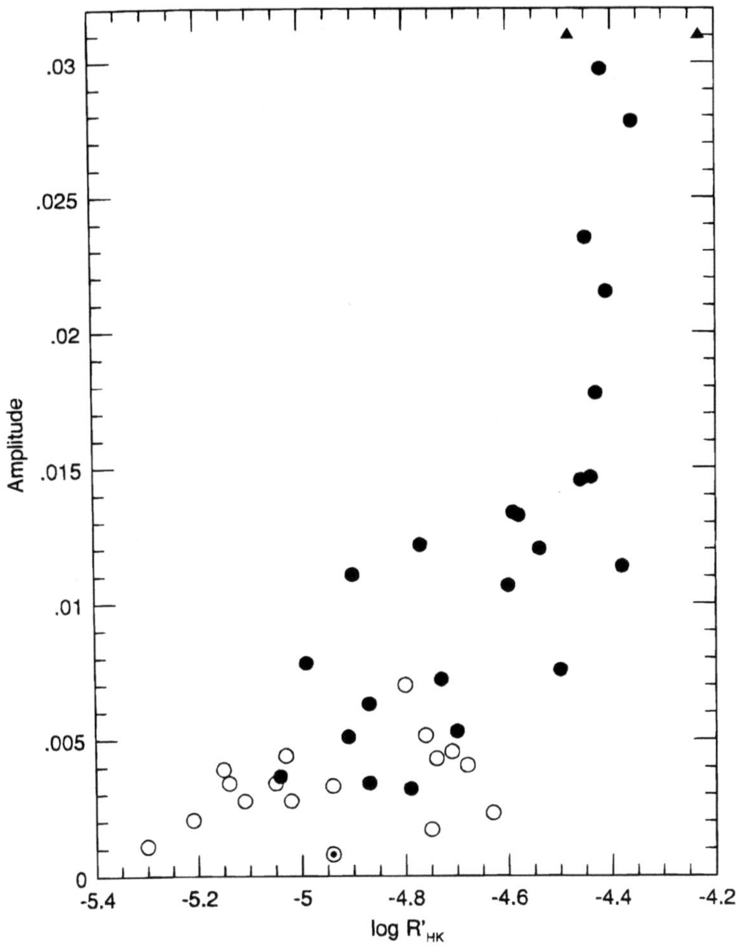

Fig. 23. The mean amplitude of brightness changes in solar-type stars as a function of mean normalized chromospheric Ca II H + K emission. The Sun is indicated. *Open* circles represent stars that appear constant to within the photometric errors. *Filled* circles are stars that are variable in their brightness (*from Lockwood et al. 1997*)

Two points are immediately clear from Fig. 23. First, the amplitude of brightness changes increases with increasing activity (i.e., R'_{HK} value). Second, the amplitude of brightness changes in this sample of solar-type stars can be considerably greater than the 0.1% variation seen in the Sun, by factors of ~10–30! Indeed, one can only imagine the profound effect such changes in brightness could have on the planetary systems, their atmospheres and climates, that may exist in this sample of stars! Another aspect of this diagram is the relative position of the Sun as indicated. In particular, the amplitude of its brightness variation appears low with respect to other stars with similar mean levels of chromospheric activity. The short-term rms variability of solar-type stars shows a similar dependence on mean chromospheric activity (Fig. 24).

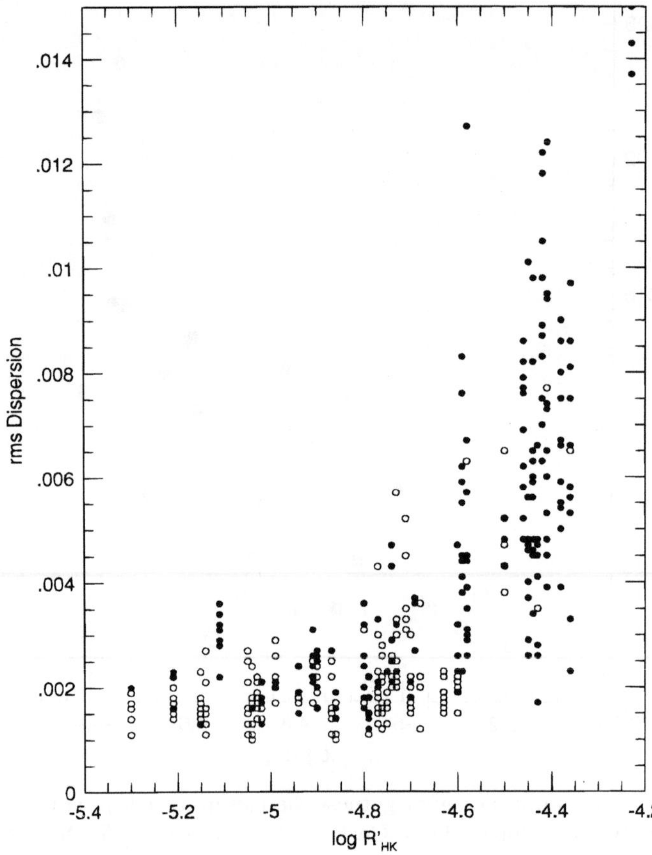

Fig. 24. Short-term rms fluctuations in brightness in solar-type stars as a function of mean normalized chromospheric Ca II H + K emission. Each program star occupies a column. *Filled* or *open* circles represent times when a star was determined to be variable or constant, respectively (*from Lockwood et al. 1997*)

In brief summary, solar-type stars with relatively higher mean chromospheric emission fraction have larger photometric variations on both short and long (i.e., cyclic) time scales. Stars with solar-like levels of chromospheric emission show low or even no photometric variation.

3.3 Nature of Correlated Variability in Brightness and Activity

The stellar observations that show a change, or evolution, from anticorrelated to correlated brightness changes with activity changes likely reflect a shift from "spot-dominated" to "faculae-dominated" irradiance variability with decreasing mean chromospheric activity level. This shift is equivalent to a shift with both stellar age and decreasing total surface (unsigned) magnetic flux.

Corroborative evidence for this hypothesis can be seen in the solar cycle variation of the ratio of the areas of white light faculae to the areas of spots, as pointed out by Foukal (1994). Inspection of Fig. 25 (from Foukal 1994) reveals that at the highest activity levels during the solar cycle, the ratio of the facular area to spot area on the disk begins to turn over. Thus, there is reduced cancellation of the decline in irradiance due to spot blockage by an increase in irradiance due to faculae. This effect is seen in active solar-type stars, giving rise to the observed anticorrelation between activity and brightness changes for stars characterized by a relatively higher mean level of chromospheric emission.

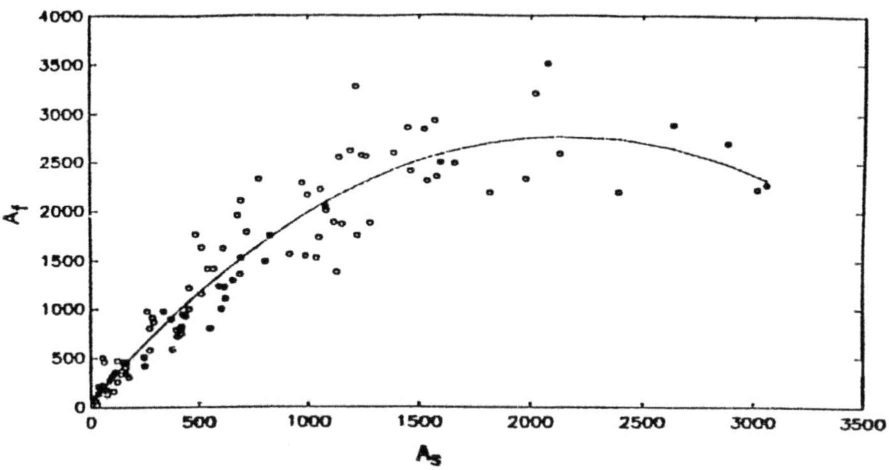

Fig. 25. Facular area as a function of spot area during the solar cycle (*from Foukal 1994*)

3.4 Solar Variability in a Stellar Context

I will now intercompare solar irradiance and chromospheric variations with that seen in solar-type stars. Figure 26 shows a comparison of both the long-term (cyclic) and short-term photometric variations of the Sun in comparison with solar-type stars as a function of mean relative chromospheric activity.

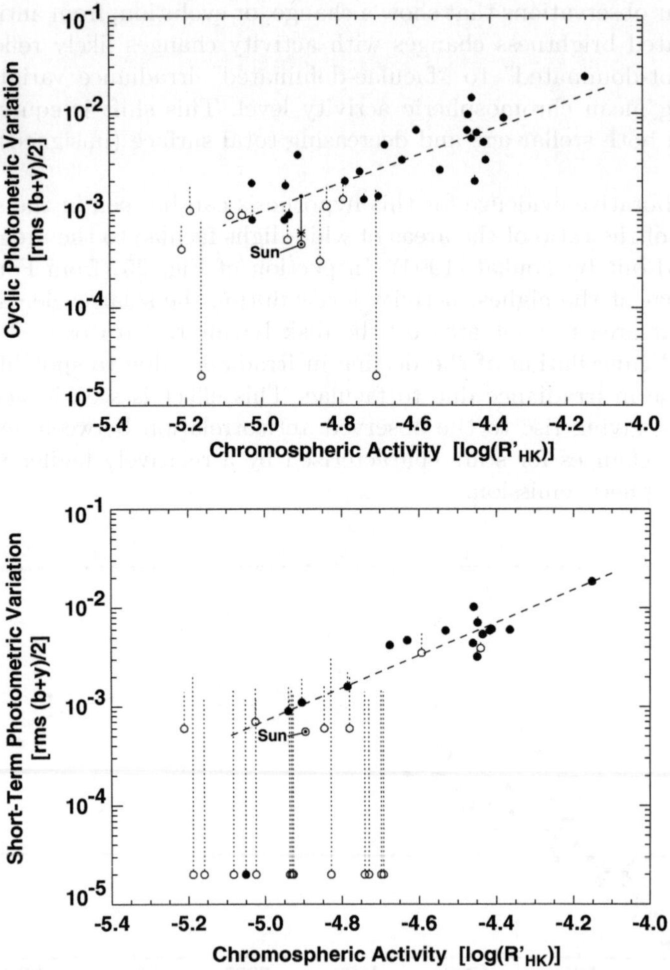

Fig. 26. Long-term, cyclic rms variations in brightness (*upper panel*) and short-term rms fluctuations in brightness (*lower panel*) in solar-type stars, each as a function of mean normalized chromospheric Ca II H + K emission. *Filled* symbols are program stars while *open* symbols are comparison stars. *Drop* lines indicate the magnitude of the small correction for any low-level, comparison star variations (*from Radick et al. 1998*)

In both cases, the brightness changes inferred for the Sun as a star appear somewhat subdued for its mean level of chromospheric activity when compared to other solar-type stars. The Sun is not a statistical outlier in these graphs but its photometric variability does appear systematically lower than that of other solar-type stars for both short-term and cyclic time scales of variability.

In the upper panel, the potential range of corrections to the stellar photometric counterpart of solar irradiance variability due to inclination is indicated by Radick et al. (1998). However, an extreme correction would be required to place the Sun on the regression line. The most likely correction is indicated in the diagram, which still places the Sun at a relatively lower amplitude of cyclic brightness change as a function of chromospheric activity.

In both panels, we may ask whether $\sim$a factor of 3 increase in the amplitude of brightness change, as suggested by a comparison of the Sun with ostensibly similar stars, could occur and thereby yield a direct and dominant impact on global climate change.

A comparison of the cyclic and short-term chromospheric variations, as a function of mean chromospheric activity level, is given in Fig. 27. In both panels, the Sun appears within the scatter of the regression curve. That is, the amplitude of both chromospheric and short-term variations of the Sun appear similar to that of solar-type stars. I only comment that from an experimental perspective, the disk-integrated solar Ca II emission is readily observable with current ground-based instrumentation, sometimes including the same spectrographs utilized for stellar observations. By contrast, it is more difficult to compare stellar photometric observations with the bolometric irradiance measurements of the Sun. Nevertheless, it has not been concluded that the relative differences between the Sun and similar stars in terms of their comparative relative changes in brightness for comparable levels of chromospheric activity is an artifact of differences in measurement technique. For now, it is considered to be a real difference in behavior between the Sun and similar stars.

Finally, we show in Fig. 28 slope of the regression of photometric brightness variation and Ca II emission change, as a function of mean activity level. The Sun appears near the dashed dividing line between direct and inverse correlation in brightness changes with chromospheric activity changes, primarily as a result of its relatively lower amplitude of brightness change for its activity level. It is interesting that the Sun is perilously close to the dividing line between inactive and active stars. In fact, Fig. 28 suggests that if the mean activity of the Sun were to increase by roughly 50% in its S-index then the Sun would be included among the active solar-type stars with their inverse correlation between activity and irradiance. Foukal (1998) states that the decrease in the ratio of facular to spot areas with increasing activity level explains the increase in photometric variability in stars beginning at values only 50% larger than the present-day Sun.

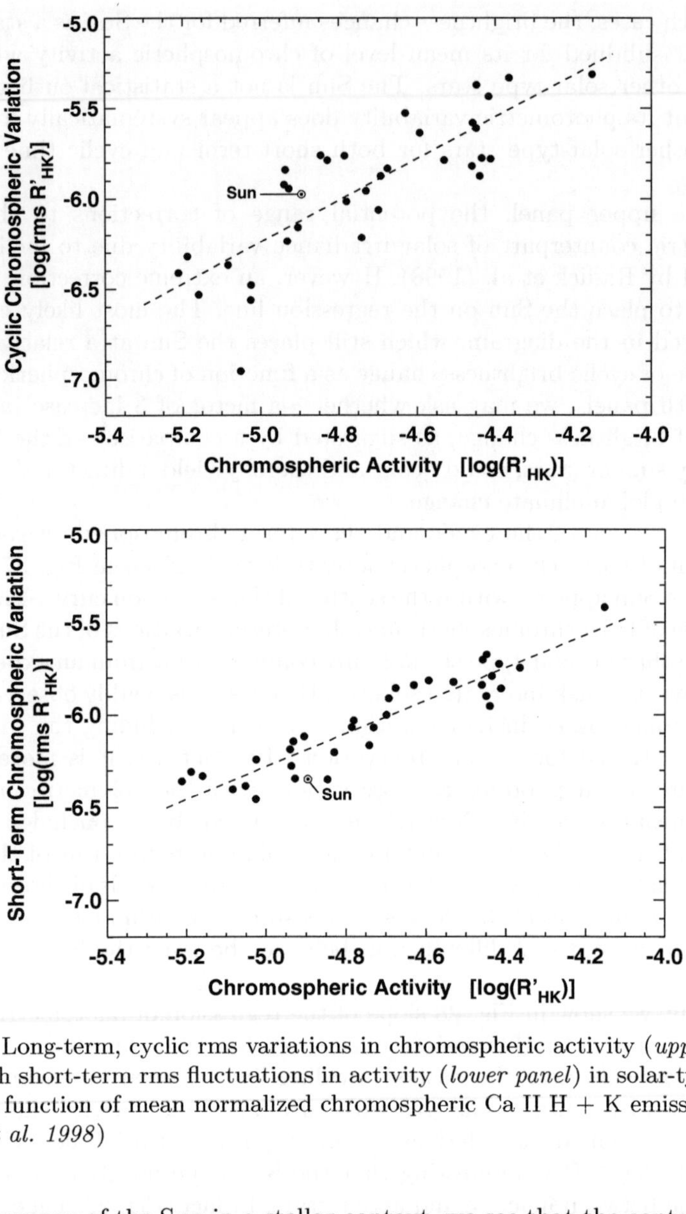

Fig. 27. Long-term, cyclic rms variations in chromospheric activity (*upper panel*) along with short-term rms fluctuations in activity (*lower panel*) in solar-type stars, each as a function of mean normalized chromospheric Ca II H + K emission (*from Radick et al. 1998*)

In summary of the Sun in a stellar context, we see that the contemporary Sun exhibits a relatively high-amplitude, well-behaved chromospheric cycle variation. The present-day Sun's total irradiance variability is somewhat subdued compared to stars of similar activity level. However, the Sun is close to the dividing line between inactive and active solar-type stars in terms of its joint variation of activity and irradiance.

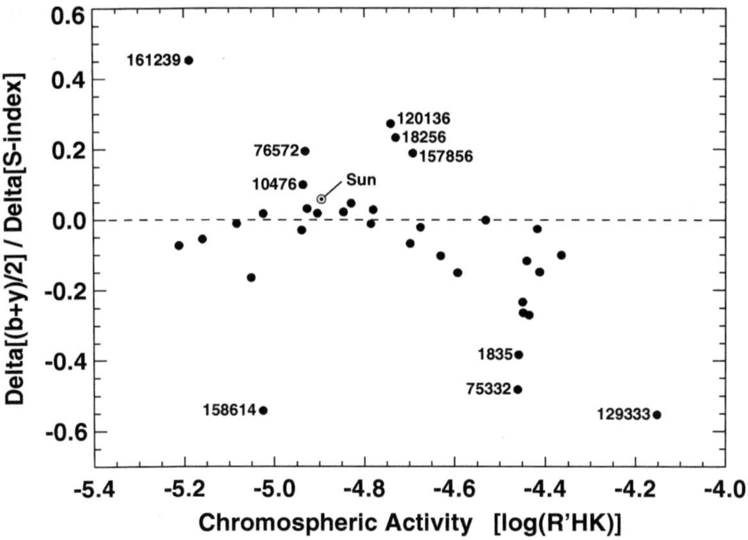

Fig. 28. Slope of the regression of photometric brightness changes per change in activity (S-index) for quiet and active solar-type stars as a function of mean normalized chromospheric Ca II H + K emission (*from Radick et al. 1998*). Stars with correlated changes in brightness and activity are above the *dashed* line while stars with anticorrelated changes are below the *dashed* line. Note that the Sun is close to the cross-over between quiet and active stars

3.5 Questions

The intercomparison of the Sun and solar-type stars in both their respective photometric and chromospheric variability raises interesting questions. Could artifacts of the different measurement techniques or properties of the stellar sample contribute significantly to apparent discrepancies between the variability of the Sun as a star and other solar-like stars? For example, inclination effects combined with the concentration of activity toward mid-latitudes in the Sun and solar-type stars could introduce geometrical factors into an intercomparison. However, the detailed study by Knaack et al. (2001) concluded from their simulations that inclination effects are not a significant factor in the solar and stellar comparisons.

The more pertinent question is whether the stellar sample is an appropriate comparison to the Sun. That is, are the stars truly counterparts of the Sun or are their characteristics, such as rotation rate, interior structure, age and metallicity, sufficiently different from those of the Sun that, in the context of dynamo theory, magnetic field generation and emergence, and the associated irradiance variability, the comparison is not appropriate. This issue is not yet resolved though we will discuss recent developments in this area in the next section on solar analogs.

In the meantime, provocative questions have been raised concerning solar and stellar irradiance variability. For example, Foukal (2003) asks not why the Sun exhibits subtle irradiance variations but rather why is the Sun so photometrically quiet? Foukal (2003) states that larger brightness variations would be expected simply from stochastic variations in the surface brightness of the millions of granules in the solar photosphere. However, Foukal (2003) argues that the enormous thermal inertia of the solar convection zone essentially suppresses any potential variations in the granule-to-granule irradiance via diffusion throughout the bulk of the convection zone. Nevertheless, it is intriguing to ask if there are examples of luminosity variations in solar-type stars that are *not* associated with changes in magnetic activity (Foukal 2003, private communication). The identification of such a phenomenon in stars would imply the occurrence of actual structural changes that give rise to brightness changes that, presumably, are uncorrelated with variations in chromospheric activity.

3.6 Future Directions

Future research directions should include programs of photometric monitoring in samples of solar-type stars that are homogeneous in age and chemical composition. Wide-field observations of clusters provide the opportunity for investigating the evolution of the joint variation of activity and irradiance in homogeneous samples of stars.

In parallel with this program, synoptic studies of brightness variability in strict solar analogs and "solar twins" should be conducted. Clearly, the prelude to such a long-term program is the identification of strict solar counterparts. The results based on observations of truly sun-like stars will provide the most reliable comparison between solar and stellar variability.

Finally, space-based bolometric measurements of stellar luminosity along with parallel, synoptic observations of activity, will yield the highest precision measurement of stellar brightness changes that can be directly compared to solar bolometric irradiance variations. Nevertheless, ground-based, high-precision differential and ensemble photometry will still serve as an excellent proxy for space-based bolometric measurements.

4 Activity in Solar Analogs

The study of the nature of activity and associated irradiance variability in solar analogs provides a unique and invaluable adjunct to the solar database. Appropriately selected samples of solar-type stars effectively extend the range in parameter space that is simply unavailable with observations of the Sun alone. For example, a snapshot of the distribution of chromospheric Ca II H & K emission strength in a sample of solar-type stars can immediately reveal the potential range of amplitudes of the solar activity cycle, assuming that the range in stellar Ca II strengths arises from cycle variations, analogous to the solar cycle. I will examine the results of a few different approaches and their implications for understanding the full range of solar variability as inferred from observations of solar-type stars. I will conclude the main part of this section with a brief review of the identification of candidate solar analogs.

4.1 Distribution of Activity in Solar-Type Stars: Baliunas & Jastrow (1990)

An especially notable study of this kind was published by Baliunas & Jastrow (1990; hereafter BJ). These investigators utilized data from the Mt. Wilson survey comprised of field solar-type stars. In particular, the BJ sample included 74 "solar-type" stars with $B - V$ colors in the range of 0.60–0.76, corresponding to a range of 0.95 to 1.10 solar masses. The stars are claimed to have ages similar to the Sun, according to their age-determination criteria. Of the 74 stars, 13 were observed monthly while the remaining 61 were observed sporadically.

Among the 13 stars with monthly HK observations, 4 were seen to be "magnetically flat", i.e., the time-series of S-index observations were essentially constant. The remaining 9 stars exhibited cycle-like variations in their S-index values, qualitatively similar to the solar cycle. The 61 stars that were observed only sporadically were included by BJ to augment their sample. The time series of S values for each star was then sampled at 0.1 year intervals to generate the histogram replicated here (Fig. 29).

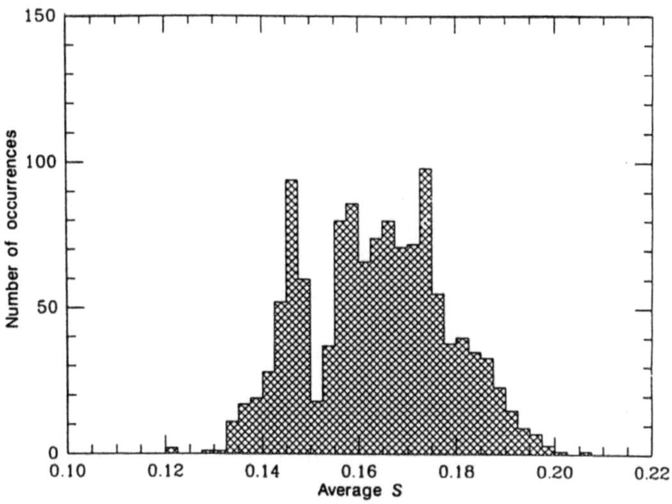

Fig. 29. Distribution of the number of occurrences of an S-index value (*abscissa*) for a sample of solar-type stars from the Mt. Wilson program (*from Baliunas & Jastrow 1990*)

The significant features of BJ's distribution of activity in solar-type and solar-age stars include a range in average S index that is broader than the corresponding range for the contemporary solar cycle. Furthermore, the distribution of activity is apparently *bimodal*, with about 1/3 of the "occurrences" of S-index values in what is identified by BJ to be a "Maunder-minimum" state of activity. The bimodal nature of the distribution seemed to confirm that the "flat-activity" stars, which were also characterized by anomalously low S-index values as compared to other solar-type stars, truly represented a distinct state of the stellar (and, by implication, the solar) dynamo in a mode that produced very little magnetic flux. Moreover, this mode occurred about one-third of the time since the low S-index mode of the bimodal distribution in Fig. 29 represents about one-third of the total number of occurrences.

The results have profound implications for our understanding of stellar/solar dynamo theory, the potential excursion in amplitude of the solar cycle of activity, and the associated range of irradiance variability that the Sun could exhibit in the future and might have occurred in the past. The bimodal distribution is qualitatively consistent with a model for the interior dynamo in the Sun and sun-like stars that "switches" between two modes. This is, in turn, reminiscent of the Vaughan-Preston Gap, suggesting that the origin of the gap is in the nature of the dynamo. Furthermore, the results imply that the Sun and solar-type stars are in a Maunder-minimum state of prolonged quiescence about 1/3 of the time.

4.2 Implications for Solar Irradiance Variability

In order to understand how the results of the stellar data presented by BJ are used to derive the possible amplitudes of solar irradiance changes, we briefly summarize the study by Lean et al. (1992). The key figure in their work is included here (Fig. 30). As described by Lean et al. (1992), the ordinate is a model of the solar irradiance that represents residual brightening in the Sun after the sunspot blocking effect and the quiet Sun are removed from the ACRIM measurements of solar irradiance variability. The estimate of the solar irradiance during the Maunder Minimum is then inferred from the Ca II HK strength reported by BJ for their non-cycling, "Maunder-minimum stars". The estimated change in solar irradiance during the Maunder Minimum deduced by Lean et al. (1992) using this approach is summarized in Table 1.

Table 1. Estimated Solar Irradiance Changes During the Maunder Minimum

Case	HK (Å)	Irradiance Change
Non-cycling stars (average)	0.145	−0.24%
Minimum possible Sun	0.130	−0.35%

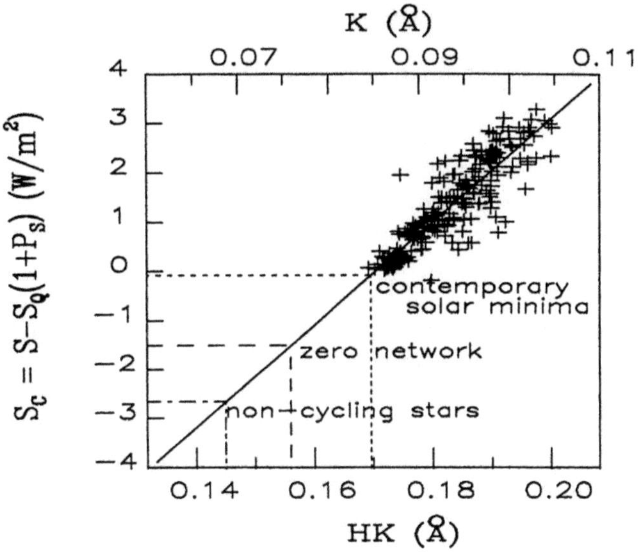

Fig. 30. A model of the solar irradiance variability as a function of Ca II HK emission (*from Lean et al. 1992*)

As inspection of Table 1 shows, Lean et al. (1992) find from BJ's results for solar-type stars that the Sun may have undergone a significant change in its irradiance that is 2.4 times the contemporary value of 0.1%. BJ cite estimates that a 0.22–0.55% change in solar irradiance could account for the estimated 0.4–0.6°C global mean temperature change in the Little Ice Age. An even larger change in the solar irradiance of –0.35% could have occurred if the solar brightness change during the Maunder Minimum corresponded to the minimum value of activity observed among the flat-activity stars in the BJ distribution (assuming that the model of residual irradiance changes in Fig. 30 is applicable).

Following this initial estimate of the solar irradiance change, Lean et al. (1995) then performed a reconstruction of the solar irradiance based on the sunspot number record modulated by a slowly varying background. The long-term component of this variation is inferred from stellar observations of the kind presented and interpreted by Baliunas & Jastrow. Utilizing this approach, Lean et al. (1995) deduce a −0.24% change in the solar irradiance during the Maunder Minimum and a −1.42% change in the solar UV (200–300 nm pass band) irradiance alone. Clearly, the stellar results yielded a critical quantitative guide in the development of the long-term record of the solar irradiance.

4.3 Baliunas & Jastrow (1990) Revisited

In view of the important inferences concerning solar variability in activity and irradiance based on observations of the solar-type stars in the BJ study, the underlying assumptions and methods of data analysis utilized for their investigation merit reexamination.

A critical question to address in Fig. 29 is whether the approach based on "number of occurrences" that is utilized to generate the histogram in Fig. 29 is valid. Recall that the Ca II light curves in the form of S-index values were sampled at 0.1 year intervals for each star. This method would seem to overweight the data for the 13 stars, including the 4 flat or "non-cycling" stars, and underweight the data for the 61 remaining stars that were sampled only sparsely. As was remarked during the lectures with respect to the sampling of the constant Ca II curve for the flat stars, this is like asking the same question of a single individual in an opinion poll ten times, getting the same answer ten times, and then recording these ten identical answers as ten independent data points! I also note that the source of data is non-uniform in the sense that about 20% (13 stars) of the sample was observed regularly while observations of the remaining 80% of the sample were obtained on an infrequent, irregular basis. It would be interesting to know simply the distribution of the *mean* S-index values for the stars in the BJ sample to see if it too is bimodal, and if the fraction of stars at unusually low values of chromospheric emission is also one-third. Unfortunately, these data are not given in the published paper.

In addition to the question of the data generation method utilized, an even more crucial question is whether the stars in the BJ sample are truly solar analogs. BJ adopt activity and projected rotational velocity (v sin$i$) as diagnostics of stellar age. As you recall from the previous sections, age and activity are correlated in the sense that activity decreases with stellar age via rotational spindown combined with the direct correlation between rotation and activity (the rotation itself declines as a result of magnetic braking). But the reasoning by BJ is circular since by these criteria, stars with S-index values that are lower or higher than the range of S-index for the contemporary solar cycle must therefore be older or younger, respectively, than the Sun and thus cannot be solar counterparts. What is the evolutionary status of the stars in the BJ sample?

This is a difficult question to answer for an inhomogeneous sample of field stars. In stellar astronomy, activity and rotation are two parameters among several others that are considered in age determinations for individual stars. Other characteristics that are considered include metallicity (i.e., the relative abundances of elements heavier than H and He) and kinematic properties. Spectroscopic estimates of gravity (luminosity) and effective temperature that enable placement in an H-R diagram are especially valuable in determining if a star can be considered a solar analog candidate. Parallax measurements, both ground-based and from *Hipparchos* data, can yield more directly the stellar luminosity when combined with accurate measures of apparent brightness. The combination of bolometric magnitude and effective temperature yields the stellar radius, which provides an important clue to the evolutionary status of an individual star. It is very possible that the flat-activity, "Maunder-minimum" stars in the BJ sample are slightly evolved sub-giants. Their low activity is a consequence of their older age (and slower rotation) relative to the Sun. If this is indeed the case then the objects that are the source of BJ's bimodal distribution are not really representative of the Sun.

4.4 Recent Results From Samples of Solar Analogs

The results from other independent studies of solar-type stars merit review and comparison with the BJ results. M. Giampapa has been leading an observational study of the solar-type stars in the solar-age and solar-metallicity open cluster, M67. The primary objective of this program is to gain insight on the possible range of *solar* chromospheric activity and the associated, potential long-term variability of the Sun through the observation of stellar analogs of the Sun. The open cluster M67 itself is an especially appropriate target of observation since it is approximately the same age (about 5 Gyr $\pm$ 1 Gyr; Demarque et al. 1992) and of the same metallicity as the Sun (Barry & Cromwell 1974).

The solar-type stars in M67, at a distance of approximately 870 pc (Montgomery et al. 1993), are relatively faint with V-magnitudes in the range of V = 14–15. Of course, they are even fainter in the deep cores of the Ca II H & K resonance features in the near UV at $\sim$400 nm. Thus, a moderate-aperture telescope is required in order to obtain spectroscopic observations of the H & K lines of acceptable quality. The advent of the 3.5-m WIYN telescope in conjunction with the Hydra multi-fiber positioner to perform multi-object spectroscopy over a 1 degree field made this project feasible. The multi-object spectrograph made it practically feasible to obtain Ca II spectra for a relatively large number of faint stars simultaneously. A conventional spectrograph would not be practical, especially in view of integration times of at least 5 hours (in very good sky conditions) that are required to obtain spectra of very good quality.

The calibration of the Hydra spectra to residual intensity and flux follows the relations given by Hall & Lockwood (1998) and Hall (1996), as functions of $(B - V)$ color. The H+K index ("HK-index") values in 1 Å bandpasses centered at the H and K lines are then deduced from the calibrated spectra. A subset of our results from the WIYN/Hydra survey of chromospheric activity in $\sim$70 solar-type stars in the solar-age and solar-metallicity cluster M67 is encapsulated in the accompanying histogram (Fig. 31).

The abscissa of the histogram for the "solar twins", the HK-index, is the sum of the residual intensities in 1 Å bandpasses centered at the Ca II H and K lines. The solar twins are defined as those stars in M67 with unreddened

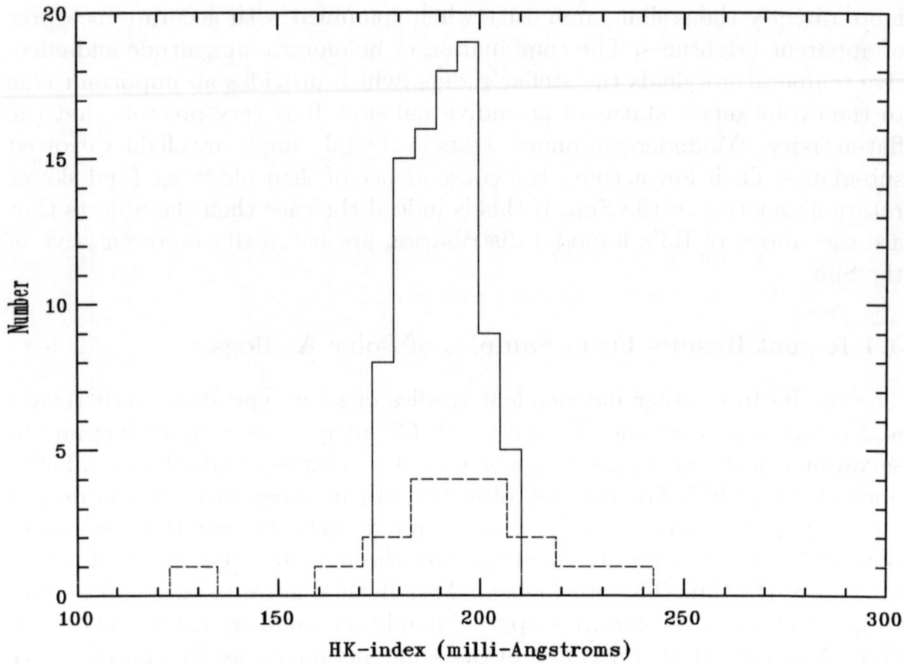

Fig. 31. The distribution of the HK-index for the solar cycle (*solid line*) and "solar twins" (*dashed line*). This index is identical to that defined by W. Livingston for his Sun-as-a-star synoptic observations at the NSO McMath-Pierce solar telescope on Kitt Peak. About 10% of the stars have HK-index values that are exceptionally quiescent, reminiscent of Maunder-minimum levels. About 20% exhibit HK-index values that exceed the value of solar maximum, at least as recorded from 1976–1994

colors in the range of $0.63 \leq B-V \leq 0.67$. This is the range of colors that has been quoted in the literature for the Sun (VandenBerg & Bridges 1984, see their Table 2). There are 20 stars in this subsample. The solar cycle in HK-index is also shown, based on data obtained by W. C. Livingston (NSO/Kitt Peak) at the McMath-Pierce Solar Telescope facility on Kitt Peak since 1976.

The broader distribution in chromospheric Ca II strength for the solar twins, compared to that of the Sun during the contemporary solar cycle, suggests that the potential excursion in the amplitude of the solar cycle is greater than what we have seen so far. The stars with HK values noticeably less than solar minimum may be in a prolonged state of quiescence analogous to the Maunder-minimum episode of the Sun during A. D. 1645–1715 when visible manifestations of solar activity vanished. This period corresponds to a time of reduced average global temperatures on the earth known as the "Little Ice Age" (Foukal & Lean 1990).

Among the solar twins we find that 10% have HK-index values less than that estimated for the Maunder Minimum while 20% exhibit HK-indices exceeding that observed at solar maximum. Thus, these preliminary results (based only on the solar twins) would indicate that the Sun can be in a state of activity outside of the envelope defined by the contemporary solar cycle at least 30% of the time. In view of the positive correlation between magnetic activity and brightness changes, our results suggest that the total solar irradiance could change by more than the 0.1% currently observed. This could, in turn, have significant implications for climate change over century-long time scales.

Conversely, the results in Fig. 31 imply that the Sun and solar-type stars are more similar than they are different in terms of their level of chromospheric activity. From this perspective, the Sun is a "normal" star. In any event, the M67 solar twins do not exhibit a bimodal distribution in Ca II strength, as in the Baliunas & Jastrow (1990) histogram of relative chromospheric emission strength in solar-type field stars. While the M67 results are not yet finalized as of this writing, an inspection of preliminary results by M. Giampapa for a broader sample of M67 solar-type stars corresponding to the same $B-V$ color range as that of Baliunas & Jastrow (1990) also show only a unimodal distribution.

A lingering question remains, however, as to whether the enhanced activity, compared to solar maximum, in some of the sun-like stars in M67 is due to an excursion in their cycles or to more rapid rotation. We are in the process of obtaining high resolution spectra from the Keck 10-m telescope and HIRES spectrograph in order to obtain v sin$i$ measurements (i.e., projected rotational velocity) for some of the M67 stars. In this way, we can determine whether the high-activity M67 stars are rotating rapidly.

Other intriguing questions are posed by these preliminary results from M67 observations. In particular, is the stellar distribution in the Fig. 31 really the result of the modulation of activity by cycles analogous to the solar cycle, or are the *relative amplitudes* of the cycles actually similar with the differences due only to differences in the *mean level* of activity among solar-type stars? Even more fundamentally, is the distribution of chromospheric H&K emission strength arising from solar-like cycles at all? The only way to address these critical issues is to obtain regular observations of a well-defined subsample of M67 "Suns" over a period of several years.

The results of this long-term program will provide fundamental empirical data on the cycle characteristics of unarguably solar-like stars. Of equal importance, they will provide valuable input for the construction of global climate change models.

4.5 Recent Results for Field Solar-Type Stars

Hall & Lockwood (2004) have just submitted for publication the results of a long-term program of spectroscopic observations of the activity and variability of 46 sun-like stars. This subset of their total sample of approximately 300 stars is characterized by colors in the same range adopted by Baliunas &

Jastrow ($0.60 \leq B - V \leq 0.76$). Hall & Lockwood's investigation differs in two principal ways from that of the M67 study described above. First, the sample is comprised of solar-type stars in the solar neighborhood, rather than objects in a single cluster of uniform age and chemical abundance. Second, long time-series of frequent observations have been acquired. The principal result is encapsulated in their histogram that is reproduced here in Fig. 32.

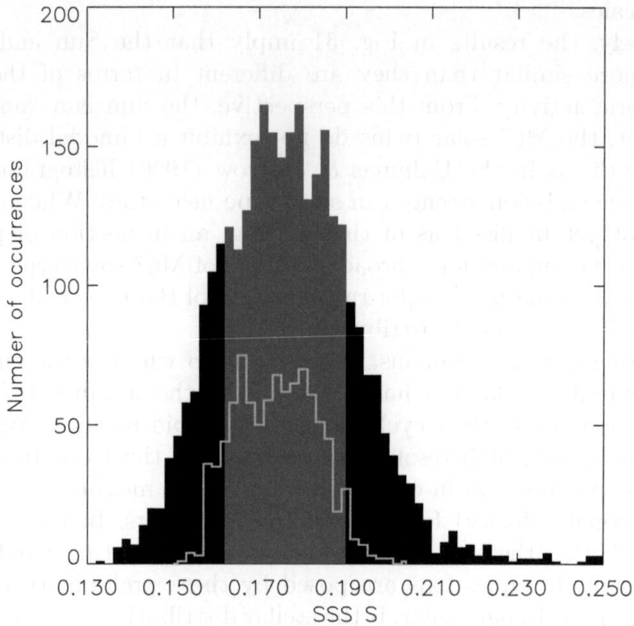

Fig. 32. The distribution of S-index values for 46 field solar-type stars from the Lowell Observatory "S-cubed" program (shown in *black*). Superimposed on the stellar distribution is the solar-cycle variation in S-index from the Mt. Wilson program (*gray*; Baliunas et al. 1995) and the Lowell program (*white outline*). The histogram is taken from Hall & Lockwood (2004)

In Fig. 32 the distribution of the Mt. Wilson S-index of Ca II H & K strengths is given for the sun-like stars (in black) compared to the equivalent distribution observed for the solar HK emission during the course of the cycle from Mt. Wilson Observatory observations and from Hall & Lockwood's data-set acquired with their Solar-Stellar Spectrograph (SSS) and converted to S index. The ordinate is the number of occurrences of each measured S-index value. In contrast to the Baliunas & Jastrow distribution, the distribution in Fig. 32 is unimodal. That is, it is not bimodal with a distinct population of stars identified as Maunder-Minimum objects. The unimodal nature of the distribution is similar to that found by Giampapa et al. (2004) for the solar-type stars in M67 in the same $B - V$ color range.

Interestingly, Hall & Lockwood find that about one-third of their sample is in a non-cycling state. This is identical to the fraction found by Baliunas & Jastrow (1990) though with an important caveat. The one-third fraction deduced by Baliunas & Jastrow is based on the fraction of the number of occurrences of S-index values–not stars–that are at Maunder-minimum-like levels of Ca II strength, i.e., fall within the lower group of their bimodal distribution. It is true that 4 of their 13 stars, or about one-third, for which they had long time series of observations were in a non-cycling state. However, there were 61 additional stars in their sample with infrequent observations and, hence, with much less well-determined cycle properties. By contrast, 14 of the 46 stars, or roughly one-third, in the Hall & Lockwood sample exhibit flat time series. However, as Hall & Lockwood (2004) emphasize, their flat-activity stars exhibit a distribution of S-index values that span the full range of mean S-index values encountered in their study. In fact, the distribution of the Ca II HK strengths of the flat-activity stars is broad and continuous over this range. Thus, the activity level of a star with a flat activity cycle is not necessarily lower than that seen in stars that exhibit cycles, analogous to the solar cycle. This result is in sharp contrast to that implied by the distribution presented by Baliunas & Jastrow (1990).

As Hall & Lockwood (2004) point out, the transition to a non-cycling state may be associated with an overall reduction in magnetic activity. Clearly, a transition occurred in the Sun that resulted in a reduction of the magnetic flux in the form of sunspots during the Maunder Minimum. But the observations of sun-like stars reveal that the absence of a chromospheric cycle is not necessarily associated with low chromospheric activity. Thus, the magnitude of any reduction in bright element and facular emission on the Sun that may have occurred during the Maunder Minimum is now difficult to assess on the basis of inferences from the current data-set of stellar observations. Moreover, the absence of the bimodal distribution in both the M67 data-set and the Hall & Lockwood data-set suggests that a Maunder-Minimum is not a distinct and separate mode of the dynamo though it may eventually be confirmed as the low-amplitude excursion of the solar dynamo. In order to gain further insight from stellar observations on the level of activity and irradiance reduction of the Maunder-minimum Sun, it will be necessary to observe the transition from a cycling to a non-cycling state in a sun-like star in order to see directly the level of any reduction in activity that might occur.

Inferences Concerning Solar Variability from Stellar Data

The estimates of solar activity and irradiance variability by J. Lean and collaborators, based on the distribution of stellar activity published by Baliunas & Jastrow (1990), have been revisited. For example, Foukal & Milano (2001) noted that the model of long-term irradiance variability associated with the sunspot cycle, as developed by Lean et al. (1992), Lean et al. (1995), implies a reduction in irradiance of 0.11% in the early part of the

20th century. In particular, Lean et al. (1992) identify a decrease of 0.11% in solar irradiance with the vanishing of the quiet network. However, Foukal & Milano (2001) do not find evidence for the disappearance of the quiet network based on their inspection of archival Ca II K spectroheliograms dating back to the early 1890's. Moreover, there is no evidence for network area variations of sufficient amplitude to produce a significant long-term component of total solar irradiance variation. Of course, the primary evidence for a long-term component was based on the report by Baliunas & Jastrow (1990) that stars with flat-activity cycles in the Mt. Wilson sample, which also exhibited extremely low values of HK emission, were bona fide solar counterparts that were in a Maunder minimum state representative of the Maunder Minimum episode of the Sun.

In recent work by Fröhlich & Lean (2002), the results of Baliunas & Jastrow (1990) are still relied on to infer facets of the long-term component of solar variability. However, these investigators now add the caveat that the Baliunas & Jastrow (1990) distribution of activity in ostensibly solar-like stars may include selection effects that would render their results less relevant to long term solar variability if some of the stars are not true solar counterparts. In particular, the distinct activity states implied by the bimodal distribution may be an artifact of an inhomogeneous sample consisting of both solar analogs and stars that are not truly analogs of the Sun. Certainly, the recent work by Giampapa et al. (2004) and Hall & Lockwood (2004) illustrate the sensitivity of the results concerning the amplitudes and behavior of cycle-related activity on the properties of the stellar sample. As pointed out by Foukal, Hall, North & Wigley (2004), the more recent results of studies of solar analogs are posing a significant challenge to–if not eliminating altogether–the scientific basis for the large-amplitude, low-frequency component of the irradiance reconstructions that have been utilized in climate models during the past decade. Clearly, the identification of true solar analogs now becomes critical for our understanding of the full range and nature of solar variability in the context of the variability of solar-type stars.

4.6 The Identification of Solar Analogs

A review of the methods and results for the identification of solar twins is given by Cayrel de Strobel (1996). The term *solar analog* refers to unevolved stars that are photometrically similar to the Sun. A star that is determined to be a *solar twin* is characterized by physical parameters (e.g., mass, effective temperature, age, metallicity, etc.) that are virtually identical to those of the Sun. The relatively bright star 18 Sco (HD146233 = HR 6060) is considered to be the best solar twin thus far identified (Porto de Mello & da Silva 1997). The position of 18 Sco in an HR diagram along with the Sun and other solar analogs is shown here in Fig. 33 from Porto de Mello & da Silva (1997). According to Porto de Mello & da Silva (1997; see their Table 1), 18 Sco (HR 6060) exhibits the smallest deviations from the Sun in key parameters

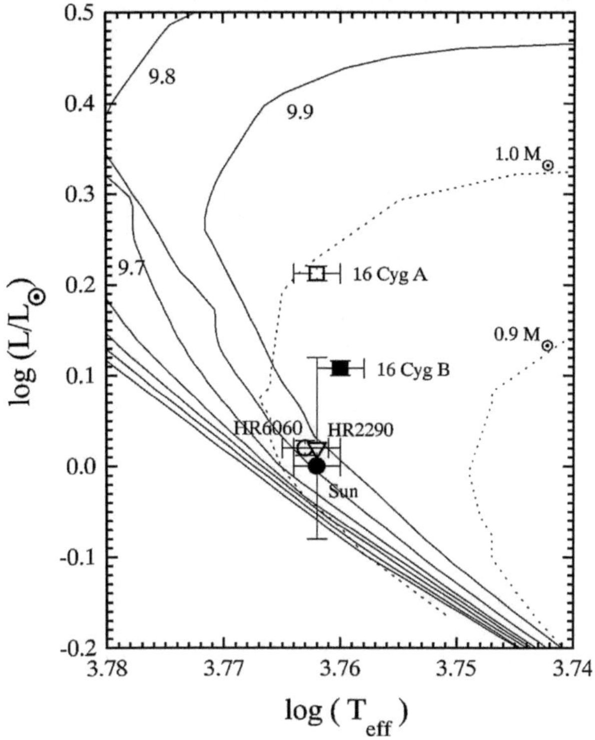

Fig. 33. HR diagram showing the position of 18 Sco (HR 6060), the Sun and other solar twin candidates. The isochrones are labeled in Gigayears (*solid lines*) and the evolutionary tracks are labeled in solar masses (*dotted lines*). The error bars are a consequence of measurement errors in the parallax (*from Porto de Mello & da Silva 1997*)

that include effective temperature, surface gravity, luminosity, metallicity and *UBV* colors.

The difficulty in identifying solar twins is highlighted by the study of Farnham et al. (2000). These investigators had as their primary motivation for identifying solar twins the goal of being able to remove the spatially integrated light of the solar spectrum from the spectra of comets, as observed with their instrumentation. In effect, Farnham et al. (2000) required a stellar counterpart of the Sun to utilize as a standard spectrum of the Sun so as to obtain the pure cometary spectrum. These investigators utilized a suite of narrow band filters that isolated various molecular and continuum bands extending from the near-UV at the atmospheric cutoff to the red near 700 nm. The essence of their results for selected candidates is seen in Fig. 34. The various candidates appear similar to each other in the visible regime. But considerable differences become evident in the near UV between 300–400 nm. The differences at short wavelengths are likely due to differences in magnetic

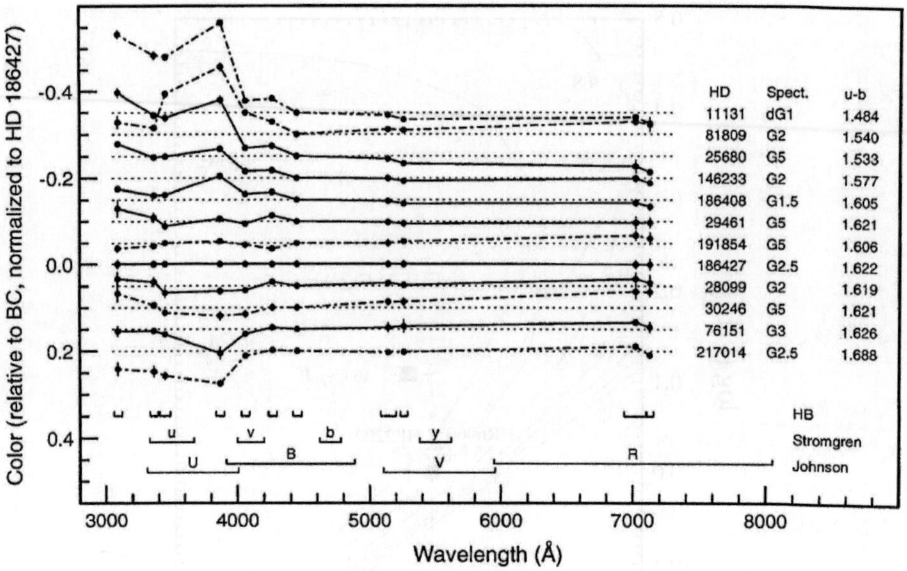

Fig. 34. Colors of solar analogs in various passbands relative to the blue continuum (BC) and normalized to HD 186427. The filter bandpasses utilized are indicated at the bottom along with standard filter bandpasses (*from Farnham et al. 2000*)

activity at the time of observation. The question that arises is the extent to which differences in activity should play a role in the identification of solar analogs and, particularly, solar twins.

The variability of the best solar twin, 18 Sco, is interesting to compare with that of the Sun. High precision differential photometry presented by G. W. Henry at the October 2003 Lowell Observatory Workshop on The Future of Stellar Cycles Research is displayed in Fig. 35. The brightness changes in 18 Sco are less than 0.05%, reminiscent of the 0.1% irradiance variability of the contemporary Sun.

The Ca II data compiled in both the Mt. Wilson and Lowell Observatory SSS programs, respectively, reveal a solar-like activity cycle with a period ~8–9 years at an amplitude of ~10% (Fig. 36). The example of 18 Sco immediately suggests that stars that are extremely similar to the Sun in their basic photospheric properties and placement in the HR diagram exhibit joint variations in activity and brightness that are also very similar to the Sun. Whether this trend persists will require the identification and monitoring of many more solar twins. This will likely require surveys at fainter magnitudes (i.e., V = 11, as recommended in the aforementioned Lowell workshop report on the future of stellar cycles research) than have thus far been performed followed by high precision spectrophotometric observations to confirm possible candidates as solar twins.

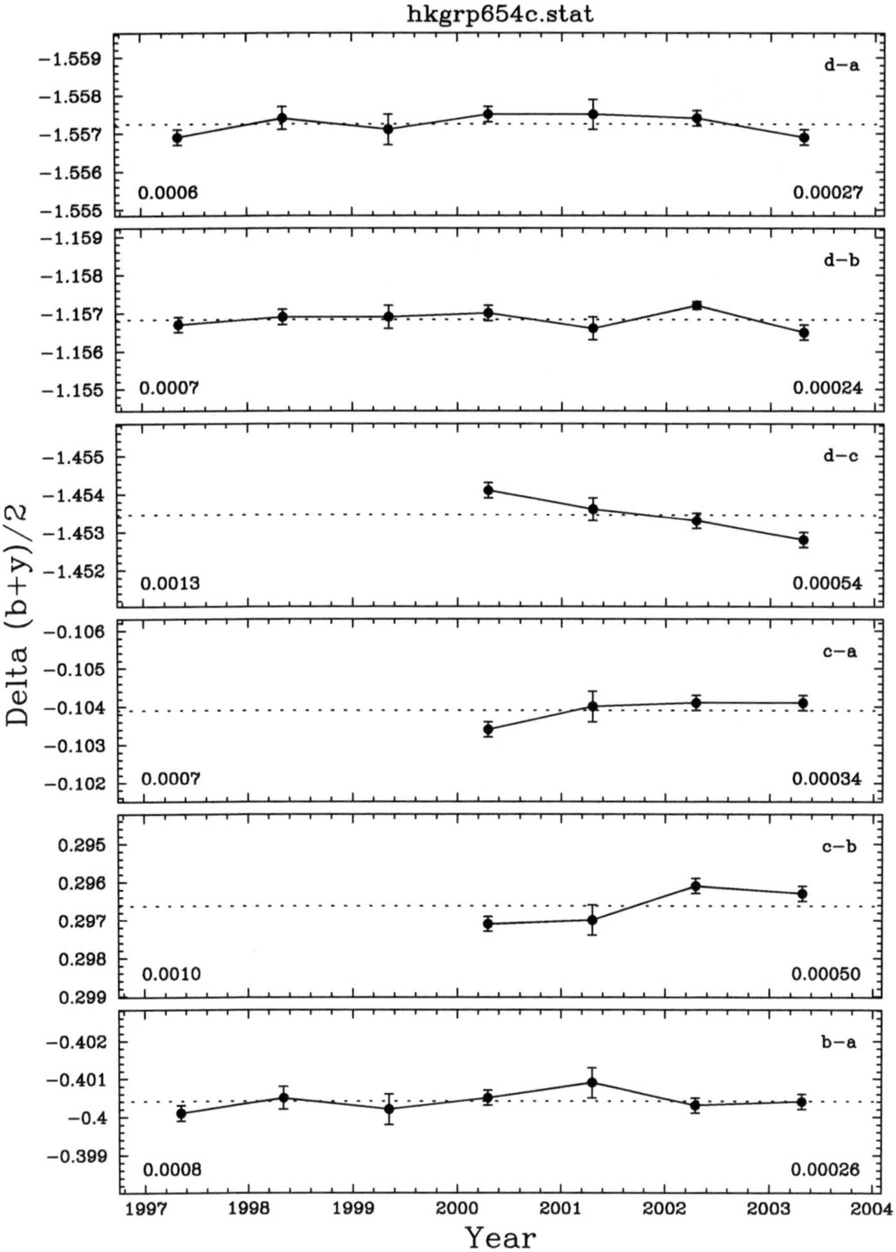

Fig. 35. High-precision, differential photometry of the solar twin, 18 Sco (star "d") obtained at the Fairborn Observatory. Variation of less than 0.05% relative to chromospherically inactive comparison stars is seen (*courtesy of G. Henry, Tennessee State University*)

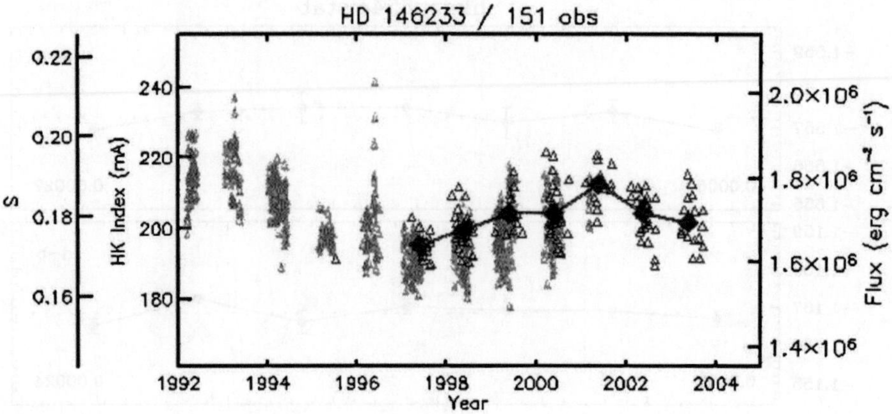

Fig. 36. The Ca II H & K cycle variation of the solar twin 18 Sco (= HD 146233) as recorded in the Mt. Wilson program (data starting in 1992) and the Lowell Observatory SSS program (beginning in about 1997). The line represents seasonal means (*courtesy of J. Hall, Lowell Observatory*)

4.7 Asteroseismology: A New Approach to Identifying Solar Twins

The emerging field of asteroseismology is a natural extension of helioseismology in which various modes of acoustic oscillations are observed to infer more directly certain fundamental properties of the interior structure of the Sun and stars. The detection of global oscillations in late-type dwarf stars will enable a more precise quantitative comparison of data with stellar models. In principle, the measurement of the frequency separations of p-mode oscillations should enable age and mass estimates of a star if its chemical abundances (metallicity) is known, say, from spectroscopic analysis.

The oscillation frequencies for the Sun and stars are described as radial eigenmodes multiplied by a spherical harmonic with two quantum numbers, n and l, or

$$\xi_{nlm}(r,\theta,\phi,t) = \xi_{nl}(r) Y_l^m(\theta,\phi) e^{-l\omega(n,l,m)t} , \quad (4)$$

where m is an azimuthal order that is normally observable in the spatially unresolved observations of stars. In (4), the radial order n is the number of nodes from the center of the star to the surface. For solar-type stars, $n \sim 20$. The quantum number l is the number of nodes around the circumference of a star. Typically, $l = 0$, 1, or 2 for stars. High orders of l are not observable in the global oscillation spectrum. Spatial resolution that thus far is only achieved in solar observations (e.g., local helioseismology) is required to observe at high-l.

The reader is referred to the review by Brown & Gilliland (1994) for a discussion of the underlying formalism in asteroseismology. I present here a summary of the interpretation of the meaning of mode frequencies. In particular, the frequency differences of modes are functions of stellar parameters. For example, frequency differences for n-modes (at a given l) are related to the sound crossing time in a star, or (Brown & Gilliland 1994):

$$\Delta\nu_0 = \left(2\int_0^{R_*} \frac{dr}{c_s}\right)^{-1} , \tag{5}$$

where R_* is the stellar radius and c_s is the local sound speed. Thus, the parameter $\Delta\nu_0$, also referred to as the *large separation*, is related to the sound crossing time through the center of the star. It can be shown that the large separation is a function of the mean density of the star (Cox 1980; Gough 1990):

$$\Delta\nu_0 \cong 135 \left(\frac{M_*}{R_*^3}\right)^{1/2} \mu Hz , \tag{6}$$

where M_* and R_* are the mass and radii, respectively, of the star in solar units. Thus, the large separation in n modes can be related directly to an important quantity in stellar structure that is a useful constraint for stellar models.

The l modes probe the conditions of the star both near the center and near the surface, with different degrees of l penetrating to different depths in the stellar interior. Modes that differ by 1 in n and +2 in l are referred to as the *small separation.* Ther frequency separations are related to the radial gradient of the sound speed according to (Däppen et al. 1988):

$$\delta\nu_{n,l} = \Delta\nu_0 \frac{(l+1)}{2\pi^2\nu_{nl}} \int_0^{R_*} \frac{dc_s}{dr}\frac{dr}{r} . \tag{7}$$

The sound speed gradients are especially sensitive to the conditions in the stellar core. As nuclear burning proceeds, the mean molecular weight in the core will change. Consequently, the gradients in sound speed change as the star evolves (as a result of the expenditure of its nuclear fuel). Thus, the small separation can reveal the evolutionary state (i.e., age) of a star.

The detection of p-mode oscillations in solar-type stars is challenging. The predominant mode in the solar oscillation frequency spectrum has a period of 5 minutes. The expected amplitudes of global oscillations in sun-like stars is ~a few meters per second. Both moderately high spectral resolution and relatively high temporal resolution are required. This is a classic photon-starved problem where every increase in telescope aperture, or throughput, yields gains. Furthermore, aliasing due to gaps in the window function of the observations reduces the precisions at which frequency differences can be measured. Multiple sites are needed to ensure as continuous a data string as possible, ideally, uninterrupted by the day-night cycle. This sort of global network is not in place yet for stellar observations. At present, restricted campaigns are carried out to establish the methods to routinely detect p-mode oscillations in solar-type stars.

Recently, Butler et al. (2004) announced the detection of p-mode oscillations in the nearby G star, α Cen A. Their detection is notable because of the unambiguous appearance of a regular pattern of oscillations in the observed time series of the Doppler signal (Fig. 37). The computed power spectrum from Butler et al. (2004) is displayed in Fig. 38. It is easy to imagine the development of semi-empirical interior models that best-fit the observed power spectrum of oscillations in order to infer fundamental stellar parameters. A solar analog candidate could quickly be confirmed or rejected simply by comparing a solar oscillation spectrum, degraded to the quality of the stellar spectrum, and determining how well they correspond to within the errors. Christensen-Dalsgaard (1992) has computed grids of models in stellar mass and age corresponding to values of the large and small separation, respectively. Though observationally intensive, asteroseismology is a promising and powerful technique to identify stars that are precise analogs of the Sun in mass, chemical composition and age.

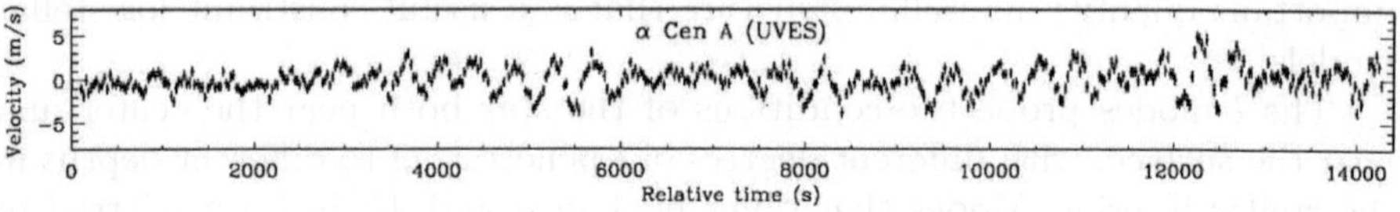

Fig. 37. The time series of observed radial velocity variations due to p-mode oscillations in α Cen A (*from Butler et al. 2004*)

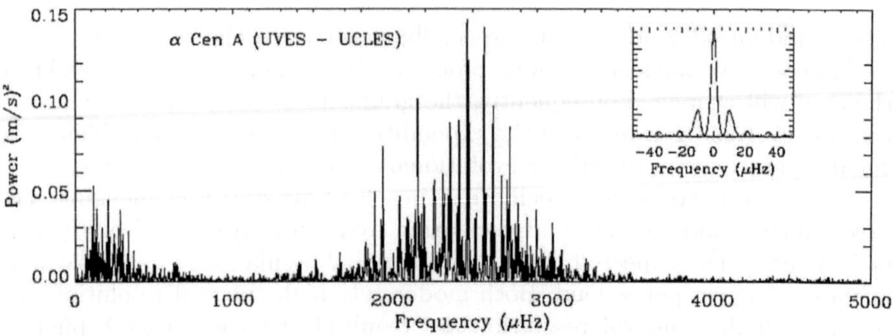

Fig. 38. Power spectrum of the time series of low-amplitude Doppler variations exhibited in Fig. 37 for α Cen A (*from Butler et al. 2004*)

5 Stellar Surface Inhomogeneities

Thermal inhomogeneities are the sites of emergent magnetic flux in the Sun and solar-type stars. It is these structures, delineated by magnetic fields, that modulate the radiative output of the Sun and sun-like stars. In this section, we will recount and discuss the evidence for inhomogeneous stellar atmospheres as manifested in key spectral and photometric diagnostics. In addition, we will discuss the salient features of model stellar chromospheres based on observed spectral line profiles and their success in reproducing the full chromospheric spectrum. Our discussion mostly will be in the context of identifying the critical properties that distinguish an "active" chromosphere star from a "quiet" chromosphere star.

5.1 Single-Component Model Chromospheres

In view of the absence of a comprehensive theory of chromospheric and coronal heating in late-type stars, the development of model chromospheres has proceeded in a semi-empirical fashion. This technique was pioneered by J. L. Linsky and his collaboratoring postdocs and students, most notably in the *Stellar Model Chromospheres* series of papers. In essence, this approach involves the acquisition of high spectral resolution line profiles, typically with echelle spectrographs. A flux-calibration is then applied to the observed profiles following their reduction to relative intensity. A model chromosphere is adopted and adjoined to an extant model photosphere appropriate for the stellar type. The latter is a radiative equilibrium model computed under the assumption of LTE (local thermodynamic equilibrium). Typically, a grid of model photospheres was computed for a range of effective temperatures and chemical abundances within a spectral class. The model chromosphere itself is operationally defined as a region above the photosphere where $dT/dh > 0$, that is, the temperature gradient is positive and, thus, the temperature is now increasing with height above the temperature minimum value. The temperature minimum, therefore, is a point of inflection that represents the onset of chromospheric heating.

In single-component models, the temperature profile as a function of column mass density (i.e., g cm^{-2}) is provided at the outset of the calculation. A microturbulent velocity height profile in the chromosphere is usually included as well. The coupled equations of radiative transfer and statistical equilibrium are then computed for the specified transitions in the model atom to ultimately yield the non-LTE source function for the line. The emergent line profile, convolved with the instrumental profile and possibly a rotational broadening profile, was then compared to the observed profile. The chromospheric temperature and microturbulent velocity distributions as a function of mass column density were then adjusted accordingly. This iterative approach continued until agreement was achieved between the computed and observed profiles. The final result was interpreted as a representation of the chromosphere of the star.

This approach to the study of solar-type, stellar chromospheres was based primarily on ground-based observations of the prominent Ca II H & K line profiles. Later, with the advent of the *International Ultraviolet Explorer (IUE)* satellite, this method was applied to the even stronger Mg II h & k lines at 280 nm. These features are characterized by higher abundance than Ca in the Sun and sun-like stars and hence are stronger. Furthermore, the Mg II resonance lines appear in a segment of the spectrum where the contribution from the background photosphere is significantly less, thus reducing uncertainties associated with the correction for the radiative equilibrium contribution to the line flux above the zero flux level. Moreover, the *IUE* satellite observatory yielded flux-calibrated line profiles directly, without the need for a more complex flux calibration procedure that is usually required with ground-based observations. However, the number of stars for which high-quality line profiles could be obtained was relatively limited because of the small aperture of the *IUE* telescope.

An example of this approach that is relevant to more nearly solar-type stars is shown in Fig. 39 from Kelch, Linsky & Worden (1979) In their models, Kelch et al. (1979) essentially fixed the temperature minimum point and adjusted the chromospheric gradient until the computed and observed Ca II resonance line profiles agreed. These investigators find that the temperature gradient in the chromosphere increases with increasing activity, i.e., increasing chromospheric Ca II H & K line strength. This result was similar to that achieved earlier by Shine & Linsky (1974) in semi-empirical models of solar plages–bright regions of enhanced magnetic flux and Ca II H & K emission (see Fig. 5) In particular, solar plages also exhibited steeper temperature gradients than did the quiet solar chromosphere. The similarity of the solar and stellar results suggested that active stars had more plage-like regions on their surfaces than did quiet stars.

By contrast, in their semi-empirical models of the chromospheres of the cool M dwarf stars, Giampapa et al. (1982) found that the temperature gradient was essentially the same for both active, dMe stars and quiet, non-dMe (dM) stars. However, the chromospheric temperature profile had to be moved inward toward high mass column densities in the dMe stars as compared to the dM stars. In either case, Kelch et al. (1979) and Giampapa et al. (1982) found that active stars essentially had more material (i.e., high column mass densities) at a given chromospheric temperature than did chromospherically quiet stars, such as the Sun.

5.2 The Applicability of Homogeneous Chromospheric Models

The computation of stellar model chromospheres has been based on the assumption of single-component, plane-parallel atmospheres in hydrostatic

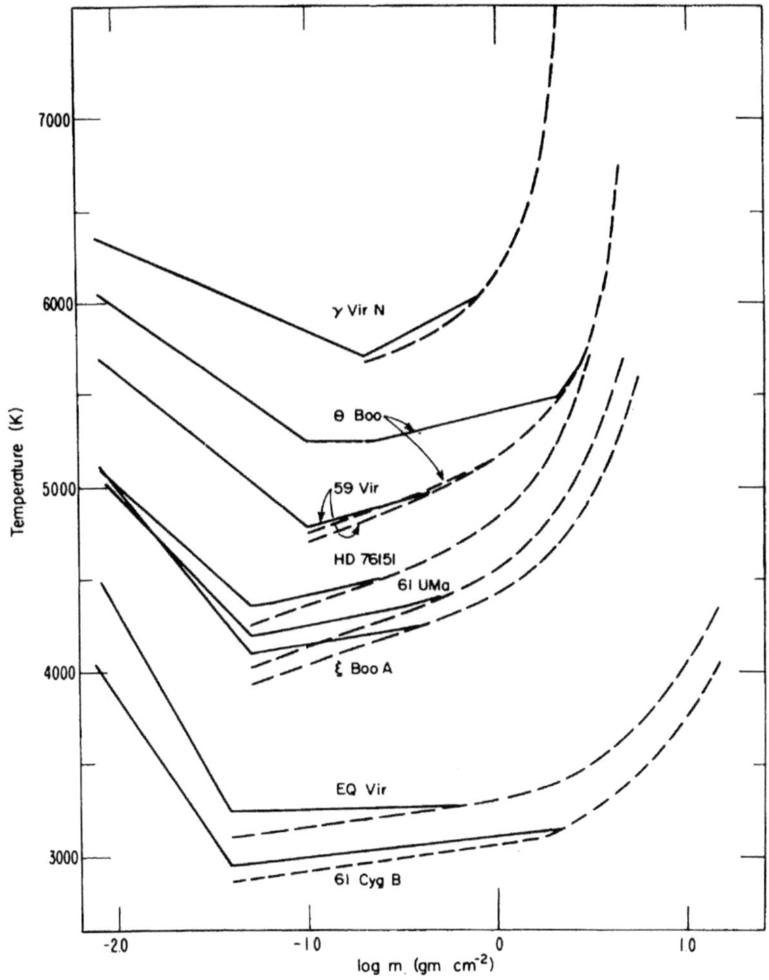

Fig. 39. Model chromospheres characterized by a chromospheric thermal gradient adjoined to a model photosphere. The chromospheric temperature gradient becomes steeper with increasing activity of the star in these models (*from Kelch et al. 1979*)

equilibrium. Of course, it is clear from solar observations that the solar chromosphere is inhomogeneous and consists of multiple components. For example, I display in Fig. 40 the Ca II K line profiles from three solar plages. The striking differences in relative strengths and shapes of the K lines reveal considerable variations in non-radiative heating rates and the K_2–K_3 velocity fields in the regions of formation in these plages. Yet, the spatially integrated K-line profile of the Sun-as-a-star typically appears as shown in Fig. 4. Evidently, the filling factor of the plage regions on the Sun is not sufficient to alter the profile so that it can exhibit the dramatic variations that can be seen

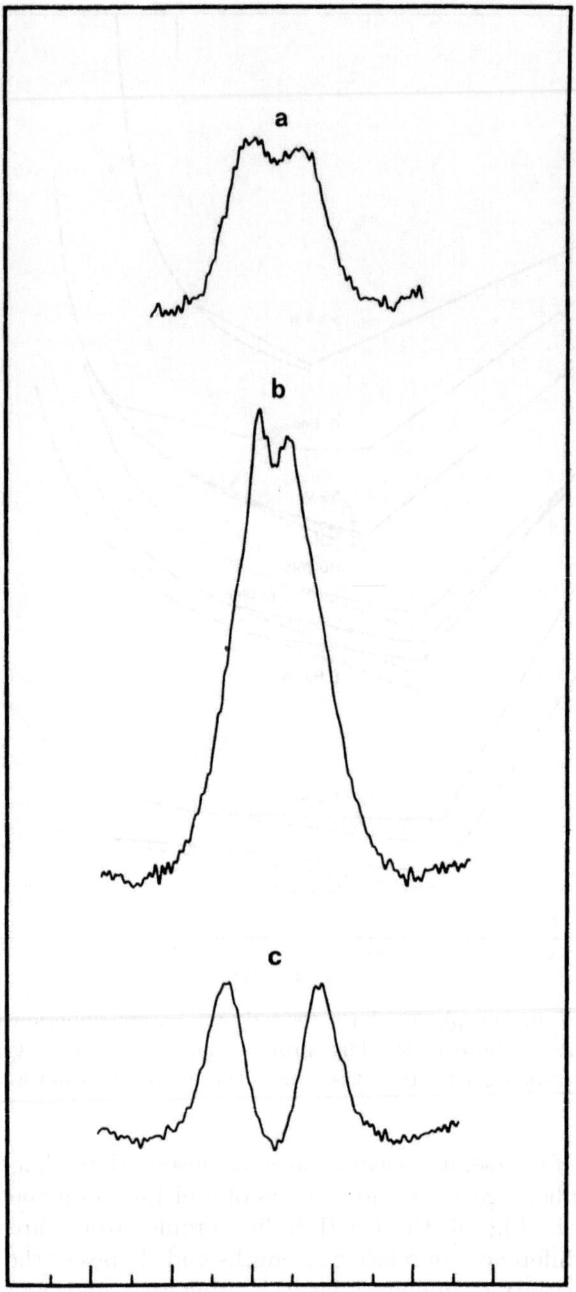

Fig. 40. The Ca II K line profile in three randomly selected plage regions on the solar disk. These observations were obtained by the author with the Fourier Transform Spectrometer at the McMath-Pierce Solar Telescope Facility of the National Solar Observatory on Kitt Peak, Arizona

among individual plage regions. I note, however, that subtle but measurable line width as well as line strength variations are seen in the integrated solar K line profile during the course of the solar cycle (Livingston & Wallace 2003).

The interpretation of single-component model chromospheres is then in the context of the average atmospheric structure. Although, in very active and exceptionally quiet chromosphere stars, the single component model can prove to be an accurate representation of the atmosphere with filling factors of plage-like regions near ~1 and zero, respectively. Nevertheless, in parallel with the development of semi-empirical model chromospheres, evidence began to emerge that multiple components were necessary to accurately account for the line spectrum of late-type stars.

In a study of the CO spectrum of the K2 giant star, Arcturus, Heasley et al. (1978) found evidence for the presence of spatial inhomogeneities. Figure 41 (from Heasley et al. 1978) shows the observed, infrared CO spectrum in Arcturus along with computed model spectra based on the range of thermal profiles given in Fig. 42. The chromospheric model (AL, from Ayres & Linsky 1975) is a semi-empirical model based on the observed Ca II resonance lines in the spectrum of Arcturus. The computed spectrum in the CO region that this model yields exhibits line depths that are too shallow with respect to the observed spectrum (Fig. 41). The radiative equilibrium model produces lines that are too deep relative to the local continuum, as compared to the observed spectrum. Ultimately, Heasley et al. (1978) conclude that the K line

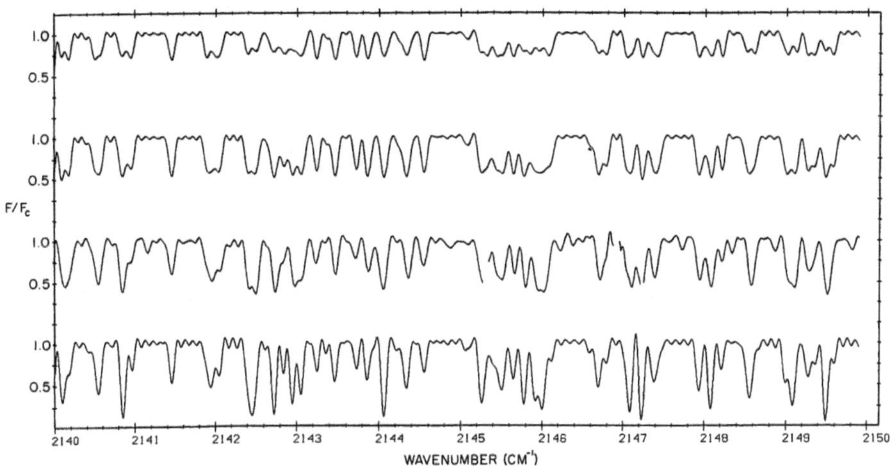

Fig. 41. A comparison of computed model chromospheres with the observed spectrum of the giant star Arcturus in the CO region. From *top* to *bottom*: synthesized spectrum from a chromospheric model (Ayres & Linsky 1975) based on Ca II K; a cool chromosphere with a plateau temperature structure; the observed spectrum; and, a pure radiative equilibrium model with no chromosphere (*from Heasley et al. 1978*)

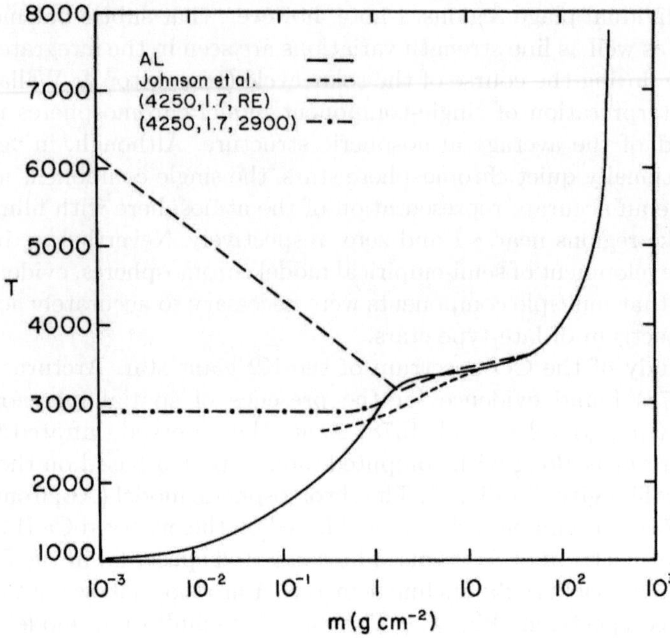

Fig. 42. Model chromospheres that correspond to the synthetic spectra in Fig. 41. (*from Heasley et al. 1978*)

and the CO fundamental cannot be simultaneously fit by a single-component atmosphere. Instead, the best-fit is a composite atmosphere comprised of the chromospheric model and the radiative equilibrium (RE) model in a ratio of 3:7. In essence, the outer atmosphere of Arcturus at chromospheric heights is composed of both hot and cold gas in this ratio.

Wiedemann et al. (1994) extended this investigation further to a study of the chromospheric structure in three bright giants. Their study disclosed the presence of optically thick CO lines at chromospheric heights. Furthermore, in this study of the applicability of a single-component model, Wiedemann et al. (1994) found that chromospheric models based jointly on the Ca II and Mg II resonance lines, and the CO lines, were completely discrepant. These investigators proposed to resolve the discrepancy by adopting a two-component model atmosphere dominated in filling factor by CO ($f \sim 90\%$) combined with a small filling factor of hot material ($f \sim 10\%$) that gives rise to the short-wavelength resonance line emission.

Similarly, McMurry & Jordan (2000) found in their study of the ultraviolet fluorescent lines of CO in cool giants, which are excited by UV lines of neutral oxygen originating in hot region, that a single-component model cannot account for the strengths of the CO features. A model with a hot region (that gives rise to the high excitation UV O I lines) in close proximity to the cool CO gas is required. Thus, McMurry & Jordan (2000) suggest that the chromospheres of these stars are bifurcated into hot magnetic regions and nearby cool regions. Alternatively, the atmospheric structure could be a result of shocks passing through a cool atmosphere that heat the gas to chromospheric temperatures. Thus, the chromosphere is a transient phenomenon at any given position on the stellar disk, consisting alternately of hot and cold gas at a given spatial location. The aforementioned stellar results are reminiscent of the thermal bifurcation model of the solar chromosphere of which T. Ayres has been a strong proponent (see Ayres 1981).

In models of dwarf M stars, Giampapa et al. (1982) attempted to fit both the observed Hα feature and the Mg II h+k observed flux with a semi-empirical model based on the observed Ca II K-line alone. Initially, it was not possible to match the observed Hα line with the single-component model. However, the addition of an inflection point in the upper chromosphere with a (steeper) temperature gradient than found in the lower chromosphere yielded a fit to both the Hα line and the K line. This was possible because the chromospheric Hα line in M dwarfs and the K line are segregated in their regions of formation. In particular, the K line formation is confined to the low chromosphere while the Hα line is strictly an upper chromospheric diagnostic. By contrast, the Mg II and Ca II resonance line have significant overlap in their regions of formation. In this case, it was not possible to develop single-component models that simultaneously matched the Ca II and Mg II total line fluxes. Giampapa et al. (1982) proposed that formation of these features in flux tubes with diverging geometry, such that the Mg II and Ca II had different filling factors at their respective levels of formation, could lead to a plausible reconciliation of the differing strengths of these features as deduced from single-component models.

As we see, a common thread that runs through the applicability of single-component chromospheric models in late-type stars is their failure once multiple diagnostics are considered. More realistic models must include the possibility of different filling factors of "hot" and "cold" material in chromospheres combined with vertical stratifications that are not constant with height but likely include line formation in diverging magnetic field structures. In the following section we will examine the diagnostics of surface inhomogeneities in dwarf stars.

5.3 Magnetic Regions in Dwarf Stars

The observational evidence summarized in §2–3 concerning the rotational and cycle modulation of chromospheric diagnostics, and the joint variation of brightness changes, represents a clear indication of spatially concentrated complexes of activity on the stellar surface. The question arises whether active and quiet chromosphere stars differ from each other mainly in the fractional area coverage (i.e., "filling factor") of otherwise similar active regions or if the differences are dominated by differing intrinsic nonradiative heating rates in these regions.

The Case of the M Dwarf Stars

The M dwarf stars span an interesting region of the H-R diagram where contributions from non-magnetic heating of the outer atmosphere, such as acoustic heating arising from the underlying turbulent convection zone (e.g., Ulmschneider 2003) are minimized. The equipartition values of the magnetic fields in their photospheres must be in the multi-kilogauss range, i.e., similar to or greater than what is seen in sunspot umbrae that are themselves similar to the photosphere of a K star (or later).

The M dwarf stars are observationally subdivided according to the appearance of Hα in their spectrum. The dMe stars exhibit Hα emission while the dM (non-dMe) stars have either an Hα absorption line or no discernible feature. Cram & Mullan (1979) investigated the origin of the Hα line in these stars through the computation of a range of schematic model chromospheres. As illustrated in Fig. 43, the Hα line is weak in the purely photospheric (radiative equilibrium) model. The dwarf M photosphere is simply too cool to excite many hydrogen atoms to the $n = 2$ state in order to build up sufficient opacity in the line. As chromospheric heating increases, the Hα responds first by becoming stronger in absorption. That is, the opacity in the $n = 2$ state increases but the line is still controlled by the background radiation temperature at this wavelength. As chromospheric densities increase in response to enhanced heating, the Hα line becomes collision dominated and it appears in emission.

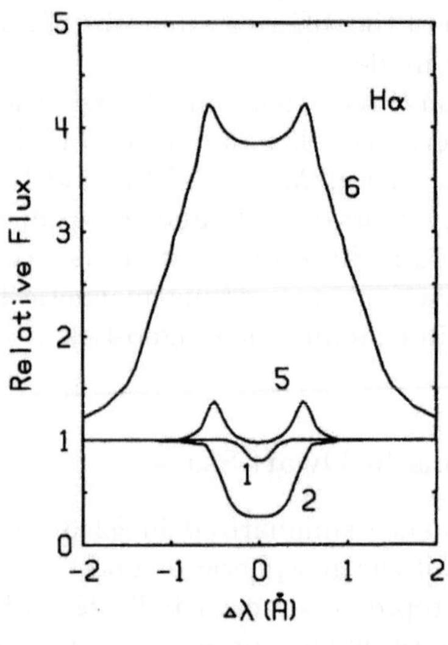

Fig. 43. The Hα profile computed using a series of model chromospheres spanning a range of activity from no chromosphere (1) to weak chromosphere (2) and strong chromosphere (6) as is found in a dMe flare star (*from Cram & Mullan 1979*)

We normally think of chromospheric lines as emission features. However, the appearance of a chromospheric *absorption* feature presents us with an opportunity to deduce some information about the filling factor of chromospheric regions, presumably associated with active (plage-like) regions on the stellar surface. In particular, the fractional area coverage or filling factor of Hα chromosphere in a dwarf M star with Hα absorption is given by (Giampapa 1985)

$$f = \frac{W_{obs} - W_q}{W_a - W_q} , \tag{8}$$

where W_{obs} is the equivalent width of the Hα line in the observed spectrum, W_q is the equivalent width of the feature in the quiet photosphere and W_a is the absorption equivalent width in the active region. For multiple active regions, W_a would be replaced with a summation over all the active regions. Of course, W_a is not directly observable for a given star (unless it can be ascertained that a single, uniform active region covers the entire visible disk). But a lower limit can be inferred if we assume that the Hα chromospheric absorption line in the surface active region is completely black, i.e., has zero flux in the line core. Recalling the definition of equivalent width, we have

$$W_a = \int_{-\Delta\lambda/2}^{+\Delta\lambda/2} \frac{I_c - I_\lambda}{I_c} \, d\lambda \, . \tag{9}$$

Since we are assuming that $I_\lambda = 0$, we have from (9) that $W_a = \Delta\lambda$. Given that Hα is weak in the quiet photosphere in the M dwarf stars, we can adopt the approximation that $W_q \approx 0$. I therefore find from (8) that

$$f \geq W_{obs}/\Delta\lambda , \tag{10}$$

where $\Delta\lambda$ is the observed base width of the line. Using this technique, Giampapa (1985) deduced from Hα line profile data in the literature (Worden et al. 1981a) *lower limits* for f in the range of 10–26%. Thus, the Hα chromosphere, that is presumably associated with magnetic field regions, in the relatively quiescent M dwarf stars without Hα emission is extensive.

A more accurate estimate for the lower limit of f can be obtained by reformulating (8) as an inequality , or

$$f \geq \frac{W_{obs} - W_q}{W_{max} - W_q} , \tag{11}$$

where W_{max} is the maximum absorption strength that can be attained in an active region on a dwarf M star of a given spectral type. The value of W_{max} is estimated from grids of model chromospheres and is therefore a model-dependent quantity. Cram & Mullan (1979) computed a value of $W_{max} = -0.69$ Å for an M dwarf star with an effective temperature of 3500 K. At approximately this temperature, Stauffer & Hartmann (1986) observed a range of Hα absorption equivalent widths of $W_{obs} = -0.27$ to -0.49 Å. Thus, inequality (10) yields a range in lower limits to the active region filling factors of 0.31–0.67 for these dwarf M stars. These estimates suggest a significant fractional area coverage of magnetic regions on even relatively quiet M dwarfs, implying that a primary difference between the dM stars and the more active dMe stars is not simply the filling factor of identical active regions but due to the intrinsic chromospheric heating rates. Of course, this was already clear by inspection since an Hα absorption line cannot be obtained from the combined contribution of plage-like, Hα emitting regions on the stellar surface. The simultaneous occurrence of more active, Hα emitting structures with Hα absorbing regions (both photospheric and chromospheric) will reduce or "dilute" the Hα absorption that would otherwise be observed in the integrated light of the star (e.g., Young et al. 1984).

Solar-Type Stars

The spectral counterparts in sun-like stars to the Hα lines in dwarf M stars are the He I lines at 5876 Å and 10830 Å, respectively. I display in Fig. 44 spectroheliograms taken in an X-ray line (at 65.5 Å) and at He I λ10830. The strong spatial coincidence of the λ10830 feature with sites of strong coronal X-ray emission on the Sun establishes the high excitation triplet lines of He I as excellent proxies for X-ray emission in the Sun and solar-type dwarfs. The He I λ5876 line is essentially absent in the quiet solar photosphere while the λ10830 is very weak. These features appear in *absorption* in the solar chromosphere and the chromospheres of active, solar-type stars. During flares, the He I triplet features can be driven into emission.

An example of the appearance of λ10830 in the spectra of an active and a quiet solar-type star is given in Fig. 45.

Given these basic characteristics of the He I lines in the (non-flare) chromosphere, we can write inequality (11) for the λ5876 and λ10830 lines as

$$f_{HeI} \geq W_{obs}/W_{max} \ , \tag{12}$$

where $W_q \simeq 0$ in the quiet photosphere. Andretta & Giampapa (1995) computed the He I λ5876 and λ10830 profiles for single-component, plane-parallel chromospheric models with solar-type photospheres (Fig. 46). The grid of models represented chromospheres of increasing activity by increasing the column mass loading at chromospheric temperatures. Both He I lines respond by increasing in absorption strength with increasing chromospheric activity. Eventually, the λ10830 is driven into emission though the relatively weaker λ5876 line remains in absorption.

The joint response of these lines to chromospheric heating is summarized in Fig. 47 for various filling factors. The U-shaped locus for each filling factor is a result of the λ10830 absorption line beginning to fill in with emission (i.e., become weaker in absorption) while the λ5876 feature continues to increase in absorption strength with increasing activity. In this interpretation, stars

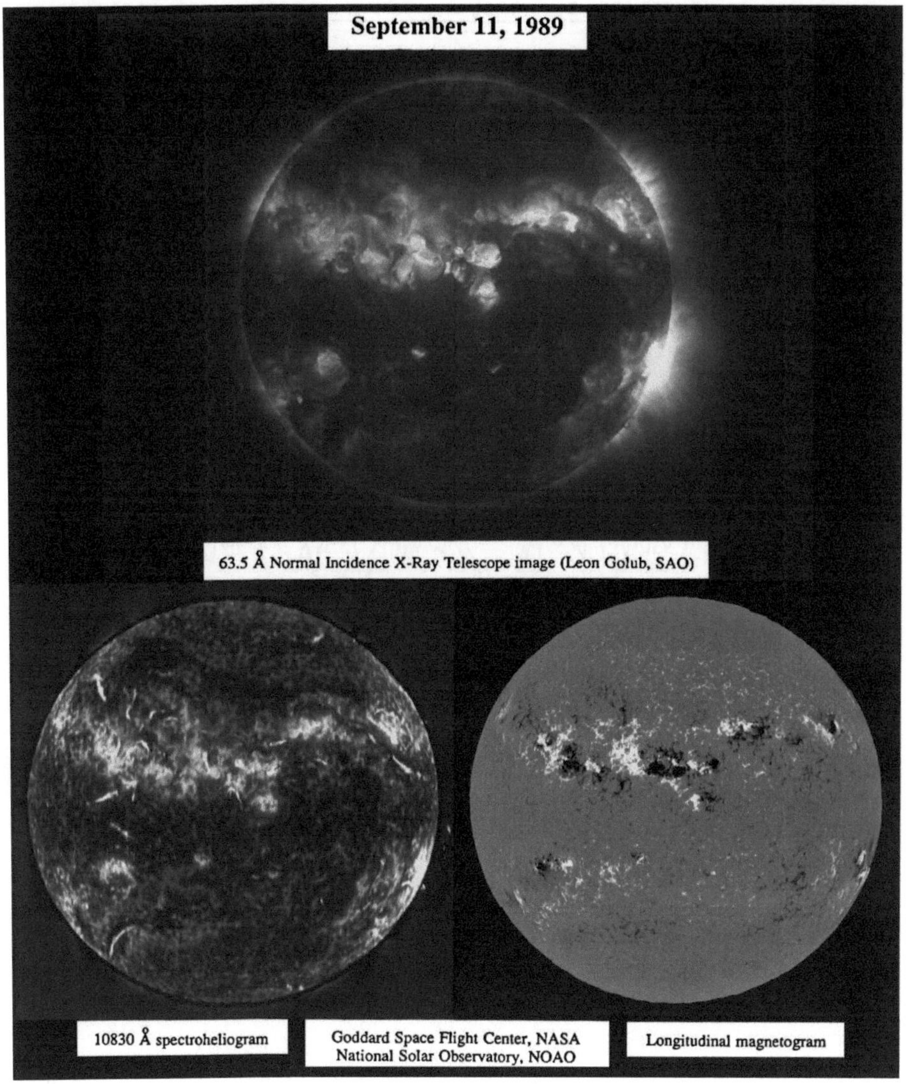

Fig. 44. Coronal X-ray and He I λ10830 spectroheliograms along with a full-disk magnetogram of the active Sun. Note the spatial coincidence of the X-ray emitting regions and the He I active regions with areas of strong magnetic flux on the Sun. Downwardly directed coronal X-ray emission likely plays a role in the excitation of the He I triplet levels (*courtesy of J. Harvey, NSO*)

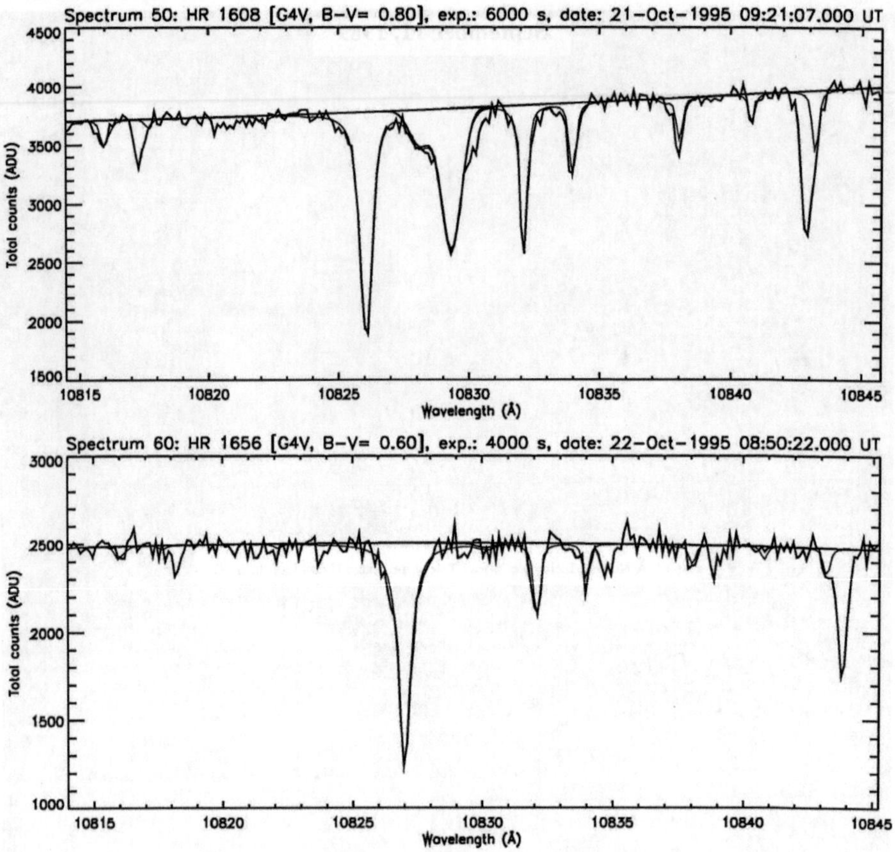

Fig. 45. The appearance of the He I $\lambda 10830$ line in an active solar-type star (*upper panel*) characterized by coronal X-ray emission that is three orders of magnitude greater than the active Sun and a quiet solar-type star of the same spectral type. In the quiet star, X-ray emission was not detected at the level of sensitivity available with the *ROSAT* satellite

with observed strengths of $\lambda 5876$ and $\lambda 10830$ that lie along one of these locii would have the same filling factor of active regions but obviously different chromospheric heating rates.

In principle, the grid of results based on plausible models as developed by Andretta & Giampapa (1995) (Fig. 47) represents a powerful method for the direct determination of active region filling factors on solar-type stars. In practice, simultaneous observations of the $\lambda 5876$ line in the visible and the $\lambda 10830$ in the near IR are required. While this is feasible with CCDs for very bright sources like the Sun, it is not practical to do with a single detector for fainter sources such as stars. Typically, two different instruments (and telescopes) observing simultaneously, one in the visible with a CCD and

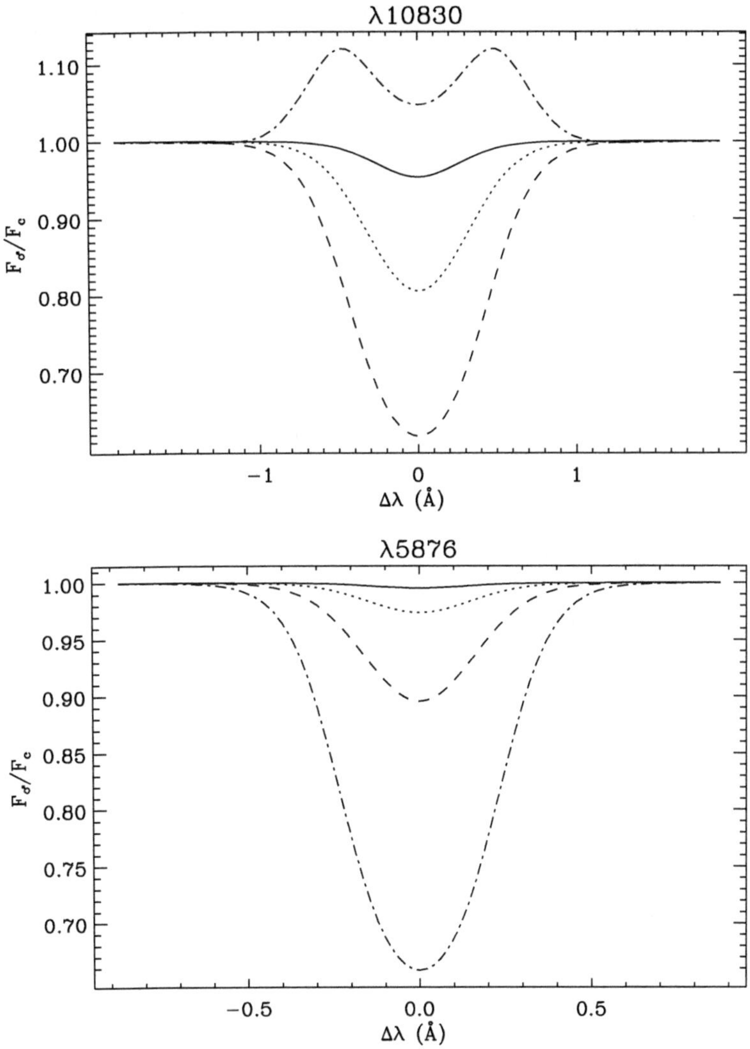

Fig. 46. The emergent profiles for the high-excitation He I triplet lines for various model chromospheres *(from Andretta & Giampapa 1995)*

one with a detector sensitive to the near infrared, are needed to obtain the requisite data. Andretta & Giampapa (2004) are in the process of reducing and analyzing spectra acquired in this way for a sample of F–G solar-type stars. Their preliminary results indicate that there are active solar-type stars with high filling factors ($f \sim 2/3$) of active regions but with different degrees of chromospheric heating.

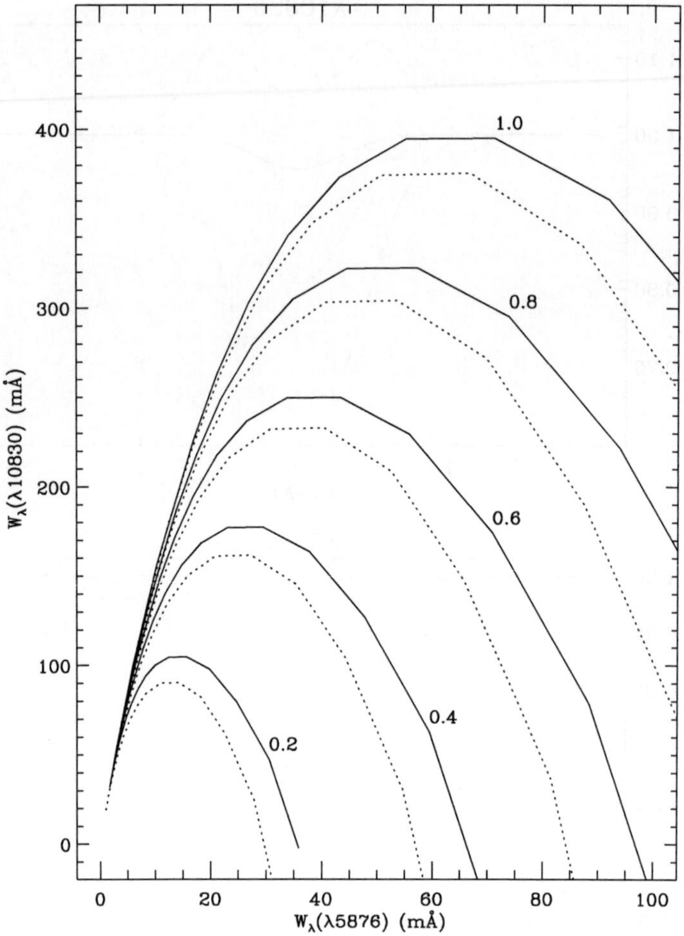

Fig. 47. The joint response of the He I triplet lines to chromospheric heating for a range of active region fractional area coverages on the visible stellar disk. The *solid* lines are based on models that include coronal X-ray irradiation while the *dotted* lines do not include coronal irradiation (*from Andretta & Giampapa 1995)*

In brief summary to this point, active and quiet chromosphere stars extending from spectral types F–M can differ in both filling factor and chromospheric heating rates. The difference is not simply due to the filling factor of basically similar active, plage-like regions on the stellar surface. Inhomogeneous, multi-component models involving both cool and hot regions are needed to account for the chromospheric and CO spectrum in cool giants (and possibly the Sun itself). In the observational arena, it would be interesting to obtain joint chromospheric and CO observations of solar-type, dwarf stars. Theoretically, the physical factors that give rise to similar active region coverages in both quiet and active stars but different non-radiative, magnetic field-related heating rates need to be identified. The influence of the differential rotation profile of a star on the generation of magnetic flux and the related outer atmospheric heating needs to be delineated.

5.4 Magnetic Field Measurements on Solar-Type Stars

Magnetic field properties in localized regions on the Sun are conventionally measured with polarimeters at high precisions. However, it is difficult to obtain useful measurements of magnetic field properties of the spatially unresolved stars with polarimetry because tangled field topologies lead to a net cancellation of opposing polarities, leading to a null result. To overcome this obstacle, Robinson et al. (1980) developed a method that utilized high signal-to-noise ratio, high spectral resolution observations of stars in unpolarized, "white-light" to directly detect the Zeeman signature in magnetically sensitive line profiles. In particular, the line profile of a magnetically sensitive line with a high Landé g-factor is compared to a nearby magnetically insensitive line observed simultaneously from the same multiplet. The residual broadening in the wings of the sensitive line is compared to the insensitive, reference line. A Fourier deconvolution of the features then yields an estimate of the magnetic field strength and fractional area coverage. Robinson (1980) gives a detailed discussion of the analysis method.

Examples of observed sensitive and insensitive reference profiles from Robinson et al. (1980) are provided in Fig. 48. Note that the Sun-as-a-star spectrum that does not show any evidence of residual Zeeman broadening; the filling factor of kilogauss-level fields is simply too small for detection. The signature of the Zeeman-split σ-components around the central, unsplit π-component in the sunspot umbra spectrum is clear. Zeeman broadening in the active G dwarf, ξ Boo A, is also apparent in Fig. 48. In this initial observation and analysis, Robinson et al. (1980) deduced field strengths $\sim$2.5 kG covering 20–45% of the surface of ξ Boo A. These kinds of field strengths are more characteristic of spots and certainly in excess of equipartition values in the normal photosphere.

This general approach to magnetic field measurements was extended to a larger sample of solar-type stars by Marcy (1984). His program included multiple observations of stars over several days, such as the example given in Fig. 49 from Marcy (1984). The crucial point conveyed by Fig. 49 is the observation of the *variability* of the magnetically sensitive line profile relative to the insensitive (actually, less sensitive) reference line. The interpretation of the subtle line broadening as magnetic in origin had been challenged as not really a manifestation of Zeeman broadening. Rather, other effects in a complex stellar atmosphere could belie the subtle signature of Zeeman broadening in the line wings. However, the detection of variability in the wings of the Zeeman-sensitive feature is decisive proof of the magnetic origin of the residual line broadening. A successful challenge to this interpretation

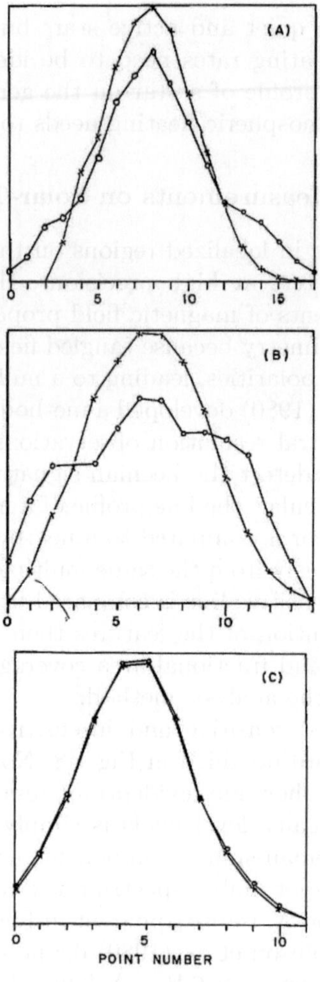

Fig. 48. Residual Zeeman broadening in the wings of the magnetically sensitive Fe I line at 684.27 nm (g = 2.5) as compared to a nearby, insensitive Fe I line at 681.027 nm (g = 1.17). (**a**) the active solar-type star ξ Boo A (G8 V), (**b**) a sunspot (equivalent to a highly active ~K star), and (**c**) the Sun seen as a star. The filling factor of strong fields on the Sun (seen as a star) is too small to give a detectable signature (*from Robinson et al. 1980*)

would be faced with the task of showing how a non-magnetic origin of the broadening could lead to variability in the magnetically sensitive feature *only* while not affecting the nearby reference line from the same multiplet.

Initial work on the detection of Zeeman broadening emphasized high resolution observations in the visible region of the electromagnetic spectrum. This was due primarily to the availability of high resolution echelle spectrographs combined with sensitive CCD arrays characterized by high dynamic range,

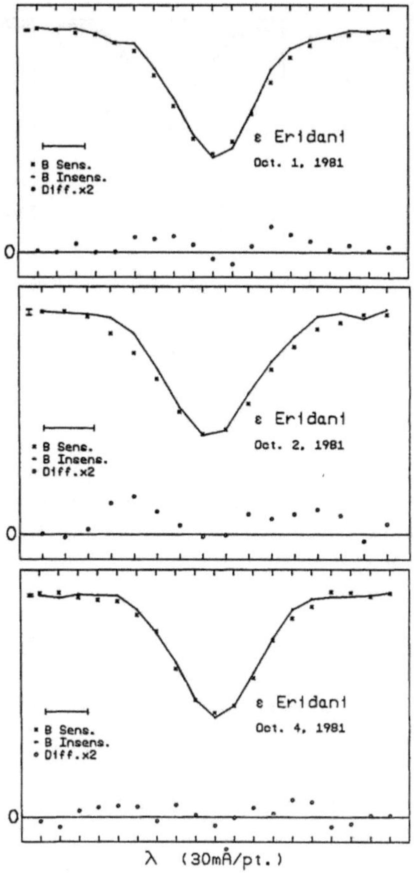

Fig. 49. Evidence of short-term magnetic flux variability in the active, solar-type dwarf ϵ Eri (*from Marcy 1984*)

thus enabling the acquisition of spectra with S/N well in excess of what was possible with photographic plates. However, observations in the infrared offer advantages for the detection and measurement of magnetic fields in late-type stars. In particular, the Zeeman effect results in a splitting of a spectral line into three components: the central (unshifted) π component and two σ components that are shifted symmetrically by an amount $\Delta\lambda_\sigma$, or

$$\Delta\lambda_\sigma = 4.67 \times 10^{-13} g\lambda^2 B \quad \text{Å} \, , \tag{13}$$

where B is in Gauss, λ is the wavelength of the line in Å, and g is the Landé g-factor that is a measure of the sensitivity of the transition to Zeeman splitting. Inspection of (13) reveals that Zeeman splitting increases with the square of the wavelength for a given field strength. Thus, it appears that Zeeman splitting is considerably enhanced in the infrared. However, we note that the

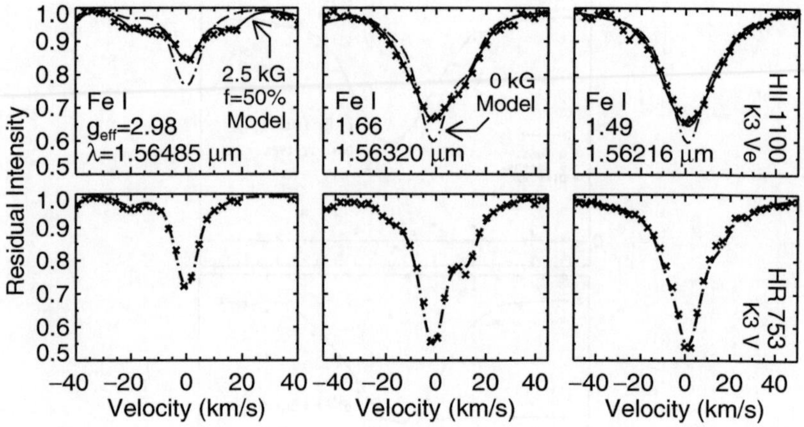

Fig. 50. Stellar magnetic field detections using the enhanced Zeeman splitting available in the infrared. The crosses are observed spectra. Fitted models include computed profiles with and without magnetic fields. Shown are an active pre-main sequence star (*upper panel*) compared to inactive dwarf in the *lower* panel (*from Valenti & Johns-Krull 2004a*)

width of the Doppler core and natural line widths increase with wavelength. In practice, therefore, the gains realized in the detection of Zeeman residual broadening are proportional to λ rather than λ^2.

An example of the enhanced Zeeman broadening encountered in the infrared is given in Fig. 50 in the infrared H band near 1.6 µm. The active, late-type pre-main sequence star in the upper panel exhibits clear signatures of Zeeman broadening, indicating the presence of strong magnetic flux in its photosphere. The broadening continues to appear even at the lower-sensitivity, $g = 1.49$ Fe I line. By comparison, the inactive bright star shows no detectable Zeeman signature. A similar example is given in Fig. 51 but for a dMe flare star (AD Leonis, dM3.5e) as recorded in the even longer-wavelength, infrared K band near 2.2 µm. Clear splitting of the σ components in the Ti I lines is evident, indicating a mean field strength of B $= 3800 \pm 260$ G covering ~73% of the surface of this active flare star (Saar & Linsky 1985).

In addition to enhanced Zeeman sensitivity, the infrared region is more sensitive to the contribution of the strong magnetic fields in cool spots (analogous to sunspots) due to the increasing brightness or, alternatively, the declining contrast, of spots at longer wavelengths. Thus the potential contribution of spot magnetic fields to the inferred mean magnetic field and flux becomes more important at infrared wavelengths. This suggests the possibility that joint optical and IR observations could be utilized to distinguish between the spot and plage-like region contributions to the stellar surface field.

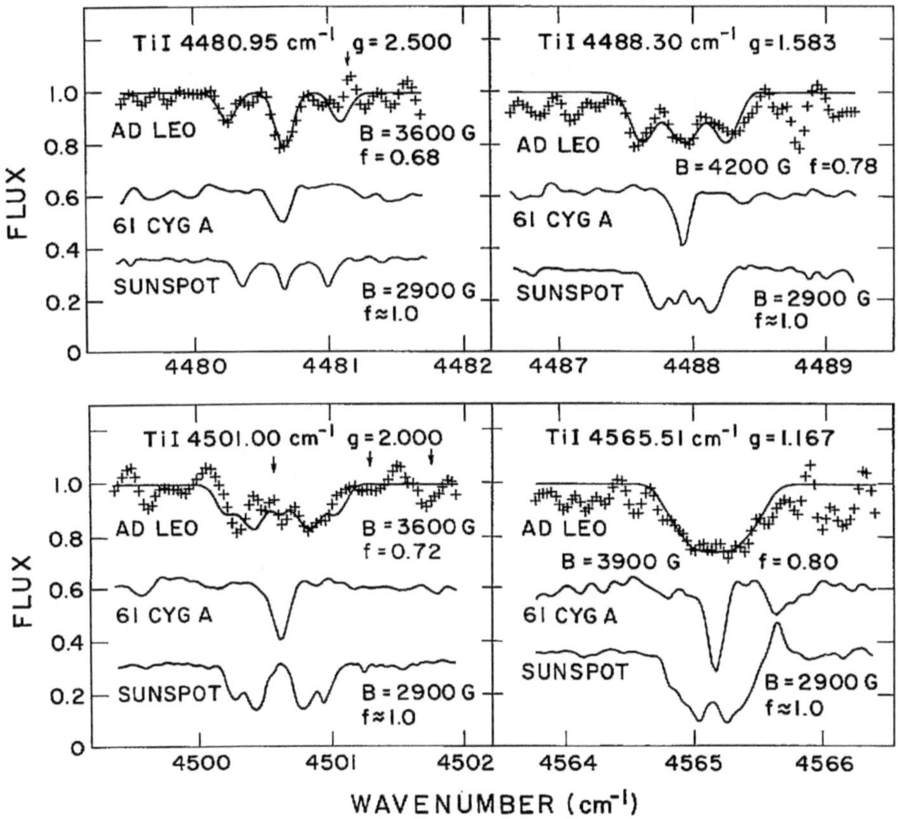

Fig. 51. Evidence of strong magnetic fields in the infrared K band in the active dMe flare star AD Leonis (dM3.5e). Zeeman splitting is clearly present in the Ti lines (*from Saar & Linsky 1985*)

Magnetic Cycles in Solar-Type Stars

The development of methods for the detection of magnetic fields in solar-type stars introduces the possibility of directly measuring the magnetic cycles of stars. Thus far, work in this area has been limited because of the difficulty of gaining access on a regular (~monthly) basis to moderate-aperture telescopes equipped with high resolution optical or infrared spectrographs. Nevertheless, some tantalizing preliminary results are emerging. For example, we display in Fig. 52 the Mt. Wilson measurements of the Ca II cycle-like variations in the solar-type star κ Ceti (G5 V). This active star has a rotation period of 9.4 days, i.e., almost 3 times faster than that of the Sun. Magnetic field measurements from Saar & Linsky (1992) are shown superimposed on the Mt. Wilson S-index values. The mean magnetic field strengths are clearly

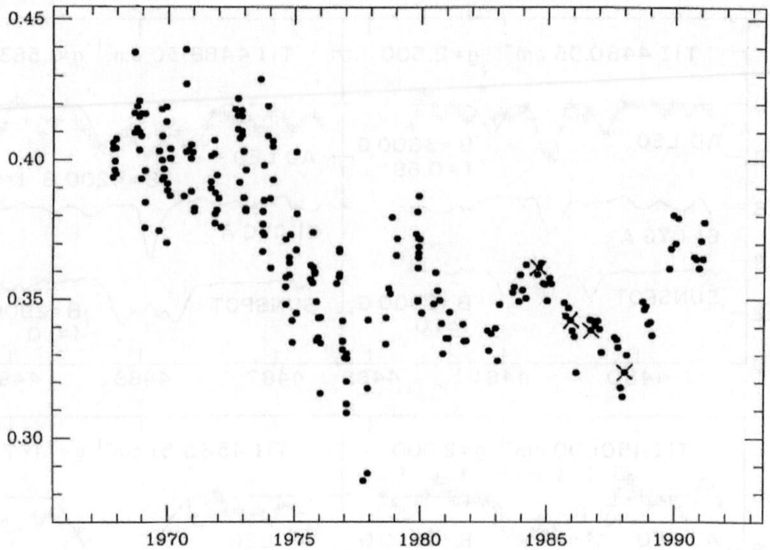

Fig. 52. Evidence for a magnetic cycle in the active solar-type star κ Ceti (G5 V). The ordinate is the Ca II measurement in terms of S-index from the Mt. Wilson program and the abscissa is year. Superimposed on the cycle variation in S-index are crosses showing magnetic flux measurements (*from Saar & Baliunas 1992*)

correlated in a direct manner with the long-term decline in chromospheric Ca II strength in this star.

The quantitative correlation between Ca II HK flux and the mean magnetic flux is given in Fig. 53. Future research in this general area should include long-term, synoptic programs of magnetic flux measurements in sun-like stars. In this way, the empirical relationships between rotation period, Rossby number, cycle period and actual magnetic flux – quantities most directly relevant to dynamo theory – can be developed. Furthermore, the origins and nature of the non-radiative heating of chromospheres and coronae via magnetic fields can be examined for a wide range of stellar parameters.

6 The Coronae of Solar Analogs, Low Mass Stars, and Brown Dwarfs

The corona of the Sun and the coronae of sun-like and other late-type stars represent the most vivid manifestations of magnetic field-related activity in stellar atmospheres. The contrast between the X-ray emission from the hot, $\sim 10^{6-7}$K corona is at its highest as seen against the background of the effectively zero X-ray emittance of the underlying $\sim 10^4$K photosphere. I note that the origin of space weather is in the solar corona. The most violent and energetic explosions in the solar system occur in the corona of the Sun.

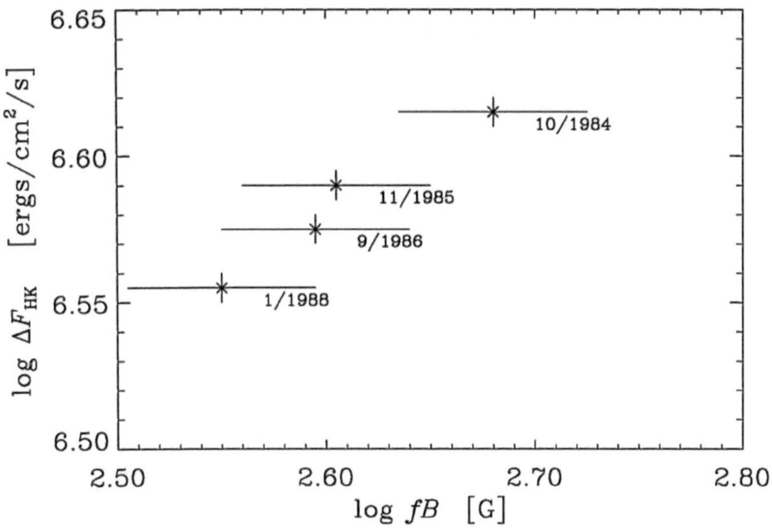

Fig. 53. The change in mean Ca II H&K emission as a function of annual mean magnetic flux on a solar-type star during its cycle (*from Saar 2002*)

6.1 Basic Properties of the Corona in Solar-Type Stars

The Sun itself when viewed as an X-ray star exhibits a pronounced cycle variation in its coronal emission. Peres et al. (2000) have estimated the X-ray luminosity of the contemporary Sun during its cycle as seen in the passband of *ROSAT*, an X-ray satellite observatory that was utilized productively for the observation of X-ray emission from stellar sources. These investigators estimated $L_x \sim 4 \times 10^{27}$ erg s^{-1} at solar maximum and a factor of roughly 20 less at solar minimum, or $L_x \sim 2\times\ 10^{26}$ erg s^{-1}, as seen in the 0.1–3 keV passband of the Position Sensitive Proportional Counter (PSPC) instrument on board the *ROSAT* satellite. The corresponding range in X-ray surface flux (assuming the emission is predominately from relatively compact regions) is $\sim 3 \times 10^3$ erg cm^{-2} s^{-1} at solar minimum to $\sim 6.5 \times 10^4$ erg cm^{-2} s^{-1} at solar maximum.

This range for the X-ray-Sun-as-a-star may be compared to the range observed in solar-type stars. Schmitt (1997) conducted a study of the X-ray properties of a complete sample of solar-type stars in the solar neighborhood. In Fig. 54, the cycle range in X-ray luminosity of the Sun – from log $L_x \sim$ 26.3 to 27.6 – overlaps the low-luminosity range of the distribution for nearby solar-type stars. In fact, the median of Schmitt's distribution (at log $L_x =$ 27.5) is similar to solar maximum levels. Thus, the Sun is not atypical in its X-ray properties though its mean coronal activity does appear somewhat below average when compared to that of a complete, volume-limited sample of nearby solar-type stars. This comparison is further illustrated in Fig. 55 in terms of X-ray surface flux as a function of $B - V$ color (i.e., spectral

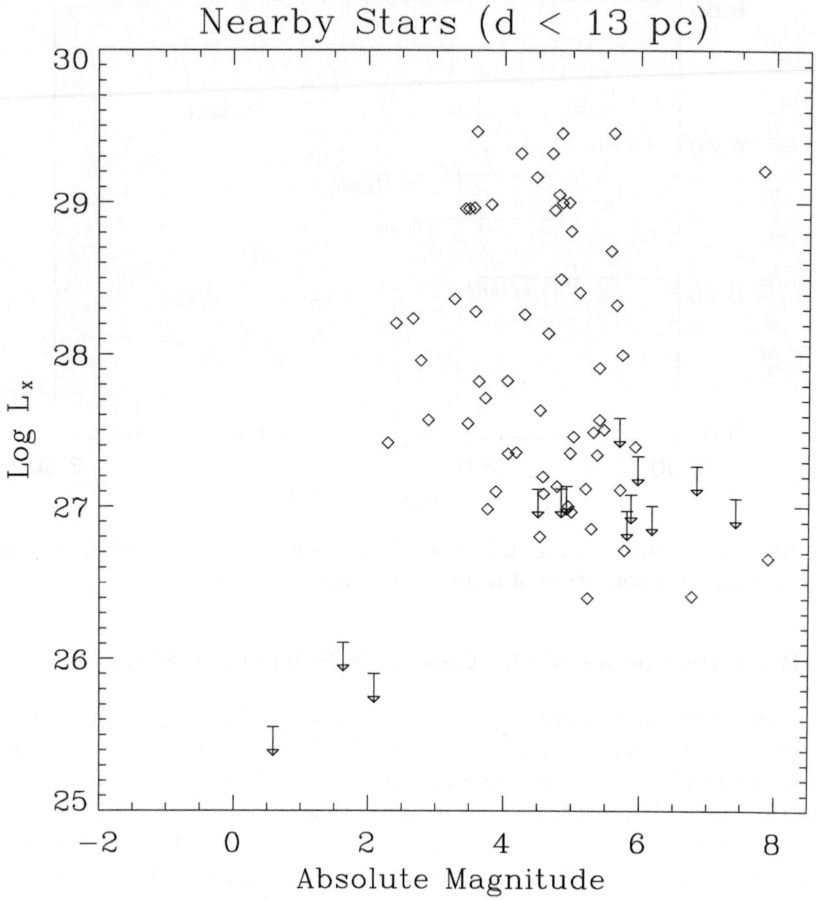

Fig. 54. The distribution of X-ray luminosity as a function of absolute visual magnitude for stars in the solar neighborhood. For reference, the absolute visual magnitude of the Sun is approximately +4.8. The cycle range of the solar X-ray luminosity is log $L_x \sim 26.3$ to 27.6 (*from Schmitt 1997*)

type). The contemporary cycle range in the solar X-ray surface flux, from log $F_x = 3.5$ to 4.8, spans the lower range of the distribution among local solar-type stars. Indeed, solar-minimum would seem to be represented by a large coronal hole covering the entire solar disk!

Among solar-type stars, the spectral hardness of the X-ray emission increases with increasing X-ray luminosity. As Schmitt (1997) points out, this result implies a corresponding increase in the emission measure weighted temperature. Those active solar type stars with harder X-ray emission require a hot component characterized by a temperature ~10^7K, which accounts for most of the inferred soft X-ray emission measure in these objects. By contrast, such a hard component is not seen in the X-ray Sun-as-a-star though it is

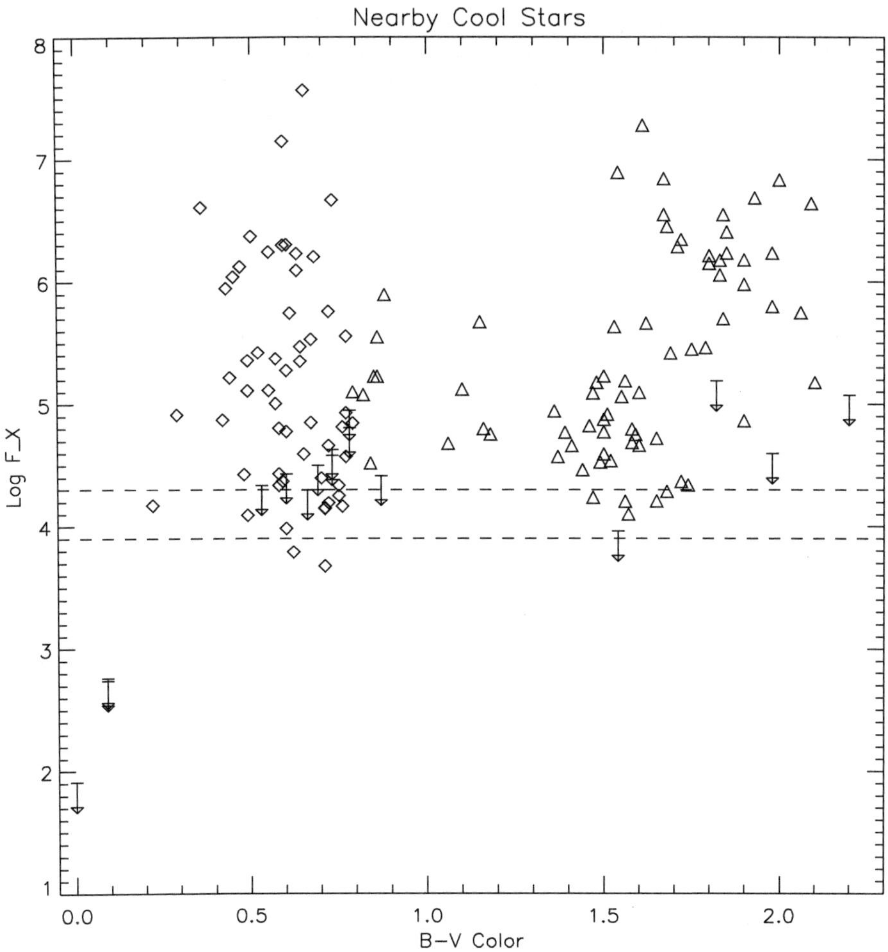

Fig. 55. The distribution of mean X-ray surface flux for solar neighborhood stars versus $B-V$ color, a measure of stellar effective temperature. The $B-V$ color of the Sun is about +0.65. The cycle range of the solar X-ray surface flux is log $F_x = 3.5$ to 4.8. The *dashed* lines delineate the typical range of X-ray surface brightness in solar coronal holes, regions of sharply reduced X-ray emission that are characterized by *open* magnetic field configurations (*from Schmitt 1997*)

observed in resolved-disk observations of solar flares. Thus, the picture of the coronae of sun-like stars that emerges consists of components comprised of coronal hole regions, lare-scale structure associated with the "quiet" corona, active regions, and a hot component that likely consists of the superposition of many transient, flare-like events. The "activity" of the star then depends on the emission measure weighted filling factor of these principal morphological components.

The action of the dynamo in solar-type stars is seen via the empirical relation between X-ray luminosity and rotation. Pallavicini et al. (1981) determined that $L_x \sim v^2$, where v is the projected rotation velocity in their sample spanning G-M stars. Alternatively, the X-ray surface flux $\sim \Omega^2$, where Ω is the stellar angular rotational velocity. As rotation velocity declines as a result of magnetic braking, the coronal X-ray luminosity will also decrease, thus leading to an evolutionary decline in total coronal emission along with a change in the overall coronal structural morphology. In particular, the contribution of the hot component will decline in importance as a sun-like star ages. While the evolution of activity in solar-type stars will be discussed in the next section, we note that X-ray surface fluxes decline about 2.5 orders of magnitude from ages $\sim 10^7$ years to solar-age ($\sim$5 Gyr). Finally, Hempelmann et al. (1996) find tentative evidence for the correlation of X-ray emission with cyclic chromospheric (Ca II K) emission in solar-type stars that exhibit regular cycle variations. The mean X-ray surface flux increases in steps of roughly 1 order of magnitude from stars with constant Ca II H+K time series to stars with regular (cycle) variations in activity to active stars with high levels of chromospheric and coronal emission combined with irregular variability in their time series of Ca II emittance.

In the next section, I will extend this discussion of coronal properties of solar-type stars into a regime that is less sun-like in order to further explore the influence of stellar parameters, such as photospheric effective temperature and fractional convection zone depth, on the nature of stellar coronae.

6.2 Coronae in Low Mass Stars and Brown Dwarfs

I discuss the implications of recent X-ray satellite observations for our understanding of the nature of the coronae and the associated dynamo in M dwarf stars and the so-called ultracool dwarfs. I discuss this aspect of activity in stars that are not solar-type because it provides unique insights on dynamo and nonradiative heating processes in a regime where the photosphere is too cool to make any appreciable contribution to chromospheric and coronal activity, i.e., the mechanisms are purely magnetic. The M dwarf stars also span a region of the H-R diagram where stars become fully convective (implying an operative dynamo that is different from the solar dynamo, at least in the interior).

Observations of stars in the X-ray during pioneering missions such as *Einstein* revealed the seemingly ubiquitous occurrence of coronae in the low-mass stars. The flare-active dMe stars are especially notable for their relatively strong X-ray emission, extending to levels near $L_x/L_{bol} \sim 10^{-2}$ and thus exceeding that of the Sun at solar maximum by roughly 3 orders of magnitude or more. While the transition in observed X-ray properties from late-type dwarfs with radiative cores to those with fully convective interiors appears seamless, the advent of even more sensitive X-ray observatories such as *ROSAT*, *Chandra*, and *XMM*, has led to new insights and deeper questions concerning the nature of coronae in stars with masses near the substellar mass limit. In particular, is steady X-ray emission, analogous to what we see in the Sun, present in the coolest stars or is it purely transient in nature?

6.3 Coronal Structure in M Dwarf Stars

I will first begin with an overview of what we have learned about the coronal structure of dMe and dM stars, based primarily on the analysis of *ROSAT* PSPC data of selected objects (Giampapa et al. 1996). In Fig. 56 we display the PSPC pulse-height spectra of the active flare star AD Leo (dM3.5e) and the quiescent dM2 star GL 411 (dM2), respectively.

A distinguishing feature of the dMe star, compared to the non-dMe star, is the relative "excess" of counts in the higher energy channels near 1 keV. By contrast, there is a paucity of counts in the 0.5–1.0 keV region in the pulse-height spectrum of GL 411. Instead, coronal emission in this quiescent object appears to be dominated by the soft component near 0.25 keV.

This observation immediately suggests the possibility that the corona of a dwarf M star is composed of two components: a soft component characterized by temperatures of a ~few $\times$ 10^6 K and a hard component with a temperature ~10^7 K. The observational distinction, therefore, between the dMe and the dM stars in the context of coronal structure is seen in the relative importance of these components in their overall level of X-ray emission. Of course, the true structure of the corona may involve the superposition of multiple thermal components that are not completely resolved in low-resolution data. In his extensive review, Güdel (2004) emphasizes that the characterization of coronal loop magnetic structures by the product of (base pressure) $\times$ (loop length) can lead to a large range of possible solutions when fit to low resolution observations and, thus, additional constraints are required.

Despite this important and valid caveat, the applicability of a two-component representation of the coronae of M dwarf stars is further supported by the nature of the variability of the X-ray light curve (see Fig. 57). In particular, time-resolved spectroscopy shows significant variability in the high-temperature (~10^7 K) component that is strongly correlated with the variable X-ray light curve. The low-temperature component of the emission measure is essentially constant and uncorrelated with the variable light curve. Thus, it is the hard component that is principally responsible for the observed X-ray variability. This is the fundamental observational justification for a model approach that treats coronae in M dwarf stars as consisting of two distinct thermal components.

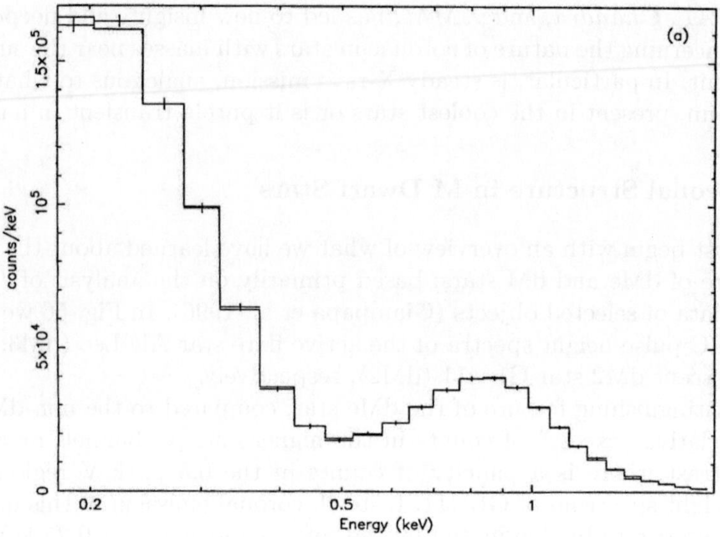

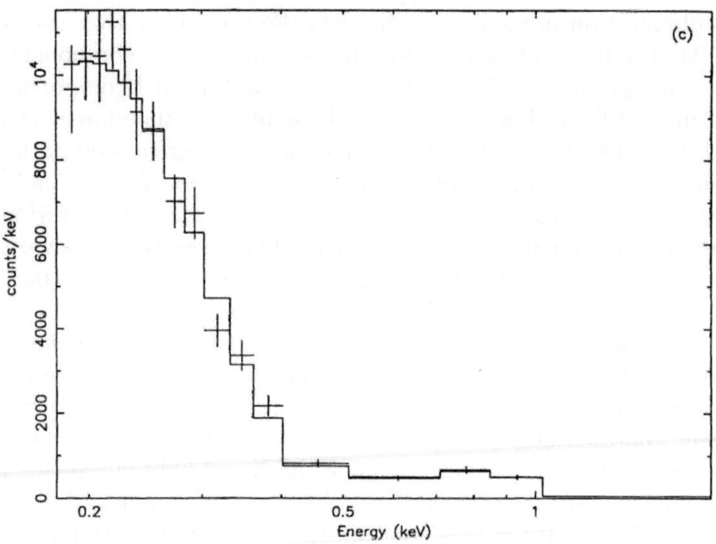

Fig. 56. The ROSAT PSPC pulse-height spectra of AD Leo and GL 411

6.4 Loop Model Results

Adopting a static loop model approach to investigate the properties of the coronae of the red dwarf stars, Giampapa et al. (1996) found that the two principal thermal components of the X-ray corona were characterized as follows: the "soft", or low-temperature component, in the temperature range of $T_L \sim (2\text{–}4) \times 10^6$ K, consists of compact loops with lengths compared to the stellar radius of $l \ll R_*$ and base pressures similar to what is inferred for qui-

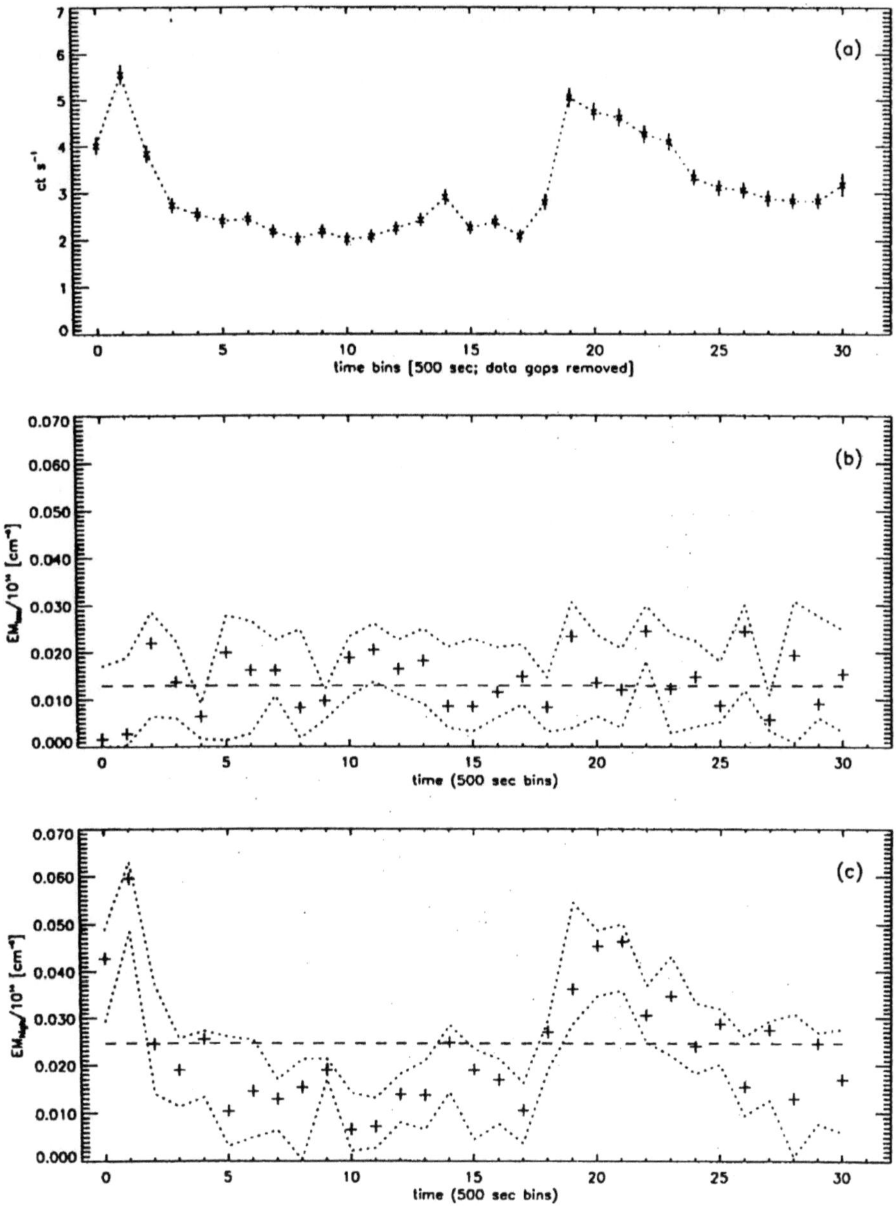

Fig. 57. (**a**) X-ray light curve for AD Leo along with the emission measure variations with time for the (**b**) low- and (**c**) high-temperature components

escent solar active regions. The "hard", or high-temperature component, with temperatures $T_H \sim 10^7$ K, requires loop model solutions with either small filling factors (~0.1), large loops ($l \gtrsim R_*$), and high base pressure ($p_0 \gtrsim p_\odot$), or very small filling factors ($f \ll 0.1$), small loops of length $l \lesssim R_*$, and very high pressure ($p_0 \gg p_\odot$). These properties are reminiscent of compact flaring structures. Thus, the coronal geometry for low-mass dwarf stars appears to be dominated by a combination of relatively compact, quiescent loop configurations and an unstable flaring component.

From an observational perspective, the hard component contributes a relatively larger fraction of the total X-ray emission in dMe stars while also giving rise to the variability in the X-ray light curve. By contrast, in the dM stars the soft component dominates the X-ray emission measure with little or no variability seen in the light curve. In the case of both the dMe and non-dMe (dM) stars, the *ROSAT* PSPC pulse-height spectra are best fit with "depleted" abundances with respect to solar. In general, thermal models based on X-ray spectroscopy – especially for active stars – could only match the observations if an underabundance of metals was assumed. Later observations at higher resolution with *XMM – Newton* and *Chandra* X-ray satellite observatories confirmed this general trend in active stars. In addition, new phenomena correlated with the first ionization potentials of elements [the so-called FIP and Inverse FIP effects] were uncovered (see Güdel 2004 for a comprehensive discussion and review of this as well as other aspects of stellar coronae).

6.5 X-Ray Emission at the End of the Main Sequence

I display in Fig. 58 the observed values of X-ray luminosity normalized by stellar bolometric luminosity, log (L_x/L_{bol}), a measure of coronal heating efficiency, versus absolute visual magnitude for late M dwarf stars from the volume-limited 7 pc sample of Fleming et al. (1995).

Inspection of Fig. 58 does not reveal any obvious change in the dynamo-related, coronal X-ray emission properties of the active M dwarfs across the range in spectral types where the onset of full interior convection has surely occurred, at least according to standard stellar interiors models (Copeland et al. 1970). A high level of coronal heating at values of $L_x/L_{bol} \sim 10^{-3}$ persists for the active dMe stars, at least to the spectral type of VB 8 (dM7e). However, *ROSAT* observations of the very low mass dwarf VB 10 (dM8e) recorded only an upper limit of log (L_x/L_{bol}) < -5.0 though the aftermath of a strong X-ray flare was detected at a peak value exceeding log (L_x/L_{bol}) $= -2.8$. The inferred differential emission measure for the flare of at least 1.8×10^{29} cm^{-5}, combined with adopted electron densities in the range of $n_e \sim 10^9$–10^{10} cm^{-3} yields a characteristic length scale for the flare in the range of $l \sim 0.3$–30 R_*, where R_* is the stellar radius. Lower electron densities would lead to even larger estimates of the scale of coronal structures.

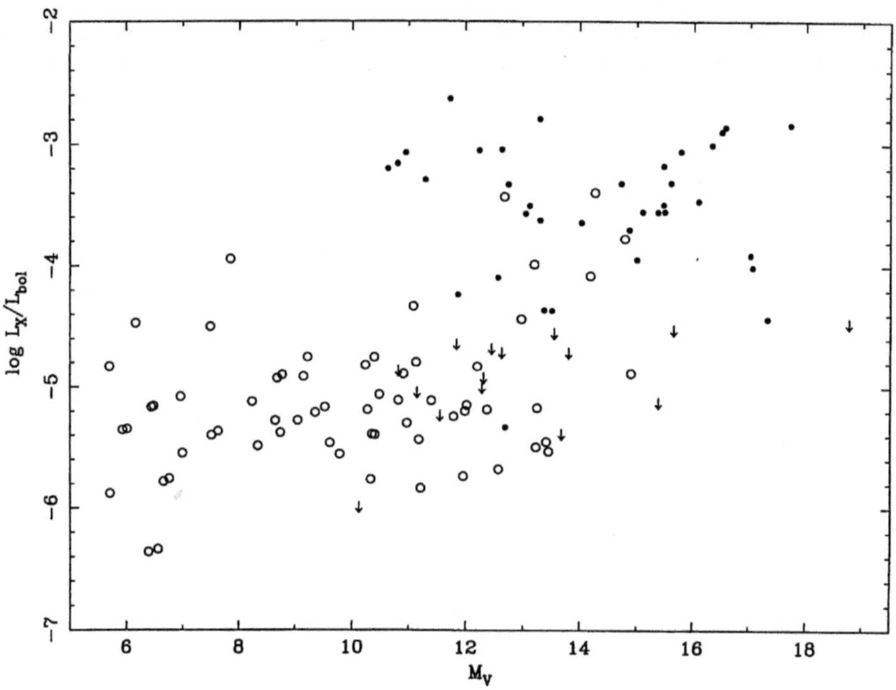

Fig. 58. Coronal heating efficiency among low-mass dwarfs (*from Fleming et al. 1995*)

The size of the source regions inferred from this simple emission measure approach needs to be verified through more rigorous modeling that will likely involve the application of loop models, at least initially. I note the caveat that flare loop models have been shown to generally overestimate the size of the flaring loops (Reale 2002). Independent of the modeling approach are time scale considerations. The minimum duration of the flare maximum of approximately 3 min implies $l \sim 0.7$ R_*, assuming the sound speed for a T $\sim 10^7$ K plasma. Thus, the flare event itself indicates the possible presence of relatively large-scale magnetic structures on VB 10.

6.6 Absence of Quiescent Coronal X-Ray Emission?

The *ROSAT* upper limit to the quiescent X-ray emission of VB 10 is at least two orders of magnitude less than that of more massive M dwarfs. This led Fleming et al. (2000) to suggest that non-flare, quiescent coronal emission, analogous to what is seen in solar active regions, may not occur in ultracool dwarfs such as VB 10. The amplitude of random motions at the footpoints of magnetic loops may be insufficient to jostle or stress magnetic fields. In addition, there are few ions in the very cool photosphere to couple the magnetic field with the surrounding photosphere which, in turn, is dominated by molecules. In particular, the ionization fraction in the upper photosphere of the Sun is $\sim 10^{-4}$ and in early M dwarfs it is $\sim 10^{-5}$. By contrast, Fleming et al. (2000) estimate the ionization fraction in the upper photosphere of VB 10 to be $\sim 10^{-7}$.

In view of the effectively neutral conditions that exist in the cool, dense photosphere of VB 10, the electrical conductivity is so low that any current system rapidly decays. Thus, the magnetic equilibrium of the field configuration is current-free everywhere. Therefore, the magnetic field is in its minimum-energy configuration with the build-up and storage of magnetic field energy excluded. Hence, no magnetic field-related heating occurs. In this scenario, the flare event would arise from more complex magnetic topologies that do not necessarily have footpoints in the cool photosphere or the flares occur as a result of reconnection events in the photosphere. This hypothesis was reinforced by the detection with *Chandra* of a flare event in the young brown dwarf LP944-20 combined with the non-detection at a sensitive upper-limit of its non-flare, quiescent emission (Rutledge et al. 2000).

6.7 Detection of Quiescent Coronal Emission in VB 10

Following these observations, Fleming et al. (2003) then obtained a 12.3 ksec integration on VB 10 using the ASCIS-S(BI) on *Chandra* in which non-flare, quiescent coronal X-ray emission was detected. The count-rate was $n = 0.00212 \pm 0.00042$ s^{-1}. Assuming a Raymond-Smith model at $T = 2 \times 10^6$ K, solar abundances, $N_H = 1.0 \times 10^{18}$ cm^{-2} yields $f_x = 6.36 \times 10^{-15}$ erg cm^{-2} s^{-1} in the 0.2–2.5 keV band. The observed flux corresponds to $L_x = (2.5 \pm 0.5) \times 10^{25}$ erg s^{-1} and $\log (L_x/L_{bol}) = -4.8$. This may be compared to the *ROSAT* upper limit of $L_x < 1.7 \times 10^{25}$ erg s^{-1} and $\log (L_x/L_{bol}) < -5.0$. VB 10 is a soft source with most source counts near 0.25 keV and nearly all the source counts below 1 keV.

Using the Raymond-Smith plasma model in the XSPEC package with $T = 2 \times 10^6$ K (i.e., a soft source) yields a differential emission measure of 1.6 $\times 10^{27}$ cm^{-5}. Adopting densities in the range of $n_e \sim 10^8$–10^9 cm^{-3} suggests large-scale magnetic structures in the range of $l \sim 0.2$–$22\ R_*$. The volume emission measure for the non-flare corona of VB 10 is $\sim 10^{48}$ cm^{-3}. As a brief comparison, the emission measure of the Sun's corona at solar minimum is $\sim 10^{49}$ cm^{-3} and $\sim 10^{50}$ cm^{-3} at solar maximum (Peres et al. 2000, see their Table 2). The quiescent X-ray emission of VB 10 is two orders of magnitude less than that of more massive M dwarfs. Furthermore, the new *Chandra* observation of VB 8 reveals non-flare X-ray emission at a level that is roughly a factor of 4 below the prior *ROSAT* detection, suggesting the likelihood of variability in the mean level of coronal emission in these cool dwarfs. I show in Fig. 59 the revised L_x/L_{bol} *vs* M_v diagram that includes the *Chandra* detection of VB 10 and the upper-limit for the non-flare X-ray emission in the brown dwarf LP 944-20. Inspection of this diagram strongly suggests

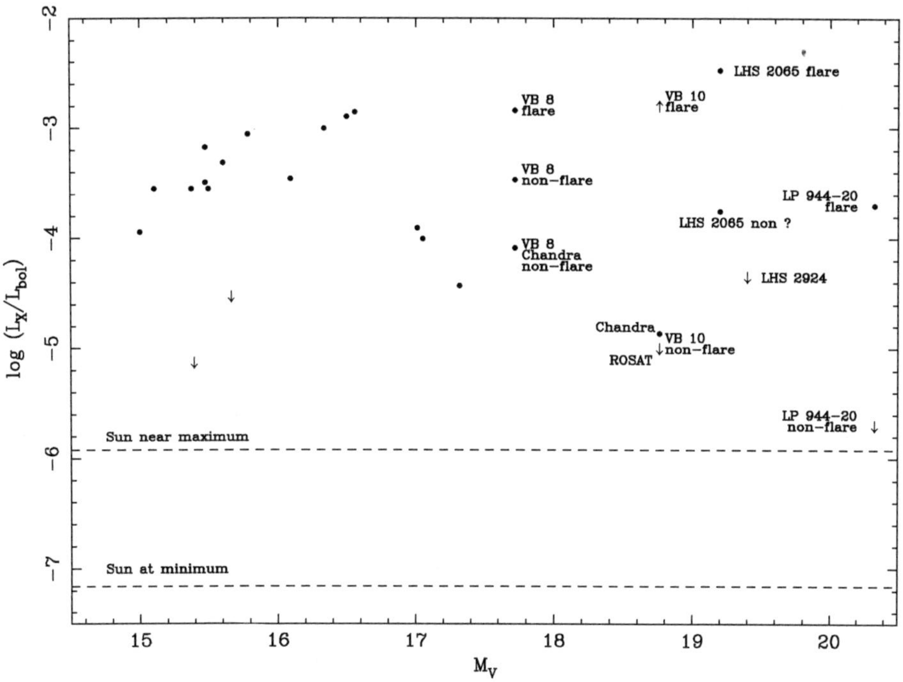

Fig. 59. The trend in coronal heating efficiency as a function of absolute visual magnitude. The new detection of quiescent coronal emission for VB 10 and the flare emission seen in the brown dwarf LP944-20 (Rutledge et al. 2000) are each shown

the onset of a sharp decline in quiescent (non-flare) coronal X-ray emission beginning near the spectral type of VB 10 (M8 V), i.e., $T_{eff} \approx 2600\,K$ and continuing to even cooler effective temperatures.

6.8 Hα and X-Ray Emission in Red Dwarfs

Given the low levels of non-flare X-ray emission in ultracool dwarfs and the corresponding feasibility of detection with *Chandra* or *XMM*, it becomes appropriate to ask how we might learn more about the coronal properties of these objects. One approach is to utilize Hα as a proxy for coronal X-ray emission. Fleming (1988) found that the mean value of the ratio of X-ray luminosity to Hα luminosity in his sample of dMe stars is $L_x/L_{H\alpha} = 6.7$. Whether this facet of the chromospheric and coronal energy balance persists in the ultracool dwarfs is not known in view of the paucity of both X-ray and Hα data for a similarly large sample of stars. In the specific case of VB 10, we infer an observed Hα flux of $f_{H\alpha} = (4.9–9.8) \times 10^{-15}$ erg cm^{-2} s^{-1} as based on the reported range of variation in Hα equivalent width combined with spectrophotometry. This may be compared to the *Chandra* detection of $f_x = 6.2 \times 10^{-15}$ erg cm^{-2} s^{-1}. Thus, the levels of Hα and X-ray emission are comparable in this object. Tinney & Reid (1998) reported for the brown dwarf LP 944-20 a value of log $(L_{H\alpha}/L_{bol}) = -5.6$. This may be compared to the X-ray upper limit of log $(L_{H\alpha}/L_{bol}) < -5.7$ (Rutledge et al. 2000).

While the data are scant, Hα may prove to be a useful guide to the X-ray properties of ultracool dwarfs. If such is the case, then the Hα results of, for example, Gizis et al. (2000) and Burgasser et al. (2002), would suggest that coronal X-ray emission undergoes a rather precipitous decline from the late M (~M8 V) dwarfs through the L and T dwarf regimes. Nevertheless, the presence of detectable Hα emission in an ultracool dwarf indicates the existence of coronal X-ray emission at a level of $f_x \gtrsim f_{H\alpha}$, following the model given by Cram (1982). In brief summary of this model, excitation of chromospheric emission is due mainly to X-ray irradiation by the overlying corona. Thus, in the context of this model, judging the practical feasibility of detection of non-flare X-ray emission by *Chandra* or *XMM* for a given object with Hα emission might therefore be guided by the aforementioned relationship between X-ray and chromospheric (Hα) emission. However, if the region of X-ray emission is in fact far from the star so that the star appears to subtend a finite solid angle (as seen from the principal site of X-ray emission) then geometrical dilution effects (i.e., $1/r^2$ reduction in observed flux) could substantially reduce the strength of any underlying chromospheric Hα emission. In such a case, we may have $f_{H\alpha} \ll f_x$.

6.9 Summary

Magnetic field-related activity occurs in ultracool dwarfs and brown dwarfs as demonstrated by the observation of (non-photospheric) Hα emission along with transient brightenings at Hα, X-ray, and radio wavelengths that are reminiscent of flares. Non-flare, quiescent coronal emission in at least one ultracool dwarf, VB 10, has now been detected. The detections and upper limits in both the X-ray and at Hα indicate the onset of a real decline in chromospheric and coronal heating efficiencies $(L_{x,\,H\alpha}/L_{bol})$ beginning in the ultracool dwarf regime.

It is also important to note that coronal emission may be completely *absent* in many ultracool and brown dwarfs, as indicated by the non-detection of Hα emission in their optical spectra. Fleming et al. (2000) tentatively suggest that $f_{H\alpha} \sim f_x$ in those very low mass stars that exhibit Hα emission. Using this as a guide to the practical feasibility of detection in the X-ray (outside of flares) by either *Chandra* or *XMM* leads to estimates of exposure times of at least 100 ksec for L and T dwarfs at distances greater than 10 pc.

7 The Early Sun

The formation and subsequent evolution of our solar system occurred in a radiative and particle environment that was predominately controlled by the early Sun. The characterization of the changing output of solar-type stars in their total irradiance and activity during the era of planet formation is needed to understand the formation and early evolution of planetary atmospheres at times less than ∼500 Myr. The study of the young precursors of solar-type stars can yield insight on the conditions that characterized our early solar system and its subsequent evolution.

I will approach this subject through a discussion of the nature of activity and variability in the young T Tauri stars. The T Tauri stars are subdivided into the Classical T Tauri stars (CTT) and the Weak T Tauri stars (WTT). The positions of a sample of both CTT and WTT stars in an H-R diagram, along with overlaid isochrones and pre-main sequence evolutionary tracks, are displayed in Fig. 60. The principal morphological difference between these two classes of late-type, pre-main sequence stars is that CTT stars have circumstellar disks and are typically characterized by massive outflows in the range of $\sim 10^{-9}$ to 10^{-8} $M_{\odot}$ yr^{-1}. By contrast, the WTT stars have little or no detectable remnant disk material and significantly lower mass loss rates. While the CTT and WTT stars largely appear coeval in Fig. 60, the two classes may represent an evolutionary sequence from CTT stars, followed by circumstellar disk dissipation, to become a WTT star. Since planet formation has occurred by the WTT stage, I will focus my discussion in this section on the nature of activity in WTT stars as likely proxies of the young Sun and its activity during the formation stage and initial evolution of planetary atmospheres.

7.1 Activity in Weak T Tauri Stars

To recapitulate, WTT stars have little or no remnant circumstellar material. Their spectra typically exhibit narrow chromospheric emission lines. The WTT stars are more common at older (pre-main sequence) ages of ∼10–30 Myr. The rotational velocities of WTT stars are generally in the range of 10–5 kms^{-1}, implying rotation periods of several days for their radii. The rotation rates of both CTT and WTT stars in the Taurus-Auriga star forming region seem to have the same distribution, suggesting that the rotation of T Tauri stars themselves is decoupled from the circumstellar disk. Rebull et al. (2004) confirmed that there is not a statistically significant difference between the distributions of rotational velocities for pre-main sequence stars with disks (CTT stars) and without disks (WTT stars). However, Rebull et al. (2004) caution that the samples of pre-main-sequence stars for which rotational data are available may not be large enough yet to investigate the potential difference in rotational velocity distributions with sufficient statistical accuracy.

The chromospheric and coronal activity of T Tauri stars is substantially enhanced relative to the Sun. This is illustrated in the case of Mg II h and k emission (in the ultraviolet near 280 nm) in Fig. 61 from Giampapa et al. (1981). Inspection of Fig. 61 reveals that the T Tauri stars shown exhibit

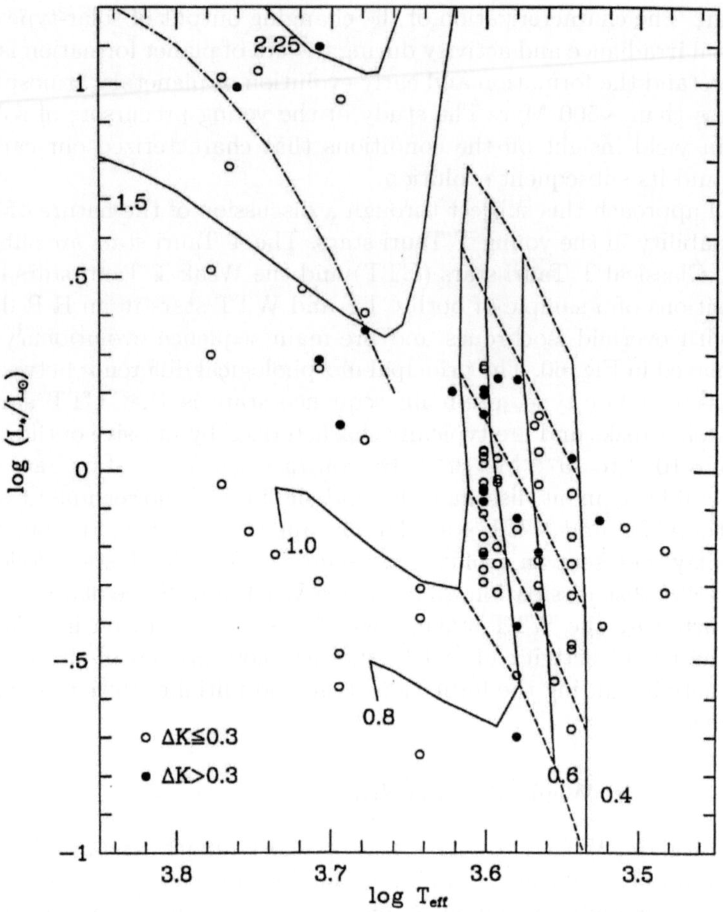

Fig. 60. An HR diagram showing classical T Tauri stars (*filled circles*) and weak T Tauri stars (*open circles*). Isochrones are given (*from K. Strom et al. 1989*)

Mg II emission levels that are typically 1.5–2 orders of magnitude greater than the quiet Sun. Their Mg II emission generally exceeds that of even the dwarf M flare stars. Of course, I note the caveat that not all the Mg II emission, which is typically associated with a compact chromosphere in normal, late-type main sequence stars, in T Tauri stars arises from a classical chromosphere. There is certainly a contribution from an extended region that is related to the outflow from these active objects. Nevertheless, the activity that is presumably magnetic field-related is significantly enhanced relative to other active stellar types or even solar active regions. This claim is further corroborated by the strong X-ray emission observed in these stars. In particular, the relative X-ray emission levels in terms of $L_x/L_{bol} \sim 1000$ times that of the quiet Sun. Powerful X-ray outbursts that are interpreted as the analogs

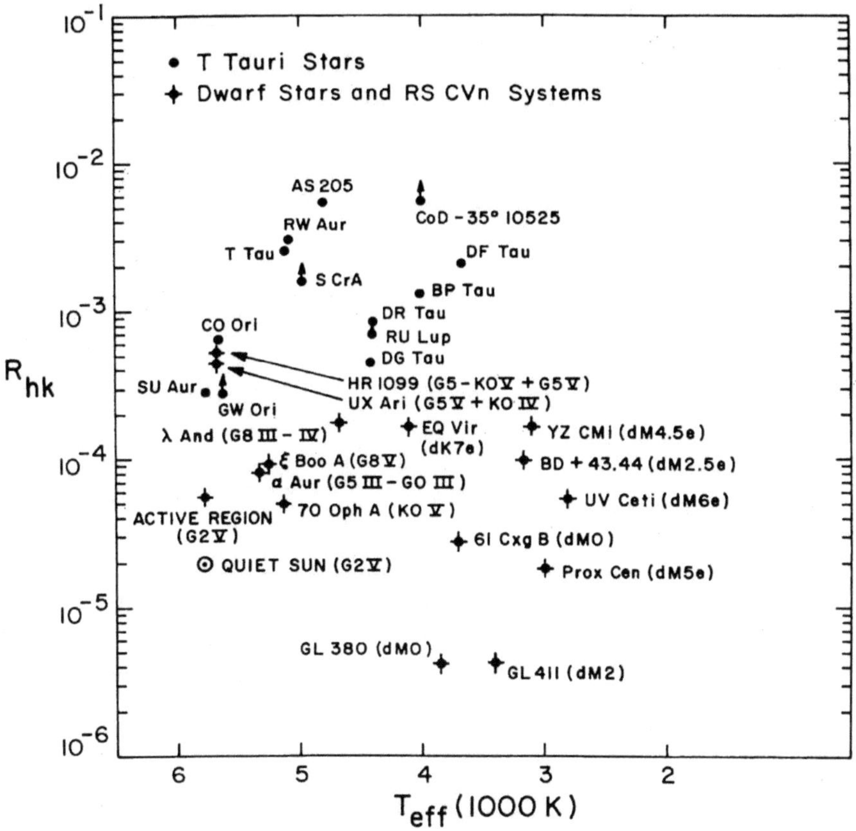

Fig. 61. Normalized Mg II h+k chromospheric emission in CTT stars. Other active stars and the Sun are shown for comparison (*from Giampapa et al. 1981*)

of solar flares are detected with X-ray flare luminosities ~1000 times more powerful than even a Class 3 solar flare. I note that in quantitative terms, a Class 3 solar flare exhibits luminosities in the 8–12 Å XUV range of ~3 $\times$ 10^{26} erg s^{-1}. The unusually energetic flare outbursts observed on the Sun during October–November 2003 attained peak intensities of at least class X7. In quantitative terms, this corresponds to ~2 $\times$ 10^{27} erg s^{-1} in the X-ray. This intense solar flare activity stimulated major geomagnetic storms. Dwarf M stellar flares can be two orders of magnitude more powerful! Examples of the chromospheric emission spectrum of WTT stars are given in Fig. 62 and Fig. 63, respectively. In Fig. 62, the Hα emission line, similar to that seen in dMe stars, is the dominant feature at these red wavelengths. Its strength is an indicator of the occurrence of significant non-radiative heating in the outer atmospheres of these stars. I note, parenthetically, the concomitant appearance of the Li I λ6707 line. Its presence in the spectrum of a late-type

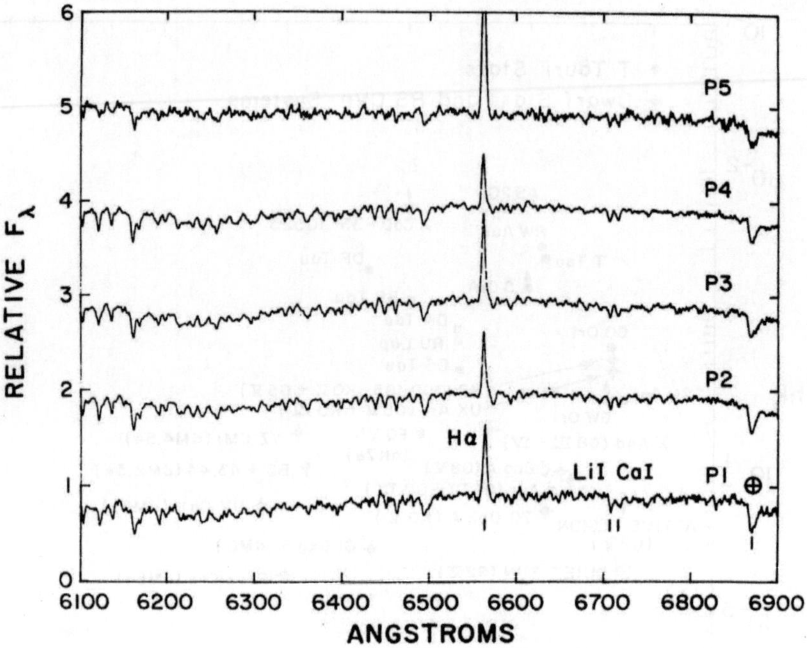

Fig. 62. The Hα region in WTT stars. The Li I λ6707, indicating relative youth and confirming the pre-main sequence nature of these stars, is also seen. The Hα emission lines are reminiscent of dMe flare stars (*from Walter 1986*)

star is an indicator of relative youth. That is, lithium is destroyed through nuclear reactions during the course of convective mixing of material to hotter regions at the base of the convection zone. Thus, the resonance line of neutral lithium at 670.7 nm decreases in strength as the element is depleted with time. Consequently, the feature is very weak or absent in older stars such as the Sun (see Brault & Müller 1971 for a discussion of the solar lithium abundance).

In Fig. 63, the Ca II H & K resonance lines dominate the visible-UV spectrum. The high Balmer lines are evident and reminiscent of the flare spectrum of dMe stars or flares on the Sun. Unlike the solar chromosphere, the Balmer lines dominate the chromospheric radiative losses in WTT stars. Their appearance in WTT stars likely arises from frequent flaring on the stellar surface.

7.2 Inhomogeneous Atmospheres of WTT Stars

The active outer atmospheres of WTT stars are inhomogeneous and consist of cool spots and hotter plages. The fractional area coverages of magnetic surface regions are in the range of ∼5–40% of the stellar surface. Examples of the evidence for thermal inhomogeneities in WTT stars are given in Fig. 64.

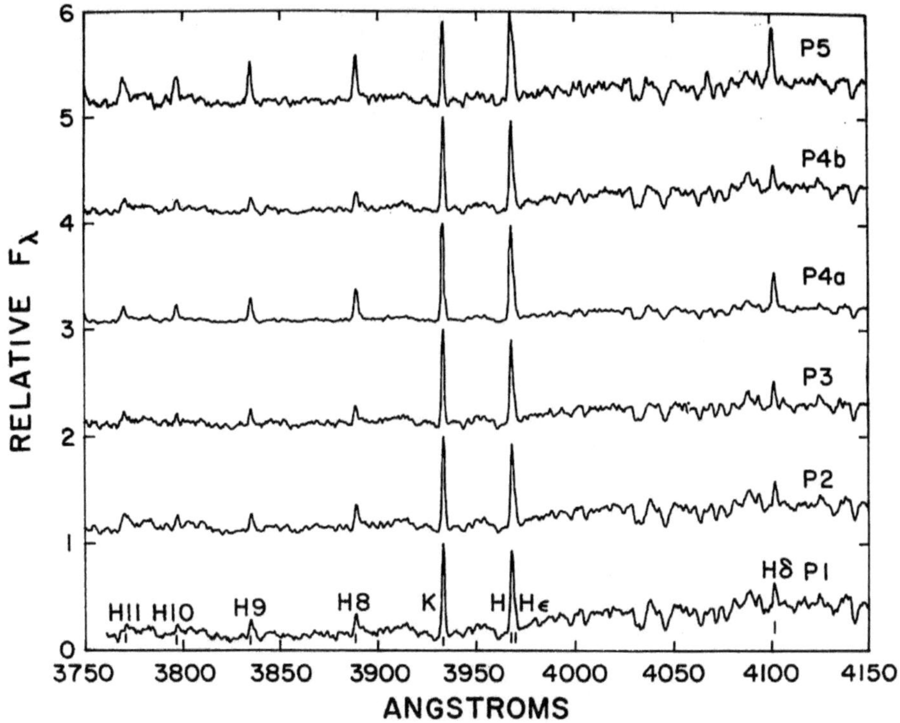

Fig. 63. The Ca II H & K spectral region in WTT stars. High Balmer lines in emission are present. The spectra are similar to that of dMe flare stars (*from Walter 1986*)

The WTT stars in Fig. 64 exhibit photometric V band variations that are periodic and generally attributed to the rotational modulation of cool spots on the stellar surface. The amplitudes of modulation are large compared to that encountered in solar-type stars (e.g., see §3), with photometric variations at the 0.1 mag level or more.

This kind of high-amplitude photometric modulation immediately suggests the presence of significant concentrations of magnetic flux on the stellar surface with field strengths in the kilogauss regime. Cryogenic echelle spectrographs with their high resolution capabilities in the infrared have provided the experimental opportunity to directly investigate this hypothesis. Recalling the discussion in §5, Zeeman splitting is proportional to λ^2 though in practice the gain in sensitivity is only directly proportional to wavelength since natural line widths increase directly with wavelength. In the T Tauri stars with their relatively higher values of v sin$i$ and, thus, larger component of rotational broadening in spectral line profiles, the infrared offers the higher sensitivity needed to compensate for the moderate rotation and

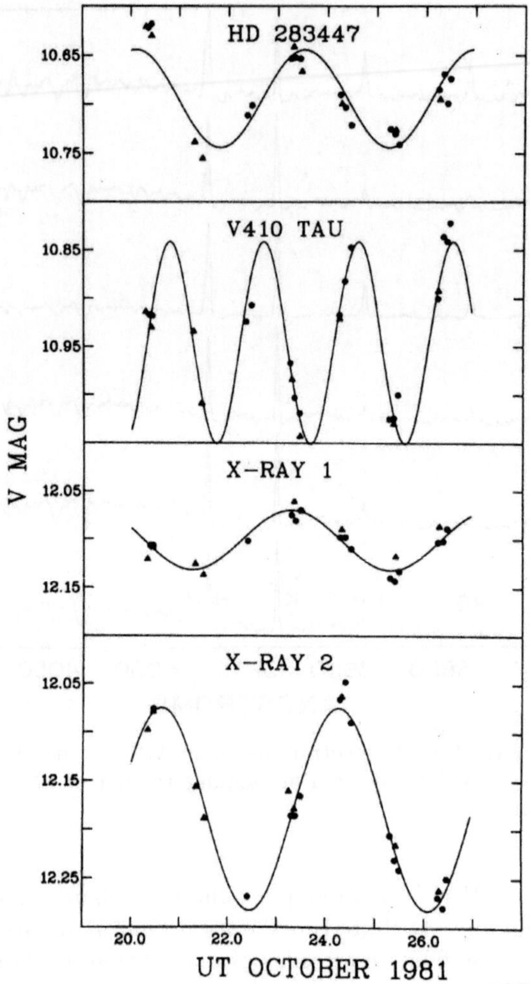

Fig. 64. Periodic photometric variations indicative of large spots on WTT stars (*from Rydgren & Vrba 1983*)

thereby enable reliable quantitative measures of field strengths and fractional area coverages.

Recent observational work by Valenti & Johns-Krull (2004a,b) confirms the widespread occurrence of kilogauss-level fields in T Tauri stars. In Fig. 65 from Valenti & Johns-Krull (2004b), the high resolution spectrum obtained in the infrared K band of the *classical* T Tauri star BP Tau reveals evidence for strong magnetic fields with a mean strength of 2.1 kG over the entire stellar surface. This field strength is inferred from the residual excess broadening of photospheric Ti I lines as compared to model spectra and assuming that the extra line broadening is due to Zeeman effects (see §5). As emphasized by

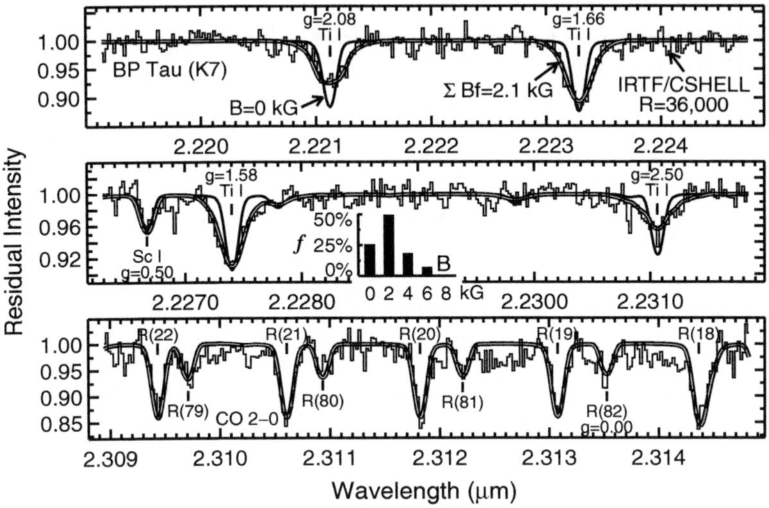

Fig. 65. Evidence of strong magnetic fields on the CTT BP Tau (*from Valenti & Johns-Krull 2004b*)

Valenti & Johns-Krull (2004b), this appears to be a reasonable assumption since the nearby, non-magnetic CO lines do not show evidence of the excess broadening. The quantity that is actually deduced, as shown in the Fig. 65, is the summed contributions of the distribution of field strengths and fractional filling factors. Evidence for similarly strong magnetic fields in *weak* T Tauri stars is also observed (Fig. 66).

In brief summary, the observational results have demonstrated that strong magnetic fields are present in the photospheres of most T Tauri stars. The magnetic fields are characterized by a distribution of strengths up to 6 kG. Furthermore, similar field strengths are present in T Tauri stars with disks (i.e., CTT stars) and without disks (i.e., WTT stars). Given these levels of field strengths and the position of the C(W)TT stars in Fig. 60 – implying surface gravities that are less than solar – it is not surprising when we say that the magnetic pressure exceeds the photospheric gas pressure in T Tauri stars. I would suggest that the multi-kilogauss fields are likely in equipartition with the cool spot photospheres rather than the "quiet" photosphere that essentially is representative of the spectral type and, hence, effective temperature of the TTS. I note that infrared observations will have increased sensitivity to the contributions of cool spot fields to Zeeman broadening because of the declining contrast (increasing relative brightness) of spots in the infrared.

The question arises whether the magnetic activity of T Tauri stars is a result of dynamo action analogous to the regenerative dynamo that operates in the Sun and cool stars. I show in Fig. 67 the observed relation between

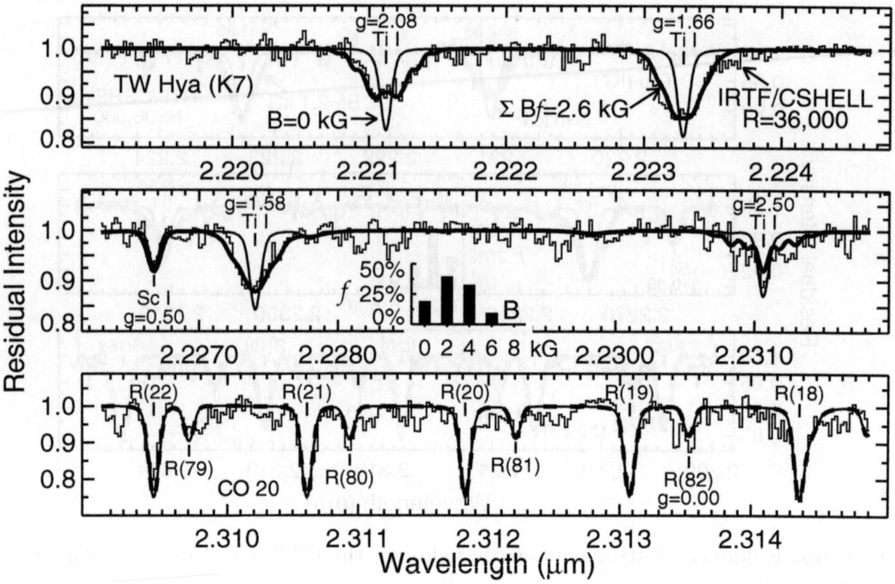

Fig. 66. Evidence of strong magnetic fields in WTT stars (*from Valenti & Johns-Krull 2004a*)

X-ray surface flux and rotation period in those WTT stars that exhibit periodic variations in their photometric light curves and for which X-ray observations have been obtained. Inspection of Fig. 67 reveals that X-ray emission in WTT stars declines with increasing rotation period. Recalling from §2, the "rotation-activity connection" is a signature of dynamo action in the late-type stars. By contrast, Feigelson et al. (2003) do not find a rotation-activity relation for the active, pre-main sequence stars in the Orion star forming region though these investigators still attribute the observed activity to dynamo action (as opposed to, say, a large-scale star-disk interaction). Thus, the precise origin of the magnetic field-related activity in T Tauri stars is still enigmatic. I speculate that a strong and continuous flaring component on these active stars may lead to apparent departures from the activity-rotation connection that has been found for most main-sequence stars that are less flare-active.

7.3 Transient Activity in T Tauri Stars

Given the enhanced levels of chromospheric and coronal emission levels, and the strong magnetic fields, that occur in classical and weak T Tauri stars, it is not surprising to see transient outbursts that are qualitatively similar to solar flares. Two such examples as seen in the X-ray in two pre-main sequence precursors to sun-like stars in the Orion star formation region are shown in Fig. 68. The X-ray light curves exhibit a sudden increase above the pre-flare

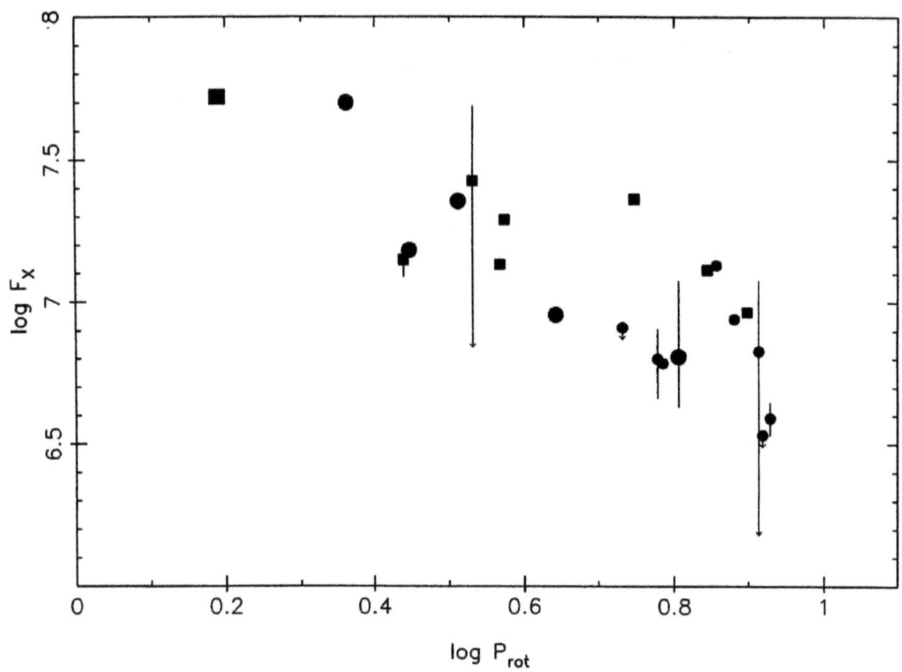

Fig. 67. The correlation between X-ray emission and rotation in CTT (*filled circles*) and WTT (*filled squares*) stars as evidence of rotation-dependent dynamo action in these objects (*from Bouvier 1989*)

levels. According to Feigelson et al. (2002), the young (age ~1Myr) solar analogs in this region exhibit X-ray flares that are typically $\sim 10^{1.5}$ times more luminous and occur $\gtrsim 300$ times more frequent than the most powerful flares thus far observed on the present-day Sun. The distribution of X-ray variability for pre-main sequence stars in the ρ Ophiuchus star formation region is shown in Fig. 69. The power law for the peaks of the distribution is similar to that found by Drake (1971) for solar X-ray flares.

Flare activity in T Tauri stars is also evident in the visible region. A flare recorded in the optical U band in a WTT is shown in Fig. 70. While the magnitude of the event is that of a strong white-light flare, the duration of onset is not sudden as is characteristic of an impulsive event. In this case, the brightening is more gradual and, thus, would be considered an "outburst" rather than a flare. Nevertheless, there are numerous examples of flare-like brightenings recorded for T Tauri stars with ground-based photometry. A notable study by Worden et al. (1981b) presented an analysis of the power spectrum of photometric fluctuations observed in a sample of CTT stars. Their fits to the low-frequency portion of the power spectra yield a frequency dependence in the U-band photometric fluctuations that is similar to solar flares. The U-band fluctuations or "flickering" implied the near-continuous

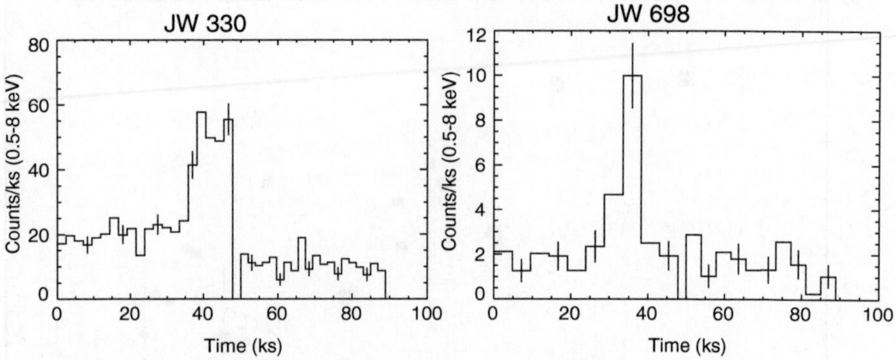

Fig. 68. Examples of X-ray flares in two young solar analogs in the Orion Nebula Cluster star formation region. These observations were obtained with the *Chandra* X-ray satellite observatory. The age of each pre-main sequence star shown here is approximately 10^6 years (*from Feigelson et al. 2002*)

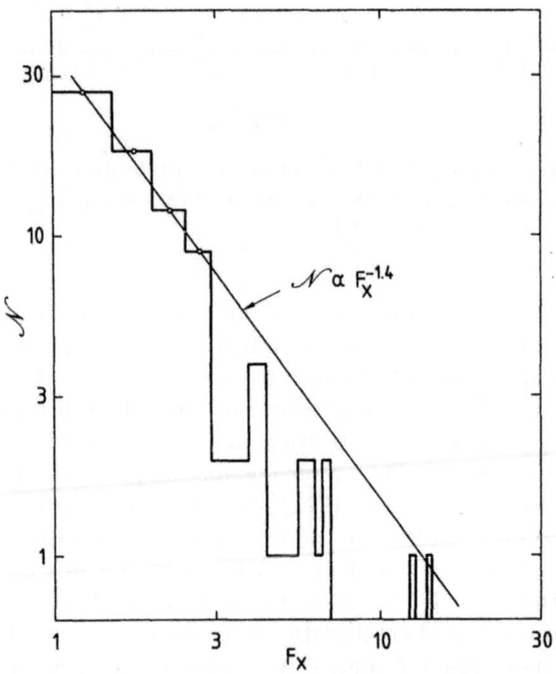

Fig. 69. The amplitude distribution of X-ray flux variations for pre-main sequence stars in the ρ Ophiuchus star formation region. The power-law fit is similar to that found for the time-integrated X-ray flux of solar flares (*from Montmerle et al. 1983*)

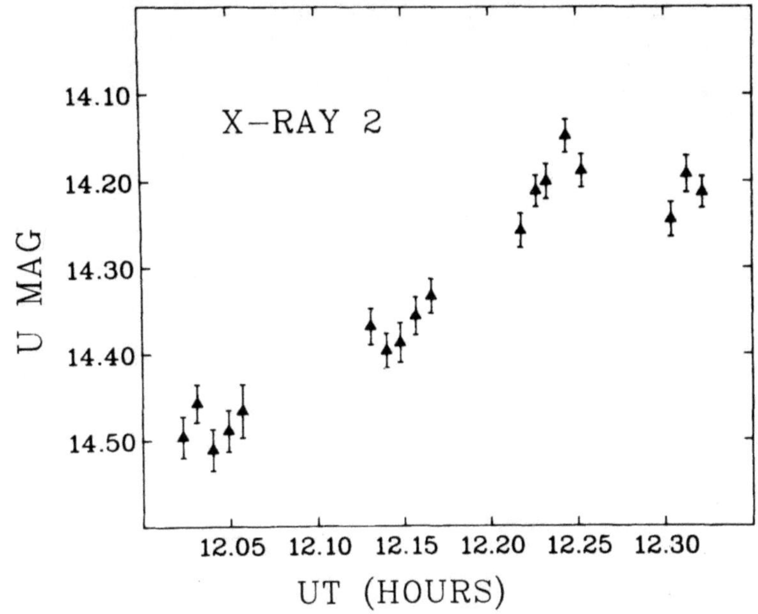

Fig. 70. A slow outburst observed in the near UV optical U band in a WTT star (*from Rydgren & Vrba 1983*)

occurrence of flaring on the surfaces of these magnetically active precursors to solar-type stars. In fact, flare activity accounted for ~5% of the U-band luminosity. Individual events were ~1000 times more energetic than large solar flares.

7.4 The Young Solar System

The conceptual illustration in Fig. 71 of a CTT conveys the principal morphological features of a precursor to a solar-type star and possibly a new solar system. These features include the surrounding, flattened disk, accretion along magnetic lines of force to the stellar surface combined with a bipolar outflow, and a central active star that exhibits both large spots and bright active regions.

A schematic of a model developed by Shu et al. (2001) that attempts to delineate the interaction between the disk and the star in quantitative terms is given in Fig. 72. Our purpose here is not to explore in detail the plethora of physical interactions involved in Shu et al.'s model. Rather, it is to convey in a schematic manner the interactions between magnetic activity on the star in the form of powerful flares and coronal winds, and the magnetic fields that thread through the accretion disk from which emanate mass outflows ("the X-Wind"). A result of these interactions is the production of rock melts from

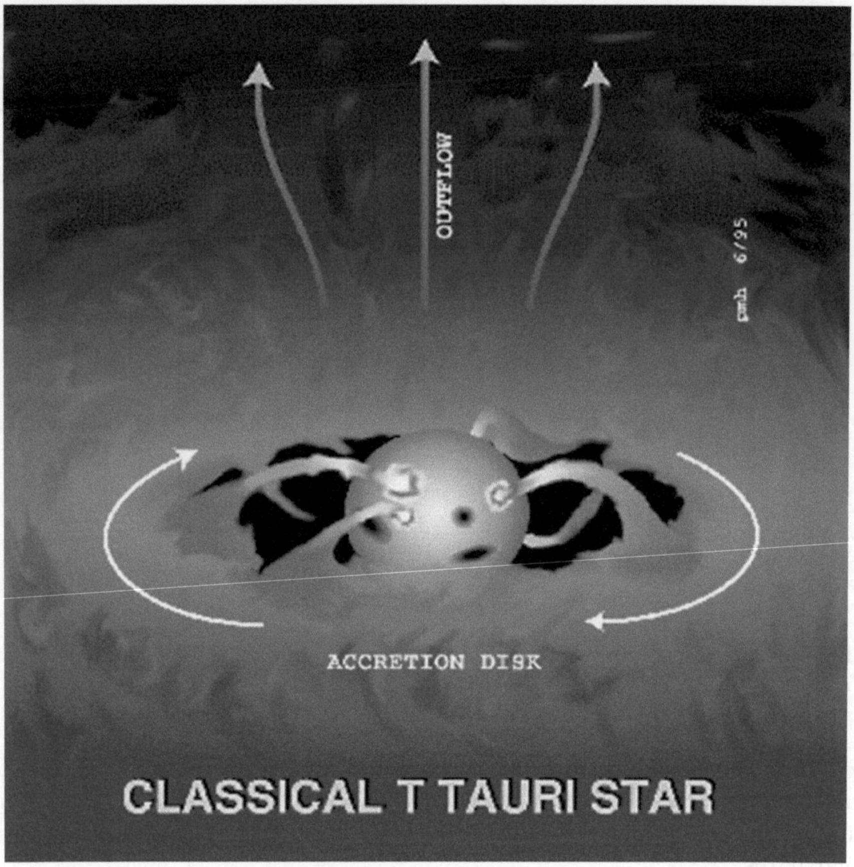

Fig. 71. A conceptual illustration of a T Tauri star showing a circumstellar disk system and polar outflows along with magnetic activity on its surface (*from http://etacha.as.arizona.edu/~eem/ttau/ by E. Mamajek*)

irradiation in the reconnection ring (see Fig. 72) that collide at low velocities and stick together, leading to the formation of small aggregates that are eventually launched via outflows to planetary distances.

Through this kind of model, we may understand the origin of isotopic anomalies in meteoritic material that constitute some of the most direct evidence for an active, early Sun. As Shu et al. (2001) point out, the nature of the carbonaceous chondrites strongly suggest irradiation by proto-solar flares near a young, active Sun. In particular, the observed isotopic excesses of ^{26}Al and ^{53}Mn found in meteorites could be produced by flare spallation in a young, active "WTT Sun". Moreover, spallation-produced ^{21}Ne is found in meteoritic grains with high track densities (e.g., see Caffee et al. 1991; Goswami 1991). The isotopic anomalies and the tracks can be explained by a 1000-fold increase in the proton fluence along with an extension of the

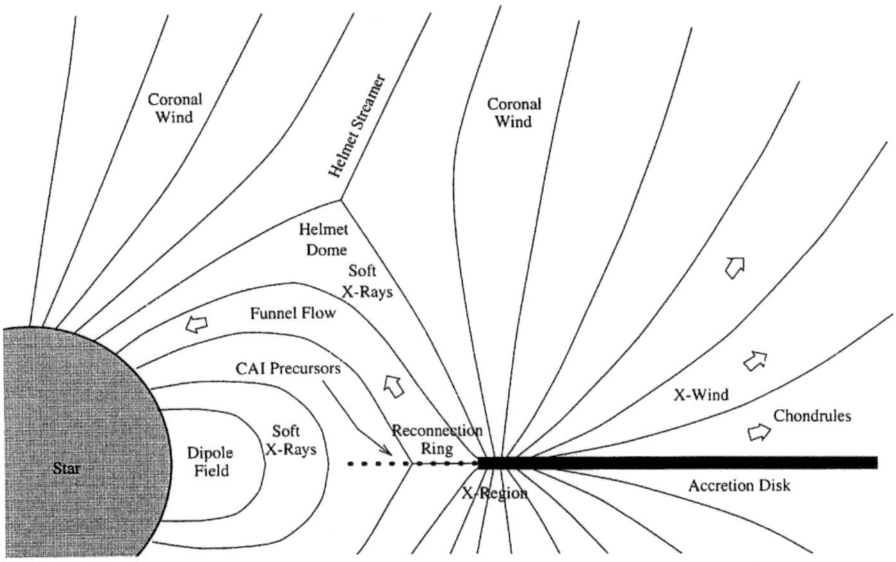

Fig. 72. The X-wind model of activity in a T Tauri star may explain the origin of isotopic anomalies in meteoritic material (*from Shu et al. 2001*)

proton energy spectrum to higher energies than what is observed today. This increase in fluence is supported by the estimates by Worden et al. (1981b) of flare energies and flare powers in T Tauri stars.

Given the abundance of evidence for an active, early Sun, we can begin to develop a picture of the Sun as WTT star. The "cartoon" in Fig. 73 from Feigelson et al. (1991) shows in a pictorial fashion the principal differences between the Sun today and as it likely was as a WTT star. The WTT Sun has larger spots, more rapid rotation and geometrically extended coronal structures, consistent with the lower surface gravity for its larger size at this stage of its pre-main sequence evolution. Some of the relevant properties of the contemporary Sun and the likely characteristics of the Sun as a WTT star are summarized in Table 2.

The enhanced radiation at chromospheric and coronal wavelengths represented in the WTT Sun serve as inputs to planetary atmospheric models during the epoch of formation. We may then ask how activity evolves with time. We know that, in general, activity in single main-sequence stars decays with time as a result of rotational spindown. The rotation rate of a solar-type star decreases via a magnetic braking mechanism first proposed by Schatzman (1962). In brief summary, plasma entrained in magnetic field lines that emanates from the star in the form of a wind carries away angular momentum. The star spins down with time and, as a consequence, the dynamo-induced activity decreases.

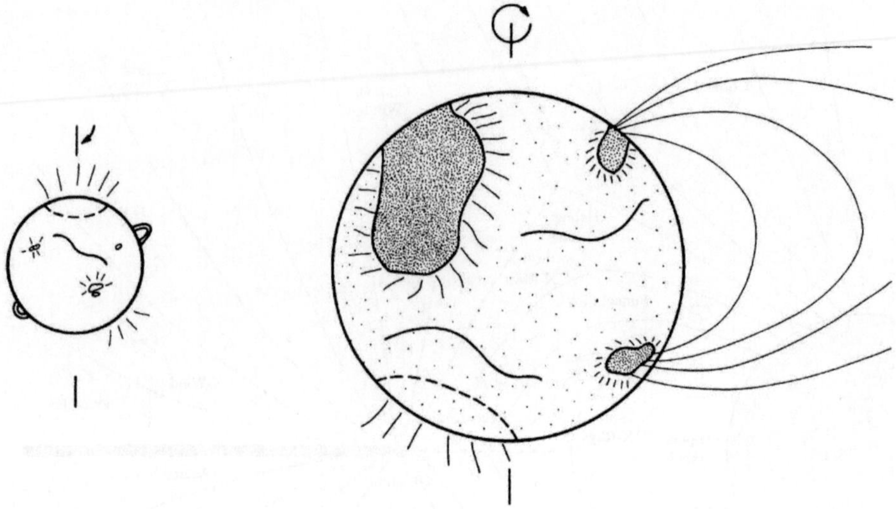

Fig. 73. A schematic illustration of the present-day Sun (*left*) and the WTT Sun (*from Feigelson et al. 1991*)

An example of this evolutionary decline in activity is depicted in Fig. 74 for X-ray emission in solar-type stars in young clusters and the Sun, from Walter & Barry (1991). Power-law decay in the form of

$$F \propto exp(A \times t^{0.5}) \ , \tag{14}$$

is fit to stellar data to describe the decay of activity for a given diagnostic F with stellar age. The constant A, which determines the rate of decay, exhibits a systematic behavior with temperature of formation of a given activity diagnostics. That is, the decay is more rapid at high temperatures of formation than at lower temperatures. For example, $A = -0.54$ for chromospheric Ca II H & K emission while for coronal X-ray emission $A = -2.20$. In the transition

Table 2. The Present-Day Sun and the WTT Sun

Property	Sun Today	WTT Sun
Mass	1 $M_\odot$	≈1 $M_\odot$
Radius	1 $R_\odot$	≈2.5 $R_\odot$
Luminosity	1 $L_\odot$	≈1.7 $L_\odot$
T_{eff}	5786 K	≈4200 K
$V_{equatorial}$	2.0 km s^{-1}	∼25 km s^{-1}
P_{rot}	≃ 25 days	∼5 days
f_{spots}	2% (sol. max.)	∼25%
L_x	$\sim 10^{26}$–$10^{27.5}$ erg s^{-1}	$\sim 10^{30}$ erg s^{-1}

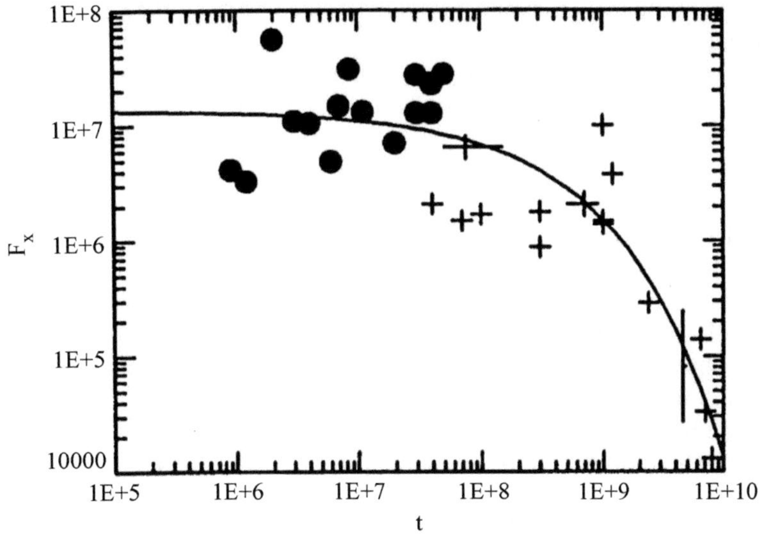

Fig. 74. The evolutionary decline in activity in the form of X-ray emission. Other magnetic field-related activity diagnostics show a similar decline with stellar age (*from Walter & Barry 1991*)

region at the temperature of N V line formation (T $\sim 10^5$ K), the constant A has an intermediate value of -1.00. Utilizing this approach, the level of emission in an activity diagnostic of a solar-type star can be estimated for a given age, up to the age of the Sun. However, this age-activity correlation is still undergoing calibration and should not be relied on exclusively at this time to provide an accurate measure of age for a given field star. For example, the solar cycle variation of X-ray emission by itself could lead to completely erroneous estimates of the age of the Sun depending on the phase of the cycle the Sun happened to be observed (i.e., X-ray emission at the maximum in the solar cycle would give a relatively younger age estimate while an older age would be inferred from the level of emission at solar minimum).

7.5 Summary

We have seen that the pre-main sequence precursors of solar-type stars are characterized by high mean levels of magnetic activity that is typically $\sim 10^2$–10^3 times that of the contemporary Sun. Both principal classes of T Tauri stars – the classical and the weak T Tauri stars – exhibit frequent and highly energetic flare activity. Meteoritic evidence is completely consistent with an early Sun that was characterized by similarly enhanced levels of magnetic activity.

The questions of relevance to studies of the history of our solar system and the evolution of extrasolar systems include:

- What is the detailed evolution of irradiance and activity in the pre-main sequence Sun?
- What is the state of planetary atmospheres during this evolution?
- How are planetary atmospheres influenced by stellar activity on short and evolutionary time scales?

8 Stellar Activity and Extrasolar Planets

The announcement in 1995 of the discovery of an extrasolar planet associated with a solar-type star began a new era of astronomical research (Mayor & Queloz 1995). We now know of many more extrasolar planets than there are planets in our own solar system!

These discoveries and the initiatives that they have spawned to detect terrestrial-size planets more similar to Earth will ultimately enable investigations in the new fields of astrobiology and comparative studies of solar systems. This will include the characterization of the nature and variability of the ambient radiation field in which planetary atmospheres form and evolve. Looking to the future, the nature of extrasolar planetary systems and their prospects as sites for life in the universe is a natural extension of the topic of solar activity and climate.

In view of the goal to detect earth-like planets with their extremely subtle signatures with current detection techniques, the influence of solar-like activity on the methods utilized for the search for extrasolar planets merits consideration. I will therefore examine how activity affects the three principal techniques used for indirect detection, namely, Doppler searches, photometric transits, and astrometric detection.

8.1 Activity and Doppler Searches

A natural asymmetry is present in the photospheric Fraunhofer lines that results from a brightness-velocity correlation between rising and falling granules. The line asymmetry is illustrated in Fig. 75 in which the characteristic "C-shape" of the line bisector is shown for both non-magnetic and magnetic regions on the Sun. The amplitude of the C-shape is lower in magnetic regions versus non-magnetic (quiet) areas on the Sun, presumably resulting from the suppression of convective motions in magnetic regions. This immediately suggests the possibility that the net line bisector amplitude in the spatially integrated spectrum of a late-type star will vary according to total surface activity. This, in turn, will cause an apparent shift in the line centroid that will contribute to the random and, possibly, systematic errors in a radial velocity determination. Obviously, periodic variations on rotational and cycle time scales could occur that mimic the radial velocity signature of an extrasolar planet.

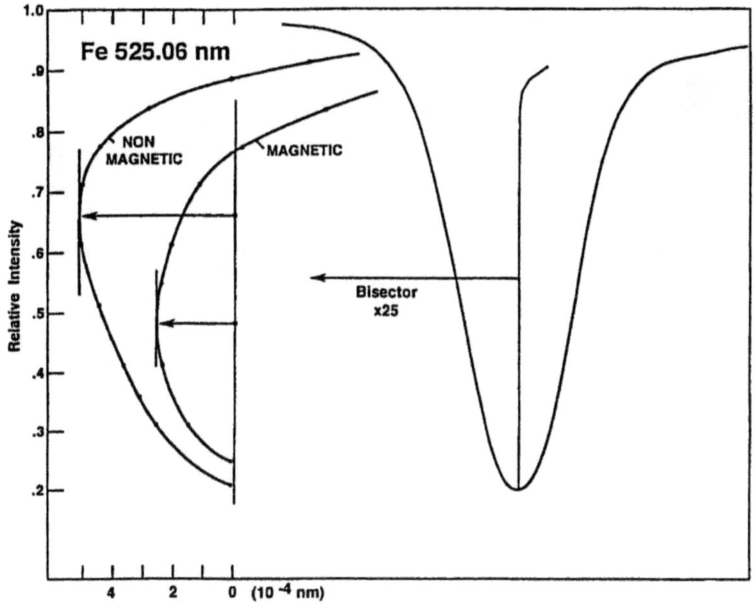

Fig. 75. A schematic illustration of the natural line asymmetry or "C-shape" in the Fraunhofer lines of the solar spectrum in quiet Sun and in magnetic active regions

In addition to the line-shape and associated line centroid position changes that can occur in response to activity, low-amplitude stellar pulsations can produce periodic changes in line positions. Gray (1997) claimed that the effects of either pulsations or activity were the actual origin of radial velocity variations in the sun-like star 51 Peg rather than the result of the reflex motion caused by a planetary companion.

Other synoptic investigations of line bisector variations in stars with putative planets did not find any significant evidence for periodic variations in line asymmetry that could mimic the radial velocity signature of a substellar, planetary companion at the reported periods (Povich et al. 2001; Gray 1998; Hatzes et al. 1998). While these findings confirmed the presence of extrasolar giant planets as the source of the sub-km s^{-1} periodic variability, they by no means excluded the possibility that activity could produce similar signatures or, at least, pose a limiting factor for Doppler detections of extrasolar planets.

In order to investigate the potential magnitude of the effect, Saar, Butler & Marcy (1998) determined the dispersion in high precision radial velocity measurements in excess of the measurement noise as a function of chromospheric activity (Fig. 76). The upshot of Fig. 76 from Saar et al. (1998) is that the error in the radial velocity measurement increases with increasing chromospheric activity (or, alternatively, the precision declines with more active stars).

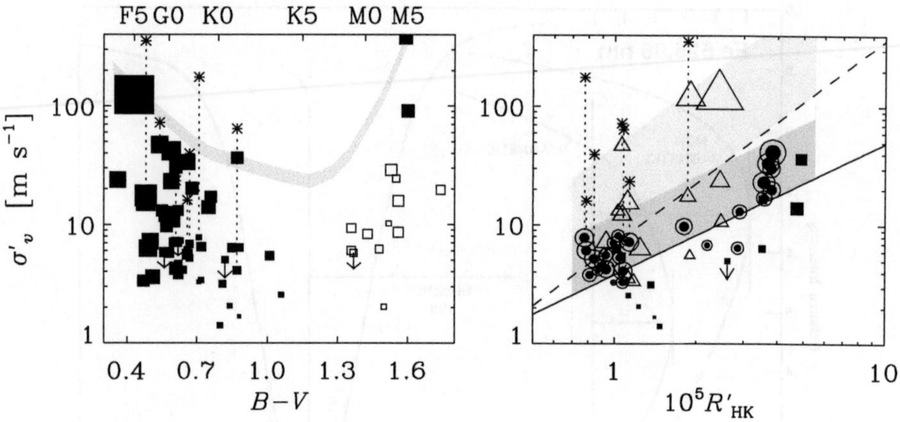

Fig. 76. Radial velocity dispersion in excess of measurement noise as a function of $B - V$ color (*left panel*) and chromospheric activity (*right panel*). Stars with planets are plotted twice–asterisks are before removal of the planetary orbit. The *drop* lines connect to the position of the star in the diagram after correcting for the influence of the planet. The different shapes correspond to different spectral types in the *right* panel along with power law fits to G and K stars (*solid line*) and F stars (*dashed line*). The *right* panel reveals a clear correlation of the scatter in radial velocity with activity (*from Saar, Butler & Marcy 1998*)

Calibration of this effect can increase the precisions attained for more chromospherically active stars, thereby enlarging the sample of targets for planet searches. Nevertheless, magnetic field-related activity can belie the presence of a planetary companion. Queloz et al. (2001) found that the periodic radial velocity variations originally attributed to a planetary companion in the G0 V star HD 166435 actually were a result of line-profile changes arising from spots on the stellar surface. A clue to the purely stellar origin of the line-position variability is provided in Fig. 77 where a clear correlation between variations in the amplitude of the line bisector and the radial velocity is seen. In addition, Queloz et al. (2001) found that both chromospheric and brightness variations were correlated with the radial velocity variations. Although, the persistent, quasi-coherence of the radial velocity variation over more than two years is unusual for a solar-type star.

Another example of a false signature of a planetary companion is given by Henry et al. (2002). In particular, Henry et al. (2002) observed the presence of rotational modulation in the activity indicators of both chromospheric emission and low-amplitude brightness changes at a period equal to that of the reported planetary orbital period of 24 days. Henry et al. therefore presume that the radial velocity variations must arise from intrinsic stellar variations related to the modulation of surface activity and not to the presence of a $\geq$ 0.75 M_{Jup} planet in a 24 day orbit.

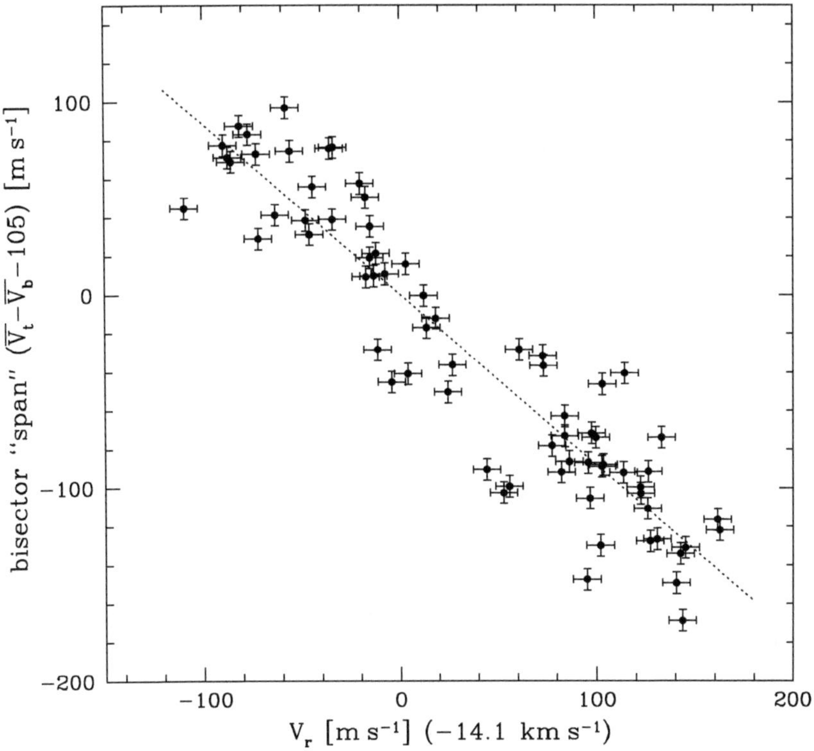

Fig. 77. Bisector amplitude variations are correlated with radial velocity variations suggesting a stellar – rather than planetary companion – origin for the low-amplitude Doppler variations (*from Queloz et al. 2001*)

In contrast to the aforementioned results, Shkolnik, Walker & Bohlender (2003) find a persistent and synchronous enhancement of chromospheric Ca II emission for over 100 orbits in the short-period (3.093 days) planetary system in HD 179949. Shkolnik et al. (2003) report that the Ca II enhancement occurs at a phase corresponding to the sub-planetary point of the 3.093-day orbit. Furthermore, these investigators attribute the synchronized increase in chromospheric activity to an interaction between the planet's magnetosphere and the extended magnetic structures coinciding with activity on the stellar surface. Ip et al. (2004) have developed a model of this kind of interaction for Jovian-size planets in close proximity to their central star (i.e., "51 Peg-like" systems; see also Cuntz et al. 2000). This interaction is illustrated schematically in Fig. 78 from Ip et al. (2004). In addition to the enhancement of chromospheric activity, Ip et al. (2004) discuss the potential enhancement of auroral activity in the giant exoplanet, with possible observable signatures in the near infrared auroral emission lines of H_3^+, analogous to what is observed in the aurora of Jupiter.

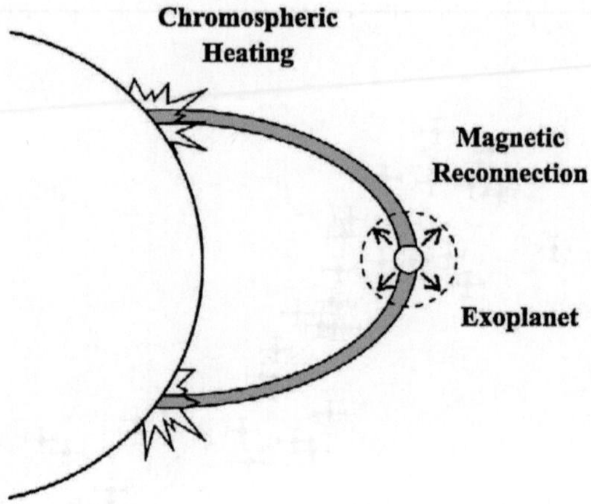

Fig. 78. Schematic model of the interaction of an exoplanet magnetosphere with localized magnetic structure on the nearby central star (*from Ip et al. 2004*)

In brief summary, stellar activity that is modulated at the rotation period can mimic the radial velocity variations of a planetary companion. However, recent observational and theoretical work that suggests the possibility of an enhancement of activity in short-period systems implies that, for chromospherically active stars, the report of a planetary companion must be interpreted cautiously with regard to the existence or non-existence of the putative planet.

8.2 Photometric Transits and Activity

A photometric transit can be detected when a planet crosses our line of sight in front of the disk of a star, as illustrated by the recent transit of Venus across the Sun's disk (Fig. 79). The result is a transient diminution of the light from the star with an amplitude that is equal to the ratio of the area of the planet disk to that of the stellar disk (Rosenblatt 1971). To date, only one such transiting star-planet system is known (Charbonneau et al. 2000) though active searches for other such systems are underway (e.g., Mallen-Ornelas et al. 2003). The photometric transit method for the discovery of extrasolar planets is observationally intensive since a detectable transit requires that the inclination of the stellar rotation axis (which is presumed to determine the inclination of the orbital plane of the planetary companion) be within a fraction of a degree of 90°. For example, the fraction of stars that can present a transit in a sample of sun-like stars, each of which has a planet located at 1 AU from the central star, is $\sim 10^{-5}$.

Despite its observational intensity, the detection of photometric transits is pursued because it is the only method that can yield a reliable estimate of the size of the extrasolar planet (assuming the size of the central star is

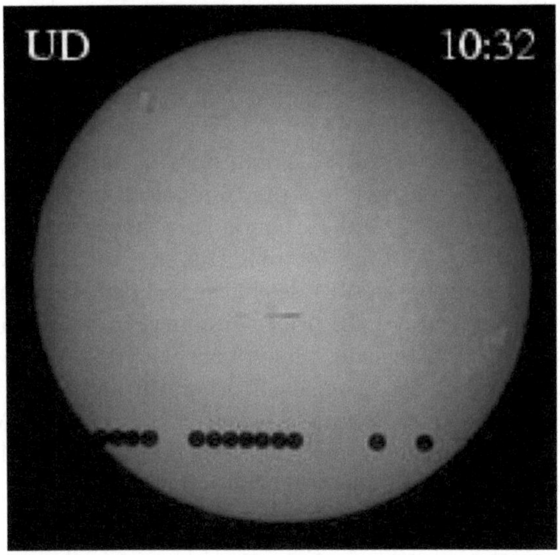

Fig. 79. The transit of Venus on 8 June 2004. This particular sequence of images of this rare event was recorded by the NSO/GONG station in Udaipur, India

known). The size along with the mass, which can be inferred from follow-up Doppler techniques without the ambiguity of inclination effects (for all practical purposes), can yield a mean density that, in turn, provides a critical constraint for the development of semi-empirical models of extrasolar planets.

As inspection of Fig. 79 suggests, there is the potential for confusion of the photometric transit signal with stellar activity, e.g., the disk passage of spots. This potential confusion is mitigated by two key considerations, namely, the time scale of the event and its color signature. In particular, photometric transit time scales are of the order of hours while the rotational modulation of spots on solar-type stars occurs on time scales of days to weeks. Of course, one may wish to know immediately if an event that is only partially seen is the result of a transit or a manifestation of stellar activity. The color signature of the event can distinguish immediately between, say, a spot transit and a planetary transit. The latter is effectively grey while the former exhibits different amplitudes at different wavelengths. I note, parenthetically, that due to limb-darkening some small color differences can occur near the beginning and near the end of a transit event (Borucki & Summers 1984). But, for the most part, a photometric transit is essentially the color-equivalent of a brick crossing in front of our line-of-sight to a star.

As an example, for a sun-like star with a spot, the ratio of continuum brightness of a spot umbra to that at disk center as seen at 0.5 μm is 0.075 (Giampapa et al. 1995). At 2.0 μm, the intensity ratio increases to 0.58 due to the declining contrast of cool spots toward longer wavelengths. Thus, for a spot filling factor of 1% on a solar-type star, the magnitude changes at 0.5 μm and 2.0 μm are 0.01 and 0.005 mag, respectively. In this case, simultaneous observations in the photometric V and infrared K bands can immediately distinguish between the disk passage of a spot and the transit of an extrasolar planet.

Since the dwarf M stars are so numerous in the Galaxy, they become attractive targets for the detection of photometric transits. For an M2 dwarf, the color difference in the V and K bands is estimated by Giampapa et al. (1995) to be 0.33 × (spot filling factor). But for an early M dwarf–brown dwarf binary, the changes in magnitudes in the visible V band and the infrared K band are virtually identical. In brief summary, in addition to time scales of the respective events, transiting planets and the disk passage of spots can be immediately distinguished from each other through joint visible and IR observations in solar-type and cooler dwarf stars. However, M dwarf-brown dwarf pairs cannot be immediately distinguished from an M dwarf-Jupiter pair. Radial velocity measurements will be required.

8.3 Astrometric Detection and Stellar Activity

The ambitious plans of NASA to detect and characterize extrasolar planets are included in the Terrestrial Planet Finder (TPF) mission. One realization of TPF is a large stellar interferometer in space that could detect the micro-arcsecond changes in the center of light of a stellar image due to the barycenter motion of a star–terrestrial (earth-like) planet system.

Transient brightness changes can shift the center of light of the stellar image and contribute noise or confusion in the interpretation of the astrometric signal. For example, the ratio of the apparent displacement of the centroid (in stellar radii) to the change in brightness in magnitudes due to a flare or spot near the limb of a star is given by

$$dr/dm = 0.92f/(1+f)\ , \tag{15}$$

where $f = \Delta I/I$ and $r = X_{CL}/R$. Here, f is the relative change in brightness and X_{CL} is the location of the center of light of the stellar image. Thus, a spot or flare at the limb of the star that produces a 1% change in brightness can shift the centroid by about 1% of the stellar radius. The relative shift in centroid position (in stellar radii) for other plausible changes in brightness (in magnitudes) is summarized in Table 3:

The upper limit to the centroid shift for compact structures at the stellar limb is 1.0 stellar radii, regardless of magnitude change. Let us explore the implications quantitatively for various Sun-planet combinations at a distance of 10 pc for a magnitude change of 5% (corresponding to a centroid shift of 0.044 stellar radii from Table 3). This change in brightness is similar to that seen in the solar-type stars in the Hyades (see §3). For the Sun at 10 pc,this would produce an apparent centroid shift of 0.021 mas. For Jupiter + Sun at 10 pc, the astrometric shift due to barycenter motion is 0.5 mas,

Table 3. Centroid Shifts

Δm (mag)	Δr (stellar radii)
0.10	0.093
1.00	0.441
0.05	0.044

or ~25 times more than the centroid shift due to a brightness change of 5% resulting from activity. However, for an Earth + Sun combination, the astrometric shift is only 3×10^{-4} mas. Thus, the apparent centroid shift due to activity in this example is approximately equivalent to a ~70 earth mass astrometric signature, which is about 74% of a Saturn mass, or about five times the mass of Neptune. At lower levels of variability of, say, Δm = 0.01 mag, the apparent centroid shift approximately corresponds to the astrometric signature of a planet with a mass of about 13 earth masses, or about the mass of Uranus.

Thus, we see that microvariability can obscure astrometric detections at sub-Jovian masses. The mitigating factors include the fact that the activity can be very transient in nature (e.g., brightening due to a stellar flare) or it may be modulated at the stellar rotation period, as in the case for spots. Moreover, the amplitude of the change in brightness due to activity is wavelength-dependent (as discussed in the previous section on photometric transits). Therefore, an astrometric approach to the detection of extrasolar planets with sub-Jovian masses, ideally, should be conducted with two-color imaging. In a single-color approach, red-infrared wavelengths should be emphasized along with repeated observations.

8.4 Summary

Stellar activity can contribute intrinsic error to the indirect methods for the detection of extrasolar planets. The adverse effects of stellar activity in these techniques can be mitigated somewhat by the distinguishing characteristics of time scale and color signature combined with a synoptic approach to detection. However, recent investigations reveal that the presence of a Jovian-size planet in close proximity to its central star can actually induce activity that is modulated at the orbital period of the planet. Thus, care must be taken not to preclude the possibility of a planetary companion just because of a coincidence of activity-related phenomenon occurring at the putative planetary period.

9 Acknowledgement

The National Solar Observatory and the NOAO are each operated by the Association of Universities for Research in Astronomy for the National Science Foundation.

References

Andretta, V., Giampapa, M.S.: ApJ **439**, 405 (1995)
Andretta, V., Giampapa, M.S.: ApJ in preparation (2004)
Ayres, T.R.: ApJ **244**, 1064 (1981)
Ayres, T.R., Linsky, J.L.: ApJ **200**, 660 (1975)
Baliunas, S., Jastrow, R.: Nature **348**, 520 (1990)
Baliunas, S.L., Donahue, R.A., Soon, W.H. et al: ApJ **438**, 269 (1995)
Barry, D.C., Cromwell, R.H.: ApJ **187**, 107 (1974)
Borucki, W.J., Summers, A.L.: Icarus **58**, 121 (1984)
Bouvier, J.: AJ **99**, 946 (1989)
Brault, J.W., E.A. Müller: Solar Phys. **41**, 43 (1971)
Brown, T. M., Gilliland, R.L.: ARA&Ap **32**, 37 (1994)
Burgasser, A.J., Liebert, J., Kirkpatrick, J.D., Gizis, J. E.: AJ **123**, 2744 (2002)
Butler, R.P., Bedding, T.R., Kjeldsen, H. et al: ApJL **600**, L75 (2004)
Caffee, M.W., Hohenberg, C.M., Nichols, R.H., Jr. et al.: Do meteorites contain irradiation records from exposure to an enhanced-activity sun?. In: *The Sun in Time*, ed by C.P. Sonett, M.S. Giampapa, M.S. Matthews (Tucson, The University of Arizona 1991) p 413
Cayrel de Strobel, G.: A&A Rev. **7**, 243 (1996)
Charbonneau, D., Brown, T.M., Latham, D.W., Mayor, M.: ApJL **529**, l45 (2000)
Christensen-Dalsgaard, J.: On the asterometric HR diagram. In: textitGONG 1992. Seismic Investigation of the Sun and Stars, ASP Conf. Ser. 42, ed. by T. M. Brown (San Francisco: ASP 1992), p. 347
Copeland. H., Jensen, J.O., Jorgensen, H.E.: A&A **5**, 12 (1970)
Cox, J.P.: *Theory of Stellar Pulsation* (Princeton, Princeton University Press 1980)
Cram, L.E.: ApJ **253**, 768 (1982)
Cram, L.E., Mullan, D.J.: ApJ **234**, 579 (1979)
Cuntz, M., Saar, S.H., Musielak, Z.E.: ApJL **533**, L151 (2000)
Däppen, W., Dziembowski, W.A., Sienkiewicz, R.: Asteroseismology–Results and Prospects. In: *Advances in Helio- and Asteroseismology, IAU Symp No. 123*, ed by J. Christensen-Dalsgaard, S. Frandsen (Dordrecht: Reidel 1988), p. 233
Demarque, P., Guenther, D.B., Green, E.M.: AJ **103**, 151 (1992)
Donahue, R. A., Saar, S.H., Baliunas, S.L.: ApJ **466**, 384 (1996)
Drake, J.F.: Solar Phys. **16**, 152 (1971)
Farnham, T. L., Schleicher, D.G., A'Hearn, M.F.: Icaurs **147**, 180 (2000)
Feigelson, E. D., Giampapa, M.S., Vrba, F.J.: Magnetic activity in pre-main sequence stars. In: textitThe Sun in Time, ed by C.P. Sonett, M.S. Giampapa, M.S. Matthews (Tucson, The University of Arizona 1991) p 658
Feigelson, E. D., Garmire, G.P., Pravdo, S.H.: ApJ **572**, 335 (2002)
Feigelson, E. D., Gaffney, J.A. III, Garmire, G.P. et al.: ApJ **584**, 911 (2003)

Fleming, T.A.: Optical Analysis of an X-Ray Selected Sample of Stars. PhD Thesis, University of Arizona, Tucson (1988)
Fleming, T.A., Schmitt, J.H.M.M., Giampapa, M.S.: ApJ **450**, 401 (1995)
Fleming, T.A., Giampapa, M.S., Schmitt, J.H.M.M.: ApJ **533**, 372 (2000)
Fleming, T.A., Giampapa, M.S., Garza, D.: ApJ **594**, 982 (2003)
Foukal, P.: Science **264**, 238 (1994)
Foukal, P.: ApJ **500**, 958 (1998)
Foukal, P.: EOS **84**, 205 (2003)
Foukal, P., Lean, J.: Science **247**, 556 (1990)
Foukal, P., Milano, L.: Geophys. Res. Lett., **28**, 883 (2001)
Foukal, P., Hall, J., North, G., Wigley, T.: Science, submitted (2004)
Fröhlich, C., Lean, J.: Astron. Nach. **323**, 203 (2002)
Giampapa, M.S.: ApJ **299**, 781 (1985)
Giampapa, M.S., Calvet, N., Imhoff, C.L., Kuhi, L.V.: ApJ **251**, 113 (1981)
Giampapa, M.S., Worden, S.P., Linsky, J.L.: ApJ **258**, 740 (1982)
Giampapa, M.S., Craine, E.R., Hott, D.A.: Icarus **118**, 199 (1995)
Giampapa, M.S., Rosner, R., V. Kashyap, V. et al.: ApJ **463**, 707 (1996)
Giampapa, M.S., Hall, J.C., Radick, R.R. et al: ApJ, in preparation (2004)
Gizis, J.E., Monet, D.G., Reid, I.N. et al.: AJ **120**, 1085 (2000)
Goswami, J.N.: Solar flare heavy-ion tracks in extraterrestrial objects. In: *The Sun in Time*, ed by C.P. Sonett, M.S. Giampapa, M.S. Matthews (Tucson, The University of Arizona 1991) p 426
Gough, D.O.: The internal structure of late-type main-sequence stars In: *Astrophysics–Recent Progress and Future Possibilities*, ed by B. Gustaffson, P.E. Nissen (Copenhagen, Kongelige Danske Videnskabernes Selskab 1990) p 13
Gray, D.F.: Nature **385**, 795 (1997)
Gray, D.F.: Nature **391**, 153 (1998)
Güdel, M.: A&AR, in press (2004)
Hall, J.C.: PASP **108**, 313 (1996)
Hall, J.C., Lockwood, G.W.: ApJ **438**, 404 (1995)
Hall, J.C., Lockwood, G.W.: ApJ **493**, 494 (1998)
Hall, J.C., Lockwood, G.W.: ApJ **541**, 436 (2000)
Hall, J.C., Lockwood, G.W.: *The Next Decade of Stellar Cycles Research*, (Lowell Observatory, Flagstaff, Arizona 2004) pp 13–21
Hall, J.C., Lockwood, G.W.: ApJ submitted (2004)
Hatzes, A.P., Cochran, W.D., Bakker, E.J.: Nature **391**, 154 (1998)
Heasley, J.N., Ridgway, S.T., Carbon, D.F. et al: ApJ **219**, 970 (1978)
Hempelmann, A., Schmitt, J.H.M.M., Stępień, K.: A&A **305**, 284 (1996)
Henry, G.W., Donahue, R.A., Baliunas, S.L.: ApJL **577**, L111 (2002)
Hoyt, D.V., Schatten, K.H., Nesmes-Ribes, E.: Geophys. Res. Lett. **21**, 2067 (1994)
Ip, W.-H., Kopp, A., Hu, J.-H.: ApJL **602**, L53 (2004)
Kelch, W.L., Linsky, J.L., Worden, S.P.: ApJ **229**, 700 (1979)
Knaack, R., Fligge, M., Solanki, S.K., Unruh, Y.C.: A&A **376**, 1080 (2001)
Lean, J., Skumanich, A., White, O.: Geophys. Res. Lett. **19**, 1591 (1992)
Lean, J., Beer, J., Bradley, R.: Geophys. Res. Lett., **22**, 3195 (1995)
Linsky, J.L., McClintock, W., Robertson, R.M., Worden, S.P.: ApJS **41**, 47 (1979)
Livingston, W., Wallace, L.: Solar Phys. **212**, 227 (2003)
Lockwood, G.W., Skiff, B.A., Radick, R.R.: ApJ **485**, 789 (1997)

Mallen-Ornelas, G., Seager, S., Yee, H.K.C. et al.: ApJ **582**, 1123 (2003)
Marcy, G.W.: ApJ **276**, 286 (1984)
Mayor, M., Queloz, D.: Nature **378**, 355 (1995)
McMurry, A.D., Jordan, C.: MNRAS **313**, 423 (2000)
Montgomery, K.A., Marschall, L.A., Janes, K.A.: **106**, 181 (1993)
Montmerle, T., Koch-Miramond, L., Falgarone, E., Grindlay, J.E.: ApJ **269**, 182 (1983)
Noyes, R.W., Hartmann, L.W., Baliunas, S.L., Duncan, D.K., Vaughan, A.H.: ApJ **279**, 763 (1984)
Pallavicini, R., Golub, L., Rosner, R., Vaiana, G.S. et al.: ApJ **248**, 279 (1981)
Peres, G., Orlando, S., Reale, F. et al.: ApJ **528**, 537 (2000)
Porto de Mello, G. F., da Silva, L.: ApJL **482**, L89 (1997)
Povich, M.S., Giampapa, M.S., Valenti, J.A. et al.: AJ **121**, 1136 (2001)
Queloz, D., Henry, G.W., Sivan, J.P. et al.: A&A **379**, 279 (2001)
Radick, R. R., Lockwood, G.W., Skiff, B.A., Baliunas, S.L.: ApJS **118**, 239 (1998)
Reale, F.: Stellar flare modeling. In: *Stellar Coronae in the Chandra and XMM-Newton Era*, ASP Conference Series Vol. 277, ed by F. Favata, J.J. Drake (ASP, San Francisco 2002) p 103
Rebull, L.M., Wolff, S.C. Strom, S.E.: AJ, in press (2004)
Robinson, R.D., Jr.: ApJ **239**, 961 (1980)
Robinson, R.D., Worden, S.P., Harvey, J.W.: ApJL **236**, L155 (1980)
Rosenblatt, F.: Icarus **14**, 71 (1971)
Rosner, R.: Stellar coronae: interpretation and modeling of stellar coronae. In: *Cool Stars, Stellar Systems, and the Sun*, SAO Spec Rept 389, ed by A. K. Dupree (Smithsonian Astrophysical Observatory 1980) pp 79–96
Rutledge, R.E., Basri, G., Martin, E.L., Bildsten, L.: ApJL **538**, L141 (2000)
Rydgren, A.E., Vrba, F.J.: ApJ **267**, 191 (1983)
Saar, S.H.: Stellar dynamos: scaling laws and coronal connections. In: *Stellar Coronae in the Chandra and XMM-Newton Era*, ASP Conference Series Vol. 277, ed by F. Favata, J.J. Drake (ASP, San Francisco 2002) p 311
Saar, S.H., Linsky, J.L.: ApJL **299**, L47 (1985)
Saar, S.H., Baliunas, S.L.: The magnetic cycle of kappa Ceti. In: *The Solar Cycle*, ASP Conf. Series, Vol. 27, ed by K.L. Harvey (ASP, San Francisco 1992) p. 197
Saar, S.H., Butler, R.P., Marcy, G.W.: ApJL **498**, L154 (1998)
Schatzman, E.: Ann. Astrophys. **25**, 18 (1962)
Schmitt, J.H.M.M.: A&A **318**, 215 (1997)
Shine, R.A., Linsky, J.L.: Solar Phys. **39**, 49 (1974)
Shkolnik, E., Walker, G.A.H., Bohlender, D. A.: ApJ **597**, 1092 (2003)
Shu, F.H., Shang, H., Gounelle, M. et al.: ApJ **548**, 1029 (2001)
Skumanich, A., Smythe, C., Frazier, E.N.: ApJ **200**, 747 (1975)
Stauffer, J.R., Hartmann, L.W.: ApJS **61**, 531 (1986)
Stimets, R.W., Giles, R.: ApJL **242**, L37 (1980)
Strom, K.M., Strom, S.E., Edwards, S. et al.: AJ **97**, 1451 (1989)
Thomas, R.N.: ApJ **125**, 260 (1957)
Ulmschneider, P.: The physics of chromospheres and coronae. In: *Lectures on Solar Physics*, Lecture Notes in Physics, Vol. 619, ed by H.M. Antia, A. Bhatnagar, P. Ulmschneider (Springer 2003) p. 232

Valenti, J.A., Johns-Krull, C.M.: Infrared Observations of Magnetic Fields on Young Stars. In: *High Resolution Infrared Spectroscopy in Astronomy*, ed by H.U. Kaufl, R. Siebenmorgen, A. MOorwood, (Springer 2004a), in press

Valenti, J.A., Johns-Krull, C.M.: Observations of magnetic fields on T Tauri stars. In: *Magnetic Fields and Star Formation: Theory versus Observations*, ed by A.I.Gomez de Castro, M. Heyer, E. Vazquez-Semadeni, R. Rebolo, M. Tagger, R.E. Pudritz, (Kluwer 2004b), Ap&SS, 291, in press

VandenBerg, D.A., Bridges, T.J.: ApJ **278**, 679 (1984)

Vaughan, A. H., Preston, G.W.: PASP **92**, 385 (1980)

Vaughan, A. H., Baliunas, S.L., Middlekoop, F. et al.: ApJ **250**, 276 (1981)

Walter, F.M.: ApJ **306**, 573 (1986)

Walter, F.M., Barry, D.C.: Pre- and main-sequence evolution of solar activity. In: *The Sun in Time*, ed by C.P. Sonett, M.S. Giampapa, M.S. Matthews (Tucson, The University of Arizona, 1991) p 633

Wiedemann, G., Ayres, T.R., Jennings, D.E., Saar, S.H.: ApJ **423**, 806 (1994)

Wilson, O.C.: ApJ **226**, 379 (1978)

Wolf, J.R.: Vierteljahrsschrift der Naturforschenden Gesellschaft in Zurich **3**, 124 (1856)

Wolf, C.A., Sumner, D.A.: Am Journal Agricultural Economics **83**, 77 (2001)

Worden, S.P., T.J. Schneeberger, M.S. Giampapa: ApJS **46**, 159 (1981a)

Worden, S.P., Schneeberger, T.J., Kuhn, J.R., Africano, J.L.: ApJ **244**, 520 (1981b)

Young, A., Skumanich, A., Harlan, E.: ApJ **282**, 683 (1984)

Index

MIX
Papier aus verantwortungsvollen Quellen
Paper from responsible sources
FSC® C105338

If you have any concerns about our products,
you can contact us on
ProductSafety@springernature.com

In case Publisher is established outside the EU,
the EU authorized representative is:
Springer Nature Customer Service Center GmbH
Europaplatz 3, 69115 Heidelberg, Germany

Printed by Libri Plureos GmbH
in Hamburg, Germany